JUVENILE HORMONES AND JUVENOIDS

Modeling Biological Effects and Environmental Fate

QSAR in Environmental and Health Sciences

Series Editor

James Devillers
CTIS-Centre de Traitement de l'Information Scientifique
Rillieux La Pape, France

Aims & Scope

The aim of the book series is to publish cutting-edge research and the latest developments in QSAR modeling applied to environmental and health issues. Its aim is also to publish routinely used QSAR methodologies to provide newcomers to the field with a basic grounding in the correct use of these computer tools. The series is of primary interest to those whose research or professional activity is directly concerned with the development and application of SAR and QSAR models in toxicology and ecotoxicology. It is also intended to provide the graduate and postgraduate students with clear and accessible books covering the different aspects of QSARs.

Published Titles

Endocrine Disruption Modeling, *James Devillers,* 2009

Three Dimensional QSAR: Applications in Pharmacology and Toxicology, *Jean Pierre Doucet and Annick Panaye,* 2010

Juvenile Hormones and Juvenoids: Modeling Biological Effects and Environmental Fate, *James Devillers,* 2013

QSAR in Environmental and Health Sciences

JUVENILE HORMONES AND JUVENOIDS

Modeling Biological Effects and Environmental Fate

Edited by

James Devillers

CRC Press is an imprint of the
Taylor & Francis Group, an **informa** business

CRC Press
Taylor & Francis Group
6000 Broken Sound Parkway NW, Suite 300
Boca Raton, FL 33487-2742

First issued in paperback 2018

ISBN-13: 978-1-4665-1321-1 (hbk)
ISBN-13: 978-1-138-38220-6 (pbk)

Library of Congress Cataloging-in-Publication Data

Juvenile hormones and juvenoids : modeling biological effects and environmental fate / [edited by] James Devillers.
pages cm. -- (QSAR in environmental and health sciences)
Summary: "Juvenile hormones play a key role in the control of larval development and metamorphosis of insects as well as the various aspects of the reproduction of adults. The book presents modeling approaches that can be used to study the mechanism of action of juvenile hormones (JHs) in insects and to estimate the adverse effects and the environmental fate of the manmade chemicals that mimic the actions of JHs. The text aims to provide a deeper understanding of the juvenile hormones mechanism of action, which may help to control the population of insects. Leading contributors address various topics that underscore the important role of natural compounds in the discovery and development of new human medicines"-- Provided by publisher.
Includes bibliographical references and index.
ISBN 978-1-4665-1321-1 (hardback)
1. Insects--Metamorphosis--Endocrine aspects. 2. Juvenile hormones. 3. Insect hormones. 4. Insect pests--Control--Environmental aspects. I. Devillers, J. (James), 1956- editor of compilation.

QL494.5J88 2013
571.8'76374--dc23 2013007332

Visit the Taylor & Francis Web site at
http://www.taylorandfrancis.com

and the CRC Press Web site at
http://www.crcpress.com

Contents

Series Introduction

The correlation between the toxicity of molecules and their physicochemical properties can be traced to the nineteenth century. Indeed, in a French thesis entitled *Action de l'alcool amylique sur l'organisme* (Action of amyl alcohol on the body), which was presented by A. Cros before the Faculty of Medicine at the University of Strasbourg in 1863, an empirical relationship was made between the toxicity of alcohols and their number of carbon atoms, as well as their solubility. In 1875, Dujardin-Beaumetz and Audigé were the first to stress the mathematical character of the relationship between the toxicity of alcohols and their chain length and molecular weight. In 1899, Hans Horst Meyer and Fritz Baum, at the University of Marburg, showed that narcosis or hypnotic activity was in fact linked to the affinity of substances to water and lipid sites within the organism. At the same time at the University of Zurich, Ernest Overton came to the same conclusion providing the foundation of the lipoid theory of narcosis. The next important step was made in the 1930s by Lazarev in St. Petersburg, who first demonstrated that different physiological and toxicological effects of molecules were correlated with their oil–water partition coefficient through formal mathematical equations in the following form: $\log C = a \log P_{oil/water} + b$. Thus, the quantitative structure–activity relationship (QSAR) discipline was born. Its foundations were definitively fixed in the early 1960s by the seminal works contributed by C. Hansch and T. Fujita. Since that period, the discipline has gained tremendous interest, and now the QSAR models represent key tools in the development of drugs as well as in the hazard assessment of chemicals. The new REACH (Registration, Evaluation, Authorization, and Restriction of Chemicals) legislation on substances, which recommends the use of QSARs and other alternative approaches instead of laboratory tests on vertebrates, clearly reveals that this discipline is now well established and is an accepted practice in regulatory systems.

In 1993, the journal *SAR and QSAR in Environmental Research* was launched by Gordon & Breach to focus on all the important works published in the field and to provide an international forum for the rapid publication of SAR (structure–activity relationship) and QSAR models in (eco)toxicology, agrochemistry, and pharmacology. Today, the journal, which is now owned by the Taylor & Francis Group and publishes three times more issues per year, continues to promote research in the QSAR field by favoring the publication of new molecular descriptors, statistical techniques, and original SAR and QSAR models. This field continues to grow very rapidly, and many subject areas that require larger developments are unsuitable for publication in a journal due to space limitation.

This prompted us to develop a series of books entitled *QSAR in Environmental and Health Sciences* to act in synergy with the journal. I am extremely grateful to Colin Bulpitt and Fiona Macdonald for their enthusiasm and invaluable help in making the project become a reality.

This book is the third of the series but the very first dedicated to the use of QSAR and other *in silico* techniques to provide insights into the mechanisms of action of juvenile hormones and their analogues as well as to estimate their environmental fate and potential adverse effects.

At the time of going to press, three books are in the pipeline—one dealing with reproductive and developmental toxicology, another focusing on the computational approaches to predict pKa values, and the last one discussing the topological descriptions of molecules.

I gratefully acknowledge Hilary Rowe for her willingness to assist me in the development of this series.

James Devillers

Acknowledgments

I am extremely grateful to the authors of the chapters for their acceptance to participate in this book and for preparing valuable contributions. To ensure the scientific quality and clarity of the book, each chapter was sent to two referees for review. I would like to thank all the referees for their useful comments. Finally, I would like to thank the publication team at CRC Press for making the publication of this book possible.

Contributors

William G. Bendena
Department of Biology
Queen's University
Kingston, Ontario, Canada

Apurba K. Bhattacharjee
Division of Experimental Therapeutics
Department of Medicinal Chemistry
Walter Reed Army Institute of Research
Silver Spring, Maryland

Grzegorz Bujacz
Center for Biocrystallographic Research
Institute of Bioorganic Chemistry
Polish Academy of Sciences
Poznan, Poland

and

Faculty of Biotechnology and Food Sciences
Institute of Technical Biochemistry
Technical University of Lodz
Lodz, Poland

Julio Caballero
Facultad de Ingeniería en Bioinformática
Centro de Bioinformática y Simulación Molecular
Universidad de Talca
Talca, Chile

Yves Carrière
Department of Entomology
University of Arizona
Tucson, Arizona

David W. Crowder
Department of Entomology
Washington State University
Pullman, Washington

Frédéric Darriet
Laboratoire de Lutte Contre les Insectes Nuisibles (IRD)
Montpellier, France

Robert Delorme
Consultant
Villiers-Saint-Frédéric, France

Xavier Deparis
Centre d'Epidémiologie et de Santé Publique des Armées
Marseille, France

Hugo Devillers
Centre de Traitement de l'Information Scientifique (CTIS)
Rillieux-La-Pape, France

James Devillers
Centre de Traitement de l'Information Scientifique
Rillieux-La-Pape, France

Jean-Pierre Doucet
ITODYS Laboratory
University Paris 7
Paris, France

Annick Doucet-Panaye
ITODYS Laboratory
University Paris 7
Paris, France

Peter C. Ellsworth
Department of Entomology
University of Arizona
Tucson, Arizona

Robert Farkaš
Laboratory of Developmental Genetics
Institute of Experimental Endocrinology
Slovak Academy of Sciences
Bratislava, Slovakia

Jerome H.L. Hui
Department of Zoology
University of Oxford
Oxford, United Kingdom

Jean-Philippe Jaeg
Department of Pharmacology and Toxicology
National Veterinary School of Toulouse
University of Toulouse
Toulouse, France

Mariusz Jaskolski
Center for Biocrystallographic Research
Institute of Bioorganic Chemistry
Polish Academy of Sciences
and
Faculty of Chemistry
Department of Crystallography
A. Mickiewicz University
Poznan, Poland

Laurent Lagadic
Institut National de la Recherche Agronomique
Équipe Écotoxicologie et Qualité des Milieux Aquatiques
Rennes, France

Christophe Lagneau
E.I.D. Méditerranée
Montpellier, France

Bruno Lapied
Laboratory of Receptors and Membrane Ion Channels
University of Angers
Angers, France

Kiyoto Maekawa
Graduate School of Science and Engineering
University of Toyama
Toyama, Japan

Steven E. Naranjo
Arid Lands Research Center
United States Department of Agriculture
Maricopa, Arizona

Agnieszka J. Pietrzyk
Center for Biocrystallographic Research
Institute of Bioorganic Chemistry
Polish Academy of Sciences
Poznan, Poland

Maja Polakovičová
Department of Drug Chemical Theory
School of Pharmacy
Comenius University
Bratislava, Slovakia

Françoise Quiniou
Consultant
Bohars, France

Bruce E. Tabashnik
Department of Entomology
University of Arizona
Tucson, Arizona

Stephen S. Tobe
Department of Cell and System Biology
University of Toronto
Toronto, Ontario, Canada

Kouhei Toga
Graduate School of Science and Engineering
University of Toyama
Toyama, Japan

Ohri Yamada
Department of Vector Control
French Agency for Food, Environmental and Occupational Health & Safety (ANSES)
Maisons-Alfort, France

André Yébakima
Centre de Démoustication/Lutte Antivectorielle (CG/ARS)
Martinique, France

Anthony J. Zera
School of Biological Sciences
University of Nebraska
Lincoln, Nebraska

Xiao-Fan Zhao
School of Life Sciences
Shandong University
Jinan, Shandong, People's Republic of China

1 Juvenile Hormones and Juvenoids

A Historical Survey

James Devillers

CONTENTS

ABSTRACT

The juvenile hormones of insects are sesquiterpenoid molecules secreted by the corpora allata. They prevent metamorphosis of larvae and play numerous other functions during the whole insect life. Juvenile hormones have prototyped the design of insecticides mimicking their activity for use in insect vector and pest control. These man-made chemicals are called juvenile hormone analogues or juvenoids. In this chapter, the conditions in which the juvenile hormones were discovered are described. The brief history of juvenoid synthesis is also recalled.

KEYWORDS

Juvenile hormone, Juvenoid, Corpora allata, Insect

1.1 INTRODUCTION

Recent estimates place the number of insect species to about 5 million, but the number of species currently catalogued is five times less important. Indeed, there are about 400,000 Coleoptera, 153,000 Diptera, 90,000 Hemiptera, 115,000 Hymenoptera, and 174,000 Lepidoptera [1] that have been described. These Arthropoda affect the living organisms including humans in a variety of ways. Insects play a crucial role in the functioning of natural and agricultural ecosystems through diverse activities ranging from decomposition of organic matter to provision of food for fishes, reptiles, birds, and so on. In fact, they act as predators, preys, parasites, hosts, parasitoids, herbivores, saprophages, and pollinators, among other things, which indicate the pervasive ecological and economic importance of these Hexapoda in both aquatic and terrestrial ecosystems [2]. For example, the annual value of the main ecological services provided by the insect in the United States has been estimated to be at least $57 billion [3]. Unfortunately, numerous insect species also impact negatively humans and the environment. Crop losses due to insect pests reduce available production of food and cash crops worldwide. Thus, one estimate indicates that about 13% of total crop production is lost annually due to insect pests in the United States [4,5]. This can be much more in developing countries [6,7]. Insects can also serve as vectors for plant, animal, and human diseases. Thus, for example, the incidence of dengue, which is a mosquito-borne infection, has grown dramatically around the world in recent decades. More than 2.5 billion people (over 40% of the world's population) are now at risk from dengue and 1.8 billion (>70%) live in Asia Pacific countries [8]. The total economic burden of the 2006 Indian dengue epidemic was approximately U.S. $27.4 million [9]. In the same way, the direct and indirect cost of recent dengue cases in Brazil, El Salvador, Guatemala, Panama, Venezuela, Cambodia, Malaysia, and Thailand was estimated to $1.8 billion, and incorporating costs of dengue surveillance and vector control would raise the amount further [10].

For all the aforementioned reasons, the insects have become organisms of choice for a variety of research endeavors to better understand their ecology, biology, physiology, and so forth.

Because the insects are encased in a cuticle, they must shed this semirigid exoskeleton periodically in order to grow. There are three major patterns of insect development. About 1% of the insects do not present a larval stage and undergo direct development. These are the ametabolous insects. They show a pronymph stage immediately after hatching, bearing the structures that have enabled it to get out of the egg. After this transitory stage, the insects begin to look like small adults; after each molt, they are bigger but unchanged in form. The springtail (Collembola) and the silverfish or fishmoth (Thysanura) are representatives of this primitive group. In the hemimetabolous insects (about 9%), the young form looks like an immature adult. At the last molt, the insect is winged and sexually mature. This group includes the grasshopper (Orthoptera), the true bug (Hemiptera), the dragonfly (Odonata), and the cockroach (Blattodea). Lastly, in the holometabolous insects (about 90%), the juvenile form undergoes a series of molts as it becomes larger. Each stage is called instar. After the last instar, the larva undergoes metamorphic molt to become a pupa and then an imago [11,12]. This evolved type of development allows to reduce larval–adult competition, but the vulnerability of the larval and pupal stages is high.

Although insect molting and metamorphosis have been scrutinized since the Greek antiquity with Aristotle through his book on the *Generation of Animals,* their physiological mechanisms have remained mysterious for a long time, even after the importance of hormones had been recognized in the vertebrates.

The purpose of this chapter is not to go into the details and intricacies of juvenile hormone (JH) and 20-hydroxyecdysone functions. Excellent recent reviews have covered these subjects [13–17]. Our goal is only to show how the JH was discovered and how this discovery was exploited by biologists and chemists to produce a new class of insecticides called JH analogues or juvenoids.

1.2 A BRIEF HISTORY OF THE DISCOVERY OF JUVENILE HORMONE

The discovery of JH has benefited undoubtedly from thoughtful and ingenious laboratory experiments. However, interpretation of the results was sometimes a notch below and even false in the early studies. Thus, for example, Conte and Vaney [18] tried to see whether acephalous caterpillars of three species of Lepidoptera were able to have a full and normal metamorphosis when they were ligatured at the junction of the head and thorax. With *Bombyx mori*, the pupae were easily obtained, but they did not give an adult, all died within one week. On dissection, they showed an advanced pupation, the silk glands were almost totally resorbed, and there was abundant adipose tissue but a limited histogenesis. The results obtained on the other species showed that generally decapitation did not have a strong effect on metamorphosis. This led the authors to claim that the integrity of animal body, including the head, was not necessary to secure the lepidopteran metamorphosis. Kopeć [19] confirmed these results, stating that the nervous system was unimportant up to the time of pupation. However, ten years later, from a rational experimental study, also performed on Lepidoptera, Kopeć [20] demonstrated the converse. More specifically, he showed that the brain (ganglion supraesophageal) of the caterpillar of *Lymantria dispar* L. was important in the general processes of metamorphosis. The presence of the brain was crucial, at least up to a certain period, for the inception of histolytical processes. Kopeć postulated that the influence of the brain in this direction was probably chemical; hence, the brain had to be considered as a gland of internal secretion. At some well-defined time before pupation, the quantity of the substance(s) secreted by the brain was already sufficient for the complete pupation of the caterpillar. Tissues of the caterpillar influenced by the brain underwent further metamorphosis independently. Other parts of the nervous system did not influence the general process of metamorphosis [20]. Koller [21] accelerated the onset of molting in normal larvae of moths by injecting the blood from other larvae in which molting had begun. Von Buddenbrock [22] repeated these experiments on a larger scale but he failed to obtain a definite acceleration of molting, but the process was not delayed. Conversely, if the blood of normal larvae was injected, molting was delayed. From this, he deducted the role of blood of the molting larvae in the acceleration process. Bodenstein [23] showed that limbs transplanted from one caterpillar to another molted at the same time as their new host. In the 1930s, Wigglesworth [24–27] brought together these findings and his own brilliant experiments on the bug *Rhodnius prolixus* to present the first synthesis that finally gave a mechanistic answer to the insect metamorphosis. *R. prolixus*, a blood-sucking triatomine, vector of the trypanosomes that causes Chagas' disease in Central America, was particularly suited to perform studies on the development of insects. An immature insect takes a large blood meal and then molts a precise number of days later. It has cuticular markers that distinguish an immature from an adult. The insect can be joined in parabiosis to others, even of very different stages [28]. In 1934, Wigglesworth wrote [25]:

> There is a 'critical period' in the molting cycle (about 7 days after feeding in the fifth nymph, about 4 days in the earlier nymphs) and removal of the head of the insect before this period prevents molting.... The blood of insects that have passed the critical period contains a factor or hormone which will induce molting in insects decapitated soon after feeding. It is suggested that this molting hormone may be secreted by the corpus allatum, since the cells of this gland show signs of greatest secretory activity during the critical period.

It is noteworthy that the corpora allata [29], first termed corpora inserta [30], and ganglia allata [31] were described in 1913 by Nabert [32] as glands of internal secretion of unknown function. In the same way, after a rather skeptical appraisal of the Wigglesworth's work [33], Bounhiol [34], through experiments on the effects of corpora allata extirpation in the silkworm (*B. mori*), confirmed the key role of the corpora allata in the insect development via the secretion of a metamorphosis-inhibiting factor. In the following years, the endocrine activity of corpora allata was verified in numerous insects belonging to different taxa [35–38].

For characterization of the corpus allatum hormone and the study of its actions, it was necessary to obtain enough active soluble extracts. Unfortunately, all early attempts to extract significant quantities from corpora allata were unsuccessful. However, a rich source of soluble JH was fortuitously discovered by Williams [39,40] in adults of *Hyalophora cecropia* moths opening the way for purification of JH. The first structural identification was made by Röller [41] and Dahm et al. [42] in *H. cercopia*, and the molecule was first named *cecropia* JH and then JH I (Figure 1.1). Meyer et al. [43] identified another sesquiterpenoid homologue in *H. cercopia* differing from JH I by the presence of a methyl group at C7. This new analogue was termed JH II (Figure 1.1). A third JH homologue, JH III (Figure 1.1), was identified in the tobacco hornworm, *Manduca sexta* [44]. JH III is the most commonly found homologue [45]. In the eggs of *M. sexta*, Bergot et al. [46] identified JH 0 and 4-methyl JH I, which is termed iso-JH I (Figure 1.1). A JH III with a second epoxide substitution at C6, C7 was identified from *in vitro* cultures of ring glands of *Drosophila melanogaster* larvae [47] and termed JH III bisepoxide or JHB_3 (Figure 1.1). A skipped bisepoxide ($JHSB_3$) was also recently identified by Kotaki et al. [48] in *Plautia stali*, a member of the family Pentatomidae (Heteroptera) (Figure 1.1). Almost simultaneously with efforts to elucidate the structure and functions of the first JHs, attempts were made to exploit them to control the development of insects. The history of this research is summarized in the next section.

1.3 DIFFERENT STEPS OF THE DISCOVERY OF THE JUVENILE HORMONE ANALOGUES

Initiation of research on juvenoids started in 1956 when Williams [39] prepared the first JH-active lipid extracts from the abdomens of male adult *cecropia* moths. He

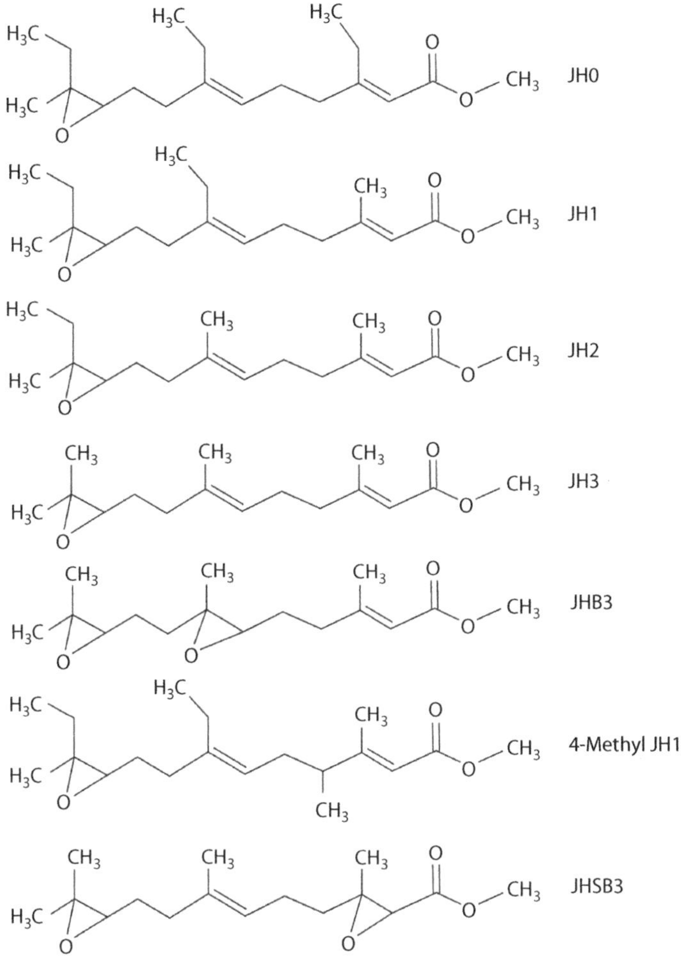

FIGURE 1.1 Structures of JHs.

pointed out that these extracts could be used as nontoxic and selectively acting insecticides. Indeed, he forewarning wrote: "It seems likely that the (juvenile) hormone, when identified and synthesized, will prove to be an effective insecticide. This prospect is worthy of attention because insects can scarcely evolve a resistance to their own hormone." Further investigations revealed the presence of extracts with positive JH responses in other invertebrates [49] as well as in microorganisms and plants [50]. This opened the door to the research of active substances in all the phyla. In the early 1960s, chemists tried to synthesize terpenoids and related structures for estimating their JH activity. Thus, for example, various terpenoid and straight-chain alcohols as well as their derivatives were synthesized, purified, and tested for their potential JH activity by Schneiderman et al. [51]. They showed that a critical size of the molecules was necessary for higher activity. It is noteworthy that at that time numerous results were controversially discussed due to the difficulty to interpret the results. Fortunately, a fortuitous event contributed to definitively install the research on juvenoids. Early in the 1960s, Dr. K. Sláma, an entomologist from Prague (former Czechoslovakia), visited the University of Harvard in order to work in Dr. Williams' laboratory. He brought with him fertile eggs from the bug *Pyrrhocoris apterus* that he had been rearing in Petri dishes without difficulty in his laboratory for 10 years. To his surprise, the bugs did not develop normally under the conditions at Harvard. Instead of metamorphosing into normal adults, at the end of the 5th larval instar, all molted into 6th instar larvae or into adultoids preserving a lot of larval characters. Some continued to grow and molted into still-larger seventh instar larvae. At the end, all insects died without completing metamorphosis or attaining sexual maturity. An audit of the culture conditions at Harvard versus Prague suggested 15 differences. Fourteen were eliminated from rational investigations. The source of JH activity was finally tracked down to exposure of the bugs to a certain paper towel that had been placed in the rearing cages. Pieces of American newspapers and journals (*New York Times*, *Wall Street Journal*, *Boston Globe Science*, and *Scientific American*) showed extremely high JH activity when placed in contact with *Pyrrhocoris*. The *London Times* and *Nature* were inactive, and so were other paper materials of European or Japanese manufacture. These findings pointed out the botanical origin of the papers. American paper products were mainly derived from the balsam fir (*Abies balsamea*). The negative tests recorded for paper samples of European and Japanese manufactures suggested that their pulp trees did not contain the active principle [52]. The active extract found in the balsam fir was called "paper factor." Sláma and Williams [52–54] and Williams and Sláma [55] were very surprised to discover that the most active extracts were without any detectable effects when injected or topically applied to previously chilled pupae of the *cecropia* or *polyphemus* silkworms. The same negative results were obtained on all the other silkworm pupae of *H. cecropia*, *H. gloveri*, *Antheraea mylitta*, and *Samia cynthia*. The extracts were without any effects on two other species of Heteroptera, *R. prolixus* and *Oncopeltus fasciatus*. These authors deducted that molecules with JH activity for one species of insect are not necessarily active on other species. They also speculated that certain plants might perhaps develop resistance against insect herbivores by evolutionary adaptations leading to synthesis of compounds with insect JH activity.

The "paper factor" found in the balsam fir was identified by Bowers et al. [56] as a methyl ester of an alicyclic sesquiterpenoid acid, called juvabione (methyl (4*R*)-4-[(2*R*)-6-methyl-4-oxoheptan-2-yl]cyclohexene-1-carboxylate, CAS RN 17904-27-7) (Figure 1.2). A structurally related chemical called dehydrojuvabione

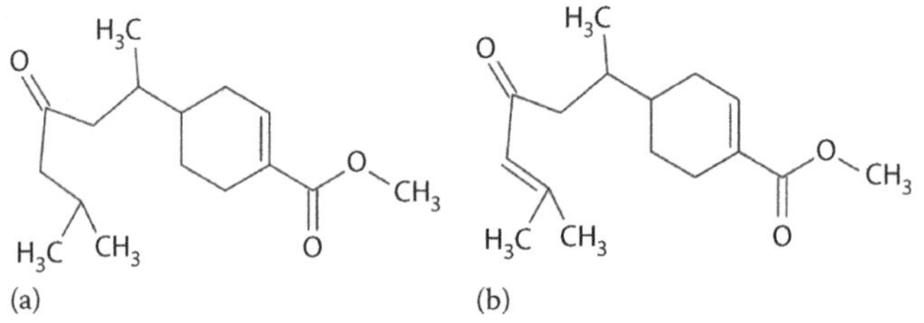

FIGURE 1.2 Structures of juvabione (a) and dehydrojuvabione (b).

((4*R*)-4-[(1*R*)-1,5-dimethyl-3-oxo-4-hexenyl]-1-cyclohexene-1-carboxylic acid methyl ester, CAS RN 16060-78-9) (Figure 1.2) was found in a balsam fir growing in Europe [57], but the common European evergreen trees did not contain JH activity at all [58]. These discoveries led to searches for both natural and synthetic substances with JH activity. At the end of the 1960s, the number of synthetic juvenoids was estimated to about 500 including acyclic terpenoids, juvabione derivatives, aromatic terpenoid ethers, thioethers, and amines [59–62].

At the end of the 1960s, Zoecon Corporation was set up by Syntex Corporation to develop the new concept that insect pests could be selectively controlled, without environmental problems, by using analogues of their natural JH. Hydroprene (ethyl (2*E*,4*E*)-3,7,11-trimethyl-2,4-dodecadienoate, CAS RN 41096-46-2) (Figure 1.3) was prepared in October 1970 and showed high activity against many insect species. Kinoprene ((*E*,*E*)-2-propynyl 3,7,11-trimethyl-2,4-dodecadienoate, CAS RN 42588-37-4) (Figure 1.3) also showed an interesting activity, but the most interesting juvenoid was the methoprene (1-methylethyl (2*E*,4*E*)-11-methoxy-3,7,11-trimethyl-2,4-dodecadienoate, CAS RN 40596-69-8) (Figure 1.3) synthesized in March 1971 at Zoecon. Methoprene was the first registered juvenoid for use in the control of floodwater mosquitoes [63–65].

Concomitantly with the registration of methoprene, epofenonane (2-ethyl-3-[3-ethyl-5-(4-ethylphenoxy)pentyl]-2-methyloxirane, CAS RN 57342-02-6), a terpenoid aromatic juvenoid (Figure 1.3), was also registered for practical use [66] by Hoffmann-La Roche AG (Basel, Switzerland). Epofenonane showed better effects than methoprene in certain field applications, but basically, its range of JH activity was more or less similar to methoprene [58].

In the early 1980s, a highly effective, nonisoprenoid type of juvenoid was discovered and developed by Dr. Maag Ltd. Dielsdorf (Switzerland) and MAAG Agrochemicals, HLR Sciences, Inc. The decisive structural innovation was the inclusion of the bicyclic, 4-phenoxyphenyl group into a classical juvenoid structure. The most active molecule in this series was fenoxycarb (ethyl [2-(*p*-phenoxyphenoxy)ethyl]carbamate, CAS RN 79127-80-3) (Figure 1.3), which was primarily registered for use in stored product insect control [67,68]. It is useful for control of fire ants, fleas, mosquitoes, cockroaches, moths, scale insects, and insects attacking vines, olives, cotton, and fruit [69]. Although fenoxycarb includes a carbamate moiety, unlike certain *N*-methyl or *N*-ethyl carbamates, it does not inhibit cholinesterase.

In the 1980s, Japanese chemists were very active in the search for new juvenoids (see Chapter 8). The most successful candidate was proposed by Sumitomo and known under the name of pyriproxyfen (2-(1-methyl-2-(4-phenoxyphenoxy)ethoxy) pyridine, CAS RN 95737-68-1) (Figure 1.3) [70]. It is a fenoxycarb derivative in which a part of the aliphatic chain has been replaced by pyridyl oxyethylene. It was shown to be a potent JH mimic affecting the hormonal balance in insects, resulting in strong suppression of embryogenesis, metamorphosis, and adult formation [71–74].

Attempts were also made at Ciba Geigy to propose new juvenoids. They prepared and tested various non-terpenoid structures, and among them, diofenolan (mixture of (2*RS*,4*SR*)-4-(2-ethyl-1,3-dioxolan-4-ylmethoxy)phenyl phenyl ether and (2*RS*,4*RS*)-4-(2-ethyl-1,3-dioxolan-4-ylmethoxy)phenyl phenyl ether, CAS RN

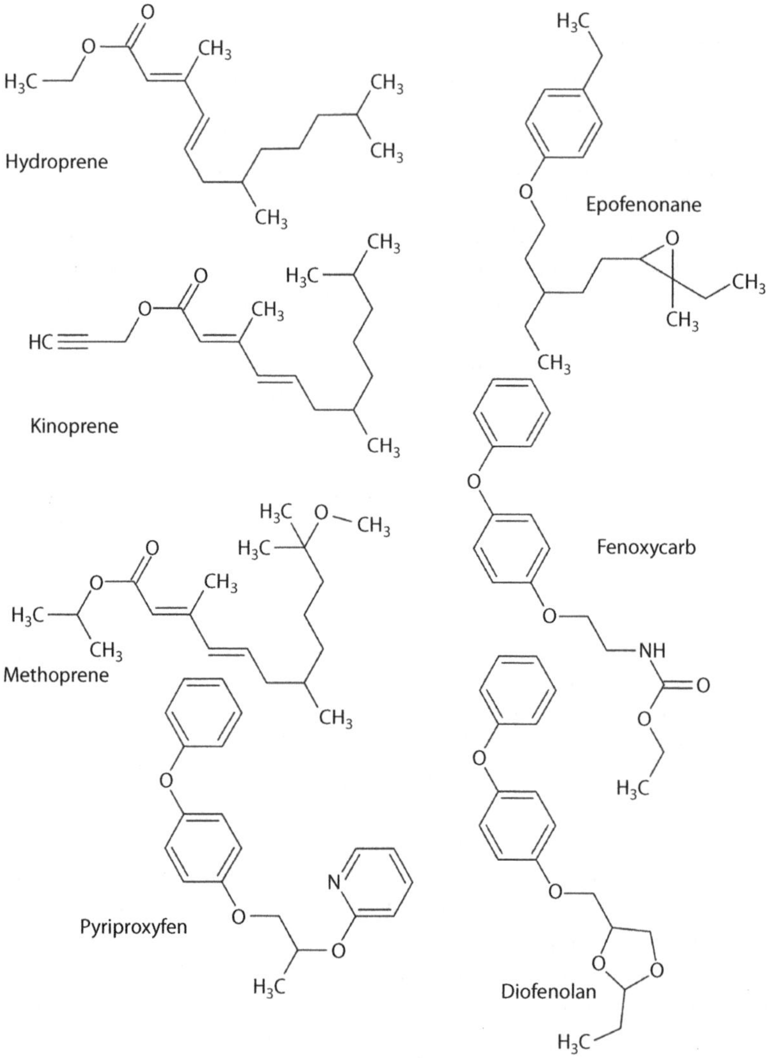

FIGURE 1.3 Structures of JH analogues.

63837-33-2) (Figure 1.3), commercialized under the name Aware, was found to be effective against lepidopteran pests in deciduous fruit, citrus, grapes, and olives and scale insects in pome fruit and citrus [75–78].

1.4 ANTI-JUVENILE HORMONE AGENTS

The discovery and subsequent development of juvenoids inspired thoughts that the reverse principle, aiming at discovering chemicals showing an anti-juvenile activity, could be also a potential strategy to find new insecticides. In the 1970s, Bowers et al. [79] isolated two simple chromenes in the plant *Ageratum houstonianum*. They induced precocious metamorphosis and sterilization in several hemipterous species. Some holometabolous species were sterilized, forced into diapause, or both. These chemicals were called antihormonal compounds precocene I (7-methoxy-2,2-dimethylchromene) and precocene II (6,7-dimethoxy-2,2-dimethylchromene). Precocene III was also identified as 7-ethoxy-6-methoxy-2,2-dimethylchoromene [80]. From studies of metabolism [81], pharmacokinetics [82], reaction with model substrates [83], and cytological investigations [84], it was shown that the precocenes were activated by oxidation to form highly reactive epoxides that destroy the parenchymal cells of corpus allatum by nucleophilic alkylation [85,86].

In 1989, a peptide causing a strong and rapid inhibitory effect against biosynthesis of JH (allatostatic effect) was found in cockroach *Diploptera punctata*. Its sequence was determined and named allatostatin (A type) [87]. Other types were identified in different insect taxa [88–90].

1.5 CONCLUSION

The insect corpus allatum hormones, known as JHs, were discovered through their role in regulating the onset of metamorphosis by stopping the expression of the adult state, and they were named accordingly. It has been also experimentally demonstrated that the JHs induced multifaceted effects during the whole insect life. As a result, it is not surprising that attempts were made to use the JHs as targets in the design of insecticides for the control of insect pests.

Since Aristotle, research in biology is rooted on the triad observations–hypotheses–experimentations and this brief history of the discovery of the JHs and juvenoids does not contradict this fact. However, over the past several decades, what has changed is our perception of the components of the biological observations and/or experiments and their corresponding outcomes that are both seen as mathematical entities with their proper characteristics that we manipulate consciously to isolate influential factors, to better interpret the results, to gain insight into a mechanism of action, and so on. The main reason for this change is that now the computer belongs to the working kit of any researcher. From the room-sized Electronic Numerical Integrator Analyzer and Computer (ENIAC) in the 1940s weighing 30 tons to the big Cray-1 system installed at Los Alamos National Laboratory in 1976 for $8.8 million to today's laptops, processing speed has increased over a millionfold and cost has dropped dramatically, allowing scientists to analyze their data and run models on their own computers rather than booking time on one of only a few "supercomputers" in the world, like in the past.

Invertebrate endocrinology has widely benefited from these changes to mine and analyze genomic data, to better understand the mechanism of action of JHs and their analogues, to model the environmental fate and effects of juvenoids, or to predict the potential JH activity of molecules from their chemical structure. Case studies related to these different modeling approaches can be found in the different chapters of this book.

ACKNOWLEDGMENT

The financial support from the French Ministry of Ecology, Sustainable Development, Transport and Housing (MEDDTL) is gratefully acknowledged (PNRPE program).

REFERENCES

1. B.R. Scheffers, L.N. Joppa, S.L. Pimm, and W.F. Laurance, *What we know and don't know about Earth's missing biodiversity,* Trends Ecol. Evol. 27 (2012), pp. 501–510.
2. D.M. Rosenberg, H.V. Danks, and D.M. Lehmkuhl, *Importance of insects in environmental impact assessment*, Environ. Manag. 10 (1986), pp. 773–783.
3. J.E. Losey and M. Vaughan, *The economic value of ecological services provided by insects*, BioScience 56 (2006), pp. 311–323.
4. J.S. Bernal, J. Prasifka, M. Sétamou, and K.M. Heinz, *Transgenic insecticidal cultivars in integrated pest management: Challenges and opportunities*, in *Integrated Pest Management: Potential, Constraints and Challenges*, O. Koul, G.S. Dhaliwal, and G.W. Cuperus, eds., CABI International, Cambridge, London, U.K., 2004, pp. 123–145.
5. D. Pimentel, L. McLaughlin, A. Zepp, B. Lakitan, T. Kraus, P. Kleinman, E. Vancini, W.J. Roach, E. Craap, W.S. Keeton, and G. Selig, Environmental and economic effects of reducing pesticide use in agriculture, Agric. Ecosyst. Environ. 46 (1993), pp. 273–288.
6. K.V.S. Reddy and U.B. Zehr, *Novel strategies for overcoming pests and diseases in India*, Proceedings of the 4th International Crop Science Congress, "New directions for a diverse planet", September 26–October 1, 2004, Brisbane, Australia, T. Fischer, N. Turner, J. Angus, L. McIntyre, M. Robertson, A. Borrell, and D. Lloyd, eds., BPA Print Group, Burwood, Australia, pp. 1–8.
7. E.C. Oerke and H.W. Dehne, *Safeguarding production—Losses in major crops and the role of crop protection*, Crop Protect. 23 (2004), pp. 275–285.
8. WHO, World Health Organization, Dengue: http://www.wpro.who.int/mediacentre/factsheets/fs_09032012_Dengue/en/index.html (Accessed on December 12, 2012).
9. P. Garg, J. Nagpal, P. Khairnar, and S.L. Seneviratne, *Economic burden of dengue infections in India*, Trans. Royal Soc. Trop. Med. Hyg. 102 (2008), pp. 570–577.
10. J.A. Suaya, D.S. Shepard, J.B. Siqueira, C.T. Martelli, L.C. Lum, L.H. Tan, S. Kongsin, S. Jiamton, F. Garrido, R. Montoya, B. Armien, R. Huy, L. Castillo, M. Caram, B.K. Sah, R. Sughayyar, K.R. Tyo, and S.B. Halstead, *Cost of dengue cases in eight countries in the Americas and Asia: A prospective study*, Am. J. Trop. Med. Hyg. 80 (2009), pp. 846–855.
11. J.W. Truman and L.M. Riddiford, *The origins of insect metamorphosis*, Nature 401 (1999), pp. 447–452.
12. J.W. Truman and L.M. Riddiford, *Endocrine insights into the evolution of metamorphosis in insects*, Annu. Rev. Entomol. 47 (2002), pp. 467–500.
13. G.R. Wyatt, *Juvenile hormone in insect reproduction—A paradox*? Eur. J. Entomol. 94 (1997), pp. 323–333.
14. L.I. Gilbert, *Insect hormones*, in *Endocrinology: Basic and Clinical Principles*, 2nd ed., S. Melmed and P.M. Conn, eds., Humana Press, Totowa, NJ, 2005, pp. 127–136.
15. R. Lafont, C. Dauphin-Villemant, J.T. Warren, and H. Rees, *Ecdysteroid chemistry and biochemistry*, in *Insect Endocrinology*, L.I. Gilbert, ed., Academic Press, Amsterdam, the Netherlands, 2012, pp. 106–176.
16. W.G. Goodman and M. Cusson, *The juveniles hormones*, in *Insect Endocrinology*, L.I. Gilbert, ed., Academic Press, Amsterdam, the Netherlands, 2012, pp. 310–365.
17. L.M. Riddiford, *How does juvenile hormone control insect metamorphosis and reproduction*? Gen. Comp. Endocrinol. 179 (2012), pp. 477–484, http://dx.doi.org/10.1016/j.ygcen.2012.06.001

18. A. Conte and C. Vaney, *Production expérimentale de lépidoptères acéphales*, C.R. Acad. Sci. Paris 152 (1911), pp. 404–406.
19. S. Kopeć, *Über die Funktionen des Nervensystems der Schmetterlinge während der sukzessiven Stadien der Metamorphose*, Zool. Anz. 40 (1912), pp. 353–360.
20. S. Kopeć, *Studies on the necessity of the brain for the inception of insect metamorphosis*, Biol. Bull. 42 (1922), pp. 323–342.
21. G. Koller, *Die innere Sekretion bei wirbellosen Tieren*, Biol. Rev. 4 (1929), pp. 269–306.
22. W. von Buddenbrock, *Untersuchungen iiber die Hautungshormone der Schmetterlingsraupen*, Zeitschr. f. vergleich. Physiol. 14 (1931), pp. 415–428.
23. D. Bodenstein, *Beintransplantationen an lepidopteren Raupen. I. Transplantationen zur Analyse der Raupen- und Puppenhäutung*, Arch. Entw. Meoh. 128 (1933), pp. 564–583.
24. V.B. Wigglesworth, *The physiology of the cuticle and of ecdysis in* Rhodnius prolixus *(Triatomidae, Hemiptera); with special reference to the function of the oenocytes and of the dermal glands*, Quart. J. Microsc. Sci. 76 (1933), pp. 269–318.
25. V.B. Wigglesworth, *The physiology of ecdysis in* Rhodnius prolixus (Hemiptera). *II. Factors controlling moulting and "metamorphosis"*, Quart. J. Microsc. Sci. 77 (1934), pp. 191–222.
26. V.B Wigglesworth, *Functions of the corpus allatum of insects*, Nature 136 (1935), pp. 338–339.
27. V.B. Wigglesworth, *The function of the corpus allatum in the growth and reproduction of* Rhodnius prolixus (Hemiptera), Quart. J. Microsc. Sci. 79 (1936), pp. 91–121.
28. J.S. Edwards, *Sir Vincent Wigglesworth and the coming of age of insect development*, Int. J. Dev. Biol. 42 (1998), pp. 471–473.
29. R. Heymons and T. Kuhlgatz, *Die Süsswasserfauna Deutschlands. Heft 7. Collembola, Neuroptera, Hymenoptera, Rhynchota*, Verlag, Jena, Germany, 1909.
30. F. Meinert, *Bidrag til de danske Myrers Naturhistorie*, K. Dan. Vidensk. Selsk. Skr. 5 (1861), pp. 273–340.
31. R. Heymons, *Die Embryonalentwickelung von Dermapteren und Orthoperen unter Besonderer Berücksichtigung der Keimblätterbildung*. Verlag, Jena, Germany, 1895.
32. A. Nabert, *Die Corpora allata der Insekten*, Z. Wiss. Zool. (Leipzig), 104 (1913), pp. 181–358.
33. J.J. Bounhiol, *Métamorphose après ablation des corpora allata chez le ver à soie (*Bombyx mori *L.)*, C.R. Acad. Sci. Paris 203 (1936), pp. 388–389.
34. J.J. Bounhiol, *Métamorphose prématurée par ablation des corpora allata chez le jeune ver à soie*, C.R. Acad. Sci. Paris 205 (1937), pp. 175–177.
35. I.W. Pfeiffer, *The influence of the corpora allata over the development of nymphal characters in the grasshopper* Melanoplus differentialis, Trans. Conn. Acad. Arts Sci. 36 (1945), pp. 489–515.
36. A. Radtke, *Hemmung der Verpuppung beim Mehlkäfer* Tenebrio molitor L., Die Naturwiss. 30 (1942), pp. 451–452.
37. R. Poisson and R. Sellier, *Brachyptérisme et actions endocrines chez* Gryllus campestris L., C.R. Acad. Sci. Paris 224 (1947), pp. 1074–1075.
38. M. Vogt, *Inhibitory effects of the corpora cardiaca and of the corpus allatum in* Drosophila, Nature 157 (1946), pp. 512.
39. C.M. Williams, *The juvenile hormone of insects*, Nature 178 (1956), pp. 212–213.
40. C.M. Williams, *The juvenile hormone. I. endocrine activity of the corpora allata of the adult* cecropia *silkworm*, Biol. Bull. 116 (1959), pp. 323–338.
41. H. Röller, K.H. Dahm, C.C. Sweeley, and B.M. Trost, *Die Struktur des Juvenilhormons*, Angew. Chem. 79 (1967), pp. 190–191.
42. K.H. Dahm, H. Röller, and B.M. Trost, *The juvenile hormone—IV. Stereochemistry of juvenile hormone and biological activity of some of its isomers and related compounds*, Life Sci. 7 (1968), pp. 129–137.

43. A.S. Meyer, H.A. Schneiderman, E. Hanzmann, and J.H. Ko, *The two juvenile hormones from the* cecropia *silk moth*, Proc. Nat. Acad. Sci. 60 (1968), pp. 853–860.
44. K.J. Judy, D.A. Schooley, L.L. Dunham, M.S. Hall, B.J. Bergot, and J.B. Siddall, *Isolation, structure, and absolute configuration of a new natural insect juvenile hormone from* Manduca sexta, Proc. Nat. Acad. Sci. 70 (1973), pp. 1509–1513.
45. K. Hartfelder, *Insect juvenile hormone: From "status quo" to high society*, Braz. J. Med. Biol. Res. 33 (2000), pp. 157–177.
46. B.J. Bergot, G.C. Jamieson, M.A. Ratcliff, and D.A. Schooley, *JH zero: New naturally occurring insect juvenile hormone from developing embryos of the tobacco hornworm*, Science 210 (1980), pp. 336–338.
47. D.S. Richard, S.W. Applebaum, T.J. Sliter, F.C. Baker, D.A. Schooley, C.C Reuter, V.C. Henrich, and L.I. Gilbert, *Juvenile hormone bisepoxide biosynthesis* in vitro *by the ring gland of* Drosophila melanogaster: *A putative juvenile hormone in the higher Diptera*, Proc. Nat. Acad. Sci. 86 (1989), pp. 1421–1425.
48. T. Kotaki,T. Shinada, K. Kaihara, Y. Ohfune, and H. Numata, *Structure determination of a new juvenile hormone from a heteropteran insect*, Org. Lett. 11 (2009), pp. 5234–5237.
49. H.A. Schneiderman and L.I. Gilbert, *Substances with juvenile hormone activity in Crustacea and other invertebrates*, Biol. Bull. 115 (1958), pp. 530–535.
50. H.A. Schneiderman, L.I. Gilbert, and M.J. Weinstein, *Juvenile hormone activity in microorganisms and plants*, Nature 188 (1960), pp. 1041–1042.
51. H.A. Schneiderman, A. Krishnakumaran, V.G. Kulkarni, and L. Friedman, *Juvenile hormone activity of structurally unrelated compounds*, J. Insect Physiol. 11 (1965), pp. 1641–1642.
52. K. Sláma and C.M. Williams, *Juvenile hormone activity for the bug* Pyrrhocoris apterus, Proc. Nat. Acad. Sci. 54 (1965), pp. 411–414.
53. K. Sláma and C.M. Williams, *"Paper factor" as an inhibitor of the embryonic development of the European bug,* Pyrrhocoris apterus, Nature 210 (1966), pp. 329–330.
54. K. Sláma and C.M. Williams, *The juvenile hormone. V. The sensitivity of the bug*, Pyrrhocoris apterus, *to a hormonally active factor in American paper-pulp*, Biol. Bull. 130 (1966), pp. 235–246.
55. C.M. Williams and K. Sláma, *The juvenile hormone. VI. Effects of the "Paper factor" on the growth and metamorphosis of the bug*, Pyrrhocoris apterus. Biol. Bull. 130 (1966), pp. 247–253.
56. W.S. Bowers, H.M. Fales, M.J. Thompson, and E.C. Uebel, *Juvenile hormone: Identification of an active compound from balsam fir*, Science 154 (1966), pp. 1020–1021.
57. V. Černý, L. Dolejš, L. Lábler, F. Šorm, and K. Sláma, *Dehydrojuvabione—A novel compound with juvenile hormone activity from balsam fir*, Tetrah. Lett. 8 (1967), pp. 1053–1057.
58. K. Sláma, *The history and current status of juvenoids*, in *Proceedings of the 3rd International Conference on Urban Pests*, W.H. Robinson, F. Rettich, and G.W. Rambo, eds., Grafické Závody Hronov, Prague, Czech Republic, 1999, pp. 9–25.
59. M. Romaňuk, K. Sláma, and F. Šorm, *Constitution of a compound with a pronounced juvenile hormone activity*, Proc. Natl. Acad. Sci. 57 (1967) pp. 349–352.
60. K. Mori and M. Matsui, *Synthesis of compounds with juvenile hormone activity-I: (±)-juvabione (methyl (±)-todomatuate)*, Tetrahedron 24 (1968), pp. 3127–3138.
61. K. Poduška, F. Šorm, and K. Sláma, *Natural and synthetic materials with insect hormone activity. 9. Structure-juvenile activity relationships in simple peptides*, Z. Naturforsch 26B (1971), pp. 719–722.
62. K. Sláma, *Insect juvenile hormone analogues*, Annu. Rev. Biochem. 40 (1971), pp. 1079–1102.

63. C.A. Henrick, G.B. Staal, and J.B. Siddall, *Alkyl 3,7,11-trimethyl-2,4-dodecadienoates, a new class of potent insect growth regulators with juvenile hormone activity*, J. Agric. Food Chem. 21 (1973), pp. 354–359.
64. C.A. Henrick, W.E. Willy, and G.B. Staal, *Insect juvenile hormone activity of alkyl (2E,4E)-3,7,11-trimethyl-2,4-dodecadienoates. Variations in the ester function and in the carbon chain*, J. Agric. Food Chem. 24 (1976), pp 207–218.
65. J.B. Siddall, *Insect growth regulators and insect control: A critical appraisal*, Environ. Health Perspect. 14 (1976), pp. 119–126.
66. W.W. Hangartner, M. Suchy, H.K. Wipf, and R.C. Zurflueh, *Synthesis and laboratory and field evaluation of a new, highly active and stable insect growth regulator*, J. Agric. Food Chem. 24 (1976), pp 169–175.
67. S. Dorn, M.L. Frischknecht, V. Martinez, R. Zurfluh, and U. Fischer, *A novel non-neurotoxic insecticide with a broad activity spectrum*, Z. Pflanzenkr. Pflanzens. 88 (1981), pp. 269–275.
68. K.J. Kramer, R.W. Beeman, and L.H. Hendricks, *Activity of Ro 13-5223 and Ro 13-7744 against stored product insects*, J. Econ. Entomol. 74 (1981), pp. 678–680.
69. S. Grenier and A.M. Grenier, *Fenoxycarb, a fairly new insect growth regulator: A review of its effects on insects*, Ann. Appl. Biol. 122 (1993), pp. 369–413.
70. M. Hatakoshi, N. Agui, and I. Nakayama, *2-[1-Methyl-2(4-phenoxyphenoxy) ethoxy] pyridine as a new insect juvenile hormone analogue: Induction of supernumerary larvae in* Spodoptera litura (*Lepidoptera: Noctuidae*), Appl. Entomol. Zool. 21 (1986), pp. 351–353.
71. W. Itaya, *Insect juvenile hormone analogue as an insect growth regulator*, Sumitomo Pyrethroid World 8 (1987), pp. 2–4.
72. H. Kawada, *An insect growth regulator against cockroach*, Sumitomo Pyrethroid World 11 (1988), pp. 2–4.
73. P. Langley, *Control of tsetse fly using a juvenile hormone mimic, pyriproxyfen*, Sumitomo Pyrethroid World 15 (1990), pp. 2–5.
74. M. Hatakoshi, Y. Shono, H. Yamamoto, and M. Hirano, *Effects of a juvenile hormone analogue pyriproxyfen on* Myzus persicae *and* Unaspis yanonensis, Appl. Entomol. Zool. 26 (1991), pp. 412–414.
75. H.P. Streibert, M.L. Frischknecht, and F. Karrer, *Diofenolan—A new insect growth regulator for the control of scale insects and important lepidopterous pests in deciduous fruit and citrus*, Brighton Crop Prot. Conf. Pests Dis. 1 (1994), pp. 23–30.
76. T.S. Dhadialla, G.R. Carlson, and D.P. Le, *New insecticides with ecdysteroidal and juvenile hormone activity*, Annu. Rev. Entomol. 43 (1998), pp. 545–569.
77. T.G. Grout, G.I. Richards, and P.R. Stephen, *Further non-target effects of citrus pesticides on* Euseius addoensis *and* Euseius citri (*Acari: Phytoseiidae*), Exp. Appl. Acar. 21 (1997), pp. 171–177.
78. S. Singh and K. Kumar, *Diofenolan: A novel insect growth regulator in common citrus butterfly*, Papilio demoleus, Phytoparasitica 39 (2011), pp. 205–213.
79. W.S. Bowers, T. Ohta, J.S. Cleere, and P.A. Marsella, *Discovery of insect anti-juvenile hormones in plants*, Science 193 (1976), pp. 542–547.
80. G.B. Staal, *Anti juvenile hormone agents*, Ann. Rev. Entomol. 31 (1986), pp. 391–429.
81. D.M. Soderlund, A. Messeguer, and W.S. Bowers, *Precocene II metabolism in insects: Synthesis of potential metabolites and identification of initial* in vitro *biotransformation products*, J. Agric. Food Chem. 28 (1980), pp. 724–731.
82. N.H. Haunerland and W.S. Bowers, *Comparative studies on pharmacokinetics and metabolism of the anti-juvenile hormone precocene II*, Arch. Insect. Biochem. Physiol. 2 (1985), pp. 55–63.
83. H. Aizawa, W.S. Bowers, and D.M. Soderlund, *Reactions of precocene II epoxide with model nucleophiles*, J. Agric. Food Chem. 33 (1985), pp. 406–411.

84. G.C. Unnithan, K.K. Nair, and W.S. Bowers, *Precocene-induced degeneration of the corpus allatum of adult females of* Oncopeltus fasciatus, J. Insect Physiol. 23 (1977), pp. 1081–1094.
85. W.S. Bowers and M. Aregullin, *Discovery and identification of an antijuvenile hormone from* Chrysanthemum coronarium, Mem. Inst. Oswaldo Cruz, 82 (1987), pp. 51–54.
86. I. Kiss, A. Fodor, T. Timar, S. Hosztafi, P. Sebok, T. Torok, E. Viragh, and M. Berenyi, *Biological activity of precocene analogues on* Locusta migratoria, Experienta 44 (1988), pp. 790–792.
87. G.E. Pratt, D.E. Farnsworth, N.R. Seigel, K.F. Fok, and R. Feyereisen, *Identification of an allatostatin from adult* Diploptera punctata, Biochem. Biophys. Res. Com. 163 (1989), pp. 1243–1247.
88. A.P. Woodhead, B. Stay, S.L. Seidel, M.A. Khan, and S.S. Tobe, *Primary structure of four allatostatins: Neuropeptide inhibitors of juvenile hormone synthesis*, Proc. Nat. Acad. Sci. 86 (1989), pp. 5997–6001.
89. B. Stay and SS. Tobe, *The role of allatostatins in juvenile hormone synthesis in insects and crustaceans*, Ann. Rev. Entomol. 52 (2007), pp. 277–299.
90. N. Audsley, H.J. Matthews, N.R. Price, and R.J. Weaver, *Allatoregulatory peptides in Lepidoptera, structures, distribution and functions*, J. Insect Physiol. 54 (2008), pp. 969–980.

2 Future Perspectives for Research on the Biosynthesis of Juvenile Hormones and Related Sesquiterpenoids in Arthropod Endocrinology and Ecotoxicology

Jerome H.L. Hui, William G. Bendena, and Stephen S. Tobe

CONTENTS

ABSTRACT

After the separation of arthropods and vertebrates from the last common ancestor of bilaterians about 550 million years ago, extant arthropods continue to share multiple conserved pathway components in the biosynthesis of peptide, steroid, and terpenoid (unsaturated hydrocarbon) hormones with those in the vertebrates. In the hope of achieving different potential applications, such as drug design or manipulation of the organisms, the regulation of the production of arthropod-specific hormones is a concern. Tremendous efforts have been invested in the last several decades to discover the actions of different hormones, with the sesquiterpenoids methyl farnesoate (MF) and juvenile hormones (JH) now well known to regulate many developmental and reproductive processes in crustaceans and insects, respectively. To date, most research on the regulation by MF and JH production has focused on the study of their interactions with other hormones. In this chapter, we highlight and summarize the latest findings on the potential rate-determining enzymes in the synthesis of MF and JH, farnesoic acid *O*-methyltransferase and juvenile hormone *O*-methyltransferase, and other key candidates to investigate and manipulate sesquiterpenoid hormone regulation in arthropods. With the recent sequencing of the first crustacean (*Daphnia*) genome, the potential of comparative genomics and microRNAs to shed light on arthropod endocrinology and ecotoxicology are also discussed.

KEYWORDS

Sesquiterpenoid, Juvenile hormone, Methyl transferase, Methyl farnesoate, Biosynthesis, Toxicology

2.1 SESQUITERPENOIDS METHYL FARNESOATE AND JUVENILE HORMONES

Hormones act as chemical signals released by programmed cells to target specific cells for responses. Such a mechanism needs to be precisely controlled in organisms, and an imbalance in the control of an established hormonal system will result in changes in responsive targets and thus alter the biological functions. The phylum Arthropoda (Greek, "jointed feet") comprises members of chelicerates (mites, scorpions, spiders), crustaceans (barnacles, crabs, lobsters, shrimps), hexapods (insects), and myriapods (centipedes, millipedes) and accounts for the majority of described living animal species in the world. The diverse body plans, behaviors, developmental modes, and reproductive and endocrine systems that have evolved in the arthropods have contributed to the habitats they have successfully populated. The reasons for studying the endocrine systems of these fascinating organisms range from conserving biodiversity, scientific curiosity, improving human health, and increasing aquaculture production to combating arthropod pests. The knowledge acquired from the study of arthropod hormonal systems is clearly beneficial to different fields of biology.

An obvious morphological characteristic differentiating the vertebrates and arthropods is the development of rigid chitinous exoskeletons in the latter. Periodic exuviation or ecdysis (shedding of the old exoskeleton) is thus required to allow the incremental increase in body size throughout arthropod life. The innovation of such developmental processes has coevolved with distinctive hormonal systems, which involve the precisely controlled biosynthesis or release of peptides, steroids, and terpenoids (unsaturated hydrocarbons, class of terpenes, e.g., sesquiterpenoid $C_{15}H_{24}$).

Actions of different families of arthropod hormones have been extensively reviewed elsewhere and will not be repeated here (e.g., arthropod neuropeptides [1–3], insect juvenile hormones (JHs) [4,5]). In brief, sesquiterpenoids (terpenoids comprising three isoprene units) including methyl farnesoate (MF) and JH are well known to control development and reproduction in many crustaceans and insects, respectively. In the last decade, to precisely monitor or manipulate the biosynthesis of these sesquiterpenoid hormones in arthropods, numerous studies have focused on their interactions with other hormones. As noted earlier, the possible rate-determining enzymes in these biosynthetic pathways include farnesoic acid *O*-methyltransferase (FAMeT) and juvenile hormone acid *O*-methyltransferase (JHAMT). In this chapter, we summarize the latest knowledge on these enzymes, discuss how comparative genomics will bring new insights to the field, and speculate regarding microRNAs as future candidates and targets to investigate the dynamics of hormonal controls and ecotoxicology. Initially, we will discuss the biosynthetic pathways for the unique arthropod sesquiterpenoids, farnesoic acid (FA), MF, and JHs.

The biosynthetic pathway from conversion of acetate (C2) to farnesyl pyrophosphate (C15) has been conserved between vertebrates and arthropods, after their separation at around the Cambrian period (~550 MYA) [4–6] (Figure 2.1). In brief, the farnesyl (C15) units result from the condensation of three isoprene (C5) units originating from acetate (C2). Unlike arthropods that do not produce cholesterol *de novo*, farnesyl units are converted to squalene (C30) by farnesyl diphosphate synthase and further processed to cholesterol in vertebrates (for details of this biosynthetic pathway, see [7,8]). In arthropods, farnesyl pyrophosphate is converted to FA via the actions of FPP phosphatase, farnesol dehydrogenase, and farnesal dehydrogenase [9].

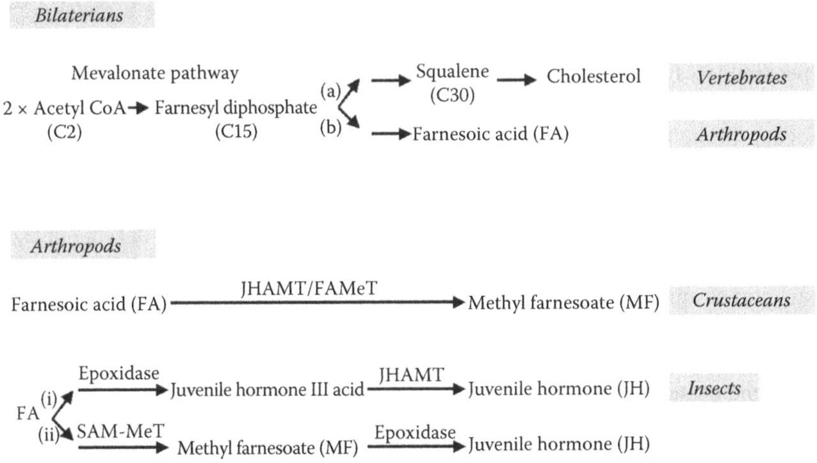

FIGURE 2.1 Biosynthetic pathway of sesquiterpenoids in insects and crustaceans. For details, see main text. (Adapted from Tobe, S.S. and Bendena, W.G., *Ann. N. Y. Acad. Sci.*, 897, 300, 1999; Hui, J.H.L. et al., *Peptides*, 31, 451, 2010.)

The end products of the biosynthetic pathway in crustaceans and insects are, however, different. In crustaceans, FA is thought to be converted to the final product MF in the presence of the cofactor *S*-adenosyl-methionine (SAM) and FAMeT (Figure 2.1). On the other hand, FA in insects can be converted by two different biosynthetic pathways: (1) FA is converted to juvenile hormone III acid (JHA) and then juvenile hormone III (JH-III) by an epoxidase and a SAM-dependent methyltransferase or (2) a similar pathway to the crustaceans whereby FA is first converted to MF and then to JH-III by the action of an epoxidase [4,5] (Figure 2.1). It should be stressed that insects in different orders use either one pathway or the other. For example, dictyopterans such as cockroaches use pathway 2, whereas the lepidopterans such as moths and butterflies use pathway 1. The organs that produce these hormones differ between crustaceans and insects, although the organs appear to have a common embryological origin. For example, FA and MF are produced by the mandibular organ (MO) in crustaceans [10,11], and all forms of JH and related sesquiterpenoids are synthesized by the corpora allata (CA) of insects [12,13]. The major functions of JH, FA, and MF are the regulation of gametogenesis/reproduction, metamorphosis, metabolic activities, and ecdysteroid secretion for molting in insects and crustaceans, respectively (for details, see [2–4]).

An outstanding perplexing issue for arthropod endocrinologists is the question of the nature of the active hormones that are capable of eliciting physiological responses in target tissues. For some groups, for example, the Hemimetabola, the ubiquitous sesquiterpenoid appears to be JH-III, with some notable exceptions, such as Hemiptera. The situation appears to be more complex for the higher insects and decapod crustaceans. In insects within the Lepidoptera, there may be up to five different but closely related JHs produced by the CA, as well as JHA. Although the activities of these sesquiterpenoids have been assessed in bioassays [13], the *in vivo* physiological actions remain to be adequately defined. The question of blends or mixtures of the various sesquiterpenoids being required to realize a physiological action has not been adequately explored. This issue may be further complicated by the possibility of synergism between the sesquiterpenoids. An equally complex situation appears to exist in the higher Diptera, in which MF, JH-III, and JH-III bisepoxide occur. Some authors have suggested that the principal hormone is MF as JH-III binds to the potential functional retinoic acid retinoid X receptor (RXR) receptor (ultraspiracle receptor (USP)) in *Drosophila* [14,15]. The possibility that one or more of these sesquiterpenoids may be active in one stage, such as the larval stage, but are not active (or less active) in another (adult) has been suggested [16]. As well, do specific sesquiterpenoids regulate specific developmental and reproductive processes? Resolution of these issues must await the definitive identification, cloning, and expression of the JH receptor. Expression of the receptor would permit stage-specific and tissue-specific assays with different sesquiterpenoid ligands. In addition, because of the occurrence of multiple forms of the sesquiterpenoids in a range of insects, it is possible that there are multiple forms of the receptor.

The situation in decapod crustaceans is equally complex and confusing. Both MF and FA exert clear physiological effects but which combination of sesquiterpenoid regulates the physiological responses is unknown (for details, see [3]). Both FA and MF are released from the MO *in vitro* although FA has not been definitively identified in hemolymph *in vivo* [11,17–20]. However, in physiological assays and *in vitro* monitoring of the production of the apolipoprotein gene product (previously incorrectly identified as the "vitellogenin" homologue in crustaceans) by the ovary and hepatopancreas, FA is considerably more active than MF [21–27]. Interestingly, ecdysteroids secreted from the Y organ can also enhance this effect. Although these studies suggest that FA is an active hormone in decapods, the major question remains as to whether FA could be rapidly degraded or converted to MF in other tissues. Since MF is also released from the MO, albeit at lower rates, it is possible that MF represents a storage form for the active hormone and that it may be converted into a related sesquiterpenoid by lipases at target tissues. At present, this is not clear, as the hydrolysis of MF to FA has not been described. Although there appears to be an MF-binding protein in decapod hemolymph [28–32], it remains unknown if hemolymph contains an FA-binding protein to aid in transport and protection of the compound. Alternatively, the receptor for FA and/or MF in crustaceans may differ from that in insects because of the differences in physiochemical properties of the sesquiterpenoids. As well, there may be contributions of ecdysone receptors and RXRs in relation to the FA and MF actions in crustaceans, given that JH-III can potentially interact with the retinoic acid RXR in insects [15,33].

Is there any evidence suggesting that knowledge of arthropod endocrinology could benefit the field of ecotoxicology and vice versa? In reality, evidence supports this notion already! For instance, in the presence of stressors such as elevated temperature, anoxia, or changes in salinity of seawater, studies have shown that the titer of MF in hemolymph increases in response [34–36]. These studies suggest that the hormonal system of arthropods is responsive to environmental stress. Furthermore, by treating gravid water fleas *Daphnia magna* with JH agonists and other contaminants that occur in the environment, the production of male offspring was promoted [37–40]. Given the interactive relationships between environmental toxicity and endocrine systems, the possibility of knowledge transfer from one field to another could be useful in the future (see further discussion as follows).

2.2 SESQUITERPENOID HORMONE CATALYTIC ENZYMES (FAMET AND JHAMT)

In light of the importance of the regulation of the hormonal systems and thus different developmental processes of arthropods, regulation of FA, JH, and MF production is another focus in the field. To date, most studies have focused on neuropeptide regulators, the allatostatins and allatotropins, on the biosynthesis of JH in insects [1,41–43] and the neuropeptides crustacean hyperglycemic hormone, MO-inhibiting hormone, pigment-dispersing hormone, and red pigment–concentrating hormone on MF production in crustaceans [44–48]. Although this line of research is unequivocally important, it should be realized that the study of rate-determining enzymes that catalyze the biosynthesis of JH, FA, and MF (i.e., JHAMT and possibly FAMeT) has been largely neglected.

Depending on the species, the final step in the sesquiterpenoid biosynthetic pathway is the conversion of either FA to MF or MF to JH, with few exceptions (e.g., higher Diptera and some Lepidoptera) (see previous section; Figure 2.1). In both cases, these SAM-dependent methylation steps could also be one of the rate-determining regulatory points in each of these pathways. Although the study of these enzymes is clearly important to our understanding and ability to manipulate the regulation of arthropod hormone production and potential applicability to other fields such as ecotoxicology (see next section), most of the research done previously has focused on either characterizing their biochemical properties or using them as rate quantifiers for FA/MF/JH production. Molecular characterization of putative FAMeTs and JHAMTs has only begun to attract attention in the last decade.

FAMeTs have been investigated in both insects and crustaceans, although they seem to play different biochemical roles. In crustaceans, FAMeTs may catalyze the conversion of FA to MF in lobster *Homarus americanus* and shrimp *Metapenaeus ensis* [49,50], although evidence for enzymatic activity is limited. Nonetheless, it appears to be a rate-determining step in this sesquiterpenoid biosynthetic pathway [48] (Table 2.1). Knockdown of FAMeT expression has also been shown to affect molting and cause lethality in shrimp *Litopenaeus vannamei* [51]. In insects, however, knockdown of FAMeT activity in the fly *Drosophila melanogaster* does not decrease sesquiterpenoid hormone production significantly [52]. Since it is now known that there are different mechanisms generating the FAMeT isoforms in insects and crustaceans (in insects, ancestral duplication followed by a loss of genes in different insect species versus multiple independent lineage-specific duplications in crustaceans) [53], whether the effect seen in *D. melanogaster* is a consequence of

TABLE 2.1
Summary of Animal Lineages Containing FAMeT/JHAMT Homologues and Their Putative Functions

Animal Lineages	Functions
FAMeT	
Crustaceans (except *Daphnia*)	FA conversion to MF?
Insects	Not clear (see *Drosophila* discussion)
Polychaetes	Unknown
Cephalochordates	Unknown
Vertebrates	No homologue
JHAMT	
Crustaceans	Identified in *Daphnia pulex* only function unknown
Insects	JHIIIA conversion to JH
Polychaetes	No homologue
Cephalochordates	No homologue
Vertebrates	No homologue

a secondary derived pathway in the drosophilid lineage or is representative of the whole order Insecta remains to be investigated.

JHAMT, on the other hand, has been demonstrated to be the enzyme responsible for the conversion of JHA or FA to JH (through MF) in a range of insects, including the silkworm *Bombyx mori*, the fly *D. melanogaster*, and the beetle *Tribolium castaneum* [54–56] (Table 2.1). Functional (knockdown/overexpression) experiments on JHAMT have suggested its essential role in maintaining the normal development of insects [54–57]. A recent study by [58] in the hemimetabolous locust *Schistocerca gregaria* has clearly demonstrated that the JHAMT is responsible for conversion of FA to MF and then JH-III, based on RNAi injection experiments, whereas knockdown of the FAMeT had no effect on MF or JH production.

2.3 LET'S RETHINK THE DISCOVERY OF JHAMT IN THE FIRST CRUSTACEAN GENOME: CAN *DAPHNIA* BE A PANCRUSTACEAN "LIVING FOSSIL" FOR ARTHROPOD ENDOCRINOLOGY AND ECOTOXICOLOGY?

With the advancement of sequencing technologies and reduction in its costs, sequencing varieties of animal genomes has become a reality. The feasibility and accessibility of next-generation sequencing has also facilitated the use of comparative genomic methods in arthropod endocrinology and ecotoxicology. Here we ask the question how has comparative genomics changed our views on FAMeT and JHAMT?

Animal lineages containing FAMeTs and/or JHAMTs have proven to be ambiguous. For example, it has long been suggested in the field that the vertebrates contain a FAMeT orthologue (e.g., [50,59]), despite the knowledge that the final steps in sterol biosynthesis are quite different from those of the sesquiterpenoid/JH pathway of the Arthropoda. The pathways diverge at farnesyl pyrophosphate that is the final common intermediate between the two groups. Arthropods do not synthesize cholesterol, whereas vertebrates do. However, vertebrates do possess farnesyl-related compounds that are covalently bound to a cysteine residue in the C-terminal regions of proteins, a process termed protein prenylation. Prenylation is believed to aid in binding of proteins to lipophilic substrates, including cell membranes. The reaction is catalyzed by at least three different enzymes, including a methyltransferase (Figure 2.1). In the first systematic phylogenetic analyses of these enzymes, it has been demonstrated that vertebrates do not contain either FAMeT or JHAMT, and the previously described "FAMeT" homologue indeed belongs to another family of largely ignored genes in the echinoderms [53] (Table 2.1 and Figure 2.2). By comparison of the FAMeT orthologues in different genomes, it is now also clear that lineages other than arthropods such as the annelids and cephalochordates contain "FAMeTs" [53] (Table 2.1). Since some species of annelids are known to produce sterols *de novo*, elucidating the precise roles and functions of these newly discovered "FAMeTs" in respective hormonal pathways will be of importance to the fundamental understanding of hormone biosynthesis. Moreover, could there be any possibility that these "FAMeTs" are not enzymes themselves, catalyzing methylation of FA, but rather, cofactors affecting or enhancing the action of JH, which in turn suggest

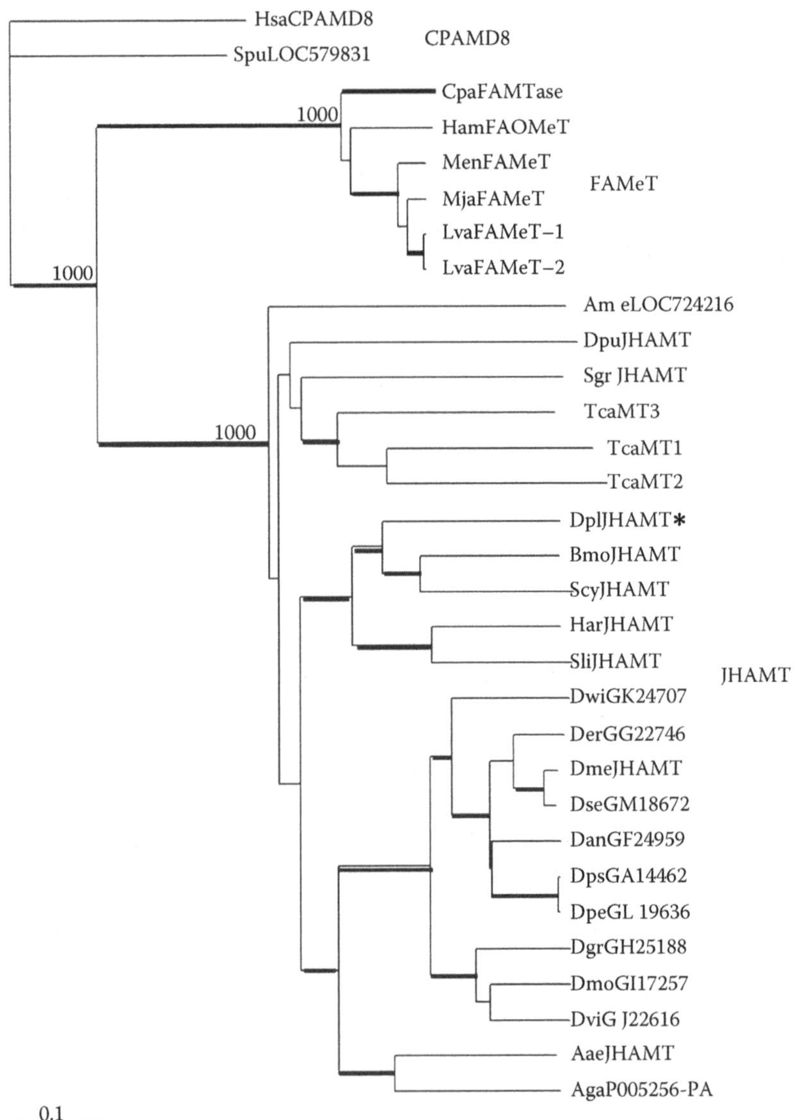

FIGURE 2.2 Neighbor-joining tree illustrating the phylogenetic relationships of JHAMT, FAMeT, and CPAMD8. In addition to the selected sequences from a previous study [53], twelve JHAMT sequences retrieved from GenBank with accession numbers ADV17350, XP_001356909, XP_002015317, EHJ73944, XP_001965404, ABE98256, XP_001976217, XP_002038170, XP_001996441, XP_002002213, XP_002052259, and XP_002066866 were included in the analysis. The asterisk depicts the position of the putative crustacean JHAMT/ FAMeT sequence from the crustacean *D. pulex*. For taxon abbreviations, please refer to [53]. Bold lines indicate clades supported by bootstrap values above 950.

other phyla could contain JH-related hormones yet to be discovered? These are some examples of the value of comparative genomics in modifying and expanding existing knowledge of arthropod endocrinology.

Before such comparative genomic analyses, most JHAMT sequences were identified by the conventional method of PCR amplification with degenerate primers designed on the basis of reported sequences. Given the limitation of this labor-intensive and cost-ineffective method, JHAMTs had only been identified in the insects. Confinement of JHAMTs to the Insecta was the common thought in the field, indicating this gene could possibly have arisen after the separation of insects and crustaceans from the Pancrustacea. With the sequencing of the first crustacean genome, *Daphnia pulex* [60], a JHAMT homologue has been unexpectedly also discovered in this animal [53] (Table 2.1 and Figure 2.2). This finding is important to both arthropod endocrinology and ecotoxicology in many respects, and its implications are discussed in the succeeding text.

To arthropod endocrinology, could the identification of the first *bona fide* crustacean JHAMT indicate that other classes of arthropods or subclasses of crustaceans contain JHAMTs in their genomes? The determination of JHAMT expression and abundance in other crustaceans and its roles in hormone biosynthesis are of immense importance to our existing knowledge of how arthropods produce their hormones. Regardless of the limited amount of existing data, this exciting finding has already suggested that the Pancrustacea possessed the JHAMT before the splitting of Arthropoda into crustaceans and insects.

On the other hand, could this simply mean that the lineage *Daphnia* has developed a different or derived hormonal system from other crustaceans? According to the analysis of *Daphnia* EST and draft genome databases, FAMeT was not identified and only a JHAMT homologue was recovered [53] (Table 2.1 and Figure 2.2). If all the other crustaceans from the Pancrustacea have lost JHAMT, *Daphnia* will then become a “living fossil” for the linking of crustacean and insect endocrinology. Since MF also has been shown to play a role in determining the sex of *D. magna* [37,38,61], perhaps revealing the relationship between JHAMT and sesquiterpenoid biosynthesis in *Daphnia* is another way to go forward. However, the MO that is responsible for producing MF in this organism is physically too small for manual dissection, and this has hindered the conventional physiological approaches to this question. Other clever experimental designs need to be developed in the future to carry out the task. It is also important to recognize that MF is a product of release from CA in many insect species, for example, *Drosophila* and other higher flies [16], again suggesting an endocrine function for this compound. Although little MF appears to be released *in vitro* from CA, with the majority of the release product being JH-III or in the case of flies, JHB_3, MF does show biological activity in insects and in fact, has been suggested to be the “authentic” JH in at least some insect species [14]. MF is also made by the CA of embryos of cockroaches early in development and precedes the ability to produce JH-III [62]. The methyltransferase is clearly present in excess at this stage since stimulation of biosynthesis by FA results in the production of large amounts of MF.

The water flea *Daphnia* is also well known as one of the most sensitive animal models for the assessment of ecotoxicity [39,40,63]. With *D. pulex* being

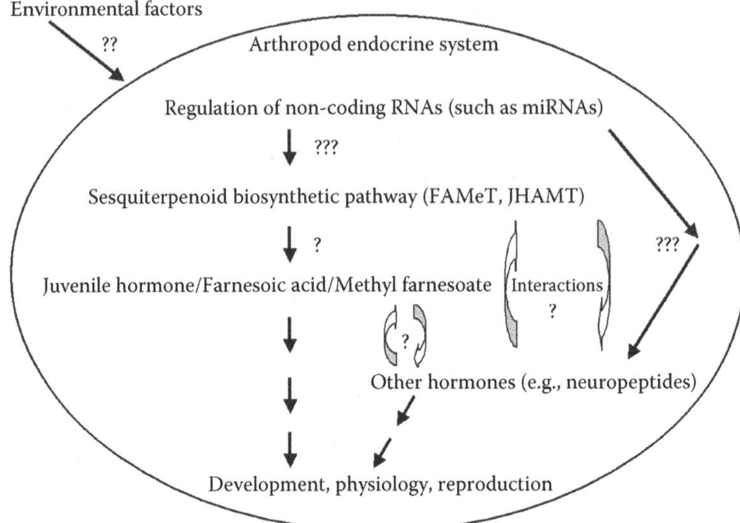

FIGURE 2.3 Possible future research directions on arthropod FAMeT, JHAMT, sesquiterpenoids, and microRNAs in response to environmental challenges. The question marks denote the current major research gaps.

the first crustacean genome sequenced, comparisons with other genomes have revealed more than one-third of its genes have no identifiable homologues [60,64]. Interestingly, *D. pulex*–specific duplicated genes are differentially expressed following environmental challenges [60]. Switching of reproductive modes between parthenogenetic and sexual reproduction has been suggested to facilitate such adaptation [65]. Irrespective of the interesting evolutionary reasons that account for such phenotypes, *Daphnia* is clearly a new genetic model for ecotoxicology. However, can the combination of understanding how JHAMT and the biosynthesis of sesquiterpenoids in response to different environmental changes make a contribution to our understanding of ecotoxicology (Figure 2.3)? Recent findings support this view. In the presence of kairomones, it has been shown that JHAMT is one of the most upregulated candidate genes in *Daphnia* [66]. Hence, we suggest that the study of the interactions of JHAMT and biosynthesis of sesquiterpenoids in different arthropods under different environmental conditions will provide useful information on ecotoxicology.

2.4 MicroRNA AS NEW TARGETS FOR REGULATION OF SESQUITERPENOID PRODUCTION AND ACTION

In recent years, it has become clear that many of the RNA transcripts from the genome do not code for "functional" proteins. These regions, which were originally called "noncoding RNAs," are now known to play important roles in many biological processes. Among the different types of noncoding RNAs, the microRNAs (miRNAs) are highly conserved in sequence and functions across animal species. Although the first animal miRNA, lin-14, was discovered in the nematode *Caenorhabditis elegans* almost two decades ago [67], the miRNA field did not garner much attention until the discovery of a second miRNA, let-7, which is highly conserved across animal species [68,69]. miRNAs are now known to be ~21 to 23 nucleotides in length that act post-transcriptionally to regulate RNA transcripts. Although the double-stranded RNAs or small-interfering RNAs (dsRNAs/siRNAs) can also change the expression of RNA transcripts and share many conserved biosynthetic pathway components with the miRNAs, there are some fundamental differences between the two. In animals, siRNAs bind with perfect complementary to the targeted mRNA, resulting in the cleavage of mRNA targets, whereas miRNAs bind with either partial or perfect complementarity to the 3'UTRs of the targeted mRNAs, resulting in posttranscriptional/translational repression [70–72]. Moreover, endogenous dsRNAs/siRNAs are relatively very rare in animals, whereas miRNAs are common and highly conserved across the majority of animals. Given these different regulatory properties, most research done on dsRNA/siRNA have utilized them exogenously, to destroy the activity of the target gene through RNA interference, whereas miRNAs seem to be ideal for the purpose of revealing and modulating the intrinsic regulation of different biological processes. Therefore, it is not surprising that the study of miRNAs has become one of the hottest research fields in the past few years.

In the last decade, although miRNAs have been found to control many different aspects of cellular function—from cancer, development, physiology, and reproduction to varieties of diseases—no study on their role in arthropod sesquiterpenoid biosynthesis has been forthcoming to our knowledge (Figure 2.3). This is likely because (1) the study of miRNAs is still relatively new and its current major focus has been biased toward the understanding of their biogenesis and regulatory functions and (2) hormonal regulation in arthropods has also focused on the study of hormonal interactions rather than other possible individual components (see earlier section). In an earlier study [73], knockdown of Dicer-1 in the hemimetabolous insect *Blattella germanica* depleted the miRNAs in the last larval instars and inhibited metamorphosis. Although no specific miRNA candidate has been identified in regulating the components of ecdysteroid and/or JH signaling, this study suggested that miRNAs do play a role in the regulation of insect metamorphosis. In our preliminary bioinformatic searches, some miRNA candidates can be identified as potential regulators of the sesquiterpenoid pathway components including the JHAMT (Hui and Tobe, unpublished data). Despite such potential miRNAs–sesquiterpenoid gene interactions, further validation is required to reveal how these miRNAs can regulate the biosynthetic pathway of sesquiterpenoids. It is very likely that miRNAs will be useful candidates to provide another dimension to our understanding of arthropod endocrinology that is largely ignored at present (Figure 2.3).

Studying miRNAs will also be of benefit to the field of ecotoxicology (Figure 2.3). The regulation of hormone biosynthesis genes following environmental challenges is very likely to be under the control of miRNAs; in other words, miRNAs can potentially be sensitive ecotoxicological markers. Furthermore, depending on the purposes of study, targeting miRNAs for manipulation of organisms following different environmental/physiological challenges is clearly another way to go forward. In summary, the combinatorial study of dynamics and relationships between environmental changes, arthropod sesquiterpenoid hormones, and miRNAs could and should become a new trend and another major line of research in the coming decade.

REFERENCES

1. W.G. Bendena, B.C. Donly, and S.S. Tobe, *Allatostatins: A growing family of neuropeptides with structural and functional diversity*, Ann. N.Y. Acad. Sci. 897 (1999), pp. 311–329.
2. S.M. Chan, P.L. Gu, K.H. Chu, and S.S. Tobe, *Crustacean neuropeptide genes of the CHH/MIH/GIH family: Implications from molecular studies*, Gen. Comp. Endocrinol. 134 (2003), pp. 214–219.
3. J.H.L. Hui, *The farnesoic acid O-methyltransferase is an essential molt regulator in the shrimp* Litopenaeus vannamei, University of Hong Kong Thesis, Hong Kong, 2005.
4. S.S. Tobe and W.G. Bendena, *The regulation of juvenile hormone production in arthropods. Functional and evolutionary perspectives*, Ann. N.Y. Acad. Sci. 897 (1999), pp. 300–310.
5. X. Bellés, D. Martín, and M.D. Piulachs, *The mevalonate pathway and the synthesis of juvenile hormone in insects*, Ann. Rev. Entomol. 50 (2005), pp. 181–199.
6. S.J. Seybold and C. Tittiger, *Biochemistry and molecular biology of de novo isoprenoid pheromone production in the Scolytidae*, Ann. Rev. Entomol. 48 (2003), pp. 425–453.
7. J.L. Goldstein and M.S. Brown, *Regulation of the mevalonate pathway*, Nature 343 (1990), pp. 425–430.
8. C. Weinberger, *A model for farnesoid feedback control in the mevalonate pathway*, Trends Endocrinol. Metab. 7 (1996), pp. 1–6.
9. A.J. Clark and K. Bloch, *The absence of sterol synthesis in insects*, J. Biol. Chem. 234 (1959), pp. 2578–2582.
10. H. Laufer, D. Borst, F.C. Baker, C.C. Reuter, L.W. Tsai, D.A. Schooley, C. Carrasco, and M. Sinkus, *Identification of a juvenile hormone-like compound in a crustacean*, Science 235 (1987), pp. 202–205.
11. S.S. Tobe, D.A. Young, and H.W. Khoo, *Production of methyl farnesoate by the mandibular organs of the mud crab,* Scylla serrata*: Validation of a radiochemical assay*, Gen. Comp. Endocrinol. 73 (1989), pp. 342–353.
12. A. Le Roux, *Description of d'organes mandibulaires nouveaux chez les crustaces décapodes*, C. R. Acad. Sci. Paris (D). 226 (1968), pp. 1414–1417.
13. S.S. Tobe and B. Stay, *Structure and regulation of the corpus allatum*, Adv. Insect Physiol. 18 (1985), pp. 305–432.
14. D. Jones and G. Jones, *Farnesoid secretions of dipteran ring glands: What we do know and what we can know*, Insect Biochem. Mol. Biol. 37 (2007), pp. 771–798.
15. G. Jones and P.A. Sharp, *Ultraspiracle: An invertebrate nuclear receptor for juvenile hormones*, Proc. Natl. Acad. Sci. USA 94 (1997), pp. 13499–13503.
16. W.G. Bendena, J. Zhang, S.M. Burtenshaw, and S.S. Tobe, *Evidence for differential biosynthesis of juvenile hormone (and related) sesquiterpenoids in* Drosophila melanogaster, Gen. Comp. Endocrinol. 172 (2011), pp. 56–61.
17. D.W. Borst, L. Kissee, and D. Ramlose, *The synthesis of methyl farnesoate (MF) by mandibular organs (MO) of two crabs*, Am. Zool. 27 (1987), p. 69a.
18. H. Laufer, M. Landau, E. Homola, and D.W. Borst, *Methyl farnesoate; its site of synthesis and regulation of secretion in a juvenile crustacean*, Insect Biochem. 17 (1987), pp. 1129–1131.
19. B. Tsukimura and D.W. Borst, *Regulation of methyl farnesoate in the hemolymph and mandibular organ of the lobster,* Homarus americanus, Gen. Comp. Endocrinol. 86 (1992), pp. 297–303.

20. S.L. Tamone and E.S. Chang, *Methyl farnesoate stimulates ecdysteroid secretion from crab Y-organs* in vitro, Gen. Comp. Endocrinol. 89 (1993), pp. 425–432.
21. A.S. Mak, C.L. Choi, S.H. Tiu, J.H. Hui, J.G. He, S.S. Tobe, and S.M. Chan, *Vitellogenesis in the red crab* Charybdis feriatus*: Hepatopancreas-specific expression and farnesoic acid stimulation of vitellogenin gene expression*, Mol. Reprod. Dev. 70 (2005), pp. 288–300.
22. S.M. Chan, A.S. Mak, C.L. Choi, T.H. Ma, J.H. Hui, and S.H. Tiu, *Vitellogenesis in the Red Crab,* Charybdis feriatus*: Contributions from Small vitellogenin transcripts (CfVg) and farnesoic acid stimulation of CfVg expression*, Ann. N.Y. Acad. Sci. 1040 (2005), pp. 74–79.
23. S.H. Tiu, J.H. Hui, J.G. He, S.S. Tobe, and S.M. Chan, *Characterization of vitellogenin in the shrimp Metapenaeus ensis: Expression studies and hormonal regulation of MeVg1 transcription* in vitro, Mol. Reprod. Dev. 73 (2006), pp. 424–436.
24. J.C. Avarre, E. Lubzens, and P.J. Babin, *Apolipocrustacein, formerly vitellogenin, is the major egg yolk precursor protein in decapod crustaceans and is homologous to insect apolipophorin II/I and vertebrate apolipoprotein B*, BMC Evol. Biol. 7 (2007), p. 3.
25. S.H. Tiu, J.H.L. Hui, B. Tsukimura, S.S. Tobe, H.G. He, and S.M. Chan, *Cloning and expression study of the lobster (*Homarus americanus*) vitellogenin: Conservation in gene structure among decapods*, Gen. Comp. Endocrinol. 160 (2009), pp. 36–46.
26. S.H. Tiu, S.M. Chan, and S.S. Tobe, *The effects of farnesoic acid and 20-hydroxyecdysone on vitellogenin gene expression in the lobster,* Homarus americanus*, and possible roles in the reproductive process*, Gen. Comp. Endocrinol. 166 (2010), pp. 337–345.
27. A. Hayward, T. Takahashi, W.G. Bendena, S.S. Tobe, and J.H. Hui, *Comparative genomic and phylogenetic analysis of vitellogenin and other large lipid transfer proteins in metazoans*, FEBS Lett. 584 (2010), pp. 1273–1278.
28. G.D. Prestwich, M.J. Bruce, I. Ujváry, and E.S. Chang, *Binding proteins for methyl farnesoate in lobster tissues: Detection by photoaffinity labeling*, Gen. Comp. Endocrinol. 80 (1990), pp. 232–237.
29. H. Li and D.W. Borst, *Characterization of a methyl farnesoate binding protein in hemolymph from* Libinia emarginata, Gen. Comp. Endocrinol. 81 (1991), pp. 335–342.
30. P. Takac, H. Laufer, and G.D. Prestwich, *Characterization of methyl farnesoate (MF) binding proteins and the metabolism of MF by some tissues of the spider crab,* Libinia emarginata, Am. Zool. 33 (1993), p. 10A.
31. L.E. King, Q. Ding, G.D. Prestwich, and S.S. Tobe, *The characterization of a hemolymph methyl farnesoate binding protein and assessment of methyl farnesoate metabolism by the hemolymph and other tissues from* Procambarus clarkii, Insect Biochem. Mol. Biol. 25 (1995), pp. 495–501.
32. S.L. Tamone, G.D. Prestwich, and E.S. Chang, *Identification and characterization of methyl farnesoate binding proteins from the crab,* Cancer magister, Gen. Comp. Endocrinol. 105 (1997), pp. 168–175.
33. S.H. Tiu, E.F. Hult, K.J. Yagi, and S.S. Tobe, *Farnesoic acid and methyl farnesoate production during lobster reproduction: Possible functional correlation with retinoid X receptor expression*, Gen. Comp. Endocrinol. 175 (2012), pp. 259–269.
34. D.L. Lovett, P.D. Clifford, and D.W. Borst, *Physiological stress elevates hemolymph levels of methyl farnesoate in the green crab* Carcinus maenas, Biol. Bull. 193 (1997), pp. 266–267.
35. D.W. Borst, J.T. Ogan, B. Tsukimura, T. Claerhout, and K.C. Holford, *Regulation of the crustacean mandibular organ*, Am. Zool. 41 (2001), pp. 430–441.
36. D.L. Lovett, M.P. Verzi, P.D. Clifford, and D.W. Borst, *Hemolymph levels of methyl farnesoate increase in response to osmotic stress in the green crab,* Carcinus maenas, Comp. Biochem. Physiol. A Mol. Integr. Physiol. 128 (2001), pp. 299–306.

37. A.W. Olmstead and G.A. Leblanc, *Juvenoid hormone methyl farnesoate is a sex determinant in the crustacean* Daphnia magna, J. Exp. Zool. 293 (2002), pp. 736–739.
38. A.W. Olmstead and G.A. LeBlanc, *Insecticidal juvenile hormone analogs stimulate the production of male offspring in the crustacean* Daphnia magna, Environ. Health Perspect. 111 (2003), pp. 919–924.
39. N. Tatarazako and S. Oda, *The water flea* Daphnia magna *(Crustacea, Cladocera) as a test species for screening and evaluation of chemicals with endocrine disrupting effects on crustaceans*, Ecotoxicology 16 (2007), pp. 197–203.
40. H.Y. Wang, A.W. Olmstead, H. Li, and G.A. Leblanc, *The screening of chemicals for juvenoid-related endocrine activity using the water flea* Daphnia magna, Aquat. Toxicol. 74 (2005), pp. 193–204.
41. R. Kwok, J.R. Zhang, and S.S. Tobe, *Regulation of methyl farnesoate production by mandibular organs in the crayfish,* Procambarus clarkii*: A possible role for allatostatins*, J. Insect Physiol. 51 (2005), pp. 367–378.
42. B. Stay and S.S. Tobe, *The role of allatostatins in juvenile hormone synthesis in insects and crustaceans*, Ann. Rev. Entomol. 52 (2007), pp. 277–299.
43. R.J. Weaver and N. Audsley, *Neuropeptide regulators of juvenile hormone synthesis: Structures, functions, distribution, and unanswered questions*, Ann. N.Y. Acad. Sci. 1163 (2009), pp. 316–329.
44. M. Landau, H. Laufer, and E. Homola, *Control of methyl farnesoate synthesis in the mandibular organ of the crayfish* Procambarus clarkii*: Evidence for peptide neurohormones with dual functions*, Invertebr. Reprod. Dev. 16 (1989), pp. 165–168.
45. D.W. Borst, B. Tsukimura, and E.S. Chang, *The regulation of methyl farnesoate levels in crustaceans, Pres. 5th International Symposium Juvenile Hormones*, La Londe les Maures, France, 1991.
46. H. Laufer, L. Liu, and F. Van Herp, *A neuropeptide family that inhibits the mandibular organ of crustacea and may regulate reproduction*, in *Insect Neurochemistry and Neurophysiology*, A.B. Barkovec and E.P. Masler, eds., Humana Press, Clifton, NJ, 1994, pp. 203–206.
47. L. Liu, H. Laufer, Y. Wang, and T. Hayes, *A neurohormone regulating both methyl farnesoate synthesis and glucose metabolism in a crustacean*, Biochem. Biophys. Res. Commun. 237 (1997), pp. 694–701.
48. G. Wainwright, S.G. Webster, and H.H. Rees, *Neuropeptide regulation of biosynthesis of the juvenoid, methyl farnesoate, in the edible crab,* Cancer pagurus, Biochem. J. 334 (1998), pp. 651–657.
49. Y.I. Gunawardene, S.S. Tobe, W.G. Bendena, B.K. Chow, K.J. Yagi, and S.M. Chan *Function and cellular localization of farnesoic acid O-methyltransferase (FAMeT) in the shrimp,* Metapenaeus ensis, Eur. J. Biochem. 269 (2002), pp. 3587–3595.
50. K.C. Holford, K.A. Edwards, W.G. Bendena, S.S. Tobe, Z. Wang, and D.W. Borst, *Purification and characterization of a mandibular organ protein from the American lobster, Homarus americanus: A putative farnesoic acid O-methyltransferase*, Insect Biochem. Mol. Biol. 34 (2004), pp. 785–798.
51. J.H.L. Hui, S.S. Tobe, and S.M. Chan, *Characterization of the putative farnesoic acid O-methyltransferase (LvFAMeT) cDNA from white shrimp,* Litopenaeus vannamei*: Evidence for its role in molting*, Peptides 29 (2008), pp. 252–260.
52. S.M. Burtenshaw, P.P. Su, J.R. Zhang, S.S. Tobe, L. Dayton, and W.G. Bendena, *A putative farnesoic acid O-methyltransferase (FAMeT) orthologue in* Drosophila melanogaster *(CG10527): Relationship to juvenile hormone biosynthesis*, Peptides 29 (2008), pp. 242–251.
53. J.H.L. Hui, A. Hayward, W.G. Bendena, T. Takahashi, and S.S. Tobe, *Evolution and functional divergence of enzymes involved in sesquiterpenoid hormone biosynthesis in crustaceans and insects*, Peptides 31 (2010), pp. 451–455.

54. T. Kinjoh, Y. Kaneko, K. Itoyama, K. Mita, K. Hiruma, and T. Shinoda, *Control of juvenile hormone biosynthesis in* Bombyx mori*: Cloning of the enzymes in the mevalonate pathway and assessment of their developmental expression in the corpora allata,* Insect Biochem. Mol. Biol. 37 (2007), pp. 808–818.
55. C. Minakuchi, T. Namiki, M. Yoshiyama, and T. Shinoda, *RNAi-mediated knockdown of juvenile hormone acid O-methyltransferase gene causes precocious metamorphosis in the red flour beetle* Tribolium castaneum, FEBS J. 275 (2008), pp. 2919–2931.
56. R. Niwa, T. Niimi, N. Honda, M. Yoshiyama, K. Itoyama, H. Kataoka, and T. Shinoda *Juvenile hormone acid O-methyltransferase in* Drosophila melanogaster, Insect Biochem. Mol. Biol. 38 (2008), pp. 714–720.
57. T. Shinoda and K. Itoyama, *Juvenile hormone acid methyltransferase: A key regulatory enzyme for insect metamorphosis*, Proc. Natl. Acad. Sci. USA 100 (2003), pp. 11986–11991.
58. E. Marchal, J. Zhang, L. Badisco, H. Verlinden, E.F. Hult, P. van Wielendaele, K.J. Yagi, S.S. Tobe, and J. Vanden Broeck, *Final steps in juvenile hormone biosynthesis in the desert locust,* Schistocerca gregaria, Insect Biochem. Mol. Biol. 41 (2011), pp. 219–227.
59. A.V. Kuballa, K. Guyatt, B. Dixon, H. Thaggard, A.R. Ashton, B. Paterson, D.J. Merritt, and A. Elizur, *Isolation and expression analysis of multiple isoforms of putative farnesoic acid O-methyltransferase in several crustacean species*, Gen. Comp. Endocrinol. 150 (2007), pp. 48–58.
60. J.K. Colbourne, M.E. Pfrender, D. Gilbert, W.K. Thomas, A. Tucker, T.H. Oakley, S. Tokishita, A. Aerts, G.J. Arnold, M.K. Basu, D.J. Bauer, C.E. Cáceres, L. Carmel, C. Casola, J.H. Choi, J.C. Detter, Q. Dong, S. Dusheyko, B.D. Eads, T. Fröhlich, K.A. Geiler-Samerotte, D. Gerlach, P. Hatcher, S. Jogdeo, J. Krijgsveld, E.V. Kriventseva, D. Kültz, C. Laforsch, E. Lindquist, J. Lopez, J.R. Manak, J. Muller, J.Pangilinan, R.P.Patwardhan, S. Pitluck, E.J. Pritham, A. Rechtsteiner, M. Rho, I.B. Rogozin, O.Sakarya, A. Salamov, S. Schaack, H. Shapiro, Y. Shiga, C. Skalitzky, Z. Smith, A. Souvorov, W. Sung, Z. Tang, D. Tsuchiya, H. Tu, H. Vos, M. Wang, Y.I. Wolf, H. Yamagata, T. Yamada, Y. Ye, J.R. Shaw, J. Andrews, T.J. Crease, H. Tang, S.M. Ucas, H.M. Robertson, P. Bork, E.V. Koonin, E.M. Zdobnov, I.V. Grigoriev, M. Lynch, and J.L. Boore, *The ecoresponsive genome of* Daphnia pulex, Science 331 (2011), pp. 555–561.
61. N. Tatarazako, S. Oda, H. Watanabe, M. Morita, and T. Iguchi, *Juvenile hormone agonists affect the occurrence of male* Daphnia, Chemosphere 53 (2003), pp. 827–833.
62. B. Stay, Z.R. Zhang, and S.S. Tobe, *Methyl farnesoate and juvenile hormone production in embryos of* Diploptera punctata *in relation to innervation of corpora allata and their sensitivity to allatostatin*, Peptides 23 (2002), pp. 1981–1990.
63. J. Martins, L.O. Teles, and V. Vasconcelos, *Assays with* Daphnia magna *and* Danio rerio *as alert systems in aquatic toxicology*, Environ. Int. 33 (2007), pp. 414–425.
64. D. Tautz and T. Domazet-Lošo, *The evolutionary origin of orphan genes*, Nat. Rev. Genet. 12 (2011), pp. 692–702.
65. D. Tautz, *Not just another genome*, BMC Biol. 9 (2011), p. 8.
66. H. Miyakawa, M. Imai, N. Sugimoto, Y. Ishikawa, A. Ishikawa, H. Ishigaki, Y. Okada, S. Miyazaki, S. Koshikawa, R. Cornette, and T. Miura, *Gene up-regulation in response to predator kairomones in the water flea,* Daphnia pulex, BMC Dev. Biol. 10 (2010), p. 45.
67. R.C. Lee, R.L. Feinbaum, and V. Ambros, *The* C. elegans *heterochronic gene lin-4 encodes small RNAs with antisense complementarity to lin-14*, Cell 75 (1993), pp. 843–854.
68. A.E. Pasquinelli, B.J. Reinhart, F. Slack, M.Q. Martindale, M.I. Kuroda, B. Maller, D.C. Hayward, E.E. Ball, B. Degnan, P. Müller, J. Spring, A. Srinivasan, M. Fishman, J. Finnerty, J. Corbo, M. Levine, P. Leahy, E. Davidson, and G. Ruvkun, *Conservation of the sequence and temporal expression of let-7 heterochronic regulatory RNA*, Nature 408 (2000), pp. 86–89.

69. B.J. Reinhart, F.J. Slack, M. Basson, A.E. Pasquinelli, J.C. Bettinger, A.E. Rougvie, H.R. Horvitz, and G. Ruvkun, *The 21-nucleotide let-7 RNA regulates developmental timing in* Caenorhabditis elegans, Nature 403 (2000), pp. 901–906.
70. D.P Bartel, *MicroRNAs: Target recognition and regulatory functions*, Cell 136 (2009), pp. 215–233.
71. M. Chekulaeva and W. Filipowicz, *Mechanisms of miRNA-mediated post-transcriptional regulation in animal cells*, Curr. Opin. Cell. Biol. 21 (2009), pp. 452–460.
72. V.N. Kim, J. Han, and M.C. Siomi, *Biogenesis of small RNAs in animals*, Nat. Rev. Mol. Cell. Biol. 10 (2009), pp. 126–139.
73. E. Gomez-Orte and X. Belles, *RNA-dependent metamorphosis in hemimetabolan insects*, Proc. Natl. Acad. Sci. USA 106 (2009), pp. 21678–21682.

3 Morph-Specific JH Titer Regulation in Wing-Polymorphic *Gryllus* Crickets

Proximate Mechanisms Underlying Adaptive Genetic Modification of JH Regulation

Anthony J. Zera

CONTENTS

ABSTRACT

A long-standing topic in insect endocrinology and evolution is the nature of adaptive variation in hormonal regulation in ecologically important polymorphisms. Studies in *Gryllus* crickets during the past two decades have been especially important in identifying genetic alterations in the hemolymph juvenile hormone titer and the regulatory mechanisms controlling these alterations in morphs (discontinuous phenotypes) adapted for flight versus reproduction. Detailed *in vitro* and *in vivo* studies in *G. rubens* and *G. assimilis* have demonstrated elevated activity of hemolymph juvenile hormone esterase (JHE) that is associated with elevated *in vivo* JH degradation and reduced JH titer during the last juvenile instar. Direct artificial selection on hemolymph JHE activity has provided a quantitative estimate of the degree to which JHE activity must evolve to produce a measurable alteration in *in vivo* JH degradation. These studies constitute the most intensive investigation of the relationship between altered JHE activity and altered *in vivo* JH degradation in morphs of any case of complex (wing, phase, caste) polymorphism in insects to date. More recent studies have identified a dramatic and unexpected morph-specific circadian rhythm for the JH titer, and JH titer regulators (e.g., biosynthetic rate) in adults of various *Gryllus* species in the laboratory and field. These data have allowed quantitative assessment of the extent to which temporal alterations in JH titer regulators account for temporal change in the JH titer. The morph-specific JH titer circadian rhythm, which occurs in the field as well as laboratory, provides a powerful experimental model for integrating endocrinology, chronobiology, ecology, and life history evolution.

KEYWORDS

Juvenile hormone (JH), Juvenile hormone esterase (JHE), Genetic variation in JH titer regulation, Wing polymorphism, Circadian rhythm, Mathematical model of JH titer regulation

3.1 INTRODUCTION AND OVERVIEW

A long-standing topic in insect physiology and evolutionary biology has been the endocrine mechanisms regulating ecologically important, complex polymorphisms, such as dispersal polymorphism, phase polymorphism in locusts, and caste polymorphism in social insects [1–6]. The word "complex" is used here to highlight the

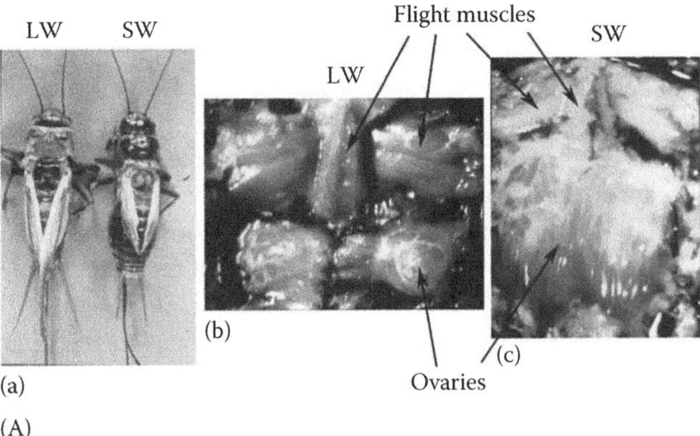

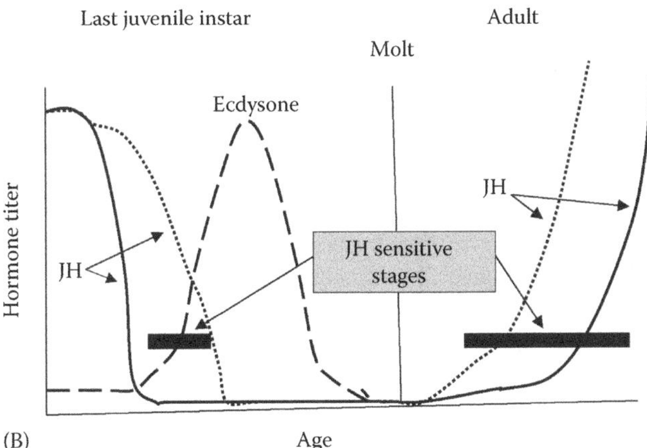

FIGURE 3.1 **(See color insert.)** (A) Flight-capable, LW and flightless, SW female morphs of *G. firmus* of the same age (day 5 of adulthood). In the panel (a), the forewings have been removed to show variation in the hind wings. The panels (b and c) illustrate dissections of the same-aged morphs showing much larger, functional flight muscles but much smaller ovaries in the LW females relative to SW females. (B) The "classic model" of JH-regulated complex polymorphism. The panel illustrates hypothetical variation in the JH titer during restricted periods (denoted by solid bars) of the juvenile and adult stages that regulates morph-specific development and reproduction. Note: "LW" is the same as "LW(f)" in some figures as follows. (From Zera, A.J., *Evol. Dev.*, 9, 499, 2007; Zera, A.J., *Integr. Comp. Biol.*, 45, 511, 2005, by permission of Oxford University Press.)

integrated multitrait basis of this type of discontinuous variation and to distinguish it from simpler single-locus cases (e.g., allozymes) [5,7]. Complex polymorphism consists of morphs (distinct phenotypes) that differ dramatically in a wide variety of anatomical, physiological, behavioral, biochemical, molecular, and life history traits, thus causing them to specialize in key roles important to the life cycle of the species (dispersers, reproductives, workers) (e.g., Figure 3.1).

The wide diversity of traits that differ between morphs, together with the known role of hormones in coordinating the expression of multiple traits, has led to the idea that modulation of endocrine regulation plays a cardinal role regulating complex polymorphism. Over the past several decades, these polymorphisms have been intensively studied, originally to identify the mechanisms by which environmental factors modulate endocrine regulatory mechanisms to produce morphs adapted for different functions [2,3]. More recently, studies of complex polymorphism have been investigated in an explicit evolutionary context to identify the proximate mechanisms underlying the evolution of development, life histories, and dispersal [5,6,8].

Juvenile hormone (JH) is a key insect hormone that regulates numerous aspects of development, reproduction, behavior, and metabolism [3,9–11]. Although other hormones have been investigated with respect to the endocrine control of insect polymorphism (e.g., ecdysteroids (EDCs), various biogenic amines [2,12,13]), most attention has centered on the role of JH. Endocrine regulation comprises numerous physiological, biochemical, and molecular aspects, such as systemic hormone titer regulation, receptor expression and hormone–receptor binding, hormone-modulated gene expression, and intracellular signaling and regulation of downstream effectors [14]. The first of these issues, the systemic JH titer and its morph-specific modulation, has been, and continues to be, a central focus of the endocrine study of complex polymorphism [3,5,6,8]. This topic has been especially well studied in dispersing and flightless morphs of *Gryllus* crickets [5–7]. Polymorphism can arise from differences in genotype (genetic polymorphism), environmental factors (environmental polyphenism), or both [8,12]. Most studies of insect polymorphism have focused on environmental polyphenism [3,6], and the *Gryllus* work is unique in its focus on genetic polymorphism for dispersal capability and its negative genetic association with reproduction. Thus, *Gryllus* studies of JH titer regulation are particularly relevant to issues pertaining to genetic variation and adaptive evolution of JH regulation, as well as to general aspects of morph-specific endocrine regulation.

Most of the *Gryllus* work has been undertaken in the author's laboratory over the past 25 years. Three separate but interrelated aspects of this research will be dealt with in the present chapter: (1) JH titer regulation in developing morphs, (2) JH titer regulation in adult morphs, and (3) experimental evolution of JH degradation by artificial selection. A particular focus of this chapter is the *in vivo* mechanisms that regulate morph-specific differences in the JH titer and the adaptive basis of these morph-specific mechanisms.

3.2 BACKGROUND

What follows is a very brief thumbnail sketch of JH endocrinology, mainly focusing on the systemic hemolymph JH titer and its regulation, to provide some background to understand the endocrine studies discussed later in this chapter. For additional details on this extensive topic, see reviews by Hardie and Lees [2], Gupta [15], Nijhout [3], Klowden [16], and, especially, Goodman and Granger [10] and Goodman and Cusson [11]. The JHs are a group of about a half-dozen structurally related (16–19 carbon) stereospecific sesquiterpenoids, essentially acyclic hydrocarbons with important epoxide and ester functional groups, which are primarily limited to insects [10,11,16,17]. JH-III is the primary JH in insects and is the sole JH in many insect groups (beetles, bees, crickets, grasshoppers, termites, and wasps). By contrast, the Lepidoptera (moths and butterflies) and "higher" Diptera (flies) contain, in addition to JH-III, other JHs, which may be the predominant JH during specific developmental stages. A unique JH has also been identified in hemipterans (true bugs [18]), and an increasing number of other JH-related compounds, such as JH metabolites (e.g., hydroxylated JHs) or precursors (e.g., methyl-farnesoate), have been identified and may have important endocrine functions [11,19].

JH, in concert with a variety of other hormones (e.g., ECDs), regulates a number of central developmental (e.g., aspects of molting and metamorphosis) and reproductive (e.g., biosynthesis and ovarian uptake of vitellogenin) aspects of the life cycle of insects in general, in addition to controlling numerous features unique to specific insect groups [3,10,11,16,20,21]. A long-standing focus in insect endocrinology has been the regulation of the systemic (hemolymph) JH titer, specifically, temporal alterations of the titer during development and adulthood, factors regulating the titer changes, and functional aspects of these titer changes [3,10,11]. While the field of insect endocrinology has certainly expanded beyond these issues to investigate such topics such as JH receptors, mechanisms of JH-regulated gene expression, and intracellular signaling [10,11,14], many fundamental aspects of the systemic JH titer and its regulation remain poorly understood (e.g., extent of rapid, large-amplitude titer change [11]). Moreover, quantifying JH titer profiles and their regulation is an especially important task in endocrine studies of complex polymorphism, due to the surprising paucity of reliable data on this topic despite decades of study [5,8]. The other important aspect of systemic JH signaling, tissue sensitivity to JH, has been substantially hampered, because the identity of the JH receptor remains elusive (reviewed in Goodman and Cusson [11]). This situation contrasts with other important insect hormones, such as ECD, where receptor expression and its relationship to systemic hormone titer have been studied in much more detail [22]. Lack of information on the JH receptor has hampered assessment of the functional significance of JH titer fluctuations.

Starting with Wigglesworth [1], the conceptual basis for hypotheses about JH control of complex polymorphism has been derived from results of basic studies of JH titer regulation [3,5,12,23]. The classic model of JH-regulated molting and metamorphosis postulates that an increase in the blood ECD titer induces a molt, whereas the JH titer during that time determines the type of molt (Figure 3.1). Before the last juvenile instar, the JH titer remains relative high (with fluctuations) and an ECD titer peak during this period induces a juvenile-to-juvenile molt. During the last juvenile instar, the JH titer declines to a low level, due to reduction in the rate of JH biosynthesis and increase in the rate of enzymatic degradation. An increase in the ECD titer during this time induces a metamorphic molt (i.e., causes expression of pupal or adult traits), in which the subsequent stage differs substantially in morphology, physiology, etc., from the preceding stage: a nymphal-to-adult molt in hemimetabolous insects (i.e., species that do not have a pupal stage: aphids, crickets, locusts, true bugs) or a juvenile-to-pupal molt in holometabolous insects (e.g., beetles, butterflies, bees). In the adult stage, JH and/or ECDs take on new functions where they regulate many aspects of reproduction, such as yolk protein biosynthesis and uptake into the developing eggs, intermediary metabolism, behavior, and flight [3,10,11].

Based on results of these studies, the classic model of JH regulation of complex polymorphism postulates that modulation of the JH titer during important periods in development and adulthood underlies development of alternate morphs that differ discontinuously in morphology and physiology (Figure 3.1). In the adult stage, differences in the JH titer are thought to regulate morph-specific ovarian growth and egg development and aspects of flight (e.g., propensity to fly) and flight capability (e.g., biosynthesis of flight fuels). Again, it should be emphasized that JH is not the only, and sometimes is not the most important, regulator of various morph differences (e.g., [24,25]). Also, the extent to which modulation of JH receptors regulates morph-specific development and adult function has largely remained an unstudied topic.

The main regulators of the hemolymph JH titer are thought to be the rate of JH biosynthesis and the rate of hormone degradation by hemolymph and non-hemolymph (e.g., fat body) juvenile hormone esterase (JHE) and JH epoxide hydrolase (JHEH) [3,10,26,27] (Figure 3.2). Dramatic changes in the JH titer during the juvenile or adult stages are often strongly associated with changes in the rates of JH biosynthesis and activity of hemolymph JHE. The relative importance of these two regulators varies from species to species and from one developmental stage to another, with modulation of the rate of JH biosynthesis appearing to play a more important or at least equivalent role in the majority of cases. Thus, studies of JH titer modulation in wing-polymorphic *Gryllus*, and most other complex polymorphisms in insects, have focused primarily on modulation of rates of JH biosynthesis and degradation (especially by hemolymph JHE) [3,5,6,8,23].

JH is biosynthesized in the corpora allata (CA), paired glands that occur behind the brain in most insects [3,10,27]. Hormone biosynthesis is regulated by direct neural input as well as by various neurohormones (e.g., allatostatins, allatotropins), other hormones (JH itself, ECDs), neurotransmitters (octopamine, dopamine), and

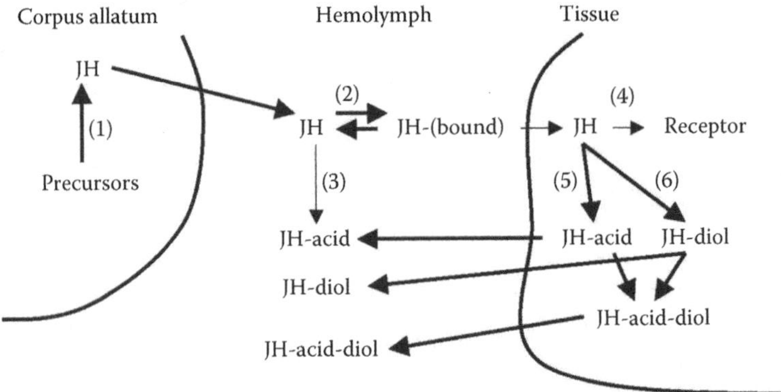

FIGURE 3.2 Major regulators of the hemolymph JH titer: (1) biosynthesis in the CA, (2) binding to JH-carrier protein, (3) hemolymph degradation by JHE, (4) intracellular receptor binding, and intracellular degradation by (5) JHE and (6) JHEH. Intracellular JH acid and JH diol are subsequently acted on by JHE and JHEH, respectively, producing JH acid–diol. See text for additional discussion.

ions (Ca^{++}, K^{+}) [10,11,27]. A relatively simple *in vitro* radiochemical assay for JH biosynthetic rate, developed by Pratt and Tobe [28], has allowed this important aspect of JH regulation to be measured in numerous insect species [10,11,27]. In many cases, *in vitro* rate of JH biosynthesis is strongly correlated with hemolymph JH titer and various developmental and reproductive traits (e.g., egg development in adult *Diploptera* [6,10,27,29]) and thus appears to regulate the JH titer in many cases. However, very little is known about the *in vivo* rate of JH biosynthesis, and several reports indicate that severing of nerve connections between the CA and the brain, required to remove the CAs from the insect for the *in vitro* assay, can alter the rate of JH biosynthesis [27,30].

JH is not stored in the CA but is directly released into the hemolymph upon biosynthesis where it is bound to a carrier protein. The carrier protein serves to prevent nonspecific binding of the lipophilic JH, to transport this hormone to its target tissues and, likely, in hormone degradation [10]. Where studied, the total number of binding sites of the carrier protein in the hemolymph compartment is greater than the concentration of hemolymph JH during all stages of the life cycle, and genetic reduction in the degree of JH binding does not appear to result in a change in the JH titer [10,11,31,32]. Thus, JH titer changes do not appear to be regulated by changes in the concentration of JH-binding protein, in contrast to the situation for some hormone carrier proteins in vertebrates [33]. On the other hand, a recent study of termites indicates that JH sequestration by hexamerin, a JH transport protein in some insects, may play a role in regulation of caste development [34].

The major JH-degrading enzyme in the hemolymph is JHE [10,26]. JHE has a very low Michaelis constant and high k_{cat}/K_M for JH, which makes this enzyme effective in scavenging JH in the hemolymph, especially during the juvenile stage when the titer is required to be very low (low or sub-nM range). JHE can hydrolyze JH even in the presence of the JH-carrier protein (JHE hydrolyzes unbound hormone that dissociates from the carrier protein) [35]. A simple radiochemical assay developed by Hammock and Sparks [36] has allowed hemolymph JHE to be quantified and correlated with the JH titer and various biological traits in numerous insects [26,37]. The JH-carrier protein has also been proposed to function in JH degradation, by removing JH from lipophilic sites that are inaccessible to JH-degrading enzymes and delivering JH to the hemolymph where it can be hydrolyzed by JHE [32,38]. In addition to hemolymph JHE, JHE and JHEHs occur in a variety of non-hemolymph tissues, such as fat body [10]. Quantifying non-hemolymph JHE and JHEH requires a more laborious thin-layer chromatographic assay and thus has not been investigated as widely in polymorphic species or species in general (reviewed in Goodman and Granger [10]).

3.2.1 Endocrine Regulation of Wing and Related Dispersal Polymorphisms

JH titer differences among morphs have been extensively studied in social insects, where reliable data on caste differences in the JH titer and rates of JH biosynthesis have been reported for a number of species (reviewed in Hardie and Lees [2], Hartfelder and Engles [39], and Hartfelder and Emlen [6,40]). However, in most other cases of complex polymorphism, such as wing and phase polymorphisms, there is a surprising paucity of data of JH titer and its regulation, even for insects that have been experimental models of complex polymorphism for decades. For example, although aphids have been an important experimental model in JH studies of wing polymorphism since the 1960s, JH titers in morphs have been measured using appropriate analytical techniques (e.g., gas or liquid chromatography coupled with mass spectrometry) in only a very few species and, with one notable exception [41], only very recently [42,43]. Most early studies of polymorphism in aphids used gross-level techniques such as hormone application, which often resulted in confusing and contradictory data (extensively reviewed in Rankin and Singer [44], Zera and Denno [12], and Zera [45]). In addition, because of their very small size, studies of titer regulation are very rare in aphids. Similarly, although JH-mediated phase polymorphism in locusts has been the focus of study since the 1960s [13,46,47], adequate, in-depth quantification of the hemolymph JH titer (i.e., titers measured at more than a few time points) has only been available to a limited degree during the past decade [48,49]. Excluding *Gryllus*, only a very few well validated studies of JH titer differences between the morphs of other wing-polymorphic species have been reported [50]. Finally, relatively recently, a number of ecologically important polymorphisms have been studied with regard to JH regulation [6]. However, many of these studies (e.g., [51,52]) used JH application (typical of the aphid studies in the 1960s–1970s) as the sole or predominant experimental technique. Thus, most inferences derived from these studies regarding JH titer differences between morphs have been criticized [45] and should be considered unsubstantiated until validated by appropriate techniques.

JH titer and titer regulation in a complex polymorphism have been most comprehensively studied in species of *Gryllus*, especially during adulthood (reviewed in Zera [5,7,53]). These studies currently comprise (1) the most detailed temporal sampling of the JH titer and its regulators in any adult insect, whether polymorphic or not; (2) the most detailed investigations of the relationship between *in vitro* and *in vivo* JH degradation for any insect; (3) investigations of JH titer regulation in both adult and juvenile stages of the same species, allowing the diversity of mechanisms of JH titer regulation to be determined for different stages of the same species; (4) parallel studies of polymorphism (end products of evolution) with the process of evolution that generates these patterns (artificial selection on hemolymph JHE activity); and (5) the only detailed investigation of JH titer circadian rhythm. Before discussing these studies, background information on wing polymorphism, focusing on *Gryllus*, will be given.

3.2.2 Wing Polymorphism, Focusing on *Gryllus*

Wing polymorphism is the most dramatic example of the fundamental trade-off between dispersal and reproduction, a common feature of species [12,54–56]. The polymorphism is found in most insect groups and is especially common in the Hemiptera/Homoptera (aphids and plant hoppers), Coleoptera (beetles), and Orthoptera (crickets and grasshoppers). While referred to as "wing polymorphism," the polymorphism actually consists of morphs that differ dramatically and discontinuously in numerous key aspects of flight capability (morphology of the wings and flight muscles, biosynthesis and allocation of flight fuel) and reproduction (ovarian growth, fecundity, protein biosynthesis, mating behavior) (Figure 3.1). Species often exhibit dramatic spatial or seasonal variation in morph frequencies, a phenomenon that has been especially well studied in water striders and plant hoppers [12,56–58]. Morphs of wing-polymorphic *Gryllus* species are typical of wing-polymorphic insects [5,7] (Figure 3.1): Relative to the flight-capable morph with fully developed wings and flight muscles (long-winged with functional muscle, LW(f)), the obligately flightless (short-winged, SW) morph (1) cannot fly because it never fully develops wings or flight muscles, (2) exhibits substantially faster ovarian growth (400% higher egg production by the end of the first week in adulthood), (3) exhibits substantially reduced accumulation of triglyceride (flight fuel), and (4) shows numerous changes in intermediary metabolism related to the production, oxidation, and transformation of lipid and protein. Thus, the flight-capable morph is adapted for flight at the expense of reproduction, while the opposite is the case for the flightless morph. Morph differences in wings, flight muscles, and ovaries are seen to a similar degree in the field as well as the laboratory in *Gryllus firmus* [59]. In addition to the obligately flightless morph described in the aforementioned text (SW), another flightless morph (LW(h)) is produced from the flight-capable morph during adulthood, by histolysis (degeneration) of flight muscles, a common phenomenon in insects [12]. When this occurs, the flightless LW(h) morph takes on many of the characteristics of the SW morph (e.g., increase in ovarian growth, decrease in lipid accumulation and biosynthesis; JH titer [7,12,60]).

Numerous ecological studies, most notably of semiaquatic water striders and salt-marsh plant hoppers, have shown that the flightless morph occurs more commonly in longer-duration habitats in which flight is not necessary for feeding or mating [12,57,58]. By contrast, the flight-capable morph is found in temporally unstable habitats, where dispersal to new habitats is important. Polymorphic species occur in heterogeneous habitats that are a mixture of stable and unstable patches, thus resulting in temporally and spatially varying selection favoring one morph or the other in different localities or seasons. The most important driving force for the evolution of flightlessness appears to be the increased fecundity that results when nutrients that are not required to construct and maintain the flight apparatus can be redirected to egg production. Extensive field information on the polymorphism provides important ecological context for interpreting the adaptive basis of variation in proximate mechanisms underlying the expression of morph traits.

3.3 ENDOCRINE REGULATION OF WING POLYMORPHISM IN *GRYLLUS*

3.3.1 JH Titer Modulation and Regulation of Morph Development

Before discussing specific endocrine studies, it is useful to first consider features of *Gryllus* that make crickets of this group useful for investigating the endocrine regulation of the JH titer and complex polymorphism. Individuals are relatively large (adults weigh 0.7–1 g), thus allowing measurement of many traits, such as the *hemolymph* JH titer and JHE activities, *in vitro* measures of JH biosynthesis on excised CAs, *in vivo* JH degradation, hemolymph volume, and organ-specific activities of JH-degrading enzymes. All of these are much more difficult or are not possible to measure in other smaller polymorphic species (e.g., aphids, plant hoppers). Individuals can be easily reared and bred in the lab, thus allowing the construction of genetic stocks, which is especially useful in evolutionary genetic studies [7,60–62]. Genetically specified phenotypes can be altered by environmental variables (density) and by hormone application, thus providing additional tools for experimental analysis of endocrine regulation, such as identifying JH-sensitive stages [63]. Finally, *Gryllus* and related genera exhibit considerable diversity with respect to degree of flightlessness [64,65]. Many species are polymorphic or monomorphic LW or SW, which provide ample opportunity for comparative studies of the endocrine regulation of flight capability/flightlessness in the laboratory or field.

Initial studies of the endocrine control of wing polymorphism in *Gryllus* primarily focused on JH regulation of genetically specified development of LW(f) versus SW morphs of *Gryllus rubens* and *G. firmus*, both of which are polymorphic in the field. Although these initial studies were reported some time ago ([63,66,67]; reviewed in Zera [8,53]), they warrant discussion here for the following reasons: (1) They still remain the most detailed investigations of JH titer modulation via JH degradation in any complex polymorphism, (2) they constitute the only case (together with studies of adult *Gryllus*; see in succeeding text) in which JH titer modulation has been extensively studied in *genetic* stocks differing in wing morph, and most importantly, (3) previous reviews did not focus in detail on *in vivo* titer regulation and the relationship between *in vitro* JHE activity and *in vivo* JH degradation, which is a central focus of this chapter.

These studies were undertaken in genetic stocks nearly purebreeding for the LW(f) or SW morphs that were derived by artificial selection from individuals collected from natural populations (Florida). The focus was on LW(f) and SW genetic stocks for two reasons: First, as mentioned earlier, the *Gryllus* studies were undertaken in an evolutionary context, where genetically specified differences in phenotypes are especially relevant. Second, presumptive LW(f) versus SW individuals cannot be distinguished morphologically until after adult emergence. Thus, genetic stocks producing known morphs, allowed comparison of endocrine events in LW(f) versus SW morphs during the juvenile stage.

Background studies demonstrated that the development of the genetically specified LW(f) morph of *G. rubens* could be redirected to the SW morph by topical application of JH or JH analogues during the penultimate or last stadium or by changing the rearing density during these same periods in development [68]. This finding implicated these last two juvenile instars as JH-sensitive periods in morph induction, that is, periods during which variation in the JH titer or tissue sensitivity differ between morphs and influence morph expression.

A major finding was the substantially (threefold) higher activity of hemolymph JHE, in the LW(f) compared with the SW morph of *G. rubens* during the last nymphal stadium (only minimal differences between morphs were found during the penultimate stadium) [63] (Figure 3.3). More detailed studies in the congener

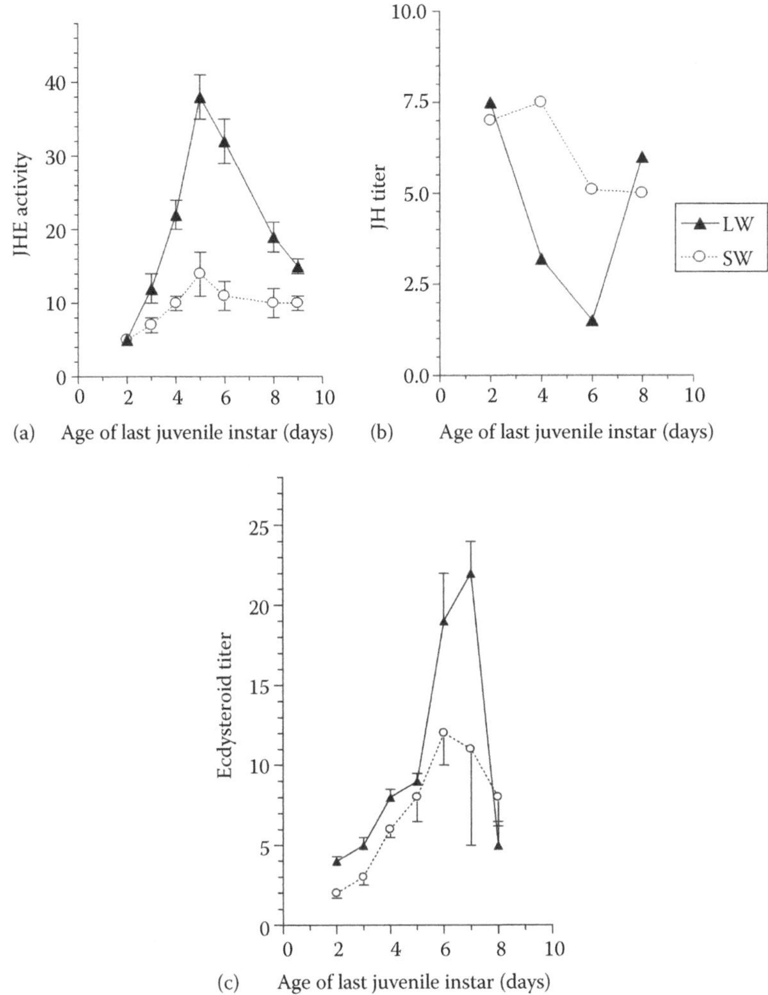

FIGURE 3.3 Differences between LW and SW female morphs of *G. rubens* in the developmental profiles of the hemolymph juvenile esterase activity ((a) nmol^{-1} min^{-1} mL hemolymph^{-1}), JH-III titer ((b) nM), and ECD titer ((c) 20-hydroxyecdysone equivalents, pmol^{-1} μL hemolymph^{-1}), measured during the last juvenile instar of females. (Data from Zera, A.J., *Comp. Biochem. Physiol. A*, 144, 365, 2006.)

G. firmus confirmed and extended this finding [7,53,62]: hemolymph JHE activity was substantially (sixfold) higher in each of three LW(f) compared with each of three SW genetic stocks (Figure 3.4). The elevated JHE activity in the hemolymph of the LW(f) morph was due to both greater whole-body JHE activity and greater (3.8-fold higher) proportion of whole-body enzyme allocated to the hemolymph. In *G. rubens*, high JHE activity strongly co-segregated with the LW(f) phenotype when the LW(f) and SW lines were crossed and backcrossed. When development of the LW(f) morph was redirected to the SW morph by changing rearing density, hemolymph JHE

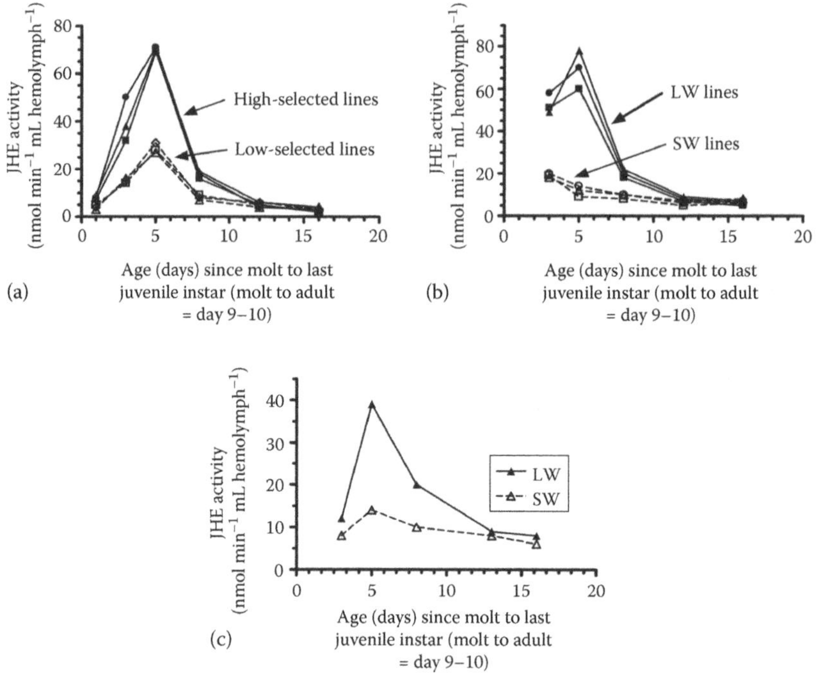

FIGURE 3.4 Hemolymph JHE activities (nmol^{-1} min^{-1} mL^{-1} hemolymph) in three species of *Gryllus* during the last female juvenile instar and first week of adulthood. (a) For *G. assimilis*, "high" and "low" refer to lines selected for high or low hemolymph JHE activity. (c) For *G. rubens* and (b) *G. firmus*, "LW" and "SW" refer, respectively, to long-wing or short-wing selected lines. (Data from Zera, A.J., *Comp. Biochem. Physiol. A*, 144, 365, 2006; Zera, A.J. and Tiebel, K.C., *J. Insect Physiol.*, 35, 7, 1989.)

activity was also reduced to values seen in the SW morph reared under standard densities. Thus, there is a very strong correlation between high hemolymph JHE activity and expression of the LW(f) phenotype. In contrast to the morph differences in hemolymph and whole-body JHE activity, no differences in whole-body JHEH activity were observed between the morphs. Nor did the rate of JH biosynthesis differ between morphs during the last stadium [69]. The JH titer was lower, albeit slightly, in the LW(f) compared with the SW morph [70]. More recently, a similar elevated JH titer was found in the SW versus LW(f) morph of a plant hopper species [50]. In addition to these differences in the hemolymph JH titer between *G. rubens* morphs, the LW(f) morph also exhibited a substantially elevated ECD titer during the last stadium [70] (Figure 3.3).

To summarize, there are dramatic morph-specific differences in hemolymph JHE activity that are negatively associated with the JH titer between nascent morphs of *G. rubens* and *G. firmus*. These data suggest a simple mechanism regulating JH titer differences between nascent morphs, namely, modulation via change in JHE activity, with no observed input by other potential JH titer regulators studied, such as JHEH activity or rate of JH biosynthesis (or JH-binding protein; see as follows). The very low JH titer, and thus large experimental error involved in its measurement (single titer points were based on pooled hemolymph from 50 or more individuals), likely precludes identifying anything other than a broad negative association between hemolymph JHE activity and JH titer. Much more detailed analyses of the functional relationship between the JH titer and JH titer regulators are discussed later in adult morphs where the JH titer is an order of magnitude higher than in juveniles. Because of space limitations, the important issue regarding the functional significance of only slight differences in the JH titer between the morphs will not be dealt with here. For extensive discussions of this topic, see [5,7,8,70].

3.3.2 *In Vivo–In Vitro* JH Degradation in *Gryllus* Morphs

If morph differences in *in vitro* JHE activity are functionally significant with regard to regulating differential morph expression by modulating the JH titer, this must occur by causing morph differences in JH degradation *in vivo*. However, various factors complicate extrapolation of morph differences in *in vitro* JHE activity to *in vivo* differences in whole-organism JH degradation: JHE activities were measured in highly diluted hemolymph, at very high substrate concentration, in the absence of potential inhibitors or activators, and under conditions that could mask important kinetic differences (e.g., substrate K_M) between the enzymes in different morphs, if present. In many cases of enzymes of intermediary metabolism, *in vitro* differences in enzyme activity are not linearly related to higher-level physiological processes, such as flux through the pathway in which the enzymes participate [71,72]. *In vitro* JHE studies also do not take into account the influence of other JH-degrading enzymes that could contribute more significantly to overall *in vivo* JH degradation, thus diminishing the physiological significance of morph differences in hemolymph JHE activity. Also JHE and JHEH assays were conducted using commercially purchased racemic hormone (mixture of 10-R and 10-S enantiomers), while the natural JH-III is the 10-R enantiomer. In some cases, the affinity of JH-binding protein differs toward the enantiomers [31,32,73], which can affect their relative rates of hydrolysis. Finally, other potential factors such as morph-specific differences in hormone excretion could augment or obviate morph differences in JHE activity.

Because of these complications, *in vivo* studies of JH degradation were undertaken: Racemic or 10R-JH-III, the naturally occurring JH enantiomer, were injected (ca. 8 pmol, 200,000 DPM in about 5 μL solvent) into the hemocoel of whole LW(f) or SW morphs during the last juvenile instar, and the rate of JH degradation was measured [66]. Prior to injection of JH, a small hemolymph sample was taken to measure JHE activity, and excretion of radiolabeled JH was also measured during the incubation period. Depending upon the method of injection (JH dissolved in oil, ethanol, or aqueous solution with BSA), the half-life of JH ranged from 24 min (injected in aqueous solution + BSA or in ethanol) to 35 min (oil). *In vivo* half-life was substantially lower in undiluted plasma (5 min). These values compare well with JH half-lives reported in other insects (10–30 min; see [66]) and JH degradation in SW female *G. rubens* exhibited first-order kinetics during the time period investigated (ca. up to 80% degradation; linear regression, $P < 0.001$). Thin-layer chromatographic analysis indicated a significant amount of JH acid–diol as well as JH acid indicating the important contribution of cellular JHEH to *in vivo* JH degradation in the hemolymph.

Two important findings resulted from these studies. First, the rate of *in vivo* JH degradation was positively associated with *in vitro* hemolymph JHE activity (Figure 3.5). Thus, the *in vitro* differences in JHE activity between morphs appear to have significant physiological consequences with respect to morph-specific *in vivo* JH degradation. For example, both *in vivo* degradation and *in vitro* JHE activity did not differ between the morphs early in the last stadium (day 2), while both were significantly higher in LW(f) versus SW morphs during the middle of the stadium (Figure 3.5). The same rank-order difference in *in vivo* degradation was found whether racemic or 10-R JH was injected, although the absolute rate of degradation was significantly higher for racemic JH (Figure 3.5). Furthermore, when *in vivo* JHE was inhibited by a trifluoroketone inhibitor, there was a correspondingly significant decrease in the *in vivo* rate of JH degradation, consistent with hemolymph JHE significantly contributing to *in vivo* JH degradation (Figure 3.5).

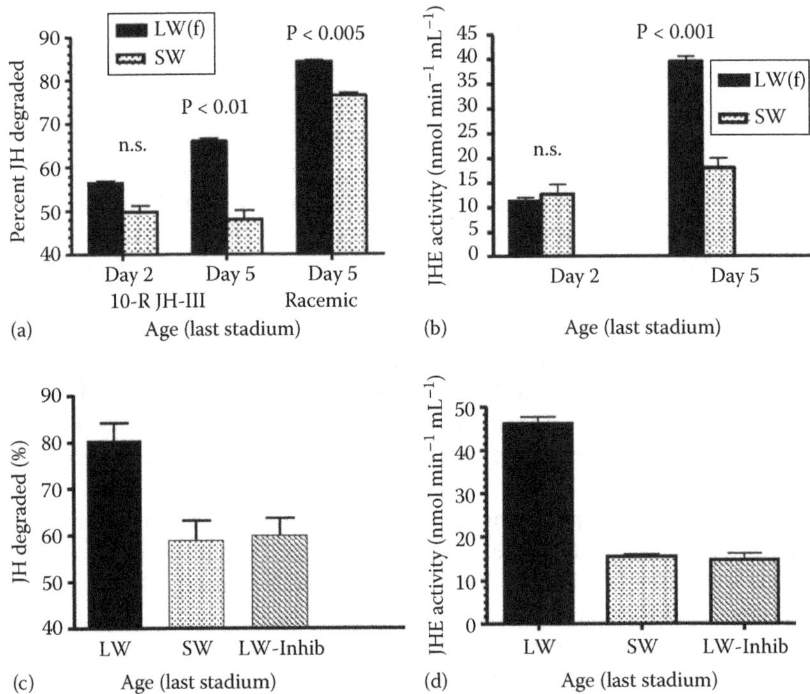

FIGURE 3.5 *In vivo* JH-III degradation (40 min after injection) and *in vitro* JHE activity in female last juvenile instar LW and SW morphs of *G. rubens*. (a) *In vivo* degradation of 10-R JH-III (natural enantiomer) or commercially available racemic JH-III in the morphs; (b) hemolymph JHE activity in the same individuals. (c) *In vivo* racemic JH-III degradation in LW, SW, and LW females that had been injected with a JHE-specific inhibitor (trifluoroketone) prior to injection of JH-III; (d) JHE activities measured in the same individuals. (Reanalyzed data of Zera, A.J. and Holtmeier, C.L., *J. Insect Physiol.*, 38, 61, 1992.)

The second important finding was the relationship between morph difference in *in vitro* JHE activity and *in vivo* rate of JH degradation: the magnitude of the difference between morphs was much greater with respect to *in vitro* JHE activity than total *in vivo* JH degradation (Figure 3.5). For example, on day 5 of the last juvenile instar of *G. rubens*, there was a two- to threefold higher *in vitro* hemolymph JHE activity in LW(f) compared to SW females but only a 1.2-fold higher rate of *in vivo* JH degradation in the LW(f) compared with the SW females [66]. The dampening of the morph difference in *in vitro* JHE activity on *in vivo* JH degradation may be due to the equivalent JHEH activity in morphs. This may have reduced the contribution of the morph difference in JHE to the overall morph difference in JH degradation *in vivo*. Morphs did not differ in the degree of JH binding to hemolymph protein during the mid-last stadium. Nor did the morphs differ with respect to the amount of JH and JH metabolites excreted, which comprised less than 2% of radiolabel injected. Thus, the reduced difference between the morphs in *in vivo* JH degradation relative to difference in *in vitro* JHE activity did not appear to arise from morph differences in hemolymph binders or JH excretion. These studies illustrate the key importance of *in vivo* studies of JH degradation to assess results obtained *in vitro*. These studies were conducted before a dramatic morph-specific JH titer circadian rhythm was identified in adults [29,74] (see as follows). However, subsequent studies indicated no significant morph-specific change in hemolymph JHE activity during the early to mid-period in the photophase (last-stadium) when the JHE activities had been measured previously (A. J. Zera, unpublished data).

In vitro and *in vivo* JH degradation were also compared between the adult female morphs, which differ dramatically in ovarian growth [67]. Hemolymph JHE activity was about 10-fold lower in adult versus last stadium *G. rubens*, *G. firmus*, or *Gryllus assimilis*, and no or only minor differences were observed between adult morphs [62,67]. Similarly, the half-life of JH in SW adult *G. rubens* (70 min; 8 pmol racemic hormone injected in 5–6 μL olive oil into day 5–7 individuals) was about 2.5-fold higher than in last stadium ($P<0.01$), consistent with the lower hemolymph JHE activity. Half-life of JH in adult *G. rubens* (and *G. firmus* [62]) was similar to that reported in other insects such as *Diploptera punctata* [27,75], *Locusta migratoria* [76], and *Manduca sexta* [77]. Degradation of 10-R JH was significantly less than racemic JH in adult *G. rubens*, similar to the situation in juveniles. However, the relative difference between morphs was the same whether 10-R or racemic JH was used, as was the case in the last nymphal stadium of *G. rubens*. The main finding was that neither *in vitro* JHE activity nor the *in vivo* rate of JH degradation differed substantially between adult morphs of the same age in *G. rubens* or *G. firmus*, irrespective of whether rates were measured using racemic or 10-R JH-III. Nor was there any difference between the morphs in the rate of JH excretion, which was very low (typically 1%–3% of injected radiolabel). On the other hand, there was a highly significant linear regression of *in vitro* JHE activity on *in vivo* JH degradation ($r^2=0.4$; $P<0.0001$) in day 5–7 adults but not day 2–3 adults independent of wing morph [67]. These data indicate that even though mean *in vitro* JHE activity and mean *in vivo* JH degradation do not differ between morphs, individual differences in hemolymph JHE activity appear to give rise to individual differences in *in vivo* JH degradation. These studies constitute the first and most comprehensive information on genetic or morph-specific differences in *in vitro/in vivo* JH degradation. These studies are especially useful for establishing the relationship between *in vitro* and *in vivo* JH degradation. However, they, unfortunately, only allow weak inferences to be made between JH degradation and the JH titer during the last juvenile instar, because of the very low hemolymph JH titer during this developmental stage.

3.3.3 Artificial Selection on JHE Activity in *G. assimilis*: Observed Evolutionary Change in JH Degradation

Studies described in the aforementioned text investigated morph differences in, and the interrelationships among, *in vitro* JHE activity, *in vivo* JH degradation, and *in vivo* JH titer in an evolved polymorphism found in natural populations. Those studies led to the hypothesis that genetic variation for hemolymph JHE activity was present in ancestral field populations and was a target of natural selection during the evolution of wing polymorphism. Selection on alleles encoding reduced JHE activity resulted in reduced *in vivo* JH degradation, elevated JH titer, and altered expression of traits involved in flight capability leading to the evolution of the flightless morph. To directly test aspects of this hypothesis, hemolymph JHE activity was subjected to direct artificial selection in *G. assimilis*, a wing-monomorphic species (i.e., all individuals winged).

Artificial selection is a powerful tool in evolutionary studies because it provides information on the extent to which phenotypic differences are due to variation in genes, as opposed to environmental factors, and the extent to which a target trait and other genetically correlated traits can respond to selection [78–80]. In addition, genetic stocks that are produced by artificial selection are very useful for subsequent identification of the variable genes that were the targets of selection that produced divergence in the selected lines. A full discussion of the JHE artificial selection project is beyond the scope of this chapter and is reviewed in detail in Zera [53]. Here I summarize basic and more recent findings of this study and focus on a central question addressed previously: what is the relationship between hemolymph JHE activity, measured *in vitro*, and *in vivo* JH degradation? Stated differently, what is the magnitude of evolutionary change in hemolymph JHE activity that is required to produce a measurable change in the *in vivo* rate of JH degradation?

G. assimilis was subjected to bidirectional selection for hemolymph JHE activity on day 3 of the last juvenile instar by selective breeding as follows [61] (Figure 3.6): In a particular population, hemolymph JHE activity was measured in a large number of individuals (n = 240), and males and females exhibiting the highest 25% of activity were bred. This was repeated over several generations to produce a high-JHE-activity line. Similarly, individuals with the 25% lowest JHE activities in a different genetic stock (derived from the same base population as the aforementioned stock) were bred in each of several generations to produce a low-activity line. Each type of selection was replicated independently in additional subpopulations, resulting in three upward selected, three downward selected, and three control (unselected) lines, all derived from the same base population. This study was the first and still constitutes the most extensive (measurements on over 10,000 individuals) artificial selection study on an endocrine trait in a non-domesticated animal or plant species.

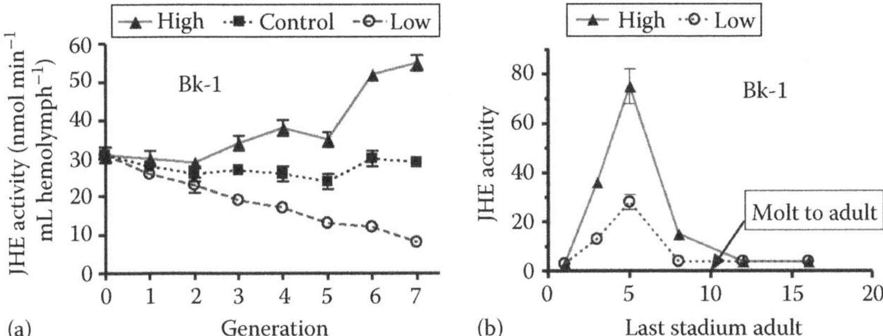

FIGURE 3.6 (a) Change in hemolymph JHE activity (*G. assimilis*) during direct artificial selection for high or low activity in a representative block (one of three independent selection trials). Data for day-3 last-instar (last stadium) females. (b) Developmental profiles of hemolymph JHE activity in females of high- and low-selected lines at generation 5 of the selection trial. (Data from Zera, A.J., *Comp. Biochem. Physiol. A*, 144, 365, 2006.)

Briefly, the main results are as follows [61,81] (Figure 3.6): Artificial selection resulted in stocks substantially differentiated in hemolymph JHE activity, demonstrating substantial variation in genes specifying JHE activity in the natural population from which *G. assimilis* were collected. Populations rapidly responded to selection with high- and low-activity lines differing six- to eightfold in hemolymph JHE by the seventh generation of selection. Activity differences were due to changes in whole-body JHE activity (about twofold) and tissue distribution: about 2.6-fold greater proportion of whole-body JHE activity was found in the hemolymph in high- versus low-activity lines. Interestingly, hemolymph JH binding was also significantly higher (1.7-fold) in high- versus low-selected lines. This suggests that genes controlling JHE activity also have positive pleiotropic effects on JH binding, probably due to genes co-regulating JHE and JH-binding protein. This finding is consistent with the hypothesis mentioned previously that a function of JH-binding protein is to deliver JH to the hemolymph for degradation by JHE.

As was the case for LW(f) and SW lines of *G. firmus*, selected lines of *G. assimilis* did not differ in whole-body, midgut, or fat body JHEH activity [82]. This demonstrates that the two major JH-degrading systems, JHE and JHEH, can evolve independently of each other. Finally, no or only minimal differences were found in JHE activity between morphs during the adult stage, similar to the situation for JHE developmental profiles in the wing-polymorphic *G. firmus* and *G. rubens* [62,67] (Figures 3.4 and 3.6). Thus, the timing of action of genes responsible for genetic differences in JHE activity during the juvenile stage is restricted to that stage. Parallel selection studies conducted on adults indicate the reverse situation: genes affecting hemolymph JHE activity in adults have little influence on JHE activity in juveniles [83]. Thus, regulation of JHE activity is modular with respect to the life cycle; evolutionary changes in juvenile JHE activity can occur independent of changes in adults and vice versa. These studies indicate that variable genes in contemporary natural populations of the wing-monomorphic *G. assimilis* that were the targets of artificial selection seem to be remarkably similar to variable genes in ancestral natural populations

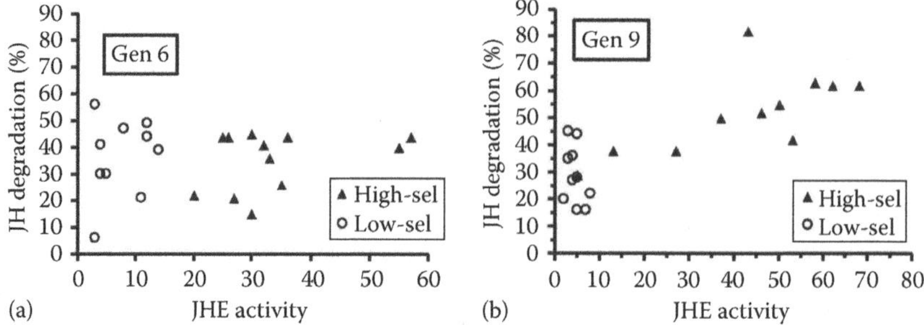

FIGURE 3.7 Relationship between *in vivo* JH-III degradation and *in vitro* hemolymph JHE activity in *G. assimilis* from lines selected for high or low hemolymph JHE activity at generations (a) 6 and (b) 9 of artificial selection. JHE activity and JH degradation were measured in the same individuals (day 3, last juvenile instar). Note the significantly elevated JHE activity and *in vivo* JH degradation in the high- versus low-selected lines at generation 9, but not generation 6. JHE activity = nmol^{-1} min^{-1} mL hemolymph^{-1}; JH degradation = % injected racemic JH-III degraded over 45 min. (Data from Zera, A.J. et al., *Arch. Insect Biochem. Physiol.*, 32, 421, 1996; reanalyzed data from Zera, A.J. and Zhang, C., *Genetics*, 141, 1125, 1995.)

that were the targets of natural selection in *G. firmus* to produce wing polymorphism. Additional discussions of these main findings of the artificial selection experiment can be found in [8,53].

As mentioned earlier, the artificial selection experiment in *G. assimilis* allowed experimental investigation regarding the degree to which hemolymph JHE activity must be altered to produce a measurable change in *in vivo* JH degradation. Studies of naturally occurring wing polymorphism discussed earlier indicated a three- to sixfold difference in *in vitro* hemolymph JHE activity between morphs but much less of a difference in *in vivo* JH degradation and even less of a difference in the hemolymph JH titer. During generation 6 of selection on *G. assimilis*, no difference in *in vivo* JH degradation was observed between two of three pairs of high- versus low-selected lines that had diverged by 2.7- to 3.7-fold in hemolymph JHE activity (see Figure 3.7 for data on one pair of selected lines). However, by generation 9, each pair of high- versus low-selected lines had diverged in hemolymph JHE activity by 5.5- to 8.2-fold, and each pair differed significantly in *in vivo* JH degradation (Figure 3.7). Thus, in juvenile *Gryllus*, it appears that a three- to fivefold change in hemolymph JHE activity is required to produce a measurable difference in *in vivo* JH degradation during that stage. This seems to explain why a relatively large-magnitude difference in hemolymph JHE activity exists between the juvenile morphs of wing-polymorphic *Gryllus*.

3.3.4 Characterization of JHE Enzyme and Morph-Specific Expression of JHE Gene(s)

To further understand the role of JHE in JH titer regulation in *Gryllus*, the JHE enzyme from high- and low-activity lines was characterized biochemically [84], a JHE gene was cloned [85], and JHE transcript abundance was compared between high- and low-activity lines of *G. assimilis* [82]. Thus far, all data are consistent with JHE activity differences between morphs or artificially selected lines of *Gryllus* species being due to differences in JHE regulation resulting in changes in enzyme concentration. No biochemical differences were observed between JHEs from LW(f) versus SW *G. rubens*, or high-activity versus low-activity selected lines of *G. assimilis*, with respect to a variety of characteristics such as Michaelis constant (K_M) for JH-III, inhibition by JHE-specific (trifluoropropanone), or general esterase (DFP, eserine) inhibitors, or thermal stability [84,87]. Multiple JHE isoforms of homogeneously purified enzyme from *G. assimilis* were identified via isoelectric focusing [86], and differences in isozyme profiles were observed between LW(f) and SW *G. rubens* [87]. However, it is not known whether the JHE isoforms differ biochemically. Kinetic properties of *Gryllus* JHEs (K_M for JH, k_{cat}, k_{cat}/K_M, K_Is for trifluoropropanones) were similar to JHEs from other insects [84,86,88]. However, the dimeric quaternary structure for the *G. assimilis* JHE differed from the monomeric quaternary structure of all studied lepidopteran and dipteran JHEs and was similar to a dimeric coleopteran enzyme [10,86].

The JHE gene from *G. assimilis* (1293 bp coding sequence of the mature protein) was amplified from a midgut–fat body cDNA library [85]. The predicted molecular weight (50 kDa) was similar to the observed molecular weight (52 kDa) of the purified protein, and the gene exhibited two motifs characteristic of JHE: the GQSAG motif around the serine of the catalytic triad and an amphipathic helix [85]. Interestingly, phylogenetic analysis of all insect JHE gene sequences available at the time, using maximum likelihood tests, indicated at least two independent origins of this gene in insects: one clade comprised all lepidopteran JHEs, while the other clade comprised JHEs in the Diptera/Coleoptera/Orthoptera clade [85], which was subsequently verified [88]. The independent origin of JHE in these two clades is consistent with structural (monomer vs. dimer) and biochemical differences between JHE of these two groups [86].

A 19 bp insertion/deletion (indel) polymorphism was found in the *G. assimilis* JHE gene, and indel phenotype was strongly associated with selected line type. That is, the insertion was found in virtually all (91%) individuals of the high-activity line and no individual of the low-activity line. Indel phenotype strongly co-segregated with JHE activity in F_2 individuals derived from line crosses/backcrosses providing strong support that the sequenced gene was, in fact, a JHE gene. These results are consistent with regulators of JHE activity, responsible for line differences in JHE activity, occurring near or within the JHE gene. JHE transcript abundance was significantly higher in fat body and midgut of the high-activity versus low-activity lines and was significantly correlated with fat body or hemolymph JHE activity. These correlations indicate that fat body appears to be the main source of JHE production, as appears to be the case for insects in general [26]. In addition, these data further support the hypothesis that JHE activity differences between the selected lines are due to selection on genes that regulate the expression and release of JHE from the fat body, rather than genes that produce biochemically differentiated JHE enzymes.

3.3.5 Morph-Specific Circadian Rhythm for the Hemolymph JH Titer

The existence and functional significance of morph-specific JH titer differences during the adult stage has been a central question in the endocrine regulation of complex polymorphism since the 1960s [3,5–7]. However, as discussed previously, morph-specific JH titers in dispersal polymorphism have been measured in only a very few species, only to a limited degree, and mainly within the past decade. Furthermore, except for the studies of JH titer regulation in adult morphs of *Gryllus* species discussed later, there is very little detailed information on the mechanisms that regulate the JH titer in a morph-specific manner.

Since the 1960s, the most widely held hypothesis regarding the endocrine regulation of morph-specific traits in adults is that the traits are expressed due to the JH titer being chronically elevated (flightless, SW morph) or reduced (flight-capable LW(f) morph) relative to some threshold [3,5,12]. This hypothesis is based on the known positive effect of JH on aspects of ovarian growth and egg development in numerous insects, coupled with the negative effect of JH on many, but not all, aspects of flight capability (e.g., degeneration of flight muscle, reduction of lipid flight fuel biosynthesis) [3,5,12]. Thus, a chronically reduced JH titer in the flight-capable morph would account for its reduced egg production, maintenance of flight muscles, and enhanced production of triglyceride flight fuel. Paradoxically, however, several studies of long-distance migrants reported that applied JH had positive effects on flight, such as increasing propensity to fly or flight duration [89,90].

The existence of JH titer differences between dispersing and reproductive morphs was directly addressed in *Gryllus* by measuring the *in vivo* hemolymph JH titer in LW(f) and SW morphs of *G. firmus*, using a well-validated JH radioimmunoassay [91]. Importantly, because the JH titer is 10-fold higher in adults than in last juvenile instar *Gryllus* [70,74], similar to the situation in insects in general, much more accurate JH titer estimates can be obtained in adults than were possible in juvenile *G. rubens*. Moreover, titers could be measured in individual adults (juvenile titer estimates required pooled hemolymph from 50 or more individuals), as could JHE activity and rate of JH biosynthesis. This, in turn, allowed more accurate investigation of the mechanisms underlying morph-specific JH titer regulation, including mathematical modeling of this phenomenon (discussed as follows).

In contrast to conventional wisdom, we found an unprecedented morph-specific JH titer circadian rhythm in adult morphs of *G. firmus* (Figure 3.8). In the LW(f) morph, the JH titer rose and fell dramatically (20- to 50-fold) during a several-hour period during the latter part of the photophase into the early part of the scotophase on each of several days of adulthood. By contrast, the JH titer was relatively constant, temporally, in the SW morph (slight diurnal fluctuation due to diurnal fluctuation in whole-body hemolymph volume) [29,74]. The JH titer daily rhythm began on day 5 of adulthood, the day on which LW(f) individuals first attain flight capability, and the cycle was lost in LW(f) individuals that histolyzed their flight muscles and thus became flightless. Thus, the JH titer daily cycle is correlated with flight capability, which is circadian in crickets (flight only occurs at night). The morph-specific patterns were observed in each of three pairs of LW(f) and SW

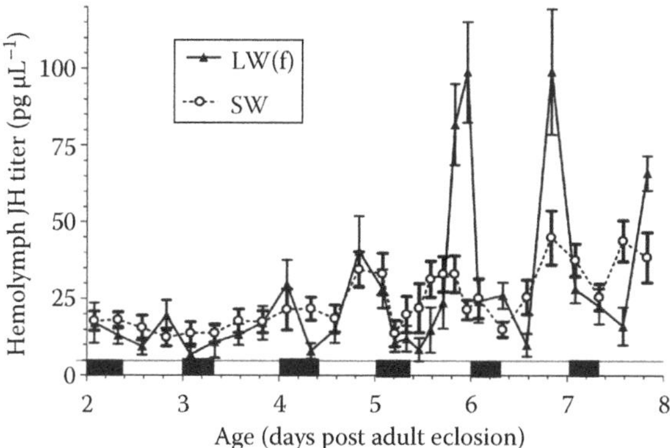

FIGURE 3.8 Morph-specific circadian rhythm for the hemolymph JH titer in adult female LW(f) and SW *G. firmus*. Dark bars: scotophase (lights off); white bars: photophase (lights on). LW(f) is the same morph (LW with functional flight muscles) as LW in previous figures. Note the dramatic daily cycle in the LW(f) morph but not in the SW morph. (Reprinted from Zhao, Z. and Zera, A.J., *J. Insect Physiol.*, 50, 93, 2004.)

artificially selected lines (same lines as those used in the JHE studies discussed earlier) and thus is a morph-associated genetic polymorphism for the circadian rhythm of the hemolymph JH titer. Finally, although the ECD titer also differs between the morphs (higher in SW), no morph-specific daily cycle was observed for this hormone. Thus, not all reproductive hormones in the hemolymph exhibit a morph-specific daily cycle.

Importantly, morph-specific patterns of *in vitro* JH biosynthesis strongly paralleled the morph-specific JH titers [92] (Figure 3.9). These strongly covarying morph-specific patterns of JH titer and JH biosynthesis provide independent confirmation of the JH titer differences between morphs, in addition to providing information on the mechanism that produces the difference. The morph-specific patterns in rate of JH biosynthesis, in turn, were inversely correlated with the concentration of allatostatin, a neurohormone found in nerves within the CA, which negatively regulates the rate of JH biosynthesis [93]. Allatostatin was high and equivalent in the LW(f) and SW CAs when rate of JH biosynthesis was low and equivalent in the morphs and dropped to a lower level in LW(f) CAs when the rate of JH biosynthesis was higher in LW(f) versus SW morphs.

By contrast, differences in *in vitro* JHE activity between the morphs were usually nonsignificant and of much lower magnitude than morph differences in rate of JH biosynthesis. Thus, results to date indicate that the morph-specific changes in the JH titer appear to be driven primarily, if not exclusively, by changes in the rates of hormone biosynthesis. An interesting finding of the adult and juvenile data taken together is the strong correlation between JH titer and rate of JH biosynthesis but not hemolymph JHE activity in adults, which contrasts with the correlation between JH titer and hemolymph JHE activity but not the rate of JH biosynthesis in juveniles (Figures. 3.3 and 3.9). These data suggest that very

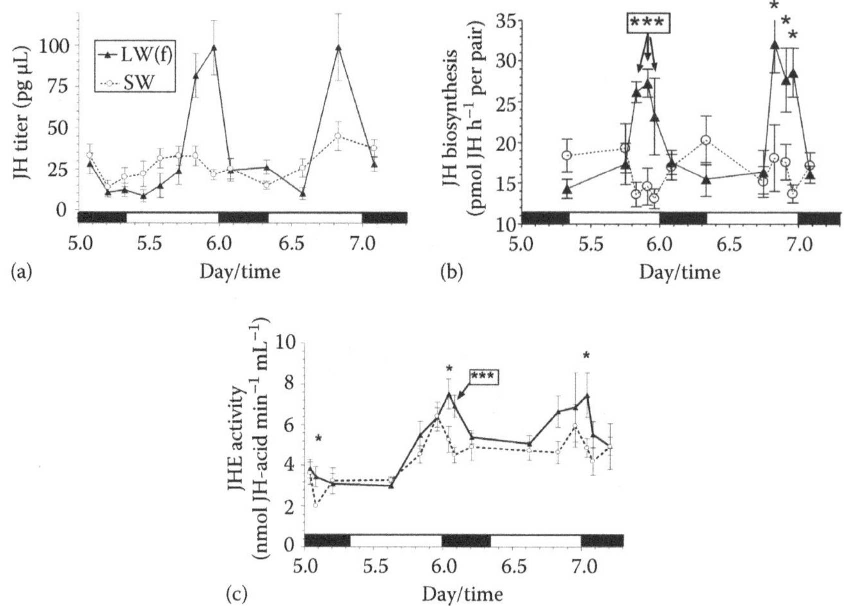

FIGURE 3.9 Morph-specific (a) JH titer (pg μL hemolymph^{-1}), rate of (b) JH biosynthesis (pmol JH h^{-1} pr. CA^{-1}), and rate of hemolymph (c) JHE activity (nmol mL^{-1} min^{-1}) in adult female *G. firmus*. Note the strong parallel patterns of JH biosynthesis and JH titer in the morphs. (Reprinted from Zhao, Z. and Zera, A.J., *J. Insect Physiol.*, 50, 965, 2004.)

different types of regulation of the JH titer can evolve, even in different life cycle stages of the same species.

Standard photoperiod experiments demonstrated that the daily rhythms of both the JH titer and rate of JH biosynthesis in the LW(f) morph were true circadian (i.e., endogenous) rhythms [94]: the cycles persisted in constant darkness and were temperature compensated, two classic pieces of information that support the circadian basis of a daily cycle. The JH titer cycle also shifted in concert with an experimental shift in the onset of the scotophase (dark portion of the cycle), and the cycle was eliminated in constant light. These effects were observed in multiple LW(f) and SW lines of *G. firmus* indicating that the morph-associated circadian rhythm is genetically based. To my knowledge, this is the first example of a morph-specific circadian rhythm in any complex polymorphism. It is also one of the first, and the most extensively described, circadian rhythms for the JH titer and its regulators in any insect to date [94,95]. Circadian rhythms have been commonly reported for many vertebrate and invertebrate hormones [95–97] and may also be common but overlooked, with respect to JH in insects. The unexpected finding of the strong JH titer circadian rhythm in *Gryllus* indicates that fundamental but neglected aspects of the JH titer pattern still await discovery [11].

The function of the morph-specific JH titer circadian rhythm is not well understood, and detailed discussion of this point is beyond the scope of this chapter.

The most likely explanation is that the circadian rhythm functions in some aspect of flight in the LW(f) morph, which, as mentioned earlier, is circadian in *Gryllus* crickets (only occurs at night). Interestingly, in the honey bees, there is an ontogenetic shift from a noncyclic (in young nurse bees) to a diurnally cyclic JH titer (in older foraging adults) that is associated with a change from within-hive activities to flight outside the hive [98]. Clearly JH regulation of aspects of flight and reproduction are more complex than previously suspected. For example, the dramatic short-duration elevation of the JH titer yet retention of flight muscles and limited ovarian growth in the LW(f) morph clearly is inconsistent with the widely held idea that an elevated hormone titer, in and of itself, induces ovarian growth and flight muscle degeneration [12,59]. The duration of elevation may be a critically important aspect of JH regulation. The short-duration elevation of the JH titer in the dispersing morph may also explain the paradoxical results as to why application of JH has positive effects on some aspects of flight (e.g., flight behavior; mentioned earlier) but negative effects on others [59]. Whether JH has positive or negative effects may depend upon the duration of elevation of the JH titer (see extensive discussion on this and related points in [59]).

3.3.6 Morph-Specific JH Titers in the Field

Excluding social insects, very little information is available on the JH titer and titer variation in natural populations of any insect [48]. In *G. firmus*, hemolymph JH titers were measured (1) in the same LW(f) and SW stocks of *G. firmus* that were studied in the laboratory but raised in the field, (2) in *G. firmus* from natural populations in which hemolymph was sampled in the field, and (3) in a few other species of crickets from natural populations [59] (Figure 3.10). Studies were undertaken near Gainesville, Florida, where natural populations of *G. firmus* occur from which individuals had been collected to initiate the LW(f) and SW selected lines discussed earlier. The morph-specific JH titer daily cycle, observed in LW(f) and SW laboratory stocks, was observed in the same stocks raised under field conditions and in *G. firmus* directly sampled from field populations. In *G. firmus* and in other cricket species, the diurnal cycle was always positively associated with the dispersing morph, while flightless individuals had a much reduced or absent cycle [59]. Titers began to rise a few hours before sunset, similar to the rise in the JH titer a few hours prior to lights off in the laboratory. The magnitude of the difference between morphs typically was not as great in the field as it was in the laboratory. This difference could be due to any of several factors such as the abrupt change in light intensity during the transition from the photophase to the scotophase (on/off) in the laboratory, as opposed to the gradual transition from day to night and vice versa, in the field. The same morph-specific JH titer cycle was also observed between male morphs in the field (males had not been previously studied in the laboratory). Similar to the situation in the laboratory, the hemolymph ECD titer was higher in SW versus LW(f) female *G. firmus* and showed no morph-specific daily fluctuation. These results are reassuring in that they indicate that morph-specific patterns in the JH and ECD titers are roughly similar in the laboratory and field.

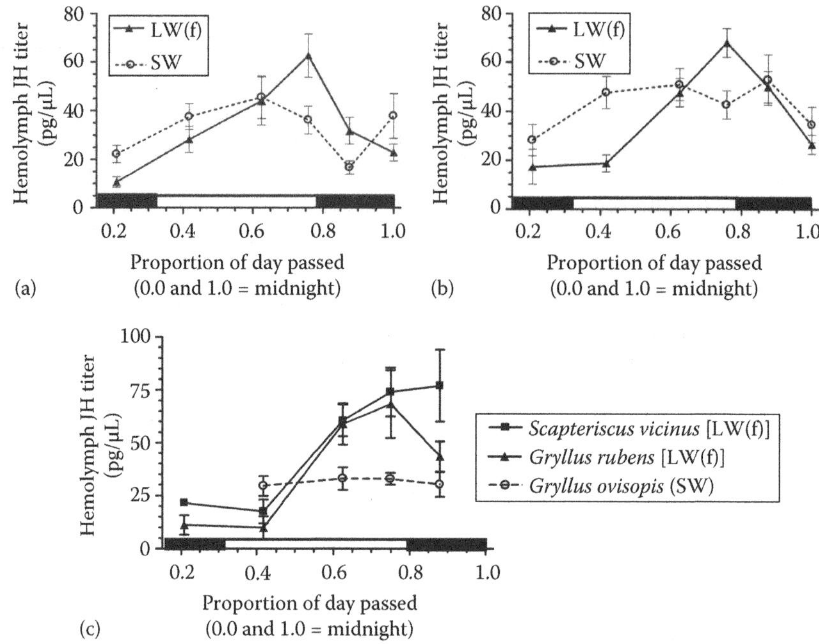

FIGURE 3.10 Morph-specific JH titer circadian rhythm in adult morphs of various cricket species measured in the field (Gainesville, Florida). JH titers in (a) male and (b) female morphs of *G. firmus* laboratory lines raised in the field. (c) JH titers in various cricket species. Note the association between more pronounced JH titer cycle and LW(f) within or between species. SW, short-winged, flightless morph; Dark bars, night; white bars, day. (Reprinted from Zera, A.J. and Zhao, Z., *J. Insect Physiol.*, 55, 450, 2009.)

3.3.7 Mathematical Modeling of Morph-Specific JH Titer Regulation

3.3.7.1 Model Description and Results

Although JH titer regulation has been a central focus of insect endocrinology since the 1970s, quantitative assessments of the extent to which experimentally determined values of JH regulation (e.g., rate of biosynthesis, *in vivo* rate of degradation, binding to carrier protein) estimate the JH titer have not often been undertaken [27,99]. Availability of estimates of the *in vivo* JH titer, *in vitro* rate of JH biosynthesis, *in vitro* JHE activity, and *in vivo* rate of JH degradation in morphs of *G. firmus* that differ dramatically in short-term temporal fluctuation in the JH titer provides an exceptional opportunity to undertake such an analysis (currently in progress, A. J. Zera and C. Brassil). Only a few aspects of our preliminary analysis of the data will be presented here for the LW(f) morph of *G. firmus*, to illustrate some major points.

Hemolymph JH titer at a particular time in adult LW(f) *G. firmus* was expressed using a simple model: $[JH]_{t+1} = [JH]_t \times k_{deg} + [JH]_{new}$. That is, the hemolymph JH titer [JH] at time ("t + 1") is equal to the hemolymph titer measured at the previous time period ($[JH]_t$) multiplied by the *in vivo* rate of JH degradation (k_{deg}), plus the amount of new JH synthesized during the previous time interval [JH-III]. This analysis was restricted to the 2-day period in adulthood (days 5–7), when values were available every 3–6 h for the *in vivo* hemolymph JH titer and *in vitro* rate of JH biosynthesis for morphs of *G. firmus* [29,74] (Figure 3.9). Estimate of the *in vivo* rate of JH degradation was not available for *G. firmus* adults, and this rate constant was obtained from [67], measured for the congener *G. rubens* (discussed earlier). Rate of JH degradation was measured during days 5–7 of adulthood in the SW morph of *G. rubens* (half-life ($t_{1/2}$) = 70 min; k_{deg} = 0.01 min, using the equation $k_{deg} = 0.693/t_{(1/2)}$). JHE activity is similar for adults of *G. rubens* and *G. firmus*. Thus, as a first approximation, the *in vivo* half-life measured in *G. rubens* is expected to be roughly similar to that of adult *G. firmus*, especially since this half-life also is similar to values reported in a number of insects (see aforementioned text). The JH half-life was measured during the mid-photophase and circadian variation in the rate of degradation was not measured (circadian cycle had not yet been discovered in *Gryllus*) and was assumed to be constant. Potential errors involved in these measures and their impact on the JH titer model are discussed as follows.

The observed and expected change in the JH titer in the LW(f) morph of *G. firmus* is given in Figure 3.11. As can be seen in Figure 3.11a, the model fairly accurately accounts for the position and height of the JH titer peaks but underestimates the amplitude of the peaks by about 2- to 2.5-fold and overestimates the basal JH titer by about 2–2.5. Given the number of potential errors in the various physiological measurements (discussed as follows), this constitutes a reasonably good fit to the model.

Since errors in measures of JH biosynthetic and degradation rates were considered more likely to be greater than errors associated with measurement of the JH titer, we altered rates of JH degradation and JH biosynthesis to determine the extent to which these values needed to be modified to produce a good fit between observed and expected change in the JH titer. Only two of these results will be presented here (Figure 3.11b and c). When the *in vivo* JH degradation rate was increased twofold (Figure 3.11b), the model more accurately predicts the position of the peaks and the basal JH titer but underestimates the height of the JH titer peaks by about twofold. Keeping this twofold elevated rate of JH degradation while increasing the rate of JH biosynthesis by 1.8-fold results in a good fit between expectations of the model and observed measurements. These twofold or less changes in *in vivo* hormone biosynthesis and degradation are modest given the likely errors involved in estimation of biosynthetic and degradative rates (see as follows). Thus, to a first approximation, our measures of rates of the JH biosynthesis, JH degradation, and JH titer provide an internally consistent picture of circadian JH titer fluctuation due primarily to change in the rate of JH biosynthesis, as reported in Zhao and Zera [29].

3.3.7.2 Potential Sources of Error in Estimates of Physiological Variables

Estimation of each of the three variables in the model discussed earlier, JH titer, *in vitro* rate of JH biosynthesis, and *in vivo* rate of JH degradation, is subject to a number of errors. A full discussion of this topic will be given in detail in Zera and Brassil (manuscript in preparation), and only a few salient points will be considered here. The JH titer in *Gryllus* was estimated using a well-validated JH radioimmunoassay [91]. Although radioimmunoassays are subject to a number of errors (see [10]),

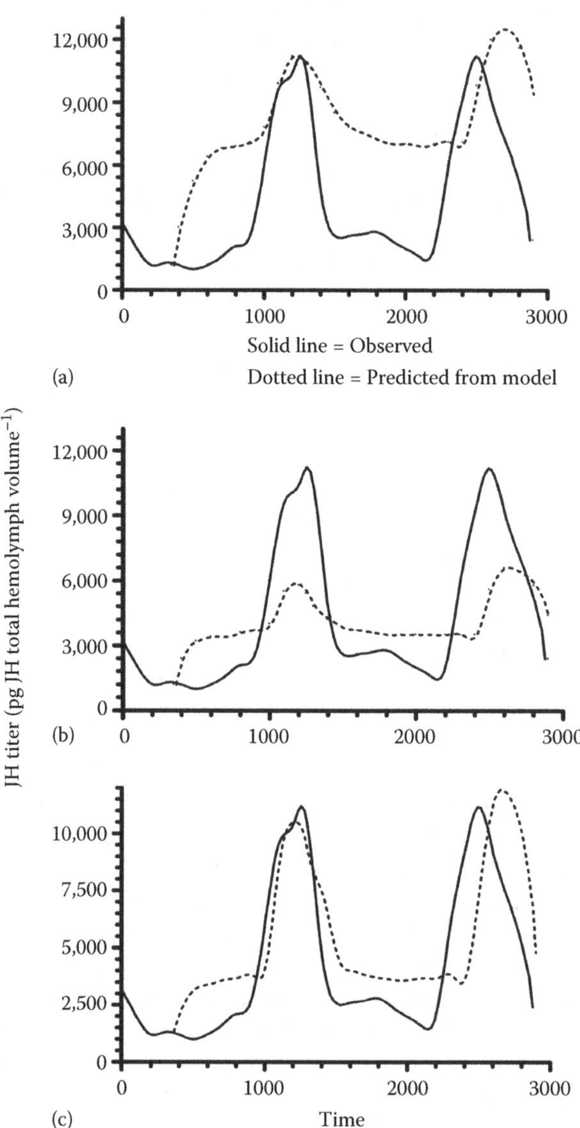

FIGURE 3.11 Preliminary results of a quantitative analysis of temporal change in the JH titer in the LW morph of *G. firmus* as a function of JH biosynthesis and JH degradation. (a) Solid line: experimentally observed change in the JH titer reported in [29] (see Figure 3.9). Dashed line: predicted change in the JH titer using the mathematical model given in the text and estimated values for rates of JH biosynthesis and JH degradation. (b) Same as in (a) but with *in vivo* rate of JH degradation increased by twofold. (c) Same as in (b) but with rate of JH biosynthesis increased by twofold. Time is in minutes with 400 min = lights on day 5 of adulthood (see Figure 3.9). Full analysis of these and other data will be published elsewhere. (Data from Zera, A.J. and Brassil, C., manuscript in preparation; Zhao, Z. and Zera, A.J., *J. Insect Physiol.*, 50, 965, 2004; see Figure 3.9.)

the JH titer estimates are probably less likely to be prone to error than the estimates of rates of JH biosynthesis and degradation.

A likely source of error in the model is the estimate of rate of JH biosynthesis, which was measured *in vitro* [29]. As mentioned previously, the JH biosynthetic rate is known to be affected by a wide variety of compounds, most notably neuropeptides that increase (allatotropins) or decrease (allatostatin) rate of biosynthesis, both of which have been reported in crickets [10,27,100]. In addition, various neurotransmitters (octopamine, glutamate, dopamine), ECDs, and JH itself (reviewed in [10]) can influence the JH biosynthetic rate. Finally, severing the neural connections between the CAs and the brain, which is required to remove the CAs from the insect for the standard *in vitro* assay, has been reported to affect the rate of JH biosynthesis in some species [10,27]. The aforementioned model indicates that particular attention should be paid to factors causing underestimation of the rate of JH biosynthesis in *Gryllus* CAs in the study of Zhao and Zera [29].

Finally, several factors might have resulted in errors in the estimated *in vivo* rate of JH degradation in *G. firmus*, using the rate constant measured in the congener *G. rubens* [67], which, for simplicity, was assumed to be temporally constant. Temporal variation in *in vitro* JHE activity (Figure 3.9) indicates that the *in vivo* rate of JH degradation may in fact vary temporally. However, the extent to which this occurs is uncertain, given that a particular difference between morphs in *in vitro* JHE activity typically results in a much smaller difference in *in vivo* JH degradation (see earlier text and Figure 3.7). To address this issue more fully, we are currently directly measuring the *in vivo* rate of JH degradation throughout the photophase and scotophase in morphs of *G. firmus*. Finally, the effects of other factors, such as the fairly substantial diurnal change in hemolymph volume (25%–30% reduction from early to late in the photophase in each morph [74]), on the estimate of JH titer have yet to be taken into account.

3.3.7.3 Comparison with Other Models

A recent analysis by Nijhout and Reed [99] indicated a far worse (several orders of magnitude) fit between rates of JH biosynthesis, JH degradation, and the JH titer in *Gryllus* as well as in *M. sexta*. A large part of the discrepancy between results of our preliminary study and those of Nijhout and Reed [99] is that these workers did not use the *in vivo* rate of JH degradation in their model as was done in our aforementioned study. Rather, Nijhout and Reed formulated a more complex model that included JHE activity and binding to JH-binding protein. They then incorporated *in vitro* JHE activities and published binding constants for JH-binding protein obtained for species other than *Gryllus* (binding constants not available for *Gryllus*) to estimate JH degradation *in vivo*. Large errors in the expected degree of *in vivo* JH sequestration due to binding to the JH-carrier protein would lead to a substantial mismatch between observed fluctuation of the JH titer and that expected by the model. However, our analysis, by contrast, was not intended to take into account the influence of JH binding per se on the JH titer in *Gryllus*, especially since some studies have indicated extensive JH binding *in vivo* to components other than the JH-carrier protein (reviewed in [27]). Rather, our use of *in vivo* JH degradation rate takes into account all *in vivo* enzymatic degradation (hemolymph and cellular JHE and JHEH) as well as the influence of JH sequestration on the rate of JH degradation. This was done because we were specifically interested in determining the extent to which variation in the rate of JH biosynthesis could account for variation in JH titer [29]. Our approach is similar to the analysis undertaken by Tobe and coworkers on the role of JH biosynthesis on JH titer regulation in the cockroach, *D. punctata* (reviewed in [27]). As mentioned earlier, a full discussion of these issues will be presented in a forthcoming publication (A. J. Zera and C. Brassil, manuscript in preparation).

3.4 SUMMARY, SYNTHESIS, AND CONCLUSIONS

Studies of *Gryllus* wing polymorphism undertaken during the past three decades have made important contributions to understanding the regulatory mechanisms underlying genetic and temporal variation in the JH titer and the evolutionary modification of these mechanisms. These studies clearly point to evolutionary modulation of JH degradation by altered JHE activity as an important factor in the evolution of wing polymorphism in *Gryllus.* Studies in *Gryllus* have been particularly important in experimentally establishing that *in vitro* differences between the morphs in hemolymph JHE activity have important *in vivo* consequences with respect to altered rates of *in vivo* JH degradation [8,66,67]. These studies are currently the most extensive investigation of the role of JH degradation and the regulation of complex polymorphism. Experimental artificial selection has augmented the aforementioned studies of natural polymorphism by providing a quantitative estimate of the degree to which hemolymph JHE must be altered to produce an observed change in *in vivo* JH degradation [53,61,81]. With the sequencing of a JHE gene, the *Gryllus* experimental model is ripe for the introduction of molecular tools (i.e., RNAi as in [34]), to probe the effect of specific gene products on JH titer regulation.

It is important to emphasize that other endocrine regulators, most notably the ECD titer [70] (Figure 3.3), also differ dramatically between nascent morphs of *Gryllus,* and thus, the precise role played by JHE and JH in regulating specific aspects of wing polymorphism in juvenile *Gryllus* morphs remains uncertain. Recent studies have demonstrated that the endocrine regulation of phase polymorphism in locusts is much more complex than previously suspected, where JH had often been considered to be the "master" phase-regulating hormone [23]. An increasing number of hormones have been identified as regulators of specific aspects of phase polymorphism in locusts, such as JH, ECDs, His^7-corazonin, and serotonin [6,13,24,49]. A similar situation likely occurs with respect to wing polymorphism in *Gryllus* and a central goal of future research should be to identify the specific aspects of morph differentiation that are regulated by JH.

In addition to the in-depth studies of JHE and JH degradation, another noteworthy aspect of the *Gryllus* work has been its focus on evolutionary genetic aspects of the endocrine regulation of complex polymorphism. This contrasts with the near exclusive focus on environmental regulation of dispersal polymorphism in other studies [6,8,53]. Evolutionary genetic studies of JHE have provided information on the nature of variable genes that influence genetic differences in hemolymph JHE activity, whole-body JHE, and whole-body JH degradation. One gene has been identified and sequenced [85], while other genes have been identified via their "quantitative genetic" phenotypes, such as differences between selected lines in whole-organism and tissue distribution of JHE activity. The picture to date, derived from biochemical genetic characterization of selected lines [8,53], is that variable genes give rise to phenotypic differences in the concentration and tissue distribution of JHE activity in contemporary populations of *G. assimilis* and that these differences in JHE activity were important targets of selection during the evolution of wing polymorphism in *Gryllus*.

Evolutionary genetic studies in *Gryllus* have transcended the issue of polymorphism by identifying the extent to which endocrine traits can evolve independently versus in concert due to genetic correlation, presumably arising from pleiotropy [8,53]. These correlations can have profound effects on the evolution of particular traits or sets of traits. For example, the absence or existence of only very weak correlations between genes regulating hemolymph JHE activity in juvenile and adult *G. assimilis* resulted in large-magnitude differences in JHE activity evolving under laboratory selection on juveniles while not producing JHE activity differences between lines in the adult stage. These results explain why dramatic differences in JHE activity exist in several species exhibiting dispersal polymorphism during the juvenile but not adult stages, and have important general implications for the evolution of life histories [80].

In a similar vein, the absence of genetic correlations between JHE and JHEH activities [53,61,81] also indicates that these two key components of JH degradation can—and have—evolved independently of each other in *Gryllus* in natural populations. By contrast, hemolymph JHE activity and JH binding are positively correlated and evolve in concert under artificial selection [81]. This result provides strong support for the hypothesis that JH-binding functions to increase JH degradation [38], rather than to protect JH from degradation as is commonly thought. It is currently unknown why certain aspects of JH regulation are genetically correlated and others are not. These data currently represent the most intensive and extensive data to date on genetic coupling or lack thereof of endocrine traits in outbred populations of any non-domesticated species.

Studies of *Gryllus* also have uncovered unanticipated, high-amplitude, morph-specific circadian rhythms for the JH titer and its regulators in both the laboratory and field [29,59,74,94]. These studies suggest that much remains to be learned about basic patterns of JH titer variation [11], as well as the extent of hormone circadian rhythms in insects, in general, and in polymorphic species, in particular [59,94]. These dramatic short-term fluctuations in the JH titer and its regulators have also provided one of the best opportunities to mathematically model JH titer regulation (Figure 3.11). While indicating a good fit between predictions of the model and experimentally derived values for the JH titer and its regulators, the model also provides insights concerning reasons for discrepancies between model predictions and experimental results (e.g., underestimation of rate of JH biosynthesis). Fascinating problems for future research in *Gryllus* are the mechanisms by which the dramatic morph-specific JH titer circadian rhythm is generated (morph-specific difference in the circadian clock? output of the clock? signal reception by the CA?) and the traits that are differentially regulated by these very different JH titer temporal patterns.

Field endocrine studies in insects have lagged far behind those in vertebrates [59]. The identification of the dramatic morph-specific JH titer circadian rhythms in *G. firmus* and several related insects in the field provides a useful model to investigate endocrine adaptation in the field. Indeed, the strong association between JH titer rhythm and wing morph, which differ dramatically in life history (e.g., reproductive) traits, provides a powerful experimental model for integrating chronobiology, endocrinology, and life history evolution.

ACKNOWLEDGMENTS

I gratefully acknowledge the National Science Foundation, which has supported my research on wing polymorphism in *Gryllus* continuously over the past 21 years (most recently, IOS-1122075, IOS-0516973, and IBN-0212486).

REFERENCES

1. V.B. Wigglesworth, *Insect polymorphism-A tentative synthesis*, in *Insect Polymorphism*, J.S. Kennedy, ed., Royal Entomological Society, London, U.K., 1961, pp. 103–113.
2. J. Hardie and A.D. Lees, *Endocrine control of polymorphism and polyphenism*, in *Comprehensive Insect Physiology, Biochemistry and Pharmacology*, Vol. 8, G.A. Kerkut and L.I. Gilbert, eds., Pergamon Press, New York, 1985, pp. 441–490.
3. H.F. Nijhout, *Insect Hormones*, Princeton University Press, Princeton, 1994.
4. M.J. West-Eberhard, *Developmental Plasticity and Evolution*, Oxford University Press, U.K., 2003
5. A.J. Zera, *The endocrine regulation of wing polymorphism: State of the art, recent surprises, and future directions*, Integr. Comp. Biol. 43 (2004), pp. 607–616.
6. K. Hartfelder and D.J. Emlen, *Endocrine control of insect polyphenism*, in *Insect Endocrinology*, L.I. Gilbert, ed., Elsevier, Amsterdam, the Netherlands, 2012, pp. 464–522.
7. A.J. Zera, *Wing polymorphism in* Gryllus *(Orthoptera: Gryllidae): Proximate endocrine, energetic and biochemical mechanisms underlying morph specialization for flight vs. reproduction*, in *Phenotypic Plasticity of Insects. Mechanisms and Consequences*, D.W. Whitman and T.N. Ananthakrishnan, eds., Science Publishers, Enfield, NH, 2009, pp. 609–653.
8. A.J. Zera, L.G. Harshman, and T.D. Williams, *Evolutionary endocrinology: The developing synthesis between endocrinology and evolutionary genetics*, Annu. Rev. Ecol. Evol. Syst. 38 (2007), pp. 793–817.
9. L. Riddiford, *Hormone action at the cellular level*, in *Comprehensive Insect Physiology, Biochemistry and Pharmacology*, Vol. 8, G.A. Kerkut and L.I. Gilbert, eds., Pergamon, Oxford, U.K., 1985, pp. 37–84
10. W.G. Goodman and N.A. Granger, *The juvenile hormones*, in *Comprehensive Molecular Insect Science*, Vol. 3, L.I. Gilbert, K. Iatrou, and S.S. Gill, eds., Elsevier, Amsterdam, the Netherlands, 2005, pp. 319–408.
11. W.G. Goodman and M. Cusson, *The juvenile hormones*, in *Insect Endocrinology*, L.I. Gilbert, ed., Elsevier, Amsterdam, the Netherlands, 2012, pp. 310–365.
12. A.J. Zera and R.F. Denno, *Physiology and ecology of dispersal polymorphism in insects*, Annu. Rev. Entomol. 42 (1997), pp. 207–231.
13. M.P. Pener and S.J. Simpson, *Locust phase polymorphism: an update*, Adv. Insect Physiol. 36 (2009), pp. 1–272.

14. L.I. Gilbert, ed., *Insect Endocrinology*, Elsevier, Amsterdam, the Netherlands, 2012.
15. A.P. Gupta, ed., *Morphogenetic Hormones of Arthropods*, Vol. 1, Rutgers University Press, New Brunswick, 1990.
16. M. Klowden, *Physiological Systems in Insects*, Academic Press, Amsterdam, the Netherlands, 2002.
17. F.C. Baker, *Techniques for identification and quantification of juvenile hormones and related compounds in arthropods*, in *Morphogenetic Hormones of Arthropods*, Vol. 1, A.P. Gupta, ed., Rutgers University Press, New Brunswick, Canada, 1990, pp. 389–453.
18. T. Kotaki, *Evidence for a new juvenile hormone in a stink bug*, J. Insect Physiol. 42 (1996), pp. 279–286.
19. D. Jones and G. Jones, *Farnesol secretions of dipteran ring glands: What we do know and what we can know*, Insect Biochem. Mol. Biol. 37 (2007), pp. 771–798.
20. L.I. Gilbert, N.A. Granger, and R.M. Roe, *The juvenile hormones: historical facts and speculations on future research directions*, Insect Biochem. Mol. Biol. 30 (2000), pp. 617–644.
21. L.I. Gilbert, K. Iatrou, and S.S. Gill, eds., *Comprehensive Molecular Insect Science*, Elsevier, Amsterdam, the Netherlands, 2005.
22. V.C. Heinrich, *The ecdysteroid receptor*, in *Comprehensive Molecular Insect Science*, Vol. 3, L.I. Gilbert, K. Iatrou, and S.S. Gill, eds., Elsevier, Amsterdam, the Netherlands, 2005, pp. 243–285.
23. H.F. Nijhout and D. Wheeler, *Juvenile hormone and the physiological basis of insect polymorphism*, Quart. Rev. Biol. 57 (1982), pp. 109–133.
24. M.L. Anstey, S.M. Rogers, S.R. Ott, M. Burrows, and S.J. Simpson, *Serotonin mediates behavioral gregarization underlying swarm formation in desert locusts*, Science 323 (2009), pp. 627–630.
25. G.L. Bloch, D.W. Borst, Z.-Y. Huang, G.E. Robinson, J. Cnaani, and A. Hefetz, *Juvenile hormone titers, juvenile hormone biosynthesis, ovarian development and social environment in* Bombus terrestris, J. Insect Physiol. 46 (2000), pp. 47–57.
26. B.D. Hammock, *Regulation of the juvenile hormone titer: Degradation*, in *Comprehensive Insect Physiology, Biochemistry and Pharmacology*, Vol. 8, G.A. Kerkut and L.I. Gilbert, eds., Pergamon Press, New York, 1985, pp. 431–472.
27. S.S. Tobe and B. Stay, *Structure and regulation of the corpora allata*, Adv. Insect Physiol. 18 (1985), pp. 305–433.
28. G.E. Pratt and S.S. Tobe, *Juvenile hormones radiobiosynthesized by corpora allata of adult female locusts*, Life Sci. 14 (1974), pp. 575–586.
29. Z. Zhao and A.J. Zera, *A morph-specific daily cycle in the rate of JH biosynthesis underlies a morph-specific daily cycle in the hemolymph JH titer in a wing-polymorphic cricket*, J. Insect Physiol. 50 (2004), pp. 965–973.
30. G. Horseman, R. Hartman, R. Virant-Doberlet, and W. Loher, *Nervous control of juvenile hormone biosynthesis in* Locusta migratoria, Proc. Natl. Acad. Sci. USA 91 (1994), pp. 2960–2964.
31. L.E. King and S.S. Tobe, *The identification of an enantioselective JH III binding protein from the hemolymph of the cockroach*, Diploptera punctata, Insect Biochem. 18 (1988), pp. 793–805.
32. C.A.D. de Kort and N.A. Granger, *Regulation of JH titers: The relevance of degradative enzymes and binding proteins*, Arch. Insect Biochem. Physiol. 33 (1996), pp. 1–26.
33. C.W. Breunner and M. Orchinik, *Plasma binding proteins as mediators of corticosteroid action in vertebrates*, J. Endocrinol. 175 (2002), pp. 99–112.
34. X. Zhou, M.R. Tarver, and M.E. Scharf, *Hexamerin-based regulation of juvenile hormone-dependent gene expression underlies phenotypic plasticity in a social insect*, Development 134 (2007), pp. 601–610.

35. Y.A.I. Abdel-Aal and B.D. Hammock, *Kinetics of binding and hydrolysis of juvenile hormone II in the hemolymph of* Trichoplusia ni *(Hubner)*, Insect Biochem. 18 (1998), pp. 743–750.
36. B.D. Hammock and T.C. Sparks, *A rapid assay for juvenile hormone esterase activity*, Anal. Biochem. 82 (1977), pp. 573–579.
37. W.G. Goodman and E.S. Chang, *Juvenile hormone cellular and hemolymph binding proteins*, in *Comprehensive Insect Physiology, Biochemistry and Pharmacology*, Vol. 8, G. Kerkut and L.I. Gilbert, eds., Pergamon Press, Oxford, U.K., 1985, pp. 491–510.
38. K. Touhara, B.C. Bonning, B.D. Hammock, and G.D. Prestwich, *Action of juvenile hormone (JH) esterase on the JH-JH binding protein complex. An* in vitro *model of JH metabolism in a caterpillar*, Insect Biochem. Mol. Biol. 25 (1995), pp. 727–734.
39. K. Hartfelder and W. Engels, *Social insect polymorphism: Hormonal regulation of plasticity in development and reproduction in the honey bee*, Curr. Topics Dev. Biol. 40 (1998), pp. 45–77.
40. K. Hartfelder and D.J. Emlen, *Endocrine control of insect polyphenism*, in *Comprehensive Molecular Insect Science*, Vol. 3, L.I. Gilbert, K. Iatrou, and S.S. Gill, eds., Elsevier, Amsterdam, the Netherlands, 2005, pp. 651–703.
41. J. Hardie, F.C. Baker, G.C. Jamieson, A.D. Lees, and D.A. Schooley, *The identification of an aphid juvenile hormone, and its titre in relation to photoperiod*, Physiol. Entomol. 10 (1985), pp. 297–302.
42. S.A. Westerlund and K.H. Hoffmann, *Rapid quantification of juvenile hormones and their metabolites in insect hemolymph by liquid chromatography-mass spectrometry (LC-MS)*, Analyt. Bioanalyt. Chem. 379 (2004), pp. 540–543.
43. E. Schwartzberg, G. Kunert, S. Westerlund, K. Hoffmann, and W. Weisser, *Juvenile hormone titres and winged offspring production do not correlate in the pea aphid*, Acyrthosiphon pisum, J. Insect Physiol. 54 (2008), pp. 1332–1336.
44. M.A. Rankin and M.C. Singer, *Insect movement: Mechanisms and effects*, in *Ecological Entomology*, C.B. Huffaker and R.L. Rabb, eds., John Wiley & Sons, New York, 1984, pp. 185–216.
45. A.J. Zera, *Endocrine analysis in evolutionary-developmental studies of insect polymorphism: Use and misuse of hormone manipulation*, Evol. Dev. 9 (2007), pp. 499–513.
46. M.P. Pener, *Endocrine aspects of phase polymorphism in locusts*, in *Endocrinology of Insects*, R.G.H. Downer and H. Laufer, eds., Alan R. Liss, New York, 1983, pp. 379–394.
47. M.P. Pener, *Locust phase polymorphism and its endocrine relations*, Adv. Insect Physiol. 23 (1991), pp. 1–80.
48. F.W. Botens, H. Rembold, and A. Dorn, *Phase-related juvenile hormone determinations in field catches and laboratory strains of different* Locusta migratoria *subspecies*, in *Advances in Comparative Endocrinology*, S. Kawashima and S. Kikuyama, eds., Monduzzi, Bologna, Italy, 1997, pp. 197–203.
49. A.I. Tawfik, K. Treiblmayer, A. Hassanali, and E.O. Osir, *Time-course haemolymph juvenile hormone titres in solitarious and gregarious adults of* Schistocerca gregaria, *and their relation to pheromone emission, CA volumetric changes and oocyte growth*, J. Insect Physiol. 46 (2000), pp. 1143–1150.
50. S. Liu, B. Yang, J. Gu, X. Yao, Y. Zhang, Y. Zhang, F. Song, and L. Zewen, *Molecular cloning and characterization of a juvenile hormone esterase gene from the brown planthopper*, Nilaparvata lugens, J. Insect Physiol. 54 (2008), pp. 1495–1502.
51. D.J. Emlen and H.F. Nijhout, *Hormonal control of male horn length dimorphism in the dung beetle* Onthophagus taurus *(Coleoptera: Scarabaeidae)*, J. Insect Physiol. 45 (1999), pp. 45–53.

52. D.J. Emlen and H.F. Nijhout, *Hormonal control of male horn length dimorphism in* Onthophagus taurus *(Coleoptera: Scarabaeidae): a second critical period of sensitivity to juvenile hormone*, J. Insect Physiol. 47 (2001), pp. 1045–1054.
53. A.J. Zera, *Evolutionary genetics of juvenile hormone and ecdysteroid regulation in* Gryllus*: A case study in the microevolution of endocrine regulation*, Comp. Biochem. Physiol. A 144 (2006), pp. 365–379.
54. R.G. Harrison, *Dispersal polymorphisms in insects*, Annu. Rev. Ecol. Syst. 11 (1980), pp. 95–118.
55. D.A. Roff, *The evolution of flightlessness in insects*, Ecol. Mono. 60 (1990), pp. 389–421.
56. A.J. Zera and J.A. Brisson, *Quantitative, physiological, and molecular genetics of dispersal and migration*, in *Dispersal Ecology and Evolution*, J. Colbert, M. Baguette, T.G. Benton, and J.M. Bullock, eds., Oxford University Press, Oxford, U.K., 2012, pp. 63–82.
57. K. Vepsalainen, *Wing dimorphism and diapause in Gerris: determination and adaptive significance*, in *Evolution of Insect Migration and Diapause*, H. Dingle, ed., Springer-Verlag, New York, 1978, pp. 218–253.
58. R.F. Denno, G.K. Roderick, M.A. Peterson, A.F. Huberty, H.G. Dobel, M.D. Eubanks, J.E. Losey, and G.A. Langellotto, *Habitat persistence underlies intraspecific variation in the dispersal strategies of planthoppers*, Ecol. Monogr. 66 (1966), pp. 389–408.
59. A.J. Zera, Z. Zhao, and K. Kaliseck, *Hormones in the field: Evolutionary endocrinology of juvenile hormone and ecdysteroids in field populations of the wing-dimorphic cricket Gryllus firmus*, Physiol. Biochem. Zool. 80 (2007), pp. 592–606.
60. A.J. Zera, J. Sall, and K. Grudzinski, *Flight-muscle polymorphism in the cricket* Gryllus *firms: Muscle characteristics and their influence on the evolution of flightlessness*, Physiol. Zool. 70 (1997), pp. 519–529.
61. A.J. Zera and C. Zhang, *Direct and correlated responses to selection on hemolymph juvenile hormone esterase activity in* Gryllus assimilis, Genetics 141 (1995), pp. 1125–1134.
62. A.J. Zera and Y. Huang, *Evolutionary endocrinology of juvenile hormone esterase: Functional relationship with wing polymorphism in the cricket*, Gryllus firmus, Evolution 53 (1999), pp. 837–847.
63. A.J. Zera and K.C. Tiebel, *Differences in juvenile hormone esterase activity between presumptive macropterous and brachypterous* Gryllus rubens*: Implications for the hormonal control of wing polymorphism*, J. Insect Physiol. 35 (1989), pp. 7–17.
64. T.J. Walker and J.M. Sivinski, *Wing dimorphism in field crickets (Orthoptera: Gryllidae:* Gryllus*)*, Ann. Entomol. Soc. Am. 79 (1986), pp. 84–90.
65. S. Masaki and T.J. Walker, *Cricket life cycles*, Evol. Biol. 21 (1987), pp. 349–423.
66. A.J. Zera and C.L. Holtmeier, *In vivo and in vitro degradation of juvenile hormone-III in presumptive long-winged and short-winged* Gryllus rubens, J. Insect Physiol. 38 (1992), pp. 61–74.
67. A.J. Zera, C. Borcher, and S.B. Gaines, *Juvenile hormone degradation in adult wing morphs of the cricket*, Gryllus rubens, J. Insect Physiol. 39 (1993), pp. 845–856.
68. A.J. Zera and K.C. Tiebel, *Brachypterizing effect of group rearing, juvenile hormone-III, and methoprene on winglength development in the wing-dimorphic cricket*, Gryllus rubens, J. Insect Physiol. 34 (1988), pp. 489–498.
69. A.J. Zera and S.S. Tobe, *Juvenile hormone-III biosynthesis in presumptive long-winged and short-winged* Gryllus rubens*: Implications for the hormonal control of wing polymorphism*, J. Insect Physiol. 36 (1990), pp. 271–280.
70. A.J. Zera, C. Strambi, K.C. Tiebel, A. Strambi, and M.A. Rankin, *Juvenile hormone and ecdysteroid titers during critical periods of wing morph determination in* Gryllus rubens, J. Insect Physiol. 35 (1989), pp. 501–511.
71. D. Fell, *Understanding the Control of Metabolism*, Portland Press, London, U.K., 1997.

72. A.J. Zera, *Microevolution of intermediary metabolism: Evolutionary genetics meets metabolic biochemistry*, J. Exp. Biol. 214 (2011), pp. 179–190.
73. M.G. Peter, *Chiral recognition in insect juvenile hormone metabolism*, in *Chirality and Biological Activity*, B. Testa, ed., Alan R. Liss, New York, 1990, pp. 111–117.
74. Z. Zhao and A.J. Zera, *The hemolymph JH titer exhibits a large-amplitude, morph-dependent, diurnal cycle in the wing-polymorphic cricket*, Gryllus firmus, J. Insect Physiol. 50 (2004), pp. 93–102.
75. S.S. Tobe, R.P. Ruegg, B.A. Stay, F.C. Baker, C.A. Miller, and D.A. Schooley, *Juvenile hormone titre and regulation in the cockroach* Diploptera punctata, Experientia 41 (1985), pp. 1028–1034.
76. R.A. Johnson and L. Hill, *Quantitative studies on the activity of the corpora allata in adult male* Locusta *and* Schistocerca, J. Insect Physiol. 19 (1973), pp. 2459–2464.
77. M.J. Fain and L.M. Riddiford, *Juvenile hormone titers in the hemolymph during late larval development of the tobacco hornworm*, Manduca sexta *(L.)*, Biol. Bull. 149 (1975), pp. 506–521.
78. M.R. Rose, T.J. Nusbaum, and A.K. Chippindale., *Laboratory evolution: the experimental wonderland and the Cheshire cat syndrome*, in *Adaptation*, M.R. Rose and G.V. Lauder, eds., Academic Press, San Diego, CA, 1996, pp. 221–241.
79. T. Garland, Jr., *Selection experiments: An under-utilized tool in biomechanics and organismal biology*, in *Vertebrate Biomechanics and Evolution*, V. Bels, J.-P. Gasc, and A. Casinos, eds., BIOS Scientific, Oxford, U.K., 2003, pp. 23–65.
80. A.J. Zera and L.G. Harshman, *Laboratory selection studies of life-history physiology in insects*, in *Experimental Evolution: Methods and Applications*, T. Garland, Jr., and M.R. Rose eds., University of California Press, Berkeley, CA, 2009, pp. 217–262.
81. A.J. Zera, J. Sall, and R. Schwartz, *Artificial selection on JHE activity in* Gryllus assimilis*: Nature of activity differences between lines and effect on JH binding and metabolism*, Arch. Insect Biochem. Physiol. 32 (1996), pp. 421–428.
82. A. Anand, E.J. Crone, and A.J. Zera, *Tissue and stage-specific juvenile hormone esterase (JHE) and epoxide hydrolase (JHEH) enzyme activities and JHE transcript abundance in lines of the cricket* Gryllus assimilis *artificially selected for plasma JHE activity: Implications for JHE microevolution*, J. Insect Physiol. 54 (2008), pp. 1323–1331.
83. A.J. Zera, T. Sanger, and G.L. Cisper, *Direct and correlated responses to selection on JHE activity in adult and juvenile* Gryllus assimilis*: Implications for stage-specific evolution of insect endocrine traits*, Heredity 80 (1998), pp. 300–309.
84. A.J. Zera and M. Zeisset, *Biochemical characterization of juvenile hormone esterases from lines selected for high or low enzyme activity in* Gryllus assimilis, Biochem. Genet. 34 (1996), pp. 421–435.
85. E. Crone, A.J. Zera, A. Anand, J. Oakeshott, T. Sutherland, R. Russell, L.G. Harshman, F. Hoffman, and C. Claudianos, *JHE in* Gryllus assimilis*: Cloning, sequence-function associations and phylogeny*, Insect Biochem. Mol. Biol. 37 (2007), pp. 1359–1365.
86. A.J. Zera, T. Sanger, J. Hanes, and L.G. Harshman, *Purification and characterization of hemolymph juvenile hormone esterase from the cricket*, Gryllus assimilis, Arch. Insect Biochem. Physiol. 49 (2002), pp. 41–55.
87. A.J. Zera, X. Gu, and M. Zeisset, *Characterization of juvenile hormone esterase from genetically-determined wing morphs of the cricket*, Gryllus rubens, Insect Biochem. Mol. Bol. 22 (1992), pp. 829–839.
88. S.G. Kamita and B.D. Hammock, *Juvenile hormone esterase: Biochemistry and structure*, J. Pest. Sci. 35 (2010), pp. 265–274.
89. M.A. Rankin, *Hormonal control of migratory flight*, in *The Evolution of Insect Migration and Diapause*, H. Dingle, ed., Springer-Verlag, New York, 1978, pp. 5–32.

90. M.A. Rankin and J.C.A. Burchsted, *The cost of migration in insects*, Annu. Rev. Entomol. 37 (1992), pp. 533–559.
91. G.A. Cisper, A.J. Zera, and D.W. Borst, *Juvenile hormone titer and morph-specific reproduction in the wing-polymorphic cricket,* Gryllus firmus, J. Insect Physiol. 46 (2000), pp. 585–596.
92. Z. Zhao and A.J. Zera, *Differential lipid biosynthesis underlies a tradeoff between reproduction and flight capability in a wing-polymorphic cricket*, Proc. Natl. Acad. Sci. USA 99 (2002), pp. 16829–16834.
93. B. Stay and A.J. Zera, *Morph-specific diurnal variation in allatostatin immunostaining in the corpora allata of* Gryllus firmus*: Implications for the regulation of a morph-specific circadian rhythm in JH biosynthetic rate*, J. Insect Physiol. 56 (2010), pp. 266–270.
94. A.J. Zera and Z. Zhao, *Morph-associated JH titer diel rhythm in* Gryllus firmus*: Experimental verification of its circadian basis and cycle characterization in artificially-selected lines raised in the field*, J. Insect Physiol. 55 (2009), pp. 450–458.
95. X. Vafopoulou and C.G.H. Steel, *Circadian organization of the endocrine system*, in *Comprehensive Molecular Insect Science*, L.I. Gilbert, I. Kostas, and S.S. Gill, eds., Vol. 3, Elsevier, Amsterdam, the Netherlands, 2005, pp. 551–650.
96. C.A. Czeisler and E.B. Klerman, *Circadian and sleep-dependent regulation of hormone research in humans*, Recent Prog. Horm. Res. 57 (1999), pp. 97–130.
97. R.J. Nelson, *An Introduction to Behavioral Endocrinology*, Sinauer, Sunderland, U.K., 1995.
98. M.M. Elekonich, D.J. Schultz, G. Bloch, and G.E. Robinson, *Juvenile hormone levels in honey bee (*Apis mellifera *L.) foragers: Foraging experience and diurnal variation*, J. Insect Physiol. 47 (2001), pp. 1119–1125.
99. H.F. Nijhout and M.C. Reed, *A mathematical model for the regulation of juvenile hormone titers*, J. Insect Physiol. 54 (2008), pp. 255–264.
100. M. Lorenz and K. Hoffmann, *Allatotropic activity in the subesophageal ganglia of crickets*, Gryllus bimaculatus *and* Acheta domesticus *(Ensifera: Gryllidae)*, J. Insect Physiol. 41 (1995), pp. 191–196.
101. A.J. Zera, *Intermediary metabolism and life history trade-offs: Lipid metabolism in lines of the wing-polymorphic cricket,* Gryllus firmus*, selected for flight capability vs. early age reproduction*, Integr. Comp. Biol. 45 (2005), pp. 511–524.

4 Soldier-Specific Organ Developments Induced by a Juvenile Hormone Analog in a Nasute Termite

Kouhei Toga and Kiyoto Maekawa

CONTENTS

ABSTRACT

Juvenile hormone (JH) is known to be an important regulator in the caste differentiation of social insects. During caste differentiation, caste-specific organ development and regression occur. To understand the proximate factors for the acquisition of each caste, it is essential to clarify the molecular developmental mechanisms underlying the formation of these organs. This chapter discusses studies examining the developmental mechanisms of soldier-specific organ formation in the nasute termite *Nasutitermes takasagoensis*. Soldiers of this species have regressed mandibles but possess a frontal tube (nasus), and defensive secretions synthesized in an exocrine gland (frontal gland) are released from a frontal pore at the tip of nasus. Among various JHs and JH analogs, hydroprene was the most effective for soldier differentiation in this species. Using an artificial induction of soldier differentiation, it was shown that organ formation was involved in the organ-specific molecular mechanisms causing the epidermal folding (nasus disc), the epidermal invagination (frontal gland) and the epithelial programmed cell death (mandible). Moreover, the limb-patterning gene *Distal-less* was involved in at least nasus formation. These results suggest that caste-specific regulation of preexisting developmental mechanisms, triggered by JH, underlie morphological modifications during the caste differentiation of termites.

KEYWORDS

Social insects, Soldier differentiation, Organ formation, Nasus, Frontal gland, Mandible, Hydroprene

4.1 INTRODUCTION

Juvenile hormone (JH) is a multifunctional hormone involved in various physiological phenomena, including molting, metamorphosis, sexual maturation, and diapause. In social insects, JH is known to be an important regulator in caste differentiation [1–3]. Namely, environmental stimuli change the physiological status, and then postembryonic processes are modified in response to the stimuli, resulting in the production of different castes. Social hymenopterans (bees, wasps, and ants) and termites are two typical species-rich social animals, and their caste systems have evolved independently [4]. Monophyly of termites within the cockroach clade was supported by substantial evidence, and termites are recognized as social cockroaches [5]. In their colonies, there are three castes (reproductive, worker, and soldier). In particular, soldiers have species-specific morphological characteristics [6], and their morphologies are highly specialized for colony defense [7] (Figure 4.1). Soldiers differentiate from workers (or late-instar larvae in some families) and always develop through intermediate instars, presoldiers [8]. Extremely dramatic morphological modifications occur during presoldier differentiation. The development of presoldiers and soldiers requires a high JH titer [9], and the application of JH and JH analogs (JHAs) to workers (or late-instar larvae in some families) has been performed for the artificial induction of presoldiers in many species [10–15].

The development and regression of certain organs is observed during soldier differentiation [7]. Defense strategies, largely categorized as mechanical defense or chemical defense, are associated with soldier-specific organ formation. Mechanical defense is achieved with the sclerotization of heads and mandibles (e.g., Figure 4.1a through d), whereas chemical defense involves the formation of an exocrine gland called the frontal gland. Defensive secretions are synthesized in the frontal gland and released from a frontal pore on the head. No structural equivalent of the frontal gland is seen in other insects [16]. Nasutitermitinae (family Termitidae), one of the most phylogenetically apical termite lineages [5], possess soldiers with a frontal tube, called a nasus, and defensive secretions are released from a frontal pore at the tip of nasus (Figure 4.1f). Moreover, in the nasutitermitid soldiers, regressive mandibular evolution is seen corresponding with the development of the nasus [17]. Clarifying the molecular developmental mechanisms underlying caste-specific organ formation

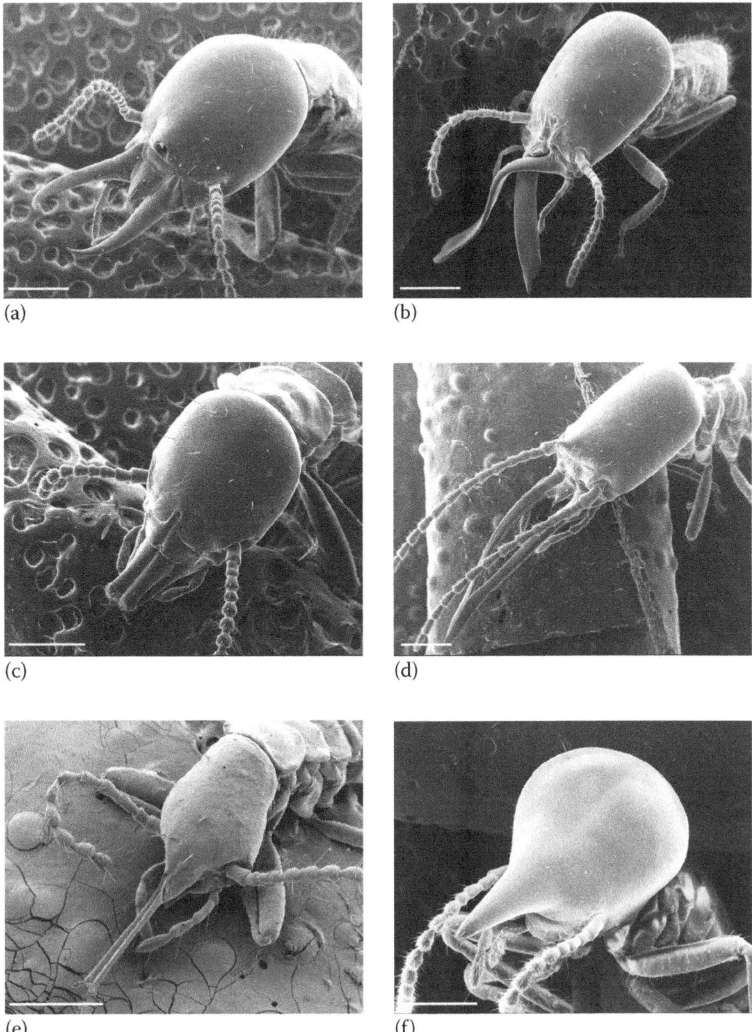

FIGURE 4.1 Soldiers of *Coptotermes* sp. (a), *Pericapritermes nitobei* (b), *Schedorhinotermes* sp. (c), *Termes rostratus* (d), *Rhinotermes* sp. (e), and *Nasutitermes takasagoensis* (f). Scale bars indicate 500 μm.

is essential for understanding the proximate factors for the acquisition of each caste in the course of termite evolution.

In this chapter, we present the findings of our previous studies, which describe the morphological and histological changes of soldier-specific morphogenesis using the artificial induction of presoldier molts in the Japanese nasutitermitid termite *Nasutitermes takasagoensis* [15,18,19]. Based on these results, we summarize the developmental alterations during the presoldier differentiation of *N. takasagoensis* and discuss future research directions.

4.2 INDUCTION OF PRESOLDIER DIFFERENTIATION BY JH AND JHA APPLICATION

4.2.1 Search for Effective JHs and JHAs

In a colony of *N. takasagoensis*, there are three types of workers (female major, male/female medium, male minor workers), and presoldiers differentiated from male minor workers in natural conditions [20] (Figure 4.2). Toga et al. [15] examined the effects on the presoldier induction of various JHs and JHAs (JH III, hydroprene, methoprene, and pyriproxyfen), which were frequently used in previous studies. The results showed that hydroprene (100 μg) induced presoldier differentiation most effectively (Table 4.1). To understand the effects of JHs and JHAs on morphogenesis during the presoldier molt, principal component analysis (PCA) on a correlation

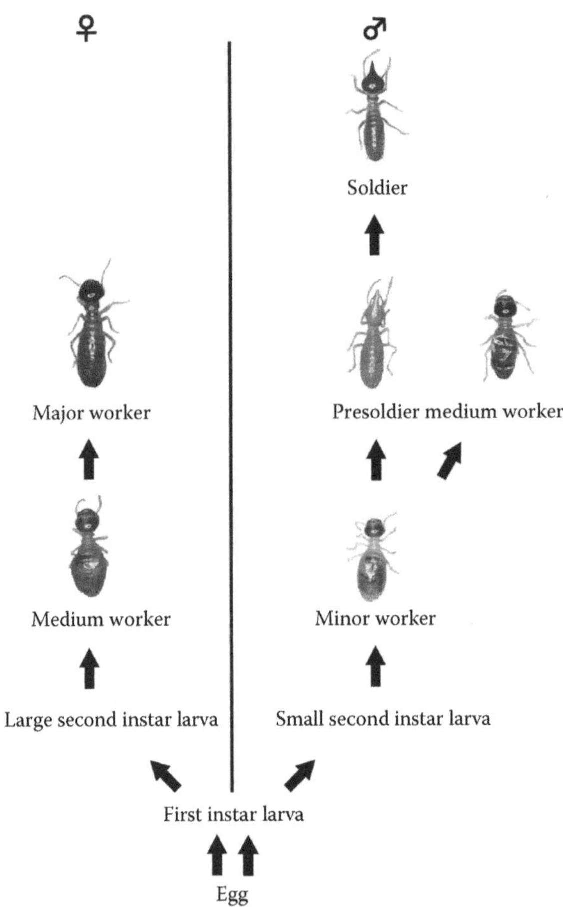

FIGURE 4.2 Developmental pathway of workers and soldiers in *N. takasagoensis*. Arrows show molts. (From Toga, K. et al., *Zool. Sci.*, 26, 382, 2009. With kind permission of the Zoological Society of Japan.)

TABLE 4.1
Overall Survival of Minor Workers (Initial $n = 20$) and Number of Induced Presoldiers per Dish

Treatment		Overall Survival (%)*	Numbers of Induced Presoldiers per Dish**
Acetone		91.7 ± 5.8 a	0.0 ± 0.0 a
JH III	10 μg	91.7 ± 5.8 a	0.0 ± 0.0 a
	100 μg	93.3 ± 5.8 a	0.0 ± 0.0 a
Hydroprene	10 μg	83.3 ± 2.9 a	0.0 ± 0.0 a
	100 μg	90.0 ± 5.0 a	4.0 ± 1.7 a
Pyriproxyfen	10 μg	90.0 ± 5.0 a	0.7 ± 1.2 a
	100 μg	70.0 ± 8.7 b	0.7 ± 1.3 a
Methoprene	10 μg	93.3 ± 2.9 a	0.0 ± 0.0 a
	100 μg	80.0 ± 5.0 ab	0.0 ± 0.0 a

Source: Toga, K. et al., *Zool. Sci.*, 26, 382, 2009. With kind permission of the Zoological Society of Japan.

Note: Sawdust (0.1 g) was treated with each JH(A) diluted in 200 μL acetone.

* Different letters denote significant differences between treatments (Tukey's multiple comparison test, $p < 0.05$).

** Different letters denote significant differences between treatments (Steel–Dwass test, $p < 0.05$).

matrix of the measured values of external morphologies was performed. PCA discriminated plots among each developmental stage except for male and female medium workers, and the presoldiers induced by JHAs (hydroprene and pyriproxyfen) were plotted with natural presoldiers (Figure 4.3). Statistical analysis (Tukey's test) of the first principal component scores showed that scores of the hydroprene-induced presoldiers were significantly lower than those of pyriproxyfen-induced presoldiers. The first principal component positively correlated with all measurements, and this component was considered to be general body size. Thus, significant differences in the first principal component indicated body size differences, whereas no significant differences in second principal component scores were found among JHA (hydroprene and pyriproxyfen)-induced and natural presoldiers. However, a pyriproxyfen-induced presoldier with a deformed nasus was observed (data not shown). As a result of these observations, hydroprene was suggested to be the most effective compound for presoldier differentiation in *N. takasagoensis*.

Just before the presoldier molt, gut contents are purged, and fat bodies are deposited within the abdomen in other species [21–23]. For example, workers of *Hospitalitermes medioflavus* (Nasutitermitinae) approaching the next molt possessed yellowish-white abdomens as a result of the gut purge and the deposition of fat bodies [21]. Using the criteria of abdomen color and the number of days from JHA application, Toga et al. [15] revealed that the gut purge and presoldier

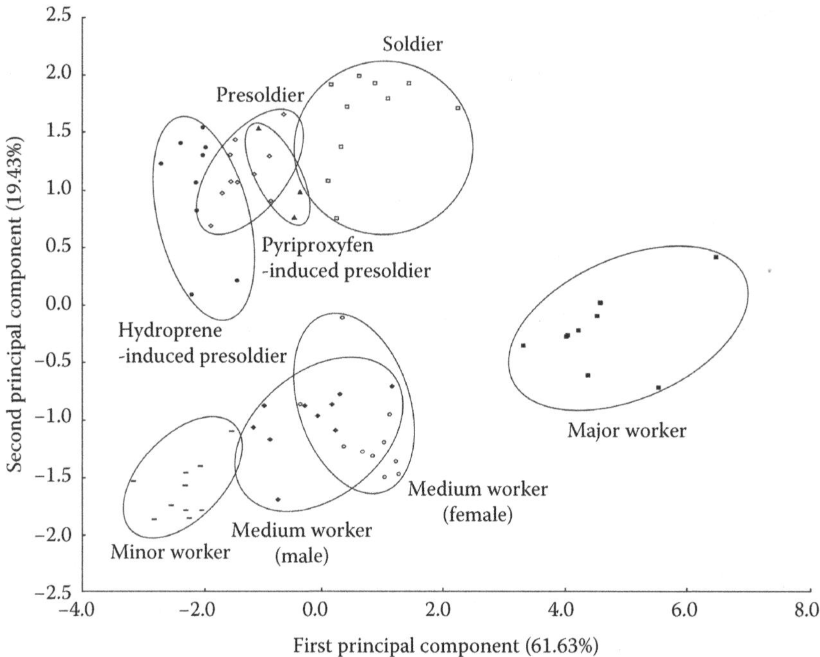

FIGURE 4.3 PCA of each developmental stage. The first principal component shows whole body size. Head height and tibia contribute to the second principal component. Contribution ratios (%) are shown in parentheses. (From Toga, K. et al., *Zool. Sci.*, 26, 382, 2009. With kind permission of the Zoological Society of Japan.)

molts occurred 8–14 days and 15–21 days after JHA application, respectively. These findings showed that it took 6–7 days from the gut purge to the presoldier molt. Furthermore, all JHA-treated minor workers with yellowish-white abdomens molted into presoldiers within 6–7 days. Hereafter, these minor workers with yellowish-white abdomens are called "day X minor workers" according to the number of days after the gut purge.

4.2.2 Effects of JHA on Nasus and Frontal Gland Formation

The nasus of the nasutitermitid soldiers is suggested to be formed by an elongation of the nasus disk, which resembles an imaginal disk of holometabolous insects and is formed under the cuticle [21]. Toga et al. [15] performed morphological and histological observations of the nasus, nasus disk, and frontal gland and examined the effects of hydroprene on organ formation during the presoldier differentiation of *N. takasagoensis*. Nasus disks were observed in the head of natural minor workers with yellowish-white abdomens with the use of a scanning electron microscope (Figure 4.4a and c). Nasus disks were also observed in day 5 minor workers (Figure 4.4b and d). These structures were not observed during molts from minor workers to male medium workers (Figure 4.4e).

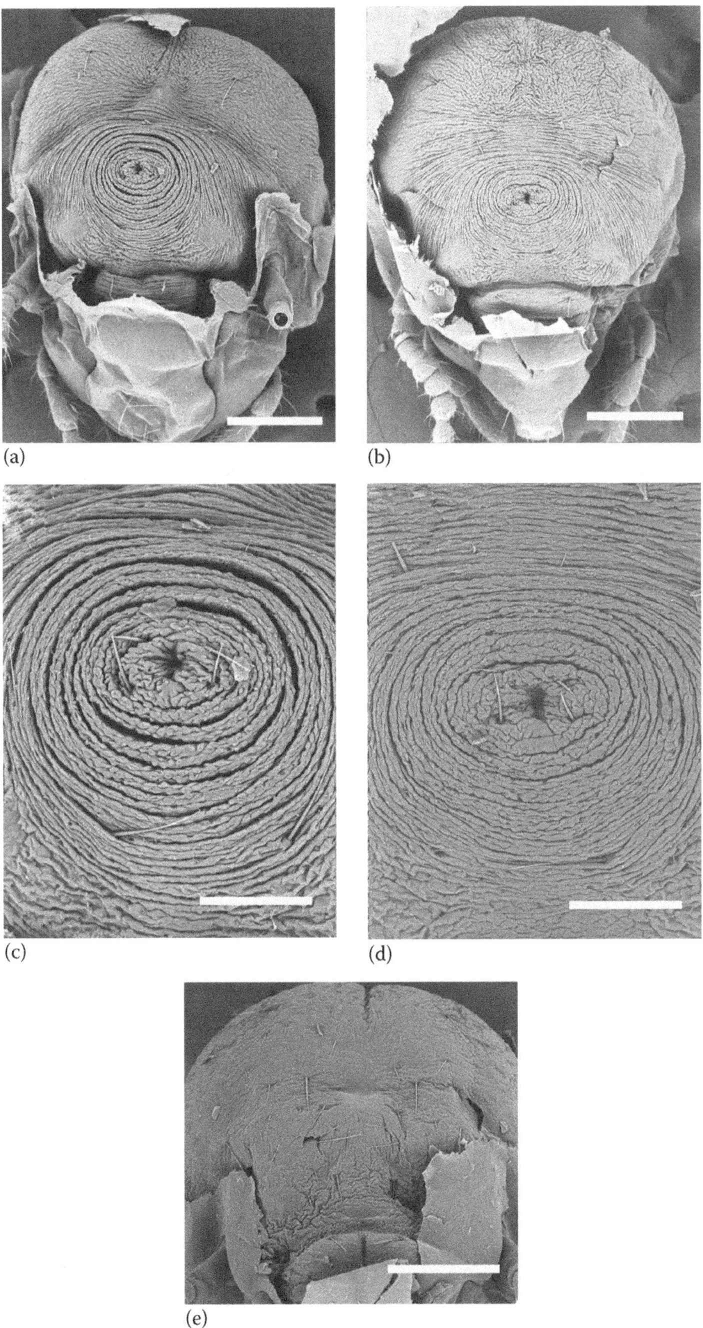

FIGURE 4.4 Nasus disks of natural minor workers (a, c) and hydroprene-treated minor workers (b, d). Nasus disks were not observed in minor workers prior to the molt into medium workers (e). Scale bars indicate 250 µm (a, b, e) and 100 µm (c, d). (From Toga, K. et al., *Zool. Sci.*, 26, 382, 2009. With kind permission of the Zoological Society of Japan.)

The nasus and frontal gland of induced presoldiers were morphologically and histologically compared with those of natural presoldiers. A layer of secretary cells was observed in the head of natural and induced presoldiers (Figure 4.5a and b, arrowheads). Both natural and induced presoldiers possessed frontal pores opening at the tip of the nasus (Figure 4.5c through f) and four bristles around the tip of nasus (Figure 4.5e and f). These bristles probably play a role in defensive behavior [24], and they were also observed in the center of the nasus disk (Figure 4.4c and d). These observations suggested that hydroprene treatment induced the differentiation of presoldiers with normal soldier-specific defensive organs.

4.3 ORGAN DEVELOPMENT DURING PRESOLDIER DIFFERENTIATION IN *N. TAKASAGOENSIS*

4.3.1 Formation of the Nasus Disk and Frontal Gland

We examined histological changes during presoldier differentiation using the JHA application methods mentioned earlier. In natural minor workers collected from nests, epithelial thickening was observed in the dorsal head (Figure 4.6a). In day 0 minor workers, epithelial proliferation resulted in increased thickness of the nasus disk, and the invagination of the frontal gland was shown to occur (Figure 4.6b). In day 3 minor workers, folded layers constituting the nasus disk were recognized, and frontal gland invaginations proceeded more than those observed in day 0 minor workers (Figure 4.6c). In day 5 minor workers, folded layers formed numerous furrows in the nasus disk, and invagination of the frontal gland progressed backward with cell proliferation (Figure 4.6d). Although little is known about the molecular developmental mechanisms of these changes, recent analyses using immunohistochemistry and RNA interference showed that the limb-patterning gene *Distal-less* was identified as a factor involved in nasus formation [19]. This suggests that the nasus was acquired through the co-option of genetic patterning, as shown in a horn development of the beetle *Onthophagus* spp. [25].

4.3.2 Regression of the Mandible

Mandibular regression during soldier differentiation was revealed by morphometric and histological analyses [18]. Morphometry of the right mandibles of minor workers (Figure 4.7a and d), presoldiers (Figure 4.7b and e), and soldiers (Figure 4.7c and f) revealed that all measured mandibular parts (shown in Figure 4.7g) regressed remarkably during the presoldier molt (Figure 4.7g). Histological observations of mandibular regression showed that apolysis of the mandibular epithelia occurred in day 0 minor workers (Figure 4.8a and b) and mandibular regression occurred in day 5 minor workers (Figure 4.8c).

Terminal deoxynucleotidyl transferase (TdT)-mediated dUTP biotin nick end labeling (TUNEL) assays showed that programmed cell death was involved in organ regression in several insect species [26–28]. Consequently, Toga et al. [18] performed TUNEL assays to examine the occurrence of programmed cell death in mandibular regression during presoldier differentiation. Spotted TUNEL signals were observed

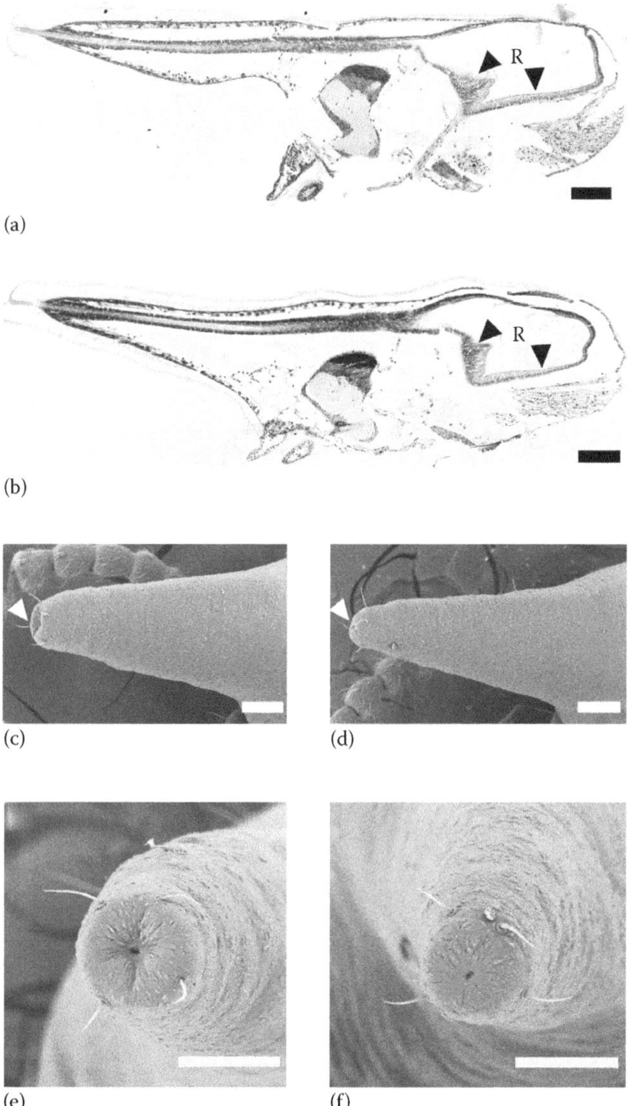

FIGURE 4.5 Frontal gland and nasus of the natural presoldiers (a, c, e) and hydroprene-induced presoldiers (b, d, f). Black arrowheads in (a) and (b) show frontal gland cells. R shows the reservoir of the defensive secretions. White arrowheads in c and d show the position of the tip of nasus. Four bristles are found around the tip of nasus (e and f). Scale bars indicate 100 μm. (From Toga, K. et al., *Zool. Sci.*, 26, 382, 2009. With kind permission of the Zoological Society of Japan.)

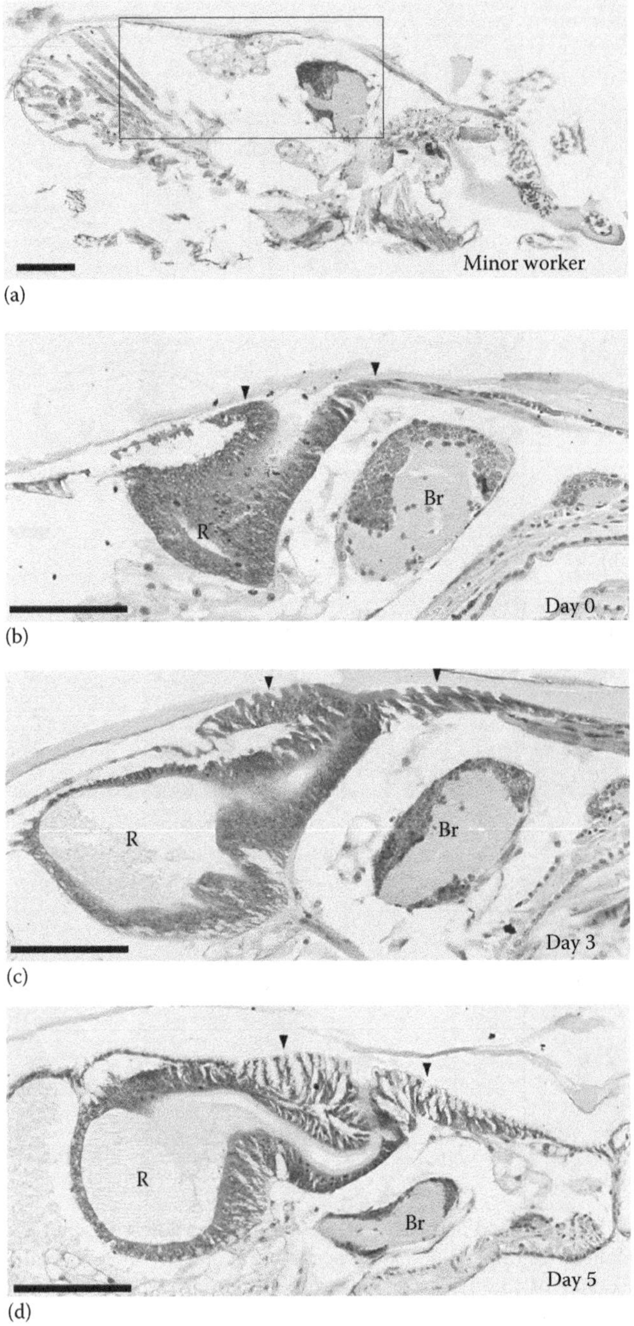

FIGURE 4.6 Parasagittal sections of minor workers (a), day 0 (b), day 3 (c), and day 5 minor workers (d). b, c, and d are magnified views of corresponding regions showing the square in a. Arrowheads show nasus disk. Br: brain, R: frontal gland reservoir. Scale bars indicate 100 μm.

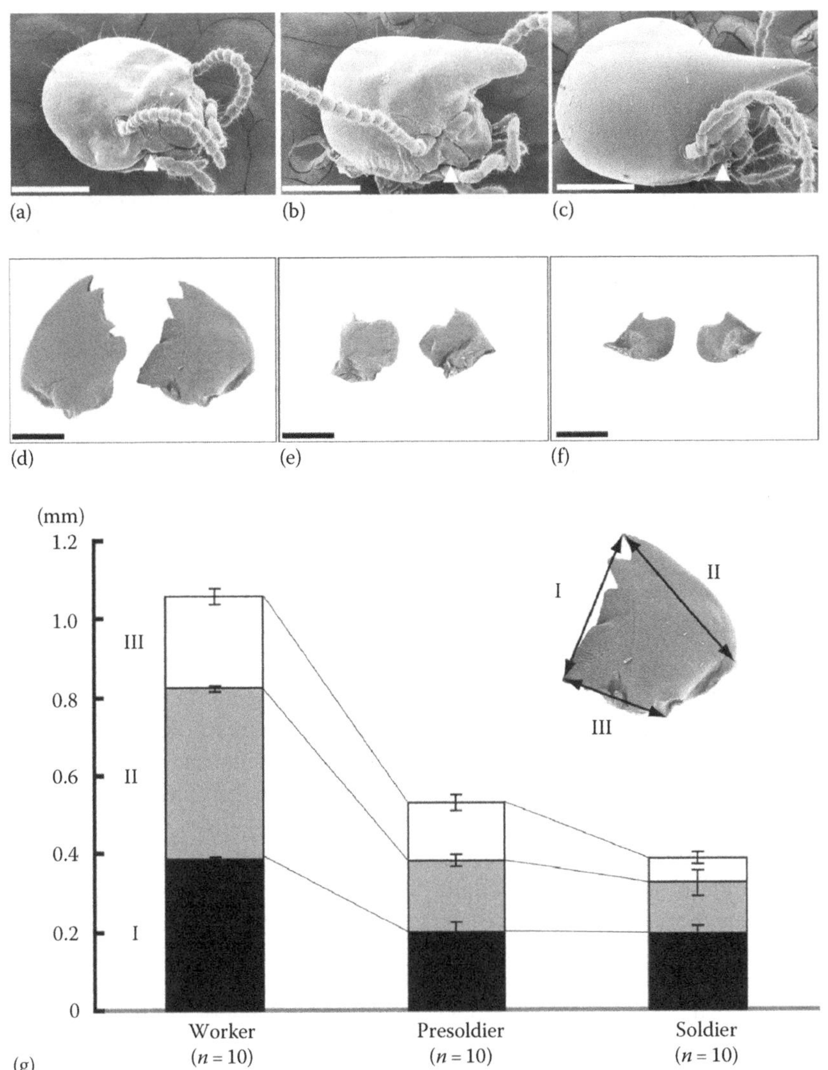

FIGURE 4.7 Scanning electron micrographs of heads and mandibles of minor workers (a, d), presoldiers (b, e), and soldiers (c, f). Arrowheads show the mandibles. Scale bars indicate 100 μm. Length of each measured region (I, II, III) is shown (g). (With kind permission from Springer Science and Business Media: *Naturwissenschaften*, The TUNEL assay suggests mandibular regression by programmed cell death during presoldier differentiation in the nasute termite *Nasutitermes takasagoensis*, 98, 801–806, 2011, Toga, K., Yoda, S., and Maekawa, K.)

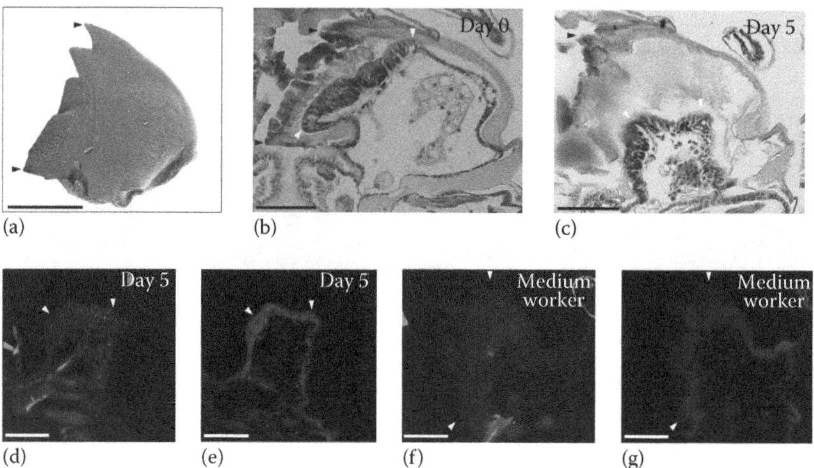

FIGURE 4.8 (See color insert.) Scanning electron micrographs of right mandible of minor workers (a). Sections of day 0 (b) and day 5 (c) minor workers. TUNEL and DAPI staining of sectioned mandibles from day 5 minor workers (d, e) and a gut-purged medium worker prior to the molting into major workers (f, g). Black and white arrowheads show corresponding regions. Scale bars indicate 100 μm. (With kind permission from Springer Science and Business Media: *Naturwissenschaften*, The TUNEL assay suggests mandibular regression by programmed cell death during presoldier differentiation in the nasute termite *Nasutitermes takasagoensis*, 98, 801–806, 2011, Toga, K., Yoda, S., and Maekawa, K.)

in the mandibular epithelia of day 5 minor workers (Figure 4.8d and e), whereas such patterns were not observed in mandibles during the molt from female medium workers to major workers (Figure 4.8f and g). These results suggested that a high JH titer triggered programmed cell death in the mandibular epithelia, resulting in mandibular regression as shown in presoldiers.

Because exaggerated organ development requires costs such as resource competition, trade-off relationships could be observed between exaggerated organ development (horns of beetles and stalked eyes of flies) and the development of other organs (antenna, eyes, wings, and genitalia) [29–31]. The nasus and frontal gland can be recognized as exaggerated organs [32]. Therefore, programmed cell death is probably involved in regulatory mechanisms for the trade-off between mandibular regression and defensive organ formation.

4.4 CONCLUSION

We have elucidated the histological changes of soldier-specific organ formation using the artificial induction of presoldier differentiation in *N. takasagoensis* (Figure 4.9). Organ development and regression already started in day 0 minor workers and continued to occur before the presoldier molt. The organ development and regression seemed to be involved in organ-specific molecular mechanisms causing the epidermal folding (nasus disk), the epidermal invagination (frontal gland), and the epithelial programmed cell death (mandible). These molecular mechanisms are predicted to

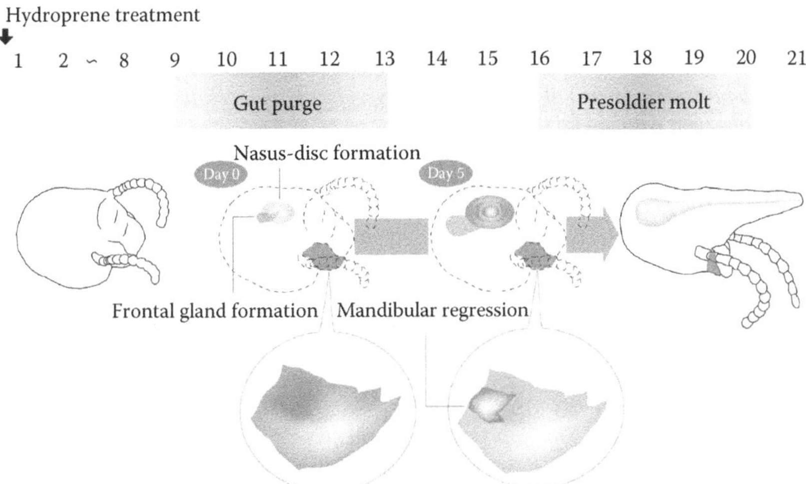

FIGURE 4.9 Time course from hydroprene treatment to presoldier differentiation in *N. takasagoensis*. Gut purge occurred after 8–14 days, and all gut-purged minor workers molted into presoldiers after 15–21 days. Developments of the nasus disk and frontal gland and regression of mandibles already started just after gut purge (day 0 minor workers). These organ formation processes proceeded until just before a presoldier molt (day 5 minor workers).

be triggered by an increase in JH titer. In other insects, JH plays an important role in polyphenism, in which two or more phenotypes are produced from a single genotype (reviewed in [33]). Moreover, some patterning genes underlying the production of polyphenic phenotypes have been identified in butterflies and beetles [25,34–36]. The limb-patterning gene *Distal-less* was shown to be involved in at least nasus formation of *N. takasagoensis* [19], but further molecular developmental approaches (e.g., analyses of other patterning genes and JH signaling) are needed to determine the regulatory mechanisms of the morphological modifications during soldier differentiation. Comparisons of these regulatory mechanisms proposed in termites with those of other insects could be important in understanding the evolution of termite castes and polyphenism in insects.

ACKNOWLEDGMENTS

We thank Drs. Osamu Kitade and Masaru Hojo for providing specimens (*Schedorhinotermes* sp. *Rhinotermes* sp. and *Termes rostratus*).

REFERENCES

1. H.F. Nijhout and D.E. Wheeler, *Juvenile hormone and the physiological basis of insect polyphenism*, Quart. Rev. Biol. 57 (1982), pp. 109–133.
2. H.F. Nijhout, *Insect Hormones*, Princeton University Press, Princeton, NJ, 1994.
3. K. Hartfelder and D.J. Emlen, *Endocrine control of insect polyphenism*, in *Comprehensive Molecular Insect Science*, Vol. 3, L.L. Gilbert, K. Iatrou, and S.S. Gill, eds., Elsevier, Amsterdam, the Netherlands, 2005, pp. 651–703.

4. E.O. Wilson, *The Insect Society*, The Belknap Press of Harvard University Press, Cambridge, MA, 1971.
5. D.J.G. Inward, A.P. Vogler, and P. Eggleton, *A comprehensive phylogenetics analysis of termites (Isoptera) illuminates key aspects of their evolutionary biology*, Mol. Phylogenet. Evol. 44 (2007), pp. 953–967.
6. F.M. Weesner, *External anatomy*, in *Biology of Termites*, K. Krishna, and F.M. Weesner, eds., Vol. I. Academic Press, New York, 1969, pp. 19–47.
7. J. Deligne, A. Quennedey, and M.S. Blum, *The enemies and defense mechanisms of termites*, in *Social Insects*, H.R. Hermann, ed., Academic Press, New York, 1981, pp. 1–76.
8. Y. Roisin, *Diversity and evolution of caste patterns*, in *Termites: Evolution, Sociality, Symbioses, Ecology*, T. Abe, D.E. Bignell, and M. Higashi, eds., Kluwer Academic Press, Dordrecht, the Netherlands, 2000, pp. 95–119.
9. R. Cornette, H. Gotoh, S. Koshikawa, and T. Miura, *Juvenile hormone titers and caste differentiation in the damp-wood termite Hodotermopsis sjostedti* (Isoptera, Termopsidae), J. Insect Physiol. 54 (2008), pp. 922–930.
10. I. Hrdy and J. Křeček, *Development of superfluous soldiers induced by juvenile hormone analogues in the termite*, Reticulitermes lucifugus santonensis, Insect. Soc. 19 (1972), pp. 105–109.
11. R.W. Howard and M.I. Haverty, *Termites and juvenile hormone analogues: A review of methodology and observed effects*, Sociobiology 4 (1979), pp. 269–278.
12. K. Ogino, Y. Hirono, T. Matsumoto, and H. Ishikawa, *Juvenile hormone analogue, S-31183, causes a high level induction of presoldier differentiation in the Japanese damp-wood termite*, Zool. Sci. 10 (1993), pp. 361–366.
13. M.E. Scharf, C.R. Ratliff, J.T. Hoteling, B.R. Pittendrigh, and G.W. Bennett, *Caste differentiation responses of two sympatric Reticulitermes termite species to juvenile hormone homologs and synthetic juvenoids in two laboratory assays*, Insect. Soc. 50 (2003), pp. 346–354.
14. M. Tsuchiya, D. Watanabe, and K. Maekawa, *Effect on mandibular length of juvenile hormones and regulation of soldier differentiation in the termite* Reticulitermes speratus *(Isoptera: Rhinotermitidae)*, Appl. Entomol. Zool. 43 (2008), pp. 307–314.
15. K. Toga, M. Hojo, T. Miura, and K. Maekawa, *Presoldier induction by a juvenile hormone analog in the nasute termite* Nasutitermes takasagoensis *(Isoptera: Termitidae)*, Zool. Sci. 26 (2009), pp. 382–388.
16. C.H. Noirot, *Glands and secretions*, in *Biology of Termites*, Vol. I, K. Krishna, F.M. Weesner, eds., Academic Press, New York, 1969, pp. 89–123.
17. W.A. Sands, *The soldier mandibles of the Nasutitermitinae* (Isoptera, Termitidae), Insect. Soc. 4 (1957), pp. 13–24.
18. K. Toga, S. Yoda, and K. Maekawa, *The TUNEL assay suggests mandibular regression by programmed cell death during presoldier differentiation in the nasute termite* Nasutitermes takasagoensis, Naturwissenschaften 98 (2011), pp. 801–806.
19. K. Toga, M. Hojo, T. Miura, and K. Maekawa, *Expression and function of a limb-patterning gene Distal-less in the soldier-specific morphogenesis in the nasute termite* Nasutitermes takasagoensis, Evol. Dev. 14 (2012), pp. 286–295.
20. M. Hojo, S. Koshikawa, T. Matsumoto, and T. Miura, *Developmental pathways and plasticity of neuter castes in* Nasutitermes takasagoensis *(Isoptera: Termitidae)*, Sociobiology 44 (2004), pp. 433–441.
21. T. Miura and T. Matsumoto, *Soldier morphogenesis in a nasute termite: Discovery of a disc-like structure forming a soldier nasus*, Proc. R. Soc. Lond. B Biol. Sci. 267 (2000), pp. 1185–1189.
22. R. Cornette, T. Matsumoto, and T. Miura, (2007) *Histological analysis of fat body development and molting events during soldier differentiation in the damp-wood termite,* Hodotermopsis sjostedti *(Isoptera, Termopsidae)*, Zool. Sci. 24 (2007), pp. 1066–1074.

23. A. Raina, Y.I. Park, and D. Gelman, (2008) *Molting in workers of the Formosan subterranean termite* Coptotermes formosanus, J. Insect Physiol. 54 (2008), pp. 155–161.
24. A. Quennedey, *Morphology and ultrastructure of termite defence glands*, in *Defensive Mechanism in Social Insects*, H. Hermann, ed., Praeger Publishers, New York, 1984, pp. 151–200.
25. A.P. Moczek and D.J. Rose, *Differential recruitment of limb patterning genes during development and diversification of beetle horns*, Proc. Natl. Acad. Sci. U.S.A. 106 (2009), pp. 8992–8997.
26. S. Sameshima, T. Miura, and T. Matsumoto, *Wing disc development during caste differentiation in the ant* Pheidole megacephala *(Hymenoptera: Formicidae)*, Evol. Dev. 6 (2004), pp. 336–341.
27. A. Gotoh, S. Sameshima, K. Tsuji, T. Matsumoto, and T. Miura, *Apoptotic wing degeneration and formation of an altruismregulating glandular appendage (gemma) in the ponerine ant* Diacamma *sp. from Japan (Hymenoptera, Formicidae, Ponerinae)*, Dev. Genes Evol. 215 (2005), pp. 69–77.
28. T. Kijimoto, J. Andrews, and A.P. Moczek, *Programmed cell death shapes the expression of horns within and between species of horned beetles*, Evol. Dev. 12 (2010), pp. 449–458.
29. D.J. Emlen, *Costs and the diversification of exaggerated animal structures*, Science 291 (2001), pp. 1534.
30. A.P. Moczek and H.F. Nijhout, *Trade-offs during the developmental primary and secondary sexual traits in a horned beetle*, Am. Nat. 163 (2004), pp. 184–191.
31. C.L. Fry, *Juvenile hormone mediates a trade-off between primary and secondary sexual traits in stalk-eyed flies*, Evol. Dev. 8 (2006), pp. 191–201.
32. T. Miura and M.E. Scharf, *Molecular basis underlying caste differentiation in termites*, in *Biology of Termites: A Modern Synthesis*, D.E. Bignell, Y. Roisin, and N. Lo, eds., Springer, Heidelberg, Germany, 2011, pp. 211–253.
33. S.J. Simpson, G.A. Sword, and N. Lo, *Polyphenism in insects*, Curr. Biol. 21 (2011), pp. 738–749.
34. P.M. Brakefield, J. Gates, D. Keys, F. Kesbeke, P.J. Wijngaarden, A. Monteiro, V. French, and S.B. Carroll, *Development, plasticity and evolution of butterfly eyespot patterns*, Nature 384 (1996), pp. 236–242.
35. B.R. Wasik, D.J. Rose, and A.P. Moczek, *Beetle horns are regulated by the Hox gene, Sex combs reduced, in a species- and sex-specific manner*, Evol. Dev. 12 (2010), pp. 353–362.
36. B.R. Wasik and A.P. Moczek, *Decapentaplegic (dpp) regulates the growth of a morphological novelty, beetle horns,* Dev. Genes Evol. 221 (2011), pp. 17–27.

5 Roles of Juvenile Hormone Analog Methoprene in Gene Transcription

Xiao-Fan Zhao

CONTENTS

ABSTRACT

Methoprene is a chemical with similar structure and function to juvenile hormone (JH) which regulates insect development. Therefore, methoprene is widely used as an insecticide and as an analog of JH in insect studies. Methoprene and retinoic acid present a common structural feature. The retinoic acid is capable of activating multiple nuclear receptor mediated pathways for gene transcription. Understanding the roles of methoprene in the gene transcription of insects and mammals is important for environmental security and public health. This chapter summarizes the role of methoprene in insect gene transcription. Methoprene upregulates the transcription of some genes in insects, but did not affect some of the other genes. A number of these genes are involved in protein synthesis, cell growth, and cell proliferation. Furthermore, methoprene is able to suppress the caspase-induced programmed cell death in insects. Methoprene and JH regulate gene transcription in insects through their candidate receptor methoprene-tolerant. Moreover, methoprene is able to bind the mammal retinoic X receptor (RXR), therefore acting as a transcriptional activator. The retinoic acid plays roles in the morphogenesis and immune response of insects. JH and retinoic acid are terpenoids. The similarity of methoprene structure to that of terpenoids might explain or indicate similarity in gene transcription functions. Nevertheless, further studies on the roles of methoprene in mammals and of retinoic acid in insects are needed.

KEYWORDS

Terpenoids, Methoprene, Juvenile hormone, Retinoic acid, Gene transcription

5.1 CHARACTERISTICS OF METHOPRENE

Methoprene is a chemical with similar structure and function to juvenile hormone (JH) [1]. JH is an insect hormone that keeps insects in larval state. Insect development involves going through larval (pupal for holometabolous insects) stage(s) before adulthood. Growth is restricted by their hard exoskeleton, which consist of chitin and proteins. Insects molt several times, including larval molting from one larval instar to the next and metamorphic molting from the last larval instar to pupae, which develops into an adult (metamorphosis). Every molting is triggered by an increase of molting hormone 20-hydroxyecdysone (20E). JH antagonizes the function of 20E and maintains the larval state by preventing metamorphosis. The hormone 20E triggers larval molting when JH amount is low and triggers metamorphic molting in the absence of JH [2]. JH contains a group of sesquiterpenoids, named as JH 0, JH I, JH II, and JH III [3]. JH 0, JH I, JH II, and JH III are isolated from lepidopteran insects, whereas JH III is found in all insects including dipterans [4]. The terpenoid backbone of JH and methoprene is the important structure recognized by insect [5].

Methoprene and retinoids also present a common structural feature (Figure 5.1). Retinoids are the derivatives of vitamin A (retinol) [6]. Retinoids are essential for cellular phenotype and in embryogenesis and apoptosis in mammals [7]. Vitamin A is obtained from food in the form of retinol, oxidized to retinal by dehydrogenases, and then oxidized to retinoic acid, which is the metabolite of vitamin A. At present, hundreds of genes are known to be regulated by retinoic acid [8]. The all-*trans*-retinoic acid (all-*trans* RA or ATRA) has been used for the treatment of various cancers such as lymphoma, leukemia, melanoma, and lung cancer [9]. However, retinoic acid can also increase the risk of cancer because it is capable of activating multiple nuclear receptor-mediated pathways [10]. In addition, retinoic acid initiates non-genomic signaling by increasing Ca^{2+} levels in the cells, but the readout of this process is not clear [11].

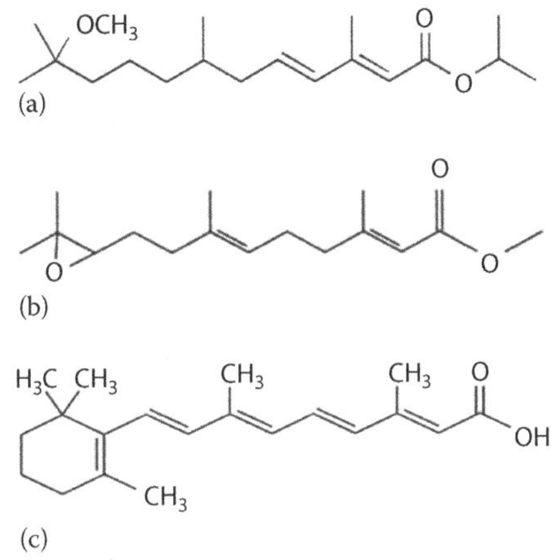

FIGURE 5.1 Structures of (a) methoprene, (b) JH III, and (c) all-trans-retinoic acid.

5.2 FUNCTIONS OF METHOPRENE IN INSECT

Methoprene mimics the function of JH in preventing various insect molting. Therefore, methoprene is widely used as an insecticide, acting as an insect growth regulator in drinking water cisterns, meat, milk, mushroom, and cattle feed, and used to control mosquitoes that cause malaria, dengue, and West Nile virus disease [12]. Methoprene is also widely used in basic research on lepidopteran insects because of the shortage of commercially available JH I and II, which are major forms of JH in lepidopteran insects.

Although some studies suggest that methoprene is nontoxic to vertebrates [13], some also indicate that the sunlight-induced photolytic products of methoprene-based pesticides cause developmental defects in zebrafish by decreasing the expression of the sonic hedgehog gene [14]. Clearly, more research is needed to elucidate further the role of methoprene in gene transcription. However, very few studies have dealt with methoprene in mammal gene transcription. Since methoprene is widely used as an insecticide and as an analog of JH in insect research, the roles of methoprene in the gene transcription of insects are summarized here as reference for further studies.

5.2.1 Methoprene Promotes Insect Lipase Transcription

The lepidopteran *Helicoverpa* expresses higher levels of lipase in the midgut during their last instar feeding period. The function of increased lipase in the midgut is considered related to digestion of lipids from food. Methoprene is able to promote lipase expression in larval tissues when injected into the hemocoel of sixth instar larvae (Figure 5.2) [15], indicating that methoprene promotes lipase transcription.

Lipases are widely distributed in various organisms and play different functions. The basic function of lipase is to catalyze the hydrolysis of dietary lipids [16]. For example, human pancreatic lipases digest fats in the gut juice. In addition, lipases

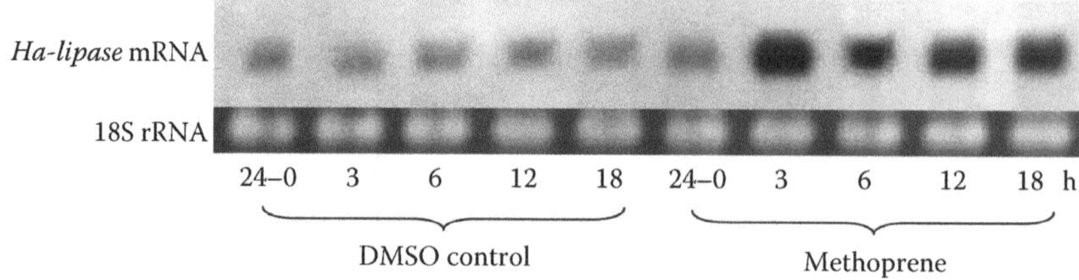

FIGURE 5.2 Methoprene regulation of the expression of *Ha-lipase* in the midgut by Northern blot analysis. The sixth instar 24 h larvae were injected methoprene (500 ng/larva). 24-0: normal sixth instar 24 h larvae. The durations (hour) after injection are 3, 6, 12, and 18. Ten micrograms of total RNA was used in each lane, and 18S ribosome RNA was used as a quantitative control.

play important roles in various cellular processes. Hepatic lipase facilitates the clearance of plasma triglyceride to avoid accumulation, thus decreasing the risk of coronary heart disease [17]. Adipose triglyceride lipase and hormone-sensitive lipase are essential for triglycerol degradation. Diacyl-sn-glycerol (DAG) lipase (DGL) catalyzes the conversion of DAG into the endocannabinoid 2-arachidonoyl-sn-glycerol (2-AG) in signal transduction [18]. Phospholipase Cs hydrolyzes phospholipids into DAG and inositol 1,4,5-trisphosphate (IP3) for signaling [19].

Problems in lipid digestion lead to diseases. Proof to this is the finding that increased lipid mobilization is an indicator of adipose tissue wasting. Also, the extensive loss of body fat is a hallmark of cancer cachexia [20]. ATRA induces upregulation of DGL in neurite of Neuro-2a cells [18]. ATRA also increases the activity of phospholipases in the release of arachidonic acid for gene transcription in neurites [21]. Methoprene is presently not known to stimulate mammalian lipase transcription.

5.2.2 Methoprene Promotes RNA-Binding Protein Gene Transcription in Insects

The eukaryotic gene expression is regulated in the gene transcription and protein translation levels. In protein translation, RNA-binding proteins (RBP) splice the mRNA [22] and locate the mRNA in the cytoplasm [23]. The RPB Y14 is a nucleocytoplasmic shuttling protein, which binds mRNA and forms a complex with other proteins [24]. Y14 imprints the mRNA in the nucleus and exports and localizes the mRNA to the cytoplasm. Y14 has participation in the enhancement of protein translation in mammalian cells [25].

Ha-RBP, the ortholog of Y14, was found in the insect *Helicoverpa armigera*. Ha-RBP is a single RNA recognition motif (RRM) protein, which is widely expressed in various tissues during larval growth, but its expression decreases and stops after larvae pupated. This evidence suggests that Ha-RBP is involved in protein translation for larval growth. Ha-RBP is located in the nucleus and cytoplasm of HaEpi cells, a cell line established from *H. armigera* epidermis [26]. Methoprene could promote the translocation of Ha-RBP from the nucleus to the cytoplasm via the methoprene-tolerant (Met) JH receptor candidate, but not by the transcription

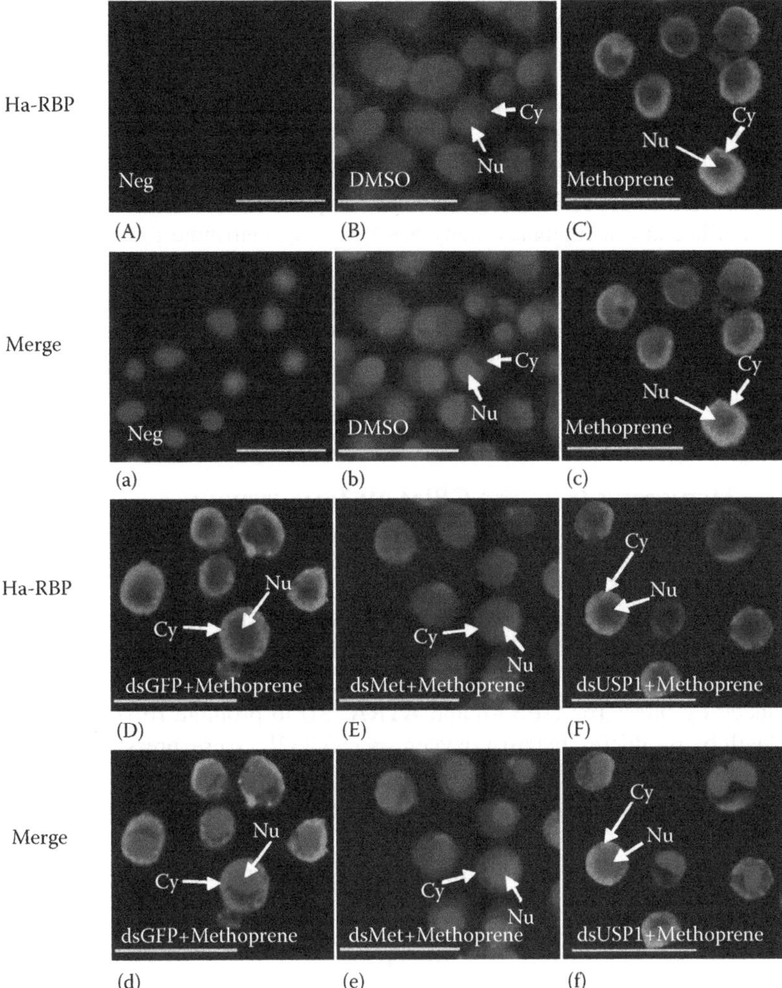

FIGURE 5.3 (See color insert.) Methoprene regulation on the subcellular translocation of Ha-RBP in HaEpi cells. Panels (A) and (a), negative control that cells treated with pre-immune serum; panels (B) and (b), cells treated with DMSO; panels (C) and (c), cells treated with methoprene for 6 h at a final concentration of 1 μM; panels (D) and (d), cells incubated with dsGFP followed by methoprene incubation; panels E and e, cells incubated with dsMet followed by methoprene incubation; panels F and f, cells incubated with dsUSP1 followed by methoprene incubation. Green portions (Alexa 488) indicate Ha-RBP detected with anti-Ha-RBP. Blue portions (DAPI) indicate nuclei. Nu, nucleus; Cy, cytoplasm. Size bars = 50 μm. At least three independent experiments were performed.

factor ultraspiracle protein (USP) (Figure 5.3). Knockdown of *Ha-RBP* decreases the level of proteins encoded by JH-responsive genes, such as USP, and the actin-binding protein, calponin, which are needed for JH signaling in gene transcription [27]. These results indicate that methoprene is able to promote RBP expression and protein translation for larval growth.

The role of RBPs in protein translation in cancer is widely recognized. RBPs bind to mRNAs of cancer-related genes to regulate their half-lives and/or translation. One of the RBPs relevant to the disease is HuR, which has three RRMs. HuR stabilizes and/or modulates the translation of many target mRNAs, such as endothelial growth factor mRNA, granulocyte/macrophage colony-stimulating factor mRNA, and mRNA of myeloid cell leukemia-1, by translocating from the nucleus to the cytoplasm. HuR is therefore recognized as an important factor in cancer-related gene expression because it regulates many mRNAs that contribute to the expression of cancer traits, such as cell proliferation, survival despite the presence of apoptotic stimuli, and local angiogenesis [28]. High levels of HuR is associated with various kinds of cancers, including breast, colon, ovarian, prostate, pancreatic, and oral cancer, and its function is tightly related to its cytoplasmic distribution [29]. The amino acid sequence of HuR in humans is different from that of Ha-RBP, which has one RRM domain, whereas HuR has three.

5.2.3 Methoprene Promotes GRIM-19 Transcription in *Helicoverpa armigera*

GRIM-19 (genes associated with retinoid–IFN-induced mortality-19) is a component of the mitochondrial respiratory complex I, which is a dual-function protein involved in promoting cell death and maintaining normal mitochondrial metabolism [30]. GRIM-19 is a pro-apoptotic factor in human cancer cell lines. GRIM-19 is induced by interferon (IFN)-β and ATRA [31] to promote IFN/ATRA-induced cell death by producing reactive oxygen species [32]. Overexpression of GRIM-19 in cancer cells suppresses cancer growth [33], whereas its downregulation promotes tumor growth via signal transducer and activator of transcription 3 (STAT3) pathway [34]. GRIM-19 is also observed to protect cells from apoptosis triggered by ultraviolet and staurosporine through the maintenance of the mitochondrial membrane potential [35].

Studies in insects showed that GRIM-19 is essential for maintaining normal cell growth. The transcription of *GRIM-19* is maintained at high levels during larval growth. Knockdown of *GRIM-19*, which is done by feeding larvae with bacteria that express dsRNA, induces programmed cell death of the larval midgut *in vivo*. The present study indicates that methoprene promotes *GRIM-19* transcription as ATRA does in mammal. Methoprene upregulates the transcription of *GRIM-19* through the JH receptor candidate *Met* but not through *USP* (Figure 5.4) [36]. GRIM-19 is important in maintaining normal cell growth, and its balanced expression is essential in keeping the normal status of cells. The increase or decrease in the expression of GRIM-19 results in cell death because of the disturbance of mitochondrial respiration. GRIM-19 protein from *Helicoverpa* is 38% similar to that of humans.

5.2.4 Methoprene Promotes Calponin Transcription

Methoprene is able to induce the transcription of some set of genes involved in insect development. One of these genes is *calponin*, which encodes for calponin

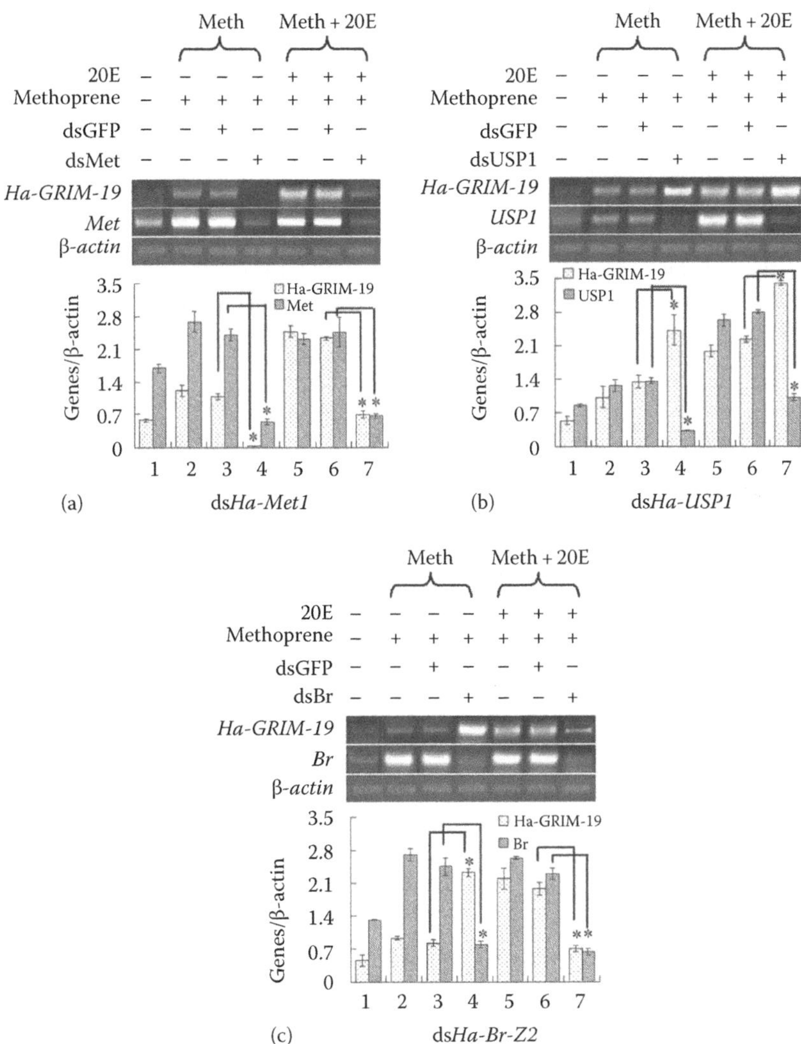

FIGURE 5.4 Semiquantitative RT-PCR to show the effects of knockdown *Ha-Met1* (a), *Ha-USP1* (b), and *Ha-Br-Z2* (c) on the expression of *Ha-GRIM-19*. Numbers 1–7 are as follows: normal cells, cells incubated with methoprene, cells treated with dsRNA of GFP and incubated with methoprene, cells treated with dsRNA of *Met1* (a)/*USP1* (b)/*Br-Z2* (c) and incubated with methoprene, cells incubated with combined methoprene and 20E, cells treated with dsRNA of GFP and incubated with combined methoprene and 20E, and cells treated with dsRNA of *Met1* (a)/*USP1* (b)/*Br-Z2* (c) and incubated with combined methoprene and 20E. Error bars represent the standard deviation in three replicates. Asterisks indicate significant differences (Student's *t*-test, $^*p < 0.05$).

protein. Calponin contains single calponin-homolog domain (Chd), also known as Chd64 in *Drosophila* and transgelin in humans. Chd is known to serve as an actin-binding domain. The ortholog of calponin in *Homo sapiens* is called transgelin. The expression level of transgelin is 25-fold higher in tumorigenic than in nontumorigenic cells. This elevated level of transgelin promotes cancer stem cell migration and invasion [37].

Calponin in *H. armigera* is 34% similar to human transgelin. *Helicoverpa* calponin is upregulated by methoprene. The calponin is expressed widely in various tissues during insect development from embryo to pupa. Methoprene promotes calponin expression and transfer to the nucleus, without being phosphorylated in the process. Similarly, 20E also promotes calponin expression and transfer to the nucleus; however, 20E induces calponin phosphorylation. Phosphorylated calponin does not bind to USP, but the unphosphorylated calponin binds to USP forming a complex for gene

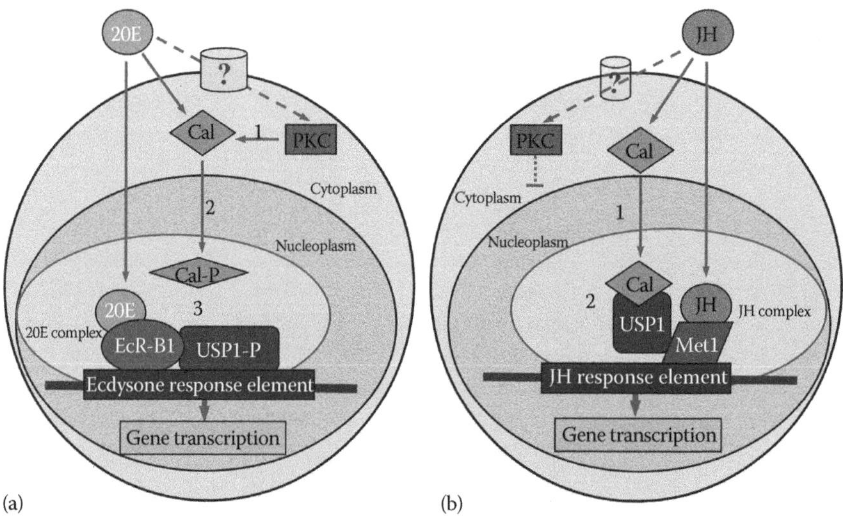

FIGURE 5.5 (See color insert.) Chart explaining the function of HaCal in the cross talk between 20E and JH pathways. (a) 20E pathway: 20E signaling leads HaCal protein phosphorylation by PKC via an unknown membrane pathway (1), HaCal is translocated into the nuclei (2), and phosphorylated HaCal does not bind with phosphorylated USP1 in 20E pathway (3). EcR binds 20E and USP and other chaperone protein to form transcription complex, combining the 20E response element initiating the gene transcription. (b) JH pathway: Methoprene maintains HaCal non-phosphorylation and translocates HaCal into the nuclei (1) and non-phosphorylated HaCal binds with non-phosphorylated USP1 (2). Met binds JH and interacts with USP and other chaperone proteins, and then this complex binds JH response element via Met to initiate JH signaling pathway. (From Li, M. et al., *Proc. Natl. Acad. Sci. USA*, 108, 638, 2011; Miura, K. et al., *FEBS J.*, 272, 1169, 2005; Li, Y. et al., *J. Biol. Chem.*, 282, 37605, 2007; Riddiford, L.M. et al., *Insect Biochem. Mol.*, 33, 1327, 2003; Antoniewski, C. et al., *Mol. Cell. Biol.*, 16, 2977, 1996; Hiruma, K. and Riddiford, L.M., *Dev. Biol.*, 272, 510, 2004; Lan, Q. et al., *Mol. Cell Biol.*, 19, 4897, 1999; Stone, B.L. and Thummel, C.S., *Cell*, 75, 307, 1993; Yao, T.P. et al., *Nature*, 366, 476, 1993; Yao, T.P. et al., *Cell*, 71, 63, 1992.)

transcription (Figure 5.5). Calponin is necessary for gene transcription in both 20E and JH pathways. The knockdown of *calponin* in *H. armigera* larvae leads to retarded larval growth and abnormal molting and results in a defective larval development [38].

In addition to calponin, many proteins contain Chd, such as the human Ras GTPase-activating proteins (Ras-GAP). Ras is a GTPase that is active in the form of Ras-GTP and inactive in the form of Ras-GDP. The activation of Ras signaling causes cell growth, cell differentiation, and cell survival. The guanine nucleotide exchange factors (GEF) regulate the conversion of Ras-GDP to Ras-GTP for the activation of Ras signaling. Ras-GAP promotes the GTPase activity of Ras-GTP to facilitate the hydrolysis of Ras-GTP to Ras-GDP to lower down Ras activity [48]. Ras-GAP is therefore a tumor suppressor [49]. Further study is still needed to investigate the effect of methoprene in Ras-GAP or Ras expression.

5.2.5 Methoprene Functions as JH Analog in Insect Studies

Methoprene is widely used in insect development studies because its structure is similar to JH. Methoprene is able to bind to Met, the candidate receptor of JH, as JH III does in *Tribolium castaneum* [50]. The mutation of *Met* results in 100-fold resistance to methoprene in *Drosophila* [51]. The exogenous JH and methoprene can induce the deposition of a second pupal cuticle in Lepidoptera and Coleoptera, but not in Diptera, because the imaginal disks of dipterans (flies) are not sensitive to JH [52]. Methoprene is shown to prevent the caspase-induced programmed cell death in *Drosophila* fat body [53], in the midgut of *Aedes aegypti* [54,55], and in *T. castaneum* [56]. Methoprene is able to upregulate the transcription of trypsin proteases that are involved in protein digestion in the midgut or in the epidermis [57,58] and suppress the carboxypeptidase A transcription, which results in the degradation of epidermal proteins in apolysis during molting in *H. armigera* [59]. In addition to the effect of methoprene in holometabolous insects, methoprene through Met and Krüppel-homolog 1 blocks the development of the hemimetabolous insect, *Pyrrhocoris apterus*, to adulthood [60], suggesting that methoprene plays roles in various insects.

Methoprene does not affect all gene expressions, for example, methoprene has no effect on the expression of protein phosphatase 6 [61] and of the small GTPase Rab32 [62]. One difference between JH and methoprene is that JH is degradable by juvenile hormone esterase (JHE), whereas methoprene is not degradable by JHE [63].

The active concentration of methoprene in insect S2 cells and Bm5 cells is from 0.1 to 10 μM. Concentrations over 10 μM may result in cell toxicity and death [64]. Mosquitoes are very sensitive to methoprene. The concentration used in *A. aegypti* larvae feeding that resulted in death is 0.41–200 ng/mL (1 μM=310 ng/mL) [55]. The concentration of methoprene in HaEpi cell line is from 1 to 10 μM, and in the sixth instar larvae, the concentration at 500 ng/larva does not show toxicity [38,65].

5.3 RECEPTORS OF METHOPRENE

Because JH receptor is not determined, the mechanism of JH in gene transcription regulation is not clear. Some studies indicate that JH might trigger signaling through a membrane receptor and protein kinase C [66], although the pathway involved is not clear yet. Other studies suggest that JH enters the nucleus and interacts with its receptor for gene transcription. One JH receptor candidate is Met, which mediates the antimetamorphic effect of JH [51]. Met is a transcription factor that belongs to the bHLH-PAS family [67]. The mutation of *Met* gene results in the tolerance for JH and methoprene [68]. This indicates that methoprene and JH require the same receptor for activation. *Tribolium* Met can directly bind JH and methoprene through a hydrophobic pocket within its PAS-B domain [50].

However, loss of Met function is not lethal in *Drosophila*, suggesting that *Met* is not a unique and vital gene in JH signaling [69]. The reason is that there is a homolog of *Met*, the germ cell-expressed (*GCE*) gene. The GCE is also a bHLH-PAS protein with high sequence identity to Met, which can compensate for Met deficiency [70]. Met and GCE can be found in *Drosophila*, but only Met can be found in Coleoptera and mosquitoes. The Met homologs from mosquitoes are more similar to GCE than to Met. The absence of a second homologous gene in mosquitoes suggests a single *Met* gene in lower Diptera [71]. *Met* and *GCE* are proposed paralogs derived from an ancestral gene [71]. *Drosophila* Met and GCE are experienced divergent evolutionary pressures following the duplication of an ancestral gce-like gene found in less derived holometabolous insects [72].

The gene encoding GCE (*gce*) is a vital gene where a deficiency causes death in *Drosophila* [4,73]. Met deficiency does not cause death in *Drosophila* [4], but it causes precocious metamorphosis in *Tribolium* [68]. Other study indicates that Met and GCE are partially redundant in transducing JH action. *Drosophila* larvae are fully viable if Met and GCE null single mutant but died during the larval–pupal transition after the Met and GCE double mutant [74]. Met and GCE use a novel nuclear receptor motif LIXXL to bind the FTZ-F1 nuclear receptor [75]. JH or JH agonists induce homodimer MET-MET and heterodimer MET-GCE formation [76].

Another possible JH receptor is the USP, a nuclear hormone receptor that is the ortholog of the mammal retinoid-X receptor (RXR), which binds to JH and methoprene with low affinity [77]; its role as a JH receptor is disputed. Studies on GRIM-19, RBP, and calponin in *Helicoverpa* suggest that methoprene, through Met, regulates gene transcription [38].

There are two classes of retinoid receptors in mammals: the retinoic acid receptor (RAR), which binds to ATRA, and the RXR, which exclusively binds to 9-*cis* retinoic acid (9*cis*RA) [78]. Both retinoic acid and JH are terpenoids that have similar structure, with the 9*cis*RA tested to best mimic JH [5]. Some studies show that methoprene is able to bind to mammal RXR and act as a transcriptional activator [79], which suggests a possible role of methoprene in mammalian cells. The metabolite of methoprene, methoprene acid, is capable of activating transcription through RAR/RXR response elements [80]. Similarly, the ATRA plays important roles in molting and immune response of the insect *Rhodnius prolixus* [5]. These evidences suggest that both methoprene and retinoic acid are active in insects and mammals.

5.4 SUMMARY

Methoprene is widely used as a pest control chemical and is structurally and functionally similar to JH and retinoic acid. Methoprene may play roles in insect or

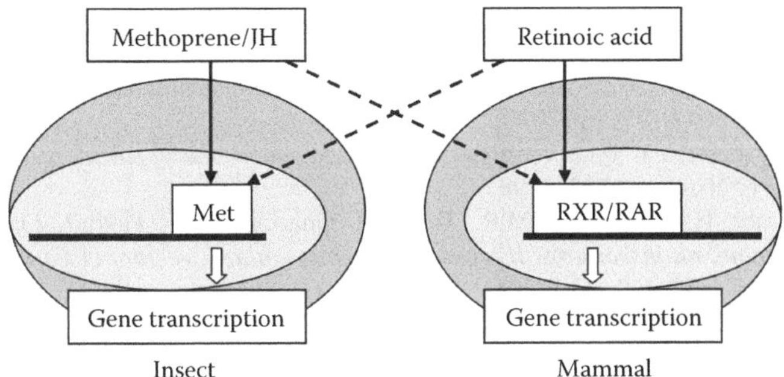

FIGURE 5.6 Schematic illustrations of the pathways and readouts of JH/methoprene and retinoic acid. Met, Methoprene-tolerant protein; RXR, retinoid-X receptor; RAR, retinoic acid receptor. The dashed line indicates the need for further study.

mammalian gene transcription. Some of the genes upregulated by methoprene may promote cell growth. Methoprene may function in mammalian cells for gene transcription, and the retinoic acid may function in insects, too. Methoprene and JH, through Met, regulate the gene transcription. The existence of a plasma membrane receptor for JH/methoprene warrants further investigation (Figure 5.6).

ACKNOWLEDGMENTS

This work was supported by grants from the National Natural Science Foundation of China (No: 31230067) and National Basic Research Program of China (973 Program, 2012CB114101).

REFERENCES

1. C.A. Henrick, *Methoprene*, J. Am. Mosq. Control. Assoc. 23 (2007), pp. 225–239.
2. L.M. Riddiford, K. Hiruma, X.F. Zhou, and C.A. Nelson, *Insights into the molecular basis of the hormonal control of molting and metamorphosis from* Manduca sexta *and* Drosophila melanogaster, Insect Biochem. Mol. 43 (2012), pp. 963–971.
3. D.S. Richard, S.W. Applebaum, T.J. Sliter, F.C. Baker, D.A. Schooley, C.C. Reuter, V.C. Henrich, and L.I. Gilbert, *Juvenile hormone bisepoxide biosynthesis* in vitro *by the ring gland of* Drosophila melanogaster*: A putative juvenile hormone in the higher Diptera*, Proc. Natl. Acad. Sci. USA 86 (1989), pp. 1421–1425.
4. A.A. Baumann and T.G. Wilson, *Molecular evolution of juvenile hormone signaling*, in *Gene Duplication*, F. Friesberg, ed., InTech, New York (2011), pp. 333–352, ISBN:978-953-307-387-3.
5. A. Nakamura, R. Stiebler, M.R. Fantappie, E. Fialho, H. Masuda, and M.F. Oliveira, *Effects of retinoids and juvenoids on moult and on phenoloxidase activity in the blood-sucking insect* Rhodnius prolixus, Acta. Trop. 103 (2007), pp. 222–230.
6. A.C. Mamede, S.D. Tavares, A.M. Abrantes, J. Trindade, J.M. Maia, and M.F. Botelho, *The role of vitamins in cancer: A review*, Nutr. Cancer 63 (2011), pp. 479–494.
7. Y. Chen and D.H. Reese, *The retinol signaling pathway in mouse pluripotent P19 cells*, J. Cell. Biochem. 112 (2011), pp. 2865–2872.
8. P.M. Amann, S.B. Eichmuller, J. Schmidt, and A.V. Bazhin, *Regulation of gene expression by retinoids*, Curr. Med. Chem. 18 (2011), pp. 1405–1412.

9. X.H. Tang and L.J. Gudas, *Retinoids, retinoic acid receptors, and cancer*, Annu. Rev. Pathol. Mech. 6 (2011), pp. 345–364.
10. N. Noy, *Between death and survival: Retinoic acid in regulation of apoptosis*, Annu. Rev. Nutr. 30 (2010), pp. 201–217.
11. N. Bushue and Y.J. Wan, *Retinoid pathway and cancer therapeutics*, Adv. Drug. Deliv. Rev. 62 (2010), pp. 1285–1298.
12. H. Darabi, H. Vatandoost, M.R. Abaei, O. Gharibi, and F. Pakbaz, *Effectiveness of methoprene, an insect growth regulator, against malaria vectors in Fars, Iran: A field study*, Pak. J. Biol. Sci. 14 (2011), pp. 69–73.
13. S.J. Degitz, E.J. Durhan, J.E. Tietge, P.A. Kosian, G.W. Holcombe, and G.T. Ankley, *Developmental toxicity of methoprene and several degradation products in* Xenopus laevis, Aquat. Toxicol. 64 (2003), pp. 97–105.
14. D.G. Smith, C. Wilburn, R.A. and McCarthy, *Methoprene photolytic compounds disrupt zebrafish development, producing phenocopies of mutants in the sonic hedgehog signaling pathway*, Mar. Biotechnol. 5 (2003), pp. 201–212.
15. Y.P. Sui, J.X. Wang, and X.F. Zhao, *Effects of classical insect hormones on the expression profiles of a lipase gene from the cotton bollworm (*Helicoverpa armigera*)*, Insect Mol. Biol. 17 (2008), pp. 523–529.
16. A. Svendsen, *Lipase protein engineering*, Biochim. Biophys. Acta 1543 (2000), pp. 223–238.
17. C. Chatterjee and D.L. Sparks, *Hepatic lipase, high density lipoproteins, and hypertriglyceridemia*, Am. J. Pathol. 178 (2011), pp. 1429–1433.
18. K.M. Jung, G. Astarita, D. Thongkham, and D. Piomelli, *Diacylglycerol Lipase-alpha and -beta control neurite outgrowth in neuro-2a cells through distinct molecular mechanisms*, Mol. Pharmacol. 80 (2011), pp. 60–67.
19. A.V. Smrcka, J.H. Brown, and G.G. Holz, *Role of phospholipase cepsilon in physiological phosphoinositide signaling networks*, Cell Signal. 24 (2012), pp. 1333–1343.
20. C. Bing, *Lipid mobilization in cachexia: Mechanisms and mediators*, Curr. Opin. Support. Palliat. Care 5 (2011), pp. 356–360.
21. A.A. Farooqui, P. Antony, W.Y. Ong, L.A. Horrocks, and L. Freysz, *Retinoic acid-mediated phospholipase A2 signaling in the nucleus,* Brain Res. Brain Res. Rev. 45 (2004), pp. 179–195.
22. L.H. Apponi, A.H. Corbett, and G.K. Pavlath, *RNA-binding proteins and gene regulation in myogenesis*, Trends Pharmacol. Sci. 32 (2011), pp. 652–658.
23. K. Shahbabian and P. Chartrand, *Control of cytoplasmic mRNA localization*, Cell Mol. Life Sci. 69 (2012), pp. 535–552.
24. N. Kataoka, J. Yong, V.N. Kim, F. Velazquez, R.A. Perkinson, F. Wang, and G. Dreyfuss, *Pre-mRNA splicing imprints mRNA in the nucleus with a novel RNA-binding protein that persists in the cytoplasm*, Mol. Cell 6 (2000), pp. 673–682.
25. A. Nott, H. Le Hir, and M.J. Moore, *Splicing enhances translation in mammalian cells: An additional function of the exon junction complex*, Gene Dev. 18 (2004), pp. 210–222.
26. H.L. Shao, W.W. Zheng, P.C. Liu, Q. Wang, J.X. Wang, and X.F. Zhao, *Establishment of a new cell line from lepidopteran epidermis and hormonal regulation on the genes*, PLoS One. 3 (2008), p. e3127.
27. X.H. Yang, P.C. Liu, W.W. Zheng, J.X. Wang, and X.F. Zhao, *The juvenile hormone analogue methoprene up-regulates the Ha-RNA-binding protein*, Mol. Cell Endocrinol. 333 (2011), pp. 172–180.
28. K. Abdelmohsen and M. Gorospe, *Posttranscriptional regulation of cancer traits by HuR*, Wiley Interdiscip. Rev. RNA 1 (2010), pp. 214–229.
29. I. Lopez de Silanes, A. Lal, and M. Gorospe, *HuR: post-transcriptional paths to malignancy*, RNA Biol. 2 (2005), pp. 11–13.

30. V. Maximo, J. Lima, P. Soares, A. Silva, I. Bento, and M. Sobrinho-Simoes, *GRIM-19 in health and disease*, Adv. Anat. Pathol. 15 (2008), pp. 46–53.
31. J.E. Angell, D.J. Lindner, P.S. Shapiro, E.R. Hofmann, and D.V. Kalvakolanu, *Identification of GRIM-19, a novel cell death-regulatory gene induced by the interferon-beta and retinoic acid combination, using a genetic approach*, J. Biol. Chem. 275 (2000), pp. 33416–33426.
32. G. Huang, Y. Chen, H. Lu, and X. Cao, *Coupling mitochondrial respiratory chain to cell death: An essential role of mitochondrial complex I in the interferon-beta and retinoic acid-induced cancer cell death*, Cell Death Differ. 14 (2007), pp. 327–337.
33. T. Okamoto, T. Inozume, H. Mitsui, M. Kanzaki, K. Harada, N. Shibagaki, and S. Shimada, *Overexpression of GRIM-19 in cancer cells suppresses STAT3-mediated signal transduction and cancer growth*, Mol. Cancer Ther. 9 (2010), pp. 2333–2343.
34. Y. Zhou, M. Li, Y. Wei, D.Q. Feng, C. Peng, H.Y. Weng, Y. Ma, L. Bao, S. Nallar, S. Kalakonda, W.H. Xiao, D.V. Kalvakolanu, and B. Ling, *Down-regulation of GRIM-19 expression is associated with hyperactivation of STAT3-induced gene expression and tumor growth in human cervical cancers*, J. Interf. Cytok. Res. 29 (2009), pp. 695–703.
35. H. Lu and X.M. Cao, *GRIM-19 is essential for maintenance of mitochondrial membrane potential*, Mol. Biol. Cell 19 (2008), pp. 1893–1902.
36. D.J. Dong, P.C. Liu, J.X. Wang, and X.F. Zhao, *The knockdown of Ha-GRIM-19 by RNA interference induced programmed cell death*, Amino Acids 42 (2012), pp. 1297–1307.
37. E.K. Lee, G.Y. Han, H.W. Park, Y.J. Song, and C.W. Kim, *Transgelin promotes migration and invasion of cancer stem cells*, J. Proteome Res. 9 (2010), pp. 5108–5117.
38. P.C. Liu, J.X. Wang, Q.S. Song, and X.F. Zhao, *The participation of calponin in the cross talk between 20-hydroxyecdysone and juvenile hormone signaling pathways by phosphorylation variation*, PLoS One 6 (2011), p. e19776.
39. C. Antoniewski, B. Mugat, F. Delbac, and J.A. Lepesant, *Direct repeats bind the EcR/USP receptor and mediate ecdysteroid responses in* Drosophila melanogaster, Mol. Cell. Biol. 16 (1996), pp. 2977–2986.
40. K. Hiruma and L.M. Riddiford, *Differential control of MHR3 promoter activity by isoforms of the ecdysone receptor and inhibitory effects of E75A and MHR3*, Dev. Biol. 272 (2004), pp. 510–521.
41. Q. Lan, K. Hiruma, X. Hu, M. Jindra, and L.M. Riddiford, *Activation of a delayed-early gene encoding MHR3 by the ecdysone receptor heterodimer EcR-B1-USP-1 but not by EcR-B1-USP-2*, Mol. Cell Biol. 19 (1999), pp. 4897–4906.
42. B.L. Stone and C.S. Thummel, *The Drosophila 78C early late puff contains E78, an ecdysone-inducible gene that encodes a novel member of the nuclear hormone receptor superfamily*, Cell 75 (1993), pp. 307–320.
43. T.P. Yao, B.M. Forman, Z. Jiang, L. Cherbas, J.D. Chen, M. McKeown, P. Cherbas, and R.M. Evans, *Functional ecdysone receptor is the product of EcR and ultraspiracle genes*, Nature 366 (1993), pp. 476–479.
44. T.P. Yao, W.A. Segraves, A.E. Oro, M. McKeown, and R.M. Evans, Drosophila *ultraspiracle modulates ecdysone receptor function via heterodimer formation*, Cell 71 (1992), pp. 63–72.
45. M. Li, E.A. Mead, and J. Zhu, *Heterodimer of two bHLH-PAS proteins mediates juvenile hormone-induced gene expression*, Proc. Natl. Acad. Sci. USA 108 (2011), pp. 638–643.
46. K. Miura, M. Oda, S. Makita, and Y. Chinzei, *Characterization of the* Drosophila *methopren -tolerant gene product. Juvenile hormone binding and ligand-dependent gene regulation*, FEBS J. 272 (2005), pp. 1169–1178.
47. Y. Li, Z. Zhang, G.E. Robinson, and S.R. Palli, *Identification and characterization of a juvenile hormone response element and its binding proteins*, J. Biol. Chem. 282 (2007), pp. 37605–37617.

48. M. Malumbres and M. Barbacid, *RAS oncogenes: The first 30 years*, Nat. Rev. Cancer 3 (2003), pp. 459–465.
49. J. Downward, *Targeting RAS signalling pathways in cancer therapy*, Nat. Rev. Cancer 3 (2003), pp. 11–22.
50. J.P. Charles, T. Iwema, V.C. Epa, K. Takaki, J. Rynes, and M. Jindra, *Ligand-binding properties of a juvenile hormone receptor, methoprene-tolerant*, Proc. Natl. Acad. Sci. USA 108 (2011), pp. 21128–21133.
51. T.G. Wilson and J. Fabian, *A* Drosophila melanogaster *mutant resistant to a chemical analog of juvenile hormone*, Dev. Biol. 118 (1986), pp. 190–201.
52. X. Zhou and L.M. Riddiford, *Broad specifies pupal development and mediates the 'status quo' action of juvenile hormone on the pupal-adult transformation in* Drosophila *and* Manduca, Development 129 (2002), pp. 2259–2269.
53. Y. Liu, Z. Sheng, H. Liu, D. Wen, Q. He, S. Wang, W. Shao, R.J. Jiang, S. An, Y. Sun, W.G. Bendena, J. Wang, L.I. Gilbert, T.G. Wilson, Q. Song, and S. Li, *Juvenile hormone counteracts the bHLH-PAS transcription factors MET and GCE to prevent caspase-dependent programmed cell death in* Drosophila, Development 13612 (2009), pp. 2015–2025.
54. J.T. Nishiura, P. Ho, and K. Ray, *Methoprene interferes with mosquito midgut remodeling during metamorphosis*, J. Med. Entomol. 40 (2003), pp. 498–507.
55. Y. Wu, R. Parthasarathy, H. Bai, and S.R. Palli, *Mechanisms of midgut remodeling: Juvenile hormone analog methoprene blocks midgut metamorphosis by modulating ecdysone action*, Mech. Dev. 123 (2006), pp. 530–547.
56. R. Parthasarathy and S.R. Palli, *Molecular analysis of juvenile hormone action in controlling the metamorphosis of the red flour beetle*, Tribolium castaneum, Arch. Insect Biochem. Physiol. 70 (2009), pp. 57–70.
57. Y.P. Sui, J.X. Wang, and X.F. Zhao, *The impacts of classical insect hormones on the expression profiles of a new digestive trypsin-like protease (TLP) from the cotton bollworm*, Helicoverpa armigera, Insect Mol. Biol. 18 (2009), pp. 443–452.
58. Y. Liu, Y.P. Sui, J.X. Wang, and X.F. Zhao, *Characterization of the trypsin-like protease (Ha-TLP2) constitutively expressed in the integument of the cotton bollworm*, Helicoverpa armigera, Arch. Insect Biochem. Physiol. 72 (2009), pp. 74–87.
59. Y.P. Sui, X.B. Liu, L.Q. Chai, J.X. Wang, and X.F. Zhao, *Characterization and influences of classical insect hormones on the expression profiles of a molting carboxypeptidase A from the cotton bollworm* (Helicoverpa armigera), Insect Mol. Biol. 18 (2009), pp. 353–363.
60. B. Konopova, V. Smykal, and M. Jindra, *Common and distinct roles of juvenile hormone signaling genes in metamorphosis of holometabolous and hemimetabolous insects*, PLoS One 6 (2011), p. e28728.
61. C.X. Wang, W.W. Zheng, P.C. Liu, J.X. Wang, and X.F. Zhao, *The steroid hormone 20-hydroxyecdysone upregulated the protein phosphatase 6 for the programmed cell death in the insect midgut*, Amino Acids 43 (2012), pp. 963–971.
62. L. Hou, J.X. Wang, and X.F. Zhao, *Rab32 and the remodeling of the imaginal midgut in* Helicoverpa armigera, Amino Acids 40 (2011), pp. 953–961.
63. S. Kamita, A. Samra, J. Liu, A. Cornel, and B. Hammock, *Juvenile hormone (JH) esterase of the mosquito* Culex quinquefasciatus *is not a target of the JH analog insecticide methoprene*, PLoS One 6 (2011), p. e28392.
64. T. Soin, L. Swevers, H. Mosallanejad, R. Efrose, V. Labropoulou, K. Iatrou, and G. Smagghe, *Juvenile hormone analogs do not affect directly the activity of the ecdysteroid receptor complex in insect culture cell lines*, J. Insect Physiol. 54 (2008), pp. 429–438.

65. W.W. Zheng, D.T. Yang, J.X. Wang, Q.S. Song, L.I. Gilbert, and X.F. Zhao, *Hsc70 binds to ultraspiracle resulting in the upregulation of 20-hydroxyecdsone-responsive genes in* Helicoverpa armigera, Mol. Cell Endocrinol. 315 (2010), pp. 282–291.
66. R. Abu-Hakima and K.G. Davey, *The action of juvenile hormone on follicle cells of* Rhodnius prolixus *in vitro: The effect of colchicine and cytochalasin B*, Gen. Comp. Endocrinol. 32 (1977), pp. 360–370.
67. M. Ashok, C. Turner, and T.G. Wilson, *Insect juvenile hormone resistance gene homology with the bHLH-PAS family of transcriptional regulators*, Proc. Natl. Acad. Sci. USA 95 (1998), pp. 2761–2766.
68. B. Konopova and M. Jindra, *Juvenile hormone resistance gene Methoprene-tolerant controls entry into metamorphosis in the beetle* Tribolium castaneum, Proc. Natl. Acad. Sci. USA 104 (2007), pp. 10488–10493.
69. T.G. Wilson and M. Ashok, *Insecticide resistance resulting from an absence of target-site gene product*, Proc. Natl. Acad. Sci. USA 95 (1998), pp. 14040–14044.
70. A. Baumann, T.G. Wilson, J. Barry, and S. Wang, *Juvenile hormone action requires paralogous genes in* Drosophila melanogaster, Genetics 185 (2010), pp. 1327–1336.
71. S. Wang, A. Baumann, and T.G. Wilson, Drosophila melanogaster *Methoprene-tolerant (Met) gene homologs from three mosquito species: Members of PAS transcriptional factor family*, J. Insect Physiol. 53 (2007), pp. 246–253.
72. A. Baumann, Y. Fujiwara, and T.G. Wilson, *Evolutionary divergence of the paralogs Methoprene tolerant (Met) and germ cell expressed (gce) within the genus* Drosophila, J. Insect Physiol. 56 (2010), pp. 1445–1455.
73. A. Baumann, J. Barry, S. Wang, Y. Fujiwara, and T.G. Wilson, *Paralogous genes involved in juvenile hormone action in* Drosophila melanogaster, Genetics 185 (2010), pp. 1327–1336.
74. M.A. Abdou, Q. He, D. Wen, O. Zyaan, J. Wang, J. Xu, A.A. Baumann, J. Joseph, T.G. Wilson, and S. Li, Drosophila *Met and Gce are partially redundant in transducing juvenile hormone action*, Insect Biochem. Mol. Biol. 41 (2011), pp. 938–945.
75. T.J. Bernardo and E.B. Dubrovsky, *The* Drosophila *juvenile hormone receptor candidates methoprene-tolerant (MET) and germ cell-expressed (GCE) utilize a conserved LIXXL motif to bind the FTZ-F1 nuclear receptor*, J. Biol. Chem. 287 (2012), pp. 7821–7833.
76. J. Godlewski, S. Wang, and T.G. Wilson, *Interaction of bHLH-PAS proteins involved in juvenile hormone reception in* Drosophila, Biochem. Biophys. Res. Commun. 342 (2006), pp. 1305–1311.
77. G. Jones, D. Jones, P. Teal, A. Sapa, and M. Wozniak, *The retinoid-X receptor ortholog, ultraspiracle, binds with nanomolar affinity to an endogenous morphogenetic ligand*, FEBS J. 273 (2006), pp. 4983–4996.
78. C. Thaller, C. Hofmann, and G. Eichele, *9-cis-retinoic acid, a potent inducer of digit pattern duplications in the chick wing bud*, Development 118 (1993), pp. 957–965.
79. M.A. Harmon, M.F. Boehm, R.A. Heyman, and D.J. Mangelsdorf, *Activation of mammalian retinoid X receptors by the insect growth regulator methoprene*, Proc. Natl. Acad. Sci. USA 92 (1995), pp. 6157–6160.
80. P.K. Schoff and G.T. Ankley, *Effects of methoprene, its metabolites, and breakdown products on retinoid-activated pathways in transfected cell lines*, Environ. Toxicol. Chem. 23 (2004), pp. 1305–1310.

6 Modeling Resistance to Juvenile Hormone Analogs

Linking Evolution, Ecology, and Management

David W. Crowder, Peter C. Ellsworth, Steven E. Naranjo, Bruce E. Tabashnik, and Yves Carrière

CONTENTS

ABSTRACT

Juvenile hormone analogs (JHAs) are insecticides that mimic insect juvenile hormone and interfere with normal insect development. JHAs disrupt a hormonal system that is specific to insects and thus kill some target pests while causing less harm to nontarget organisms than broad-spectrum insecticides. JHAs have become increasingly important in agriculture worldwide, where their specificity and efficacy has been used selectively to reduce target pest populations while conserving key natural enemies. Evolution of resistance by target pests, however, can reduce the effectiveness of JHAs. This chapter reviews how models have been used to analyze the evolution of pest resistance to JHAs and to develop strategies to delay pest resistance. We describe results of general mathematical models and a case study showing how simulation, conceptual, and spatially explicit statistical models have been applied to better understand and manage evolution of resistance to the JHA pyriproxyfen by the whitefly, *Bemisia tabaci*. Our results show how genetic, ecological, and human factors affect evolution of pest resistance to JHAs. Integrating knowledge of these factors into models can help to produce useful predictions about pest resistance to JHAs and to improve management strategies for preserving the effectiveness of this important functional class of insecticides.

KEYWORDS

Integrated pest management, Insect resistance management, Modeling, Population genetics, Pyriproxyfen, Resistance evolution, Whitefly

6.1 INTRODUCTION

Controlling pests with selective insecticides that have relatively small environmental impacts and conserve beneficial species is a cornerstone of integrated pest management (IPM) [1–3]. Juvenile hormone analogs (JHAs) are one functional class of selective insecticides that are used to control a variety of insect pests including mosquitoes [4–6] and whiteflies [7–10]. JHAs mimic the action of juvenile hormone in insects, disrupting natural hormonal balance and interfering with growth and development [7,8,11,12]. JHAs are generally pest specific and therefore have minimal impacts on beneficial or other nontarget species, particularly when compared to broad-spectrum alternatives [11,13,14]. The selectivity of JHAs, and their high efficacy, has led to their widespread use in pest management programs worldwide.

The sustained effectiveness of JHAs in agricultural ecosystems could be threatened by the evolution of resistance in target pests [4,9,15–24]. Evolution of insecticide resistance is a global problem that provides some of the most compelling examples of adaptation by natural selection [25–29]. Multiple insect species have evolved resistance to JHAs in the laboratory and the field, including mosquitoes [4,15], whiteflies [16–18], stored-grain beetles [19], fruit flies [20,21], codling moth [22], and houseflies [23,24]. However, the sustained efficacy of JHAs could be preserved by insect resistance management (IRM) practices, which aim to delay or prevent the evolution of resistance in target pests [26,29].

Analytical, conceptual, and simulation models are often used as part of IRM programs to evaluate the potential for resistance evolution in pests. Insecticide resistance models integrate knowledge of behaviors, genetics, life history, and population dynamics to make predictions about the rates of resistance evolution and the sustainability of insecticides under various environmental and management conditions [26,29–41]. Such models vary widely in their complexity and their aims. Population genetics models are used to determine how genetic factors may influence resistance evolution [37,40]. Such models are often not built on a foundation of population dynamics, but nonetheless can sometimes provide robust predictions on relative rates of resistance evolution in particular pests [40]. Conceptual IPM models demonstrate how practitioners of IRM can manage pest resistance by limiting insecticide usage, by diversifying their insecticide use, and by partitioning insecticides across space and time [42,43]. These models often visually represent relationships between pest avoidance, pest sampling, biological control, and chemical control while also showing feedbacks between biological and management practices [44]. Simulation models integrate population dynamics with genetics across implicit or spatially explicit landscapes [31–36,38,39]. In general, all three types of models are often useful in guiding real-world IPM strategies.

Geographic information system (GIS)-based statistical analyses are also well suited for the development of resistance management strategies. In contrast to simulation models that use sensitivity analyses to explore the effects of specific factors on resistance evolution, statistical approaches use field data to assess the global effects of many biological processes. For example, statistical analyses have been used to examine the effectiveness of refuges in delaying resistance evolution. Refuges are habitats where an insecticide considered for resistance management is not applied, which promotes the survival of susceptible insects that can mate with rare resistant insects. Spatially explicit statistical analyses can provide information on the area over which refuges can reduce the frequency of resistance and therefore directly contribute to the development of resistance management strategies [45,46].

In this chapter, we demonstrate the role of models in analyzing factors associated with the evolution of resistance to JHAs. In Sections 6.2 and 6.3, we detail factors associated with population genetics and ecology of resistance and present theory and results from general IRM and IPM models for a generic JHA insecticide. Through these examples, we aim to present the theoretical principles underlying the evolution of resistance and detail the conditions under which resistance can be delayed or prevented. In Section 6.4, we build on this foundation to describe how resistance and pest management models have been used as part of IRM and IPM programs in a system involving the JHA pyriproxyfen targeting whiteflies in the *Bemisia tabaci* species complex. This system is amenable to a broad review as it has been well characterized from a population genetics, ecological, and pest management perspective. In addition, the dynamics of whiteflies and pyriproxyfen have been studied with models that examine genetics, ecology, and anthropogenic factors, unlike any other system involving a JHA to date. By reviewing this case study, we demonstrate how models combined with extensive real-world field data provide a solid foundation for sustaining the efficacy of a JHA insecticide.

6.2 POPULATION GENETICS OF RESISTANCE

Evolution of insecticide resistance involves a genetically based decrease in susceptibility to a toxin [40,47,48]. Resistance occurs when a population that has experienced selection from an insecticide becomes significantly less susceptible than a conspecific population with less exposure [40]. Carrière et al. [41] detailed three conditions necessary for the evolution of insecticide resistance (see also [49]): (1) genetic variation in resistance to an insecticide, (2) heritability of resistance, and (3) selective advantage of resistance. In the following sections, we present results of general IRM models showing how variation in these factors is expected to affect resistance evolution to a generic JHA insecticide. We also focus on the role of pest reproductive strategy, which could have broad implications for the evolution of resistance to JHAs [9,50,51].

6.2.1 Genetic Models

Simple population genetics models of insecticide resistance often assume that resistance is controlled by a single locus with two alleles (*S* for susceptibility, *R* for resistance) [9,37,40,50–53]. Although the single-locus model may be simplistic compared with polygenic models of resistance [54,55], it is appropriate in cases where selection pressure is high and favors single mutations of relatively large effect to increase survival to a toxin [41]. For example, resistance to the JHA methoprene in fruit flies and mosquitoes is affected by mutations at a single locus, the methoprene-tolerant (Met) bHLH-PAS gene [15,20,21]. Similarly, Horowitz et al. [56] used a population genetics approach to demonstrate that whitefly resistance to the JHA pyriproxyfen was likely conferred by resistance alleles at a single locus. Horowitz and colleagues showed that mortality curves for haploid males produced by virgin heterozygous females (whiteflies are haplodiploid) displayed a broad plateau at 50% mortality. This indicated that males had only two genotypes (*S* or *R*), as would be the case under the single-locus model [56]. The assumption of monogenic resistance thus provides an effective foundation for predictions of resistance evolution to a generic JHA.

The single-locus model is typically applied to diploid pests (but see Section 6.2.2) that are assumed to have infinitely large populations and mate randomly across an implicit landscape. In the model, the change in frequency of the *R* allele in each generation is [50–53]

$$\Delta q = \frac{pq(W_R - W_S)}{W_M} \tag{6.1}$$

where

p is the frequency of the *S* allele
q is the frequency of the *R* allele

W_S is the marginal fitness of the S allele
W_R is the marginal fitness of the R allele
W_M is the mean fitness of all the genotypes

In this model, the direction of change is determined solely by the term $W_R - W_S$, such that the resistance allele frequency (q) increases when $W_R > W_S$. Under Hardy–Weinberg equilibrium, $W_S = pW_{SS} + qW_{RS}$, while $W_R = qW_{RR} + pW_{RS}$, where W_{SS}, W_{RS}, and W_{RR} are the fitness of the SS, RS, and RR genotypes, respectively [50–53]. Thus, q increases under the following conditions:

$$qW_{RR} + pW_{RS} > pW_{SS} + qW_{RS} \tag{6.2}$$

Thus, resistance evolution is affected by the intensity of selection on susceptible, heterozygous, and resistant genotypes; the allele frequencies; and the dominance of resistance. The dominance of resistance (h) relates the survival of heterozygotes to homozygotes and can be calculated as [57,58]

$$h = \frac{W_{RS} - W_{SS}}{W_{RR} - W_{SS}} \tag{6.3}$$

Using this model, we can generate predictions about the evolution of insecticide resistance to a generic JHA insecticide. For simplicity, we assume that the JHA is applied to 100% of the landscape (such a landscape might represent a single field). This assumption is relaxed in Section 6.3.1. For the purposes of this chapter, we assume that the initial resistance allele frequency (q) is 0.001 and that the JHA causes either relatively high or low mortality (99% or 50% mortality to homozygous susceptible insects, respectively) and that homozygous-resistant insects incur no mortality. We report the number of generations for q to increase from the initial value to 0.5. Results of this model are shown in Table 6.1.

The model described here yields several important predictions about the evolution of resistance to a typical JHA insecticide. Increased efficacy of the toxin leads to a greater selective advantage for resistant individuals and faster resistance evolution. The potential trade-off, of course, is that JHAs with lower efficacy may not be as effective at controlling pests. Furthermore, resistance evolution occurs faster in pests when resistance is inherited as an additive or dominant trait. The effect of dominance is magnified when toxin concentrations are low. Experiments should therefore attempt to quantify both the efficacy of JHA insecticides and the dominance of resistance to make realistic predictions about the sustainability of any novel JHA (see Section 6.4).

6.2.2 Pest Reproductive Strategy

Several studies have highlighted the role of reproductive strategy on the evolution of resistance to pesticides [9,50,51]. JHAs commercialized to date target several diploid pests such as mosquitoes [4–6], codling moths [22], and flies [20,21,23,24]. Other pests targeted by JHAs, however, have alternative reproductive strategies.

TABLE 6.1
Generations for Resistance to Evolve (Resistance Allele Frequency Increased from 0.001 to 0.5) in a Hypothetical Model of a JHA Insecticide That Causes High or Low Mortality

JHA Efficacy	W_{SS}	W_{RS}	W_{SS}	h	Generations
High	0.01	0.01	1	0	15
High	0.01	0.505	1	0.5	3
High	0.01	1	1	1	3
Low	0.5	0.5	1	0	1012
Low	0.5	0.75	1	0.5	18
Low	0.5	1	1	1	12

Models were conducted with varying survival of susceptible homozygotes (W_{SS}), heterozygotes (W_{RS}), and resistant homozygotes (W_{RR}) and varying dominance of resistance (h). Dominance values of 0, 0.5, and 1 indicate recessive, additive, or dominant resistance, respectively.

For example, whiteflies, which are commonly targeted by the JHA pyriproxyfen [7–10], have a haplodiploid reproductive strategy [58]. Furthermore, many species of aphids and soft scale insects, which are also affected by JHAs [59–63], have a parthenogenetic reproductive strategy for most or all of their life cycle. Here, we present models that demonstrate the role of pest reproductive strategy on resistance evolution to our generic JHA insecticide.

In haplodiploid species, females contribute twice as much to the gene pool as males [64]. In turn, the change in resistance allele frequency in each generation can be calculated as [51]

$$\Delta q = \left[\frac{2pq(W_{R/Fem} - W_{S/Fem})}{3W_{M/Fem}}\right] + \left[\frac{pq(W_{R/Mal} - W_{S/Mal})}{3W_{M/Mal}}\right] \tag{6.4}$$

where

- $W_{S/Fem}$ and $W_{R/Fem}$ are the marginal fitness of the S and R alleles, respectively, in females
- $W_{S/Mal}$ and $W_{R/Mal}$ are the fitness of the S and R genotypes in males
- $W_{M/Fem}$ is the mean fitness in females
- $W_{M/Mal}$ is the mean fitness in males

Under these conditions, q declines when [51]

$$2qW_{RR/Fem} + 2pW_{SS/Fem} + W_{R/Mal} > 2qW_{RS/Fem} + 2pW_{SS} + W_{S/Mal} \tag{6.5}$$

In parthenogenetic organisms, selection proceeds based on the fitness of each genotype relative to the mean fitness. In this case, each genotype is a distinct lineage, and the change in the resistant allele frequency each generation is [51]

$$\Delta q = F'_{RR} + 0.5F'_{RS} - q \tag{6.6}$$

where F'_{RR} and F'_{RS} are the frequency of the *RS* and *RR* genotypes, respectively, after selection and q is the frequency of the R allele before selection. Here, q declines under the following condition [51]:

$$2F_{RR}W_{RR} + F_{RS}W_{RS} > 2F_{RR}W_M + F_{RS}W_M \tag{6.7}$$

In Figure 6.1, we iterate Equations 6.1, 6.4, and 6.6 over time to show the effects of reproductive strategy on the evolution of resistance to a high-dose JHA, when resistance is recessive and the JHA is sprayed on 100% of the landscape. Using this model, resistance evolves faster in haplodiploid and parthenogenetic pests than in diploid pests. Thus, JHAs targeting whiteflies and aphids may be more prone to the evolution of resistance than JHAs targeting mosquitoes, flies, or other diploid pests (all else being equal).

Although results here represent a single scenario, Crowder et al. [51] showed over a broad range of conditions that resistance to JHAs evolves faster in non-diploid compared with diploid pests. Thus, the sustainability of any novel JHA will likely be strongly affected by the reproductive strategy of the target pest. Although pest reproductive strategy has received limited attention in regard to its effects on resistance evolution, the results demonstrate that a detailed understanding of pest reproduction is important for predicting the evolution of resistance to any JHA.

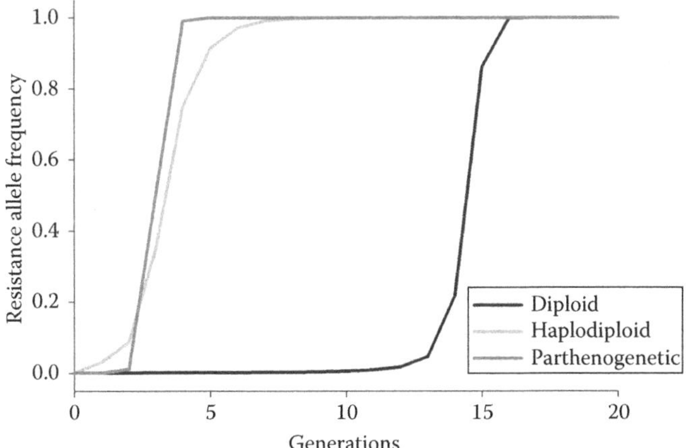

FIGURE 6.1 Frequency of the allele conferring resistance to the typical JHA over 20 generations in with a model of a JHA insecticide that causes high mortality when resistance is recessive (1% survival of homozygous susceptible females, heterozygous females, and susceptible males).

6.3 ECOLOGY OF RESISTANCE

The evolution of resistance to insecticides is affected by pest population dynamics. Fields that are intensively sprayed with insecticides may act as sinks where populations decline, while unsprayed refuges may act as sources [41]. In turn, gene flow between source and sink habitats affects the evolution of resistance by affecting the selective advantage of resistant genotypes across landscapes [41,65,66]. In this section, we detail a few key factors that play a role in shaping these dynamics and that might affect resistance to JHAs.

6.3.1 Pest Dynamics

As detailed in Section 6.2, the selective advantage for resistant genotypes is affected by pesticide efficacy and dominance of resistance, among other factors. Using a simple model, we demonstrated how these factors affect resistance to a generic JHA sprayed on 100% of the landscape. This situation is roughly analogous to a single crop field being treated repeatedly with a JHA. Here, we expand this model to include two crop fields (one treated and one untreated) to demonstrate how incorporating simple pest population dynamics and dispersal into models can influence the rate of resistance evolution to a generic JHA insecticide.

To demonstrate the role of dispersal, we relax the assumption that the generic JHA is sprayed in 100% of the landscape. For many insecticides, refuges of unsprayed crops are used to delay the evolution of resistance [31–40,50–53]. The refuge strategy involves leaving habitats of untreated crops throughout the landscape, which will produce large numbers of susceptible insects that can disperse and mate with rare resistant pests surviving on insecticide-treated crops [41,67]. When resistance is inherited as a recessive trait, heterozygous offspring from such matings are killed by the insecticide, and resistance evolution can be delayed. The refuge strategy can also be effective when resistance is additive or dominant, particularly when refuges are large, as refuges decrease the selective advantage of resistant phenotypes across landscapes regardless of dominance.

The refuge strategy depends critically on insect dispersal between patches of treated and untreated crops. Models that consider population dynamics are thus more likely to be realistic than models that do not consider this parameter. In Table 6.2, we show results of our diploid model developed in Section 6.2.1 with the added assumption that the landscape contains two patches. The first patch is treated with the insecticide and makes up 90% or 50% of the total landscape. The second patch is the untreated refuge and makes up the remaining 10% or 50% of the landscape. We assume two levels of insect dispersal: (1) Dispersal is high and random mating between individuals from the refuge and treated crop patches occurs or (2) dispersal is low and there is no mating between individuals from refuges or treated patches (i.e., all mating occurs in the natal patch). Results of this model clearly show that maintaining refuges in even a small proportion of the landscape can substantially delay the evolution of resistance (Table 6.2). However, the effectiveness of refuges declines with low levels of pest dispersal and additive or dominant resistance. As with genetically engineered insecticidal crops, where the refuge strategy has been extensively deployed to combat the evolution of

TABLE 6.2
Generations for Resistance to Evolve (Resistance Allele Frequency Increased from 0.001 to 0.5) in a Hypothetical Model of a JHA Insecticide That Causes High or Low Mortality

JHA Efficacy	W_{SS}	W_{RS}	W_{SS}	h	Refuge	Dispersal	Generations
High	0.01	0.01	1	0	10%	High	129
High	0.01	0.505	1	0.5	10%	High	5
Low	0.5	0.5	1	0	10%	High	1235
Low	0.5	0.75	1	0.5	10%	High	21
High	0.01	0.01	1	0	10%	Low	15
High	0.01	0.505	1	0.5	10%	Low	3
Low	0.5	0.5	1	0	10%	Low	1012
Low	0.5	0.75	1	0.5	10%	Low	18
High	0.01	0.01	1	0	50%	High	1032
High	0.01	0.505	1	0.5	50%	High	18
Low	0.5	0.5	1	0	50%	High	3022
Low	0.5	0.75	1	0.5	50%	High	46
High	0.01	0.01	1	0	50%	Low	1032
High	0.01	0.505	1	0.5	50%	Low	18
Low	0.5	0.5	1	0	50%	Low	1023
Low	0.5	0.75	1	0.5	50%	Low	41

Models were conducted with varying survival of susceptible homozygotes (W_{SS}), heterozygotes (W_{RS}), and resistant homozygotes (W_{RR}) and recessive or additive resistance (h) and varying refuge size (proportion of landscape left unsprayed). We also vary the proportion of individuals that disperse between habitat patches from high (insects move and mate randomly across the landscape) to low (no insects move across the landscape; mating is entirely in the natal field). Dominance values of 0 and 0.5 recessive or additive resistance, respectively.

resistance [31–41,51–53], results here suggest that the refuge strategy is also likely to be effective in delaying resistance to JHA insecticides (see also Section 6.4.4).

6.3.2 Landscape Configuration

As in Section 6.2.1, the model developed in Section 6.3.1 is admittedly simple, containing only two habitat patches (treated and treated). Of course, it is possible to develop a wide array of models that assume more complex, spatially explicit landscapes [35,38,68,69]. In these cases, refuges may be modeled as separate crop patches or mixed into fields containing areas of treated crops. Furthermore, models may contain a mosaic of different crop types [70], with various crops in the landscape serving as source/sink habitats and differentially impacting resistance evolution. Such spatially explicit models may be useful to predict how growers might modify the landscape to naturally control pests and delay resistance evolution.

Yet, evidence from studies of resistance to genetically modified crops suggests that simple models like the ones developed in Section 6.2.1 and 6.3.1 can also yield important insights for understanding the evolution of resistance across landscapes [40,41]. Thus, as spatially explicit models have not been considered in the context of resistance to JHAs, we do not expand further on these ideas here, although the literature on genetically modified crops contains many examples of spatially explicit models if readers wish to pursue this topic further.

6.4 CASE STUDY: WHITEFLIES AND PYRIPROXYFEN

In the previous sections, we describe how population genetics and ecological models can be applied to examine the evolution of resistance to a generic JHA insecticide. In this section, we build on these principles to describe the use of models to examine resistance by the whitefly *B. tabaci* to the JHA pyriproxyfen. This system has been well studied using ecological, conceptual, population genetics, and statistical models and therefore is an excellent example of how models can be used to develop IRM and IPM strategies for a pest–JHA complex.

6.4.1 Natural History

Whiteflies in the *B. tabaci* species complex include several of the world's most destructive crop pests [10,58,71,72]. The *B. tabaci* species complex consists of over 20 species [73] that attack field and greenhouse crops across subtropical and tropical regions [10,58, 71–75]. The species now known as Middle East–Asia Minor I (MEAM1) in the *B. tabaci* complex [73] was introduced to the United States at the end of the 1980s [10,74] and is a key pest of cotton and other crops. This species was previously referred to as the B biotype of *B. tabaci* [63], although we use the new MEAM1 designation here.

After establishing in the United States, MEAM1 quickly became a pest of field and horticultural crops. In the U.S. Southwest, where we focus our analysis, MEAM1 is a pest of cotton, melon, vegetable, and ornamental crops [10,76]. Throughout the 1990s, broad-spectrum insecticides such as pyrethroids, organophosphates, and carbamates were heavily applied for control of this pest [10,76]. Unfortunately, broad-spectrum insecticides eventually failed to provide effective whitefly control in many regions due to the evolution of insecticide resistance [10,76]. This was followed by growers applying more insecticides at higher concentrations yet still failing to obtain effective control and produce marketable yields of several crops [76].

The JHA pyriproxyfen was introduced for control of MEAM1 in 1996 [10,76]. Use of this insecticide to control MEAM1 exemplifies integration of a selective JHA into an IPM system [9,10,76]. A single, timely application of pyriproxyfen (or an insect growth regulator (IGR) buprofezin, a chitin inhibitor, which was also introduced in 1996) led to dramatic reductions in broad-spectrum insecticide use, helped conserve natural enemies, and restored farmer's profits, concurrent with a reduction in pest problems due to whiteflies [10,14,76]. However, laboratory bioassays revealed an area-wide decline in susceptibility of field-collected MEAM1 populations to pyriproxyfen over time [18,77]. Thus, there was significant concern that resistance evolution could threaten the long-term sustainability of pyriproxyfen as a management tool.

Throughout this section, we detail how models have been used to explore the biological, environmental, and anthropogenic factors that affect resistance to pyriproxyfen in MEAM1. We start with simple population genetics models and build models of increasing complexity. Understanding the role of each of these factors, and their interactions, has proven essential for implementing effective and economic management strategies for this pest and provides a foundation for similar studies in related systems.

6.4.2 Genetics and Ecology of Pyriproxyfen Resistance

Crowder et al. [78,79] developed a model of the population genetics and ecology of pyriproxyfen resistance. The model assumed resistance is controlled by a single locus with two alleles (*S* and *R*, see Section 6.2.1). Additionally, the model assumed there were only two crop patches, one treated with pyriproxyfen and one untreated refuge. However, there are several differences between this model and the two-patch models described earlier. First, the present model was based on a solid foundation of MEAM1 population dynamics. Population growth was controlled by several factors, including fecundity, development time, mortality, and contained functions relating these factors to environmental conditions such as temperature. Second, the model contained more complex dynamics for exposure to the JHA. Pyriproxyfen was only applied to fields when MEAM1 populations exceeded an action threshold of three adults per leaf, based on IPM guidelines [80,81]. Additionally, pyriproxyfen decayed in the environment, and selection pressure favoring resistant genotypes was reduced as the toxin degraded. Thus, while the model was built on the principles developed in Sections 6.2 and 6.3, it provides an example of how this foundation can be expanded with detailed knowledge of pest biology and JHA characteristics. Here, we review the key findings of this model and point out the similarities and dissimilarities from the models developed earlier.

Initial simulations of this model, which were conducted before detailed experiments on whitefly resistance to pyriproxyfen had been performed, suggested that pyriproxyfen resistance would evolve faster with increases in toxin concentration and dominance of resistance (Figure 6.2a) [78]. These results are in line with the predictions of the analytical models described in Section 6.2. Similarly, resistance evolved faster when the initial frequency of the resistance allele increased or the proportion of the region treated with pyriproxyfen increased [78]. Factors that delayed the evolution of resistance included greater susceptibility of males compared to females and fitness costs associated with resistance. Greater susceptibility in males delayed resistance because whiteflies are haplodiploid, and greater male susceptibility prevents the rapid buildup of resistance alleles in haploid males [50,78]. Fitness costs occur when resistant individuals have lower fitness in refuges compared to susceptible individuals, which often occurs because mutations that confer resistance to a toxin may disrupt the normal gene function and decrease fitness in the absence of the toxin [80–82]. Fitness costs, therefore, reduce the selective advantage of resistant alleles across landscapes and slow resistance evolution.

These initial simulations were subsequently used to inform empirical investigations of whitefly resistance to pyriproxyfen. Based on model results, Crowder and

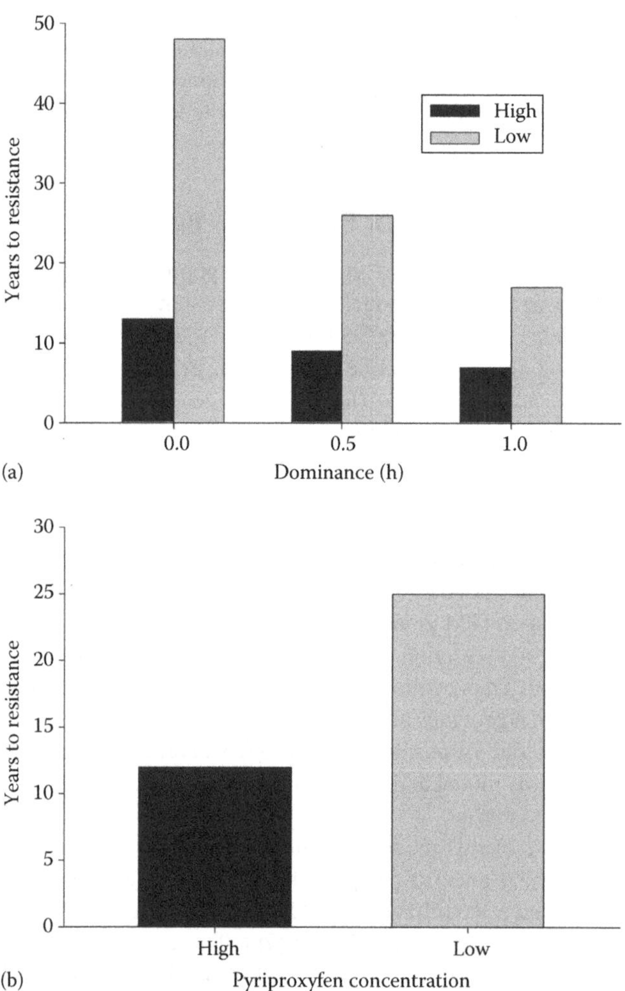

FIGURE 6.2 Years for resistance to pyriproxyfen to evolve in (a) simulations with variable dominance and pyriproxyfen concentration and (b) simulations with variable concentrations, where dominance varied across concentrations. In (a), the high and low toxin concentrations killed 95% and 50% of susceptible insects, respectively, and concentration was varied independently from dominance. In (b), the concentrations were the same, although the dominance of resistance was 0.25 with a high concentration and 0.7 with a low concentration. In both panels, the years to resistance report the number of years for the resistance allele frequency to increase from 0.001 to 0.5, in a landscape with 20% pyriproxyfen-treated fields and 80% refuges. (From Crowder, D.W. et al., *J. Econ. Entomol.*, 99, 1396, 2006; Crowder, D.W. et al., *Environ. Entomol.*, 37, 1514, 2008.)

colleagues investigated the toxicity of pyriproxyfen to whiteflies, the dominance of resistance, susceptibility in males compared to females, and fitness costs [83–85]. In field-collected populations selected or not for resistance to pyriproxyfen, fitness costs were not associated with pyriproxyfen resistance, and males and females were equally susceptible to the toxin [83–85]. Thus, factors initially identified to delay the evolution of resistance in simulations appeared not to occur under field conditions. Furthermore, the dominance of resistance was affected by toxin concentration [83]. When concentrations were low (killing 50% of whiteflies), resistance was inherited as a semidominant trait (estimated $h=0.7$); when concentrations were high (killing >95% of whiteflies), resistance was inherited as a semirecessive trait (estimated $h=0.25$). When this change in dominance was incorporated into simulations, the difference between high- and low-dose simulations declined (Figure 6.2b) [79].

The models of Crowder et al. [78,81] also demonstrated how population genetics of resistance to a JHA are linked with pest population dynamics (see also Section 6.4.3). In simulations, whitefly densities increased as pyriproxyfen resistance evolved. This occurred because pyriproxyfen became less effective at controlling resistant compared to susceptible whitefly populations. However, large-scale changes in resistance allele frequency preceded measurable changes in population densities. For example, when pyriproxyfen concentrations were low (killing 50% of susceptible insects), consistent with field efficacy of this JHA [83], the predicted resistance allele frequency increased from 0.001 to 0.4 in 20 years under normal growing conditions [79]. Over this same time period, however, simulated maximum yearly whitefly densities only increased from 8 to 12 adults per leaf [79]. Such a small increase in densities may be undetected by growers, despite the fact that resistance allele frequencies had increased 400-fold. This may be a realistic prediction because pest densities are typically affected by a myriad of factors, including host plant quality, weather, insecticide use, and natural enemies [10,14]; year-to-year variability in densities might be attributed to these factors rather than the evolution of resistance. These simulation results may partially explain why bioassays detected measurable changes in pyriproxyfen resistance from field populations in the first 10 years after pyriproxyfen was introduced [18,77], but pyriproxyfen remained effective in the field and no failures due to resistance were reported over the same period [86]. Yet, simulations suggest that the progressive buildup in resistance would eventually lead to measurable increases in whitefly abundance, if pyriproxyfen remained a key insecticide for whitefly control. For example, when resistance allele frequencies increased to 0.9, whitefly populations were predicted to exceed 25 adults per leaf [79]. In this case, simulations predicted a critical threshold of the frequency of resistance above which population densities would be significantly impacted.

These results suggest that for MEAM1 and pyriproxyfen, and likely other pest–JHA complexes, models can reconcile differences between bioassays and observed efficacy of JHAs in the field. Similar to other cases of insecticide resistance [36,87,88], simulation models of whiteflies and pyriproxyfen revealed that resistance evolution leads to increased pest densities under field conditions [78,79]. Such models linking resistance with pest population dynamics could increase the usefulness of resistance bioassays as a predictor of future pest densities and provide a tool for growers to decide when to implement preventative IRM strategies [79]. Furthermore, the

MEAM1–pyriproxyfen system shows how models can be used to inform experiments and in turn how experimental data can be reincorporated into future models. This back-and-forth process is important for investigating resistance to JHAs and other insecticides, as models provide testable hypotheses, while experimental data improve and challenge the realism of model predictions.

6.4.3 Integrated Pest Management and Pyriproxyfen Resistance

6.4.3.1 Relationship between IPM and IRM

Resistance management is an important aspect of IPM that helps preserve the efficacy of key modes of action (MOAs). As such, IRM is a component of IPM. At the same time, a central tenet of IRM is to limit the use of MOAs to the lowest practical level. As a mutual goal of IPM—limiting the use of control agents to the lowest practical level—IPM is a component of IRM. This apparent strange loop [89] or self-referential paradox is reminiscent of Escher's famous lithograph, Drawing Hands, where it appears that each hand is drawing the image of the other. In this way, IPM and IRM are intimately and inextricably linked. However, despite large gains in the research and understanding of resistance mechanisms, ecological dynamics, and insights gained through modeling efforts, our conceptual model for practitioners of IRM for JHAs and other insecticides suggests there are only three basic tools at their disposal: (1) limiting usage of MOAs, (2) diversifying MOAs, and (3) partitioning MOAs in space or time so as to segregate their usage [42,43].

The central tenet of IPM to apply control agents only when needed feeds directly into limiting key MOAs in IRM and thus reducing the potential for resistance evolution (see Section 6.4.3.2). As Arizona prepared to recover from the devastating infestations of pyrethroid-resistant whitefly populations in the mid-1990s, growers consented to and pursued a Section 18 Emergency Exemption from registration for pyriproxyfen and buprofezin that placed extraordinary limits on the usage of these IGRs [10,90]. For the years when these exemptions were in place, growers were limited to just one use of each IGR per cotton season. This constraint remains in place today for pyriproxyfen use in U.S. cotton. In turn, Arizona cotton growers were faced with diversifying MOAs as much as possible both before and after the availability of the JHA pyriproxyfen. Prior to 1996, growers followed a rotation of non-pyrethroid mixtures and synergized pyrethroid mixtures, dependent on just four classes of chemistry (pyrethroids, organophosphates, carbamates, and a cyclodiene) [91]. This limited rotation of related chemistries with severe natural enemy destruction was not sustainable. Starting in 1996, cotton growers had access to these same chemistries as well as pyriproxyfen and a second IGR buprofezin. As effective options were developed thereafter, new classes of chemistry and MOAs became available and were incorporated into IPM and IRM guidelines [92].

The final basic tactic in IRM is the partitioning of MOAs through space or time so as to segregate usage of key MOAs. Before 1996, MOAs were organized into a two-stage IRM program that encouraged a temporal partitioning of chemistry, non-pyrethroid mixtures first followed later by pyrethroid mixtures as needed [93]. Once pyriproxyfen became available in 1996, a three-stage IRM temporal partitioning became possible with the two IGRs used first and in rotation with each other as needed and followed by broader-spectrum MOAs, as needed [10,17,94]. This plan was updated as new selective and partially selective MOAs became available for use in cotton [92].

The initial exemption and first labels for pyriproxyfen in the United States were confined to cotton usage only. However, over time the USEPA Section 3 label was expanded to include many other crops, some of which are grown in Arizona. This created additional risks for selection of resistance in this mobile, multi-host pest. First, in an effort to rationalize buprofezin usage and later to accommodate the multi-crop registrations of neonicotinoids, Arizona researchers partnered with industry to create a plan for partitioning MOAs over ecological space. These voluntary cross-commodity guidelines included key agreements among growers of melons, vegetables, and cotton [17,95]. Today's labels for pyriproxyfen effectively confine its use to cotton in our multi-crop system, because of Arizona only restrictions preventing usage on brassica leafy vegetables and cucurbits. So, pyriproxyfen benefits by the most extreme IRM measures: limited to one use per season, generally followed by rotations to other chemistries as needed, and confined in ecological space to cotton only.

6.4.3.2 Role of Natural Enemies in IPM and IRM

Many arthropod predators, insect parasitoids, and several groups of fungi attack *B. tabaci* [96–98]. In Arizona, about 20 species of arthropod predators and five parasitoid species are known to attack this pest in cotton [14,99–102], and the overall contribution of natural enemies to *B. tabaci* population dynamics has been well studied and defined [103–105]. Infield life table studies have shown that natural levels of mortality to immature stages of *B. tabaci* routinely exceed 90% in cotton [103]. Sucking predators contribute the highest levels of mortality followed by dislodgement, which results from a combination of chewing predation and the effects of weather events. Various species of specialist parasitoids contribute a small amount of mortality, but predation has consistently been identified as the key factor governing changes in generational mortality and associated population dynamics [103–105]. This key role played by predators has further been demonstrated by the ease with which *B. tabaci* population resurgence and outbreak can be precipitated by the use of broad-spectrum insecticides that disrupt the natural enemy community [106,107].

The IPM and IRM strategies in place for *B. tabaci* and other key pests in Arizona cotton were designed to take advantage of the biological control provided by the natural enemy complex, particularly predators. The central role of biological control is realized through routine scouting and adherence to economic thresholds to ensure that insecticides are used only as a last resort [92,108,109]. Further, when insecticides are needed, recommendations are for the use of highly selective insecticides such as the JHA pyriproxyfen. Although this insecticide has been associated with harm to certain natural enemy species in some systems [44], exhaustive field-based evaluations have clearly demonstrated its high selectivity [14,110] and high efficacy in our cropping system [106]. The success of the current cotton IPM program rests on a broad base of understanding and tactics that largely enable us to avoid pest problems in the first place [10,110]. One key element is recognition of the seasonal cycle of this mobile multi-voltine, multi-crop pest and the important role it plays in regional pest dynamics and in guiding management decisions. Life table studies on multiple crops and plants hosting *B. tabaci* have shown that, much like in cotton,

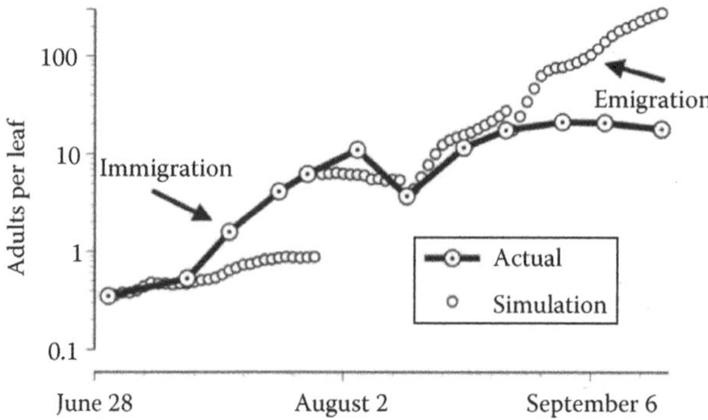

FIGURE 6.3 Comparison between actual and simulated population dynamics of *B. tabaci* in cotton. A temperature-dependent, stage-structured model was used to simulate densities and was initiated and reset for each generation with actual field densities and observed levels of mortality from life tables. (From Naranjo, S.E. and Ellsworth, P.C., *Entomol. Exp. Appl.*, 116, 93, 2005.)

natural levels of generation mortality >90% are common [14,105]. The one notable exception is spring cantaloupes where mortality is typically less than 70%. This low level of mortality on a key spring crop along with the crop's juxtaposition with cotton in the seasonal cycle heavily influences IPM and IRM considerations.

Comparison of outputs from a simple temperature-dependent population dynamics model with field data shows the pulses of dispersal into cotton in the early season and out of cotton in late season (Figure 6.3). Close correspondence of model and data in the middle portion of the season shows that dynamics during this period are largely controlled by endogenous factors such as predation [103]. This indicates that crops and their management are connected in the landscape. This understanding is the basis of the cross-commodity IRM plan discussed earlier, which strives to limit the exposure of adjacent *B. tabaci* generations to the same MOA. It would appear that movement occurs in pulses only at the boundaries of seasonal overlaps between different crops and perhaps between fields of the same crop differing in maturity. Thus, an assumption of random genetic mixing among insects from targeted and refuge crops over time may not accurately represent true system behavior.

The general role of biological control in mediating resistance evolution remains unclear. The rate of resistance evolution is expected to be affected by natural enemies if they differentially attack individuals with and without resistance alleles [111]. In general, resistance is expected to be slowed by natural enemies if they disproportionally attack individuals with resistance alleles, while resistance is expected to be accelerated if natural enemies disproportionally attack susceptible individuals. Empirical and modeling studies, however, have shown that there is no uniform outcome relative to the positive or negative role of biological control in resistance evolution [112–114], suggesting that each system needs to be examined separately. While biological control operates at a high level in the management of *B. tabaci* in cotton, specific experimental or modeling studies have not been done

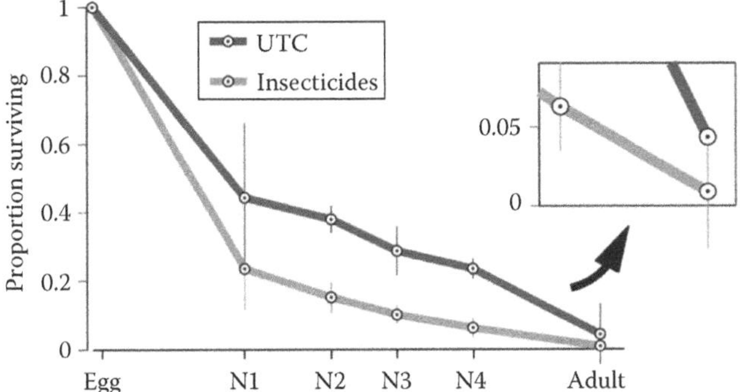

FIGURE 6.4 Mean survivorship (+SEM) curves for *B. tabaci* in cotton fields treated with IGR (insecticides) or left untreated (UTC) based on life table studies, 1997–1999 Maricopa, AZ, USA. The inset shows the small difference in mortality that separates outbreak populations from well-managed populations of the pest. The low level of survivorship under any conditions is a result of high mortality of larvae, in part due to control provided by natural enemies. (From Naranjo, S.E. and Ellsworth, P.C., *Entomol. Exp. Appl.*, 116, 93, 2005; Naranjo, S.E. and Ellsworth, P.C., *Biol. Control*, 51, 458, 2009.)

so far to assess its role in moderating insecticide resistance evolution. However, biological control in the Arizona cotton system likely played a positive role in preserving the efficacy of pyriproxyfen and ostensibly buffering against insecticide resistance evolution. Natural enemies are effective in limiting populations of *B. tabaci* [104], leading to low levels of larval survival even when insecticides are not applied (Figure 6.4). The use of selective compounds such as pyriproxyfen preserves natural enemy populations, which can provide population control after insecticides are applied [10,104]. Thus, selection for resistance could be diminished, because whitefly individuals surviving exposure to pyriproxyfen are then vulnerable to high levels of natural control, which could diminish the fitness advantage provided by resistance alleles. In contrast, this bioresidual is much smaller or nonexistent with the use of broader-spectrum insecticides resulting in more sprays and greater opportunity for resistance selection to those MOAs. Indeed, measurable levels of pyriproxyfen resistance in bioassays have not been associated with any observed levels of performance reduction in the field [86]. This may well be the result of active biological control and a strong bioresidual component in the control system.

6.4.3.3 IPM and IRM and Their Impacts on Modeling Resistance

Comprehensive sampling and threshold systems that guide the precise usage of selective JHAs like pyriproxyfen are key to achieving the goal of limited usage [92,94]. The active role that bioresidual plays in selective systems to further limit the whitefly population growth both serves to limit the need for intervention with additional MOAs and potentially buffer or mitigate incipient resistances to pyriproxyfen and other chemistries. Therefore, as a broad goal to limit the usage of all MOAs, Arizona has achieved remarkable gains, spraying on average just 1.5 times for all arthropod pests in the cotton system in recent years, compared with greater than 10 times in some years before the registration of pyriproxyfen.

Prevention of resistance to any chemistry including JHAs is supported by the existence and strategic usage of a diversity of MOAs. Pyriproxyfen was introduced to our cotton system along with buprofezin; however, growers often favored pyriproxyfen over buprofezin, which placed additional pressure on this key JHA. By 2002, however, acetamiprid was registered in cotton and proved to be the most active foliar neonicotinoid against whiteflies. Since 2003, it has been the most popular whitefly control agent in cotton, and pyriproxyfen usage remains very steady but low at ca. 0.1–0.4 sprays. Four major MOAs are now routinely used to control whiteflies in a three-stage program that attempts to temporally partition usage based on efficacy and selectivity of compounds [92]. Many other prevention tactics serve to avoid or lower whitefly densities including the adoption of smooth leaf cultivars, planting date management, and water and nitrogen inputs optimized for plant health. Together with other tactics practiced widely by growers, this created opportunity for broad-scale or area-wide lowering of pest densities [17]. In an effort to partition MOAs in space and isolate key chemistry to certain crops and cropping scenarios, cotton growers agreed to forego the usage of neonicotinoids when they were growing within multi-crop communities (as defined by cotton fields grown together with melons and vegetables within a 2 mile radius) [95]. These voluntary restrictions on neonicotinoid use placed further pressure on pyriproxyfen for growers of cotton in these multi-cropped areas. Nevertheless, there has been an area-wide lowering of whitefly densities even in untreated crops, largely due to broad-scale adoption of soil-applied neonicotinoids in vegetables and adoption of the cotton IPM program that emphasized use of IGRs first (J. Palumbo, personal communication).

The combined result of the Arizona whitefly management programs is much smaller populations and fewer sprays deployed in cotton and other crops to control whiteflies. This is an extraordinary example of cross-commodity organization for resistance management (see [115]) and a reversal of the classic pesticide treadmill: fewer and more selective sprays lead to better performance of natural controls that in turn lead to fewer primary and secondary pest problems in turn leading yet again to fewer sprays. Yet, monitoring data show a progressive decline in pyriproxyfen susceptibility in statewide whitefly populations starting around 2001 (Figure 6.5). Similarly, modeling results [78,79] suggest risks of resistance are high and progressive in our system. These facts, at least for now, seem to contradict the great successes currently seen in the Arizona cotton whitefly system and the continued performance of pyriproxyfen as a whitefly control agent [86]. As susceptibility in laboratory assays has declined, pyriproxyfen usage has either declined as a proportion of whitefly sprays made or stayed about the same as a proportion of all insect sprays made (Figure 6.5).

The causes underlying differences between these empirical outcomes from laboratory bioassay and modeling results and current sustained effectiveness of pyriproxyfen in the field remain unclear. As mentioned earlier (Section 6.4.2), sustained effectiveness of pyriproxyfen may be due to the relatively small increase in population density of whiteflies resulting from initial increases in resistance allele frequency. Other reasons could include complexity in the control dynamic that has not been sufficiently accounted for in models (such as the importance of natural control, see Section 6.4.3.2) or other facts on the ground that violate basic

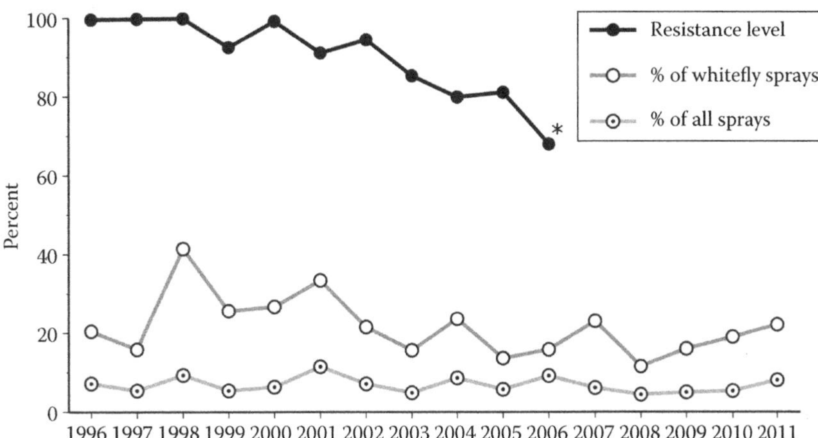

FIGURE 6.5 Statewide resistance levels (1996–2006; % mortality at 0.1 μg/mL) and pyriproxyfen usage as a percentage of whitefly or all arthropod foliar sprays (1996–2011). Empirical and modeling data suggest declines in susceptibility of whiteflies to pyriproxyfen. Yet statewide data would suggest that pyriproxyfen usage has not changed significantly over time. Furthermore, field performance assays showed continued efficacy of pyriproxyfen, similar to the period of its introduction in Arizona cotton. (From Dennehy, T.J., DeGain, B.A., Harpold, V.S., Brown, J.K., Morin, S., Fabrick, J.A., Byrne, F.J., and Nichols, R.L., *New challenges to management of whitefly resistance to insecticides in Arizona*, in *1999 Vegetable Report, Series P-144*, D. Byrne ed., University of Arizona, College of Agriculture, Tucson, AZ, 2006; Ellsworth, P.C. et al., *Arizona Cotton Insect Losses*, University of Arizona, College of Agriculture and Life Sciences, Cooperative Extension, Tucson, AZ, 2012; Ellsworth, P.C., Barkley, V., Dennehy, T., Degain, B., Ellingston, B., Naranjo, S., and Sims, M., *Assessment of Knack field performance through precision field and laboratory bioassays in cotton,* Arizona Cotton Report (2007), pp. 167–182.)

assumptions included in models of our system. For example, the bioresidual of our system whereby pyriproxyfen selection pressures act but are then masked by the overwhelming buffering potential of the natural control system may be key in future modeling efforts. Nonetheless, the relationships among evolution, ecology, and management in the Arizona pyriproxyfen–cotton whitefly system have provided a detailed understanding useful and supportive of our conceptual model for the practice of IRM: limit, diversify, and partition MOAs through space and time to the maximum practical extent.

6.4.4 Predicting Variation in Resistance across Crop Landscapes

As described earlier in this chapter, analytical, conceptual, and simulation models can provide insights for resistance management to JHAs and other insecticides; yet, they often rely on parameters that are difficult to measure or have not been estimated, which introduces uncertainty into their predictions. For example, Crowder et al. [79] used simulation models to investigate how refuges of cotton, melons, and other crops might delay the evolution of resistance to pyriproxyfen in *B. tabaci*. They used sensitivity analysis to investigate the effects of a range of plausible values for the movement of susceptible individuals from refuges of various crops to treated cotton fields. Although the simulation results indicated that both cotton and non-cotton refuges could delay resistance, the lack of data on movement led to uncertainty about the capacity of refuges of different crops to delay resistance.

A complementary modeling approach entails spatially explicit statistical models, which can use empirical data to explore the relationship between spatial variation in resistance and the abundance and distribution of potential refuges and treated fields [45,46]. A prerequisite for such spatial analyses is mapping the relevant habitat patches near "focal sites" where the frequency of resistance is known. Using GIS technology, one can measure the area of putative refuges and treated fields in rings of increasing radius around each focal site. Because refuges provide susceptible individuals, a negative association is expected between resistance frequency at the focal sites and the area of refuges. However, susceptible insects moving from refuges to treated fields are diluted in space as the distance between refuges and focal sites increases. Accordingly, at some distance from a given type of refuge, the refuges will have no statistically significant effect on resistance at the focal sites. The greatest distance (i.e., ring width) at which the association between a type of refuge and resistance at the focal sites is significant defines the zone of influence of this refuge type [45,46]. The same logic applies for analyzing the effects of treated fields on resistance at various spatial scales, but because treated fields select for resistance, we expect a positive association between resistance frequency at focal sites and the area of treated fields.

We applied this spatially explicit statistical modeling approach to analyze *B. tabaci* resistance to pyriproxyfen, which has been monitored in Arizona cotton fields since 1996 [18,116]. The monitoring data include the proportion of whitefly eggs surviving a diagnostic concentration of 0.1 μg pyriproxyfen/mL in laboratory bioassays [18]. This diagnostic concentration separates individuals with and without resistance alleles and thus provides estimates of the frequency of resistance. In our statistical analyses, we used data on resistance from 84 focal sites sampled in four counties of central Arizona between 2002 and 2008 [46].

We conducted our analyses at distances of up to 3 km from focal sites, with this distance based on previously reported data on whitefly movement [118–120]. Both laboratory and field data indicate that *B. tabaci* dispersal is bimodal, with most adults moving short distances when suitable host plants are present and a small percentage moving longer distances. In laboratory flight chambers, most individuals responded to vegetative cues and landed rapidly, although 6% of individuals remained airborne for long periods [120]. In the field in Arizona, the number of whiteflies captured outside an infested cantaloupe field declined steadily as distance from the field increased to 750 m but rose between 1.5 and 2.5 km from the cantaloupe field [119]. In California, a significant positive association occurred between the area of spring melons within a distance of 2.5 km from cotton fields and whitefly density in the cotton fields [118].

To test the hypothesis that the distribution of refuges and treated cotton affects the evolution of resistance, we used GIS technology and remote sensing data to map the distribution of four types of whitefly host plants at up to 3 km from the focal sites monitored for resistance [46]. The habitats considered were treated cotton and three types of putative refuges that were not sprayed with pyriproxyfen: alfalfa, melons, and untreated cotton. We analyzed the first 4 years of data to formulate statistical models of the association between resistance to pyriproxyfen and abundance of the four host plant types near each of 46 sites sampled from 2002 to 2005 and to determine the maximum spatial scale at which each habitat affected resistance. In the first set of analyses, among-year variation in resistance to pyriproxyfen and abundance of the four host plant types was removed statistically before pooling the data. Because there was also variation among the four counties in pyriproxyfen use and in the distribution of the whitefly host plants, we also performed a second set of analyses after statistically removing the effects of county and year. Results from both sets of analyses showed that the areas of melon and alfalfa fields were not significantly associated with spatial variation in pyriproxyfen resistance [46]. However, both sets of analyses revealed that the areas of untreated cotton were negatively associated with resistance and areas of treated cotton were positively associated with resistance. In both sets of analyses, the zone of influence of cotton refuges was 3.0 km, while the maximum distance at which treated cotton acted as a source of resistance was 2.75 km.

We tested the predictive power of the analyses as follows: we used models developed from the data of 2002 to 2005 to make predictions about resistance to pyriproxyfen in 38 populations sampled from 2006 to 2009. For predictions, we used the multiple regression models from 2002 to 2005 at the spatial scale that explained the highest proportion of spatial variation in resistance for those years. To calculate the predicted values for resistance to pyriproxyfen, we substituted the areas of treated and untreated cotton near the 38 sites sampled during 2006–2009 into these multiple regression models. The association between predicted and observed resistance to pyriproxyfen was positive and significant in both analyses, confirming that cotton fields treated with pyriproxyfen selected for resistance, while untreated cotton fields delayed resistance [46].

The statistical approach described here complemented previous simulation models assessing the role of cotton and non-cotton refuges in delaying the evolution of resistance [78,79]. While simulation models demonstrated that non-cotton refuges could delay resistance if they provided a sufficient number of susceptible individuals, the statistical analyses showed that non-cotton refuges did not affect the evolution of resistance in central Arizona [46]. This lack of concordance might be due in part to the periods of immigration of whiteflies into and out of cotton field (Figure 6.3, see Section 6.4.3). Empirical evidence for the refuge strategy had been limited to small-scale experiments and historical comparisons across species and regions [40,48,121,122]. The results of the statistical approach described here provide more direct support for effectiveness of the refuge strategy in the field based on large-scale data for a single species in a particular region [46].

6.5 CONCLUSIONS

This chapter shows how models can be used to make predictions about the evolution of resistance by pests to JHAs. Simple mathematical models like those described in Sections 6.2 and 6.3 might be particularly useful in predicting the sustainability of JHAs before they are widely deployed or in cases where limited information is available on pest–JHA interactions. Section 6.4 describes how more complex simulation models (Section 6.4.2) and conceptual models (Section 6.4.3) can link detailed knowledge of pest biology with real-world IRM and IPM strategies. Such models may be particularly useful when more extensive information is available, as is the case for the MEAM1–pyriproxyfen system. Complex simulation models may be useful in guiding IRM and IPM when they link the evolution of resistance to population dynamics under field conditions, especially when used in conjunction with observational studies and experiments to predict the sustainability of JHAs. Conceptual models may be useful when examining links between insecticide use, natural enemies, pest biology, and management strategies. In Section 6.4.4, we also show that statistical models linking spatial variation in resistance to the abundance and distribution of various types of host plants can contribute in designing precise resistance management strategies for JHA.

However, even in well-developed study systems, models may not always capture all elements of pest biology, and predictions can differ from the evolution of resistance in field populations. The differences may be explained by difficulties in accurately measuring resistance allele frequencies in the field (Section 6.4.2), the lack of spatial resolution for depicting changes in resistance dynamics (Section 6.4.4), and unexplained complexity in field conditions affecting resistance evolution (Section 6.4.3), among other factors. In these cases, models can be used to highlight gaps in knowledge, while experiments can inform models. This feedback between models and empirical data is imperative for studying the long-term dynamics of resistance to JHAs and other insecticides. As many pests have evolved resistance to JHAs, the management of resistance is likely to be an ongoing problem, particularly as the use of this functional class of insecticides increases worldwide. We hope that models can help to address this problem by increasing the ability to understand and manage pest resistance to JHAs.

REFERENCES

1. W.J. Lewis, J.C. van Lenteren, S.C. Phatak, and J.H. Tumlinson III, *A total approach to sustainable pest management*, Proc. Natl. Acad. Sci. USA 94 (1997), pp. 12243–12248.
2. P.A. Matson, W.J. Parton, A.G. Power, and M.J. Swift, *Agricultural intensification and ecosystem properties*, Science 277 (1997), pp. 504–509.
3. A.R. Horowitz and I. Ishaaya, *Insect Pest Management: Field and Protected Crops*, Springer, Berlin, Germany, 2004.
4. M. Ashok, C. Turner, and T.G. Wilson, *Insect juvenile hormone resistance gene homology with the BHLH-PAS family of transcriptional regulators*, Proc. Natl. Acad. Sci. USA 95 (1998), pp. 2761–2766.
5. T.L. Russell and B.H. Kay, *Biologically based insecticides for the control of immature Australian mosquitoes: A review*, Aust. J. Entomol. 47 (2008), pp. 232–242.
6. G.J. Devine, E.Z. Perea, G.F. Killeen, J.D. Stancil, S.J. Clark, and A.C. Morrison, *Using adult mosquitoes to transfer insecticides to* Aedes aegypti *larval habitats*, Proc. Natl. Acad. Sci. USA (2009), pp. 11530–11534.
7. I. Ishaaya and A.R. Horowitz, *Novel phenoxy juvenile-hormone analog (pyriproxyfen) suppresses embryogenesis and adult emergence of sweetpotato whitefly (Homoptera: Aleyrodidae)*, J. Econ. Entomol. 85 (1992), pp. 2113–2117.
8. I. Ishaaya and A.R. Horowitz, *Pyriproxyfen, a novel insect growth regulator for controlling whiteflies: Mechanisms and resistance management*, Pestic. Sci. 43 (1995), pp. 227–232.

9. I. Denholm, M. Cahill, T.J. Dennehy, and A.R. Horowitz, *Challenges with managing insecticide resistance in agricultural pests, exemplified by the whitefly* Bemisia tabaci, Phil. Trans. R. Soc. Lond. B 353 (2008), pp. 1757–1767.
10. P.C. Ellsworth and J. L. Martinez-Carillo, *IPM for* Bemisia tabaci*: A case study from North America*, Crop Prot. 20 (2001), pp. 853–869.
11. K.G. Palma, S.M. Meola, and R.W. Meola, *Mode of action of pyriproxyfen and methoprene on eggs of* Ctenocephais felis *(Siphonaptera, Pulicidae)*, J. Med. Entomol. 30 (1993), pp. 421–426.
12. H. Oberlander and D.L. Silhacek, *Mode of action of insect growth regulators in lepidopteran tissue culture*, Pestic. Sci. 54 (1998), pp. 300–302.
13. J. Miyamoto, M. Hirano, Y. Takimoto, and M. Hatakoshi, *Insect growth-regulators for pest-control, with emphasis on juvenile-hormone analogs—Present status and future-prospects*, ACS Symp. Ser. 524 (1993), pp. 144–168.
14. S.E. Naranjo, P.C. Ellsworth, and J.R. Hagler, *Conservation of natural enemies in cotton: Role of insect growth regulators for management of* Bemisia tabaci, Biol. Control 30 (2004), pp. 52–72.
15. T.G. Wilson and M. Ashok, *Insecticide resistance resulting from an absence of target-site gene product*, Proc. Natl. Acad. Sci. U.S.A. 95 (1998), pp. 14040–14044.
16. A.R. Horowitz, Z. Mendelson, M. Cahill, I. Denholm, and I. Ishaaya, *Managing resistance to the insect growth regulator, pyriproxyfen, in* Bemisia tabaci, Pestic. Sci. 55 (1999), pp. 272–276.
17. J.C. Palumbo, A.R. Horowitz, and N. Prabhaker, *Insecticidal control and resistance management for* Bemisia tabaci, Crop Prot. 20 (2001), pp. 739–765.
18. A. Li, T.J. Dennehy and R.L. Nichols, *Baseline susceptibility and development of resistance to pyriproxyfen in* Bemisia argentifolii *(Homoptera: Aleyrodidae) in Arizona*, J. Econ. Entomol. 96 (2003), pp. 1307–1314.
19. J.G. Daglish, *Impact of resistance on the efficacy of binary combinations of spinosad, chlorpyrifos-methyl and s-methoprene against five stored-grain beetles*, J. Stored Prod. Res. 44 (2008), pp. 71–76.
20. T.G. Wilson, S.L. Wang, M. Beno, and R. Farkas, *Wide mutational spectrum of a gene involved in hormone action and insecticide resistance in* Drosophila melanogaster, Mol. Genet. Genom. 276 (2006), pp. 294–303.
21. C. Turner and T.G. Wilson, *Molecular analysis of the methoprene-tolerant gene region of* Drosophila melanogaster, Arch. Insect Biochem. Phys. 30 (1995), pp. 133–147.
22. A. Brun-Barale, J.C. Bouvier, D. Pauron, J.B. Berge, and S. Benoit, *Involvement of a sodium channel mutation in pyrethroid resistance in* Cydia pomonella *L, and development of a diagnostic test*, Pest Manag. Sci. 61 (2005), pp. 549–554.
23. L. Zhang, K. Harda, and T. Shono, *Cross resistance to insect growth regulators in pyriproxyfen-resistant housefly*, Appl. Entomol. Zool. 33 (1998), pp. 195–197.
24. L. Zhang, S. Kasai, and T. Shono, in vitro *metabolism of pyriproxyfen by microsomes from susceptible and resistant housefly larvae*, Arch. Insect Biochem. Phys. 37 (1998), pp. 215–224.
25. L.B. Brattsten, C.W. Holyoke, J.R. Leeper, and K.F. Raffa, *Insecticide resistance: Challenge to pest management and basic research*, Science 231 (1985), pp. 1255–1260.
26. R.T. Roush and B.E. Tabashnik, *Pesticide Resistance in Arthropods*, Chapman & Hall, New York, 1990.
27. I. Denholm and M.W. Rowland, *Tactics for managing pesticide resistance in arthropods—Theory and practice,* Ann. Rev. Entomol. 37 (1992), pp. 91–112.
28. I. Denholm, G.J. Devine, and M.S. Williamson, *Evolutionary genetics—Insecticide resistance on the move*, Science 297 (2002), pp. 2222–2223.
29. D.W. Onstad, *Insect Resistance Management: Biology, Economics, and Prediction*, Elsevier, New York, 2007.

30. F. Gould, *Potential and problems with high-dose strategies for pesticidal engineered crops*, Biocontrol Sci. Technol. 4 (1994), pp. 451–461.
31. M.A. Caprio, Bacillus thuringiensis *gene deployment and resistance management in single-tactic and multi-tactic environments*, Biocontrol Sci. Technol. 4 (1994), pp. 487–497.
32. D.W. Onstad and F. Gould, *Modeling the dynamics of adaptation to transgenic maize by European corn borer (Lepidoptera: Pyralidae)*, J. Econ. Entomol. 91 (1998), pp. 585–593.
33. D.W. Onstad and F. Gould, *Do dynamics of crop maturation and herbivorous insect lifecycle influence the risk of adaptation to toxins in transgenic host plants?* Environ. Entomol. 27 (1998), pp. 517–522.
34. S.L. Peck and S.P. Ellner, *The effect of economic thresholds and life-history parameters on the evolution of pesticide resistance in a regional setting*, Am. Nat. 149 (1997), pp. 43–63.
35. N.P. Storer, S.L. Peck, F. Gould, J.W. Van Duyn, and G.G. Kennedy, *Spatial processes in the evolution of resistance in* Helicoverpa zea *(Lepidoptera: Noctuidae) to* Bt *transgenic corn and cotton in a mixed agroecosystem: A biology-rich stochastic simulation model*, J. Econ. Entomol. 96 (2003), pp. 156–172.
36. D.W. Onstad, D.W. Crowder, P.D. Mitchell, C.A. Guse, J.L. Spencer, E. Levine, and M.E. Gray, *Economics versus alleles: Balancing IPM and IRM for rotation-resistant western corn rootworm*, J. Econ. Entomol. 96 (2003), pp. 1872–1885.
37. B.E. Tabashnik, F. Gould, and Y. Carrière, *Delaying evolution of insect resistance to transgenic crops by decreasing dominance and heritability*, J. Evol. Biol. 17(4) (2004), pp. 904–912.
38. M.S. Sisterson, Y. Carrière, T.J. Dennehy, and B.E. Tabashnik, *Evolution of resistance to transgenic crops: Interactions between insect movement and field distribution*, J. Econ. Entomol. 98 (2005), pp. 1751–1762.
39. D.W. Crowder and D.W. Onstad, *Using a generational time-step model to simulate the dynamics of adaptation to crop rotation and transgenic corn by western corn rootworm (Coleoptera: Chrysomelidae)*. J. Econ. Entomol. 98 (2005), pp. 518–533.
40. B.E. Tabashnik, A.J. Gassmann, D.W. Crowder, and Y. Carrière, *Insect resistance to* Bt *crops: Evidence versus theory*, Nat. Biotechnol. 26 (2008), pp. 199–202.
41. Y. Carrière, D.W. Crowder, and B.E. Tabashnik, *Evolutionary ecology of insect adaptation to* Bt *crops*, Evol. Appl. 3 (2010), pp. 561–573.
42. P.C. Ellsworth and J.W. Diehl, *Lygus in cotton no 2: An integrated management plan for Arizona*, (1998) 2 pp. Available at http://ag.arizona.edu/crops/cotton/insects/lygus/lygus2.pdf.
43. A.R. Horowitz, P.C. Ellsworth, and I. Ishaaya, *Biorational pest control: An overview*, in *Biorational Control of Arthropod Pests: Application and Resistance Management*, A.R. Horowitz and I. Ishaaya, eds., Springer, London, U.K., 2007, pp. 1–20.
44. S.E. Naranjo and P.C. Ellsworth, *50 Years of the integrated control concept: Moving the model and implementation forward in Arizona*, Pest Manag. Sci. 65 (2009), pp. 1267–1286.
45. Y. Carrière, P. Dutilleul, C. Ellers-Kirk, B. Pedersen, S. Haller, L. Antilla, T.J. Dennehy, and B.E. Tabashnik, *Sources, sinks, and the zone of influence of refuges for managing insect resistance to* Bt *crops*, Ecol. Appl. 14 (2004), pp. 1615–1623.
46. Y. Carrière, C. Ellers-Kirk, K. Hartfield, G. Larocque, B. Degain, P, Dutilleul, T.J. Dennehy, S.E. Marsh, D.W. Crowder, X. Li, P.C. Ellsworth, S.E. Naranjo, J.C. Palumbo, A. Fournier, L. Antilla, and B.E. Tabashnik, *Large-scale, spatially-explicit test of the refuge strategy for delaying insecticide resistance*, Proc. Natl. Acad. Sci. USA 109 (2012), pp. 775–780.
47. National Research Council, *Pesticide Resistance: Strategies and Tactics for Management*, National Academy Press, Washington, DC, 1986.
48. B.E. Tabashnik, J.B.J. Van Rensburg, and Y. Carrière, *Field-evolved insect resistance to* Bt *crops: Definition, theory, and data*, J. Econ. Entomol. 102 (2009), pp. 2011–2025.

49. J.A. Endler, *Natural Selection in the Wild*, Princeton University Press, Princeton, NJ, 1986.
50. Y. Carrière, *Haplodiploidy, sex, and the evolution of pesticide resistance*, J. Econ. Entomol. 96 (2003), pp. 1626–1640.
51. D.W. Crowder and Y. Carrière, *Comparing the refuge strategy for managing the evolution of insect resistance under different reproductive strategies*, J. Theor. Biol. 261 (2009), pp. 423–430.
52. Y. Carrière and B.E. Tabashnik, *Reversing insect adaptation to transgenic insecticidal plants*, Proc. R. Soc. Lond. B 268 (2001), pp. 1475–1480.
53. B.E. Tabashnik, T.J. Dennehy, and Y. Carrière, *Delayed resistance to transgenic cotton in pink bollworm*, Proc. Natl. Acad. Sci. USA 43 (2005), pp 15389–15393.
54. F.R. Groeters and B.E. Tabashnik, *Roles of selection intensity, major genes, and minor genes in evolution of insecticide resistance*, J. Econ. Entomol. 93 (2000), pp. 1580–1587.
55. A.J. Gassmann, D.W. Onstad, and B.R. Pittendrigh, *Evolutionary analysis of herbivorous insects in natural and agricultural environments*, Pest Manag. Sci. 65 (2009), pp. 1174–1181.
56. A.R. Horowitz, K. Gorman, G. Ross, and I. Denholm, *Inheritance of pyriproxyfen resistance in the whitefly,* Bemisia tabaci *(Q biotype)*, Arch. Insect Biochem. Phys. 54 (2003), pp. 177–186.
57. Y.B. Liu and B.E. Tabashnik, *Inheritance of resistance to* Bacillus thuringiensis *toxin Cry1C in diamondback moth*, Appl. Environ. Microbiol. 63 (1997), pp. 2218–2223.
58. D.N. Byrne and T.S. Bellows, *Whitefly biology*, Annu. Rev. Entomol. 36 (1991), pp. 431–457.
59. T.X. Liu and T.Y. Chen, *Effects of a juvenile hormone analog, pyriproxyfen, on the apterous form of* Lipaphis erysimi, Entomol. Exp. Appl. 98 (2001), pp. 295–301.
60. M.L. Richardson and D.M. Lagos, *Effects of a juvenile hormone analogue, pyriproxyfen, on the apterous form of soybean aphid (*Aphis glycines*)*, J. Appl. Entomol. 131 (2007), pp. 297–302.
61. J. Kuldova, I. Hardy, and Z. Wimmer, *Response of the hop aphid,* Phorodon humuli *(Homoptera: Aphididae), to the application of juvenile hormone analogue in field trials*, Crop. Prot. 17 (1998), pp. 213–218.
62. W. Wakgari and J. Giliomee, *Effects of some conventional insecticides and insect growth regulators on difference phonological stages of the white wax scale,* Ceroplastes destructor *Newstead (Hemiptera: Coccidae), and its primary parasitoid,* Aprostocetus ceroplastae *(Girault) (Hymenoptera: Eulophidae)*, Int. J. Pest Manag. 47 (2001), pp. 179–184.
63. M. Eliahu, D. Blumberg, A.R. Horowitz, and I. Ishaaya, *Effect of pyriproxyfen on developing stages and embryogenesis of California red scale (CRS),* Aonidiella aurantii, Pest Manag. Sci. 63 (2007), pp. 743–746.
64. D.L. Hartl, *A fundamental theorem of natural selection for sex linkage or arrhenotoky*, Am. Nat. 106 (1971), pp. 516–524.
65. H.N. Comins, *The development of insecticide resistance in the presence of migration*, J. Theor. Biol. 64 (1977), pp. 177–197.
66. G.P. Georghiou and C.E. Taylor, *Genetic and biological influences in the evolution of insecticide resistance*, J. Econ. Entomol. 70 (1977), pp. 319–323.
67. F. Gould, *Sustainability of transgenic insecticidal cultivars: Integrating pest genetics and ecology*, Annu. Rev. Entomol. 43 (1998), pp. 701–726.
68. S. Peck, F. Gould, and S.P. Ellner, *Spread of resistance in spatially extended regions of transgenic cotton, implications for management of* Heliothis virescens *(Lepidoptera, Noctuidae)*, J. Econ. Entomol. 92 (1999), pp. 1–16.
69. M.A. Caprio, *Source-sink dynamics between transgenic and nontransgenic habitats and their role in the evolution of resistance*, J. Econ. Entomol. 94 (2001), pp 698–705.

70. D.W. Onstad, D.W. Crowder, S.A. Isard, E. Levine, J.L. Spencer, M.E. O'Neal, S.T. Ratcliffe, M.E. Gray, L.W. Bledsoe, C.D. DiFonzo, J.B. Eisley, and C.R. Edwards, *Does landscape diversity slow the spread of rotation-resistant western corn rootworm (Coleoptera: Chrysomelidae)?* Environ. Entomol. 32 (2003), pp. 992–1001.
71. I.D. Bedford, R.W. Brighton, J.K. Brown, R.C. Rosel, and P.G. Markham, *Geminivirus transmission and biological characterization of* Bemisia tabaci *(Gennadius) biotypes from different geographic regions*, Ann. Appl. Biol. 125 (1994), pp. 311–325.
72. M.M. Viscaret, I. Torres-Jerez, E. Agostini de Maneo, S.N. Lopez, E.E. Botto, and J.K. Brown, *Mitochondrial DNA evidence for a distinct New World group of* Bemisia tabaci *(Gennadius) (Hemiptera: Aleyrodidae) indigenous to Argentina and Bolivia, and presence of the old world B biotype in Argentina*, Ann. Entomol. Soc. Am. 96 (2003), pp. 65–72.
73. P.J. De Barro, S.-S. Liu, L.M. Boykin, and A.B. Dinsdale, Bemisia tabaci*: A statement of species status*, Annu. Rev. Entomol. 56 (2011), pp. 1–19.
74. J.K. Brown, D.R. Frohlich, and R.C. Rosell. *The sweetpotato or silverleaf whiteflies: Biotypes of* Bemisia tabaci *or a species complex?* Annu. Rev. Entomol. 40 (1995), pp. 511–534.
75. T.M. Perring, *The* Bemisia tabaci *species complex*, Crop Prot. 20 (2001), pp. 725–737.
76. T.J. Dennehy and L. Williams III, *Management of resistance in* Bemisia *in Arizona cotton*, Pestic. Sci. 51 (1997), pp. 398–406.
77. T.J. Dennehy, B.A. Degain, V.S. Harpold, and S.A. Brink, *Whitefly resistance to insecticides in Arizona: 2002 and 2003 results*, Proceedings of Beltwide Cotton Conference, 2004, San Antonio, TX, USA, pp. 1926–1938.
78. D.W. Crowder, Y. Carrière, B.E. Tabashnik, P.C. Ellsworth, and T.J. Dennehy, *Modeling evolution of resistance to pyriproxyfen by the sweetpotato whitefly (Homoptera: Aleyrodidae)*, J. Econ. Entomol. 99 (2006), pp. 1396–1406.
79. D.W. Crowder, P.C. Ellsworth, B.E. Tabashnik, and Y. Carrière, *Effects of operational and environmental factors on evolution of resistance to pyriproxyfen in the sweetpotato whitefly (Hemiptera: Aleyrodidae)*, Environ. Entomol. 37 (2008), pp. 1514–1524.
80. Y. Carrière, J.-P. Deland, D.A. Roff, and C. Vincent, *Life history costs associated with the evolution of insecticide resistance*, Proc. R. Soc. Lond. B 58 (1994), pp. 35–45.
81. F.R. Groeters, B.E. Tabashnik, N. Finson, and M.W. Johnson, *Fitness costs of resistance to* Bacillus thuringiensis *in the diamondback moth (Plutella xylostella)*, Evolution 48 (1994), pp. 197–302.
82. A.J. Gassmann, Y. Carrière, and B.E. Tabashnik, *Fitness costs of insect resistance to* Bacillus thuringiensis, Annu. Rev. Entomol. 54 (2009), pp. 147–163.
83. D.W. Crowder, T.J. Dennehy, C. Ellers-Kirk, C.M. Yafuso, P.C. Ellsworth, B.E. Tabashnik, and Y. Carrière, *Field evaluation of resistance to pyriproxyfen in* Bemisia tabaci *(B biotype)*, J. Econ. Entomol. 100 (2007), pp. 1650–1656.
84. D.W. Crowder, C. Ellers-Kirk, C.M. Yafuso, T.J. Dennehy, B.A. Degain, V.S. Harpold, B.E. Tabashnik, and Y. Carrière, *Inheritance of resistance to pyriproxyfen in* Bemisia tabaci *males and females (B biotype),* J. Econ. Entomol. 101 (2008), pp. 927–932.
85. D.W. Crowder, C. Ellers-Kirk, B.E. Tabashnik, and Y. Carrière, *Lack of fitness costs associated with pyriproxyfen resistance in the B biotype of* Bemisia tabaci, Pest Manag. Sci. 65 (2009), pp. 235–240.
86. P.C. Ellsworth, V. Barkley, T. Dennehy, B. Degain, B. Ellingston, S. Naranjo, and M. Sims, *Assessment of Knack field performance through precision field and laboratory bioassays in cotton*, Arizona Cotton Report (2007), pp. 167–182. Available at http://cals.arizona.edu/pubs/crops/az1437/az14374b.pdf.
87. D.W. Onstad and C.A. Guse, *Economic analysis of the use of transgenic crops and nontransgenic refuges for management of European corn borer (Lepidoptera: Pyralidae)*, J. Econ. Entomol. 92 (1999), pp. 1256–1265.

88. D.W. Crowder, D.W. Onstad, M.E. Gray, P.D. Mitchell, J.L. Spencer, and R.J. Brazee, *Economic analysis of dynamic management strategies utilizing transgenic corn for control of western corn rootworm Coleoptera: Chysomelidae)*, J. Econ. Entomol. 98 (2005), pp. 961–975.
89. D.R. Hofstadter, *Gödel, Escher, Bach: An Eternal Golden Braid*, Basic Books, New York, 1979.
90. P.C. Ellsworth and J.W. Diehl, *Whiteflies in Arizona No. 5: Insect Growth Regulators*, (1996), 2 pp. Available at http://ag.arizona.edu/crops/cotton/insects/wf/wfly5.pdf.
91. T.J. Dennehy, P.C. Ellsworth, and T.F. Watson, *Whiteflies in Arizona: Pocket guide*, Cooperative Extension #195009 (1995). Available at http://ag.arizona.edu/crops/cotton/insects/wf/wfly795.pdf.
92. P.C. Ellsworth, J.C. Palumbo, S.E. Naranjo, T.J. Dennehy, and R.L. Nichols, *Whitefly management in Arizona cotton*, IPM Series No. 18., Publ. # AZ1404 (2006), 4 pp. Available at http://cals.arizona.edu/pubs/insects/az1404.pdf.
93. T.J. Dennehy, P.C. Ellsworth, and R.L. Nichols, *Whitefly management in Arizona cotton 1995*, IPM Publication Series No. 3, Publication #195008 (1995), 2 pp.
94. P.C. Ellsworth, T.J. Dennehy, and R.L. Nichols, *Whitefly management in Arizona cotton 1996*, IPM Series No. 3. Publication #196004 (1996), 2 pp. Available at http://ag.arizona.edu/crops/cotton/insects/wf/cibroch.html.
95. J.C. Palumbo, P.C. Ellsworth, T.J. Dennehy, and R.L. Nichols, *Cross-commodity guidelines for neonicotinoid Insecticides in Arizona*, IPM Series 17, Publ. No. AZ1319 (2003), 4 pp. Available at http://cals.arizona.edu/pubs/insects/az1319.pdf.
96. M. Faria and S.P. Wraight, *Biological control of Bemisia tabaci with fungi*, Crop Prot. 20 (2001), pp. 767–778.
97. D. Gerling, O. Alomar, and J. Arnó, *Biological control of* Bemisia tabaci *using predators and parasitoids*, Crop Prot. 20 (2001), pp. 779–799.
98. J.G.R. Arno, T. X. Liu, A.M. Simmons, and D. Gerling, *Natural enemies of* Bemisia tabaci*: Predators and parasitoids*, in Bemisia*: Bionomics and Management of a Global Pest*, P.A. Stansly and S.E. Naranjo eds., Springer, Dordrecht, the Netherlands, 2010, pp. 385–421.
99. J.R. Hagler and S.E. Naranjo, *Determining the frequency of heteropteran predation on sweetpotato whitefly and pink bollworm using multiple ELISAs*, Entomol. Exp. Appl. 72 (1994), pp. 63–70.
100. J.R. Hagler and S.E. Naranjo, *Qualitative survey of two Coleopteran predators of* Bemisia tabaci *(Homoptera, Aleyrodidae) and* Pectinophora gossypiella *(Lepidoptera, Gelechiidae) using a multiple prey gut content ELISA*, Environ. Entomol. 23 (1994), pp. 193–197.
101. J.R. Hagler and S.E. Naranjo, *Use of a gut content ELISA to detect whitefly predator feeding activity after field exposure to different insecticide treatments*, Biocontrol Sci. Technol. 15 (2005), pp. 321–339.
102. S.E. Naranjo, *Establishment and impact of exotic Aphelinid parasitoids in Arizona: A life table approach*, J. Insect Sci. 8 (2008), p. 36.
103. S.E. Naranjo and P.C. Ellsworth, *Mortality dynamics and population regulation in* Bemisia tabaci, Entomol. Exp. Appl. 116 (2005), pp. 93–108.
104. S.E. Naranjo and P.C. Ellsworth, *The contribution of conservation biological control to integrated control of* Bemisia tabaci *in cotton,* Biol. Control 51 (2009), pp. 458–470.
105. S.E. Naranjo, P.C. Ellsworth, and L. Cañas, *Mortality and populations dynamics of* Bemisia tabaci *within a multi-crop system*, Proceedings of the Third International Symposium on Biological Control of Arthropods, P.G. Mason, D.R. Gillespie, and C.D. Vincent eds., Christchurch, New Zealand, 2009, pp. 202–207.
106. P.C. Ellsworth, J.W. Diehl, I.W. Kirk, and T.J. Henneberry, Whitefly growth regulators: Large-scale evaluation, (1997), pp. 279–293. Available at http://arizona.openrepository.com/arizona/bitstream/10150/211092/1/370108-279-293.pdf.

107. P. Asiimwe, *Relative influence of plant quality and natural enemies on population dynamics of* Bemisia tabaci *and* Lygus hesperus *in cotton*. Ph.D. diss., University of Arizona, Tucson, AZ.
108. P. Ellsworth, J. Diehl, T. Dennehy, and S. Naranjo, *Sampling sweetpotato whiteflies in cotton*, IPM Series No. 2, Univ. Arizona Coop. Ext. Publ. 194023 (1994), 2 pp.
109. P.C. Ellsworth, J.W. Diehl, and S.E. Naranjo, *Sampling sweetpotato whitefly nymphs in cotton*, Cooperative Extension, IPM Series No. 6, University of Arizona (1996). Available at http://ag.arizona.edu/crops/cotton/insects/wf/ipm6.html.
110. S.E. Naranjo, J.R. Hagler, and P.C. Ellsworth, *Improved conservation of natural enemies with selective management systems for* Bemisia tabaci *(Homoptera: Aleyrodidae) in cotton*, Biocontrol Sci. Technol. 13 (2003), pp. 571–587.
111. F. Gould, G.G. Kennedy, and M.T. Johnson, *Effects of natural enemies on the rate of herbivore adaptation to resistant host plants,* Entomol. Exp. Appl. 58 (1991), pp. 1–14.
112. S. Arpaia, F. Gould, and G. Kennedy, *Potential impact of* Coleomegilla maculata *predation on adaptation of Leptinotarsa decemlineata to Bt-transgenic potatoes*, Entomol. Exp. Appl. 82 (1997), pp. 91–100.
113. M.T. Johnson, F. Gould, and G.G. Kennedy, *Effect of natural enemies on relative fitness of* Heliothis virescens *genotypes adapted and not adapted to resistant host plants*, Entomol. Exp. Appl. 83 (1997), pp. 219–230.
114. N. Mallampalli, F. Gould, and P. Barbosa, *Predation of Colorado potato beetle eggs by a polyphagous ladybeetle in the presence of alternate prey: Potential impact on resistance evolution*, Entomol. Exp. Appl. 114 (2005), pp. 47–54.
115. G. Head and C. Savinelli, *Adapting insect resistance management programs to local needs*, in *Insect Resistance management: Biology, Economics, and Prediction*, D.W. Onstad ed., Academic Press, London, U.K., 2008, pp. 89–106.
116. T.J. Dennehy, B.A. DeGain, V.S. Harpold, J.K. Brown, S. Morin, J.A. Fabrick, F.J. Byrne, and R.L. Nichols, *New challenges to management of whitefly resistance to insecticides in Arizona*, in *1999 Vegetable Report, Series P-144*, D. Byrne ed., University of Arizona, College of Agriculture, Tucson, AZ (2006). Available at http://cals.arizona.edu/pubs/crops/az1382/az1382_2.pdf.
117. P.C. Ellsworth, L. Brown, A. Fournier, and T.D. Smith, Arizona *cotton insect losses*, University of Arizona, College of Agriculture and Life Sciences, Cooperative Extension, Tucson, AZ (2012). Available at http://cals.arizona.edu/crops/cotton/insects/cil/cil.html.
118. R. Brazzle, K.M. Heinz, and M.C. Parrella, *Multivariate approach to identifying patterns of* Bemisia argentifolii *(Homoptera: Aleyrodidae) infesting cotton*, Environ. Entomol. 26 (1997), pp. 995–1003.
119. N. Byrne, R.J. Rathman, T.V. Orum, and J.C. Palumbo, *Localized migration and dispersal by the sweet potato whitefly,* Bemisia tabaci, Oecologia, 105 (1996), pp. 320–328.
120. J.L. Blackmer and D.N. Byrne, *Flight behavior of* Bemisia tabaci *in a vertical flight chamber: Effect of time of day, sex, age and host quality*, Physiol. Entomol. 18 (1993), pp. 223–232.
121. Y.-B. Liu and B. E. Tabashnik, *Experimental evidence that refuges delay insect adaptation to* Bacillus thuringiensis, Proc. R. Soc. Lond. B 264 (1997), pp. 605–610.
122. J.-Z. Zhao, J. Cao, H.L. Collins, S.L. Bates, R.T. Roush, E.D. Earle, and A.M. Shelton, *Concurrent use of transgenic plants expressing a single and two* Bacillus thuringiensis *genes speeds insect adaptation to pyramided plants*, Proc. Natl. Acad. Sci. USA. 102 (2005), pp. 8426–8430.

7 Population Dynamics Models for Assessing the Endocrine Disruption Potential of Juvenile Hormone Analogues on Nontarget Species

James Devillers and Hugo Devillers

CONTENTS

ABSTRACT

A number of xenobiotics released into the environment have the potential to disturb the normal functioning of the endocrine system of invertebrates and vertebrates. The juvenile hormone analogues have been designed specifically for such an activity against insect pest populations and for vector control. Unfortunately, their adverse effects are also observed on nontarget species. The goal of this chapter was to review the different population dynamics models derived for estimating the endocrine disruption potential of juvenoids on nontarget aquatic and terrestrial organisms. Advantages and limitations of the different modeling approaches were discussed.

KEYWORDS

Endocrine disruptor, Nontarget species, Deterministic model, Stochastic model, Honey bee

7.1 INTRODUCTION

In both invertebrates and vertebrates, endocrine signaling is involved in pivotal physiological functions such as reproduction, embryo development, and growth [1]. While the basic endocrine strategy to regulate biological processes has been widely conserved throughout evolution, specific components of the endocrine system used in the various systematic groups have undergone significant evolutionary divergence resulting in distinct differences between taxa [2]. This is especially true for invertebrates exhibiting a wide range of different chemical signaling systems, with some of them which are highly specific, while others are not far from those found in vertebrates [3]. The endocrine systems of invertebrates generally regulate the same processes that are found in vertebrates, but they present a higher diversity.

In the last 60 years, a wide range of xenobiotics have been shown to adversely affect wildlife at both the individual and population level by altering the normal functioning of their reproductive, growth, and immune systems. Thus, for example, until the late 1950s, sparrowhawks (*Accipiter nisus* L.) were common and widespread in Britain, but they suddenly showed a marked decline in numbers, almost disappearing from some districts. Their population crash followed the widespread introduction of pesticides, namely, aldrin, dieldrin, and DDT [4]. Such declines in bird populations were also observed in other countries where hydrophobic organochlorine pesticides were commonly used [4–6]. Another well-known example of endocrine disruption in wildlife, also discovered accidentally, deals with the effect of tributyltin (TBT) on the mollusks. While neogastropod mollusks are gonochoristic (i.e., sexes are separate), in 1970, Blaber [7] observed that many females of dog whelks (*Nucella lapillus*) had a penis-like structure behind the right tentacle. Shortly thereafter, Smith [8] reported similar reproductive abnormalities in the American mud snail (*Ilyanassa obsoleta*, formerly *Nassarius obsoletus*) on the Connecticut coast and coined the term imposex to describe this superimposition of male characters onto females [9]. At this stage, no link had been made with pollution but 10 years later, Smith [10,11] showed that levels of imposex in *I. obsoleta* were elevated close to marinas where the mollusks were exposed to antifouling paints including TBT. These chemicals that are able to disturb the normal physiology and endocrinology of organisms are called endocrine disruptors. An endocrine-disrupting chemical (EDC) is an exogenous substance or mixture that alters the functions of the endocrine system and consequently causes adverse health effects in an intact organism, its progeny or (sub)population [12]. It is obvious that the juvenile hormone analogues belong to EDCs.

There is a considerable body of observational and experimental evidence of individual-level effects produced by exposure of invertebrates to EDCs including juvenile hormone analogues [3,7–11,13–21]. However, to fully understand the ecological impacts of EDCs on the ecosystems, these adverse effects must be obligatorily analyzed at the population level. This can be done through the use of models specifically designed for such a task. A mathematical model is a simplified representation of a system. Generally, a model has an information input, an information processor, and an output of expected result(s). Its construction requires making some assumptions about the essential structure of the studied system and about the relationships existing between its constitutive elements. This means that to make a model, certain compromises have to be made with reality. In other words, a model will be never able to fully reproduce a real-world system [22]. Various modeling methods exist to extrapolate adverse effects from individual to population levels [23–25]. In this context, an attempt was made to review the different population dynamics models aiming at simulating the endocrine disruption potential of juvenile hormone analogues on nontarget aquatic and terrestrial invertebrates.

7.2 AQUATIC COMPARTMENT

7.2.1 Simple Equation Model

By virtue of their nature and position between marine and terrestrial environments, estuaries are the focal point for a wide variety of human activities, and consequently they suffer from an intensive pollution [26,27]. Hill [28] studied the effects of *S*-methoprene on *Eurypanopeus depressus*, a xanthid mud crab susceptible to be exposed in an estuarine setting during surface water runoff or pesticide-spraying events. Randomly selected larvae were exposed in a dynamic, flow-through system with nominal concentrations of 2, 4, 8, 16, 32, and 64 µg/L of methoprene and compared against control exposures without juvenile hormone analogue. Stage-specific survival and intermolt duration were monitored for each of the four zoeal stages (Z^{I} to Z^{IV}) through the metamorphosis of the megalopae (Meg) to the postlarval first crab stage (C-1). Subsamples of each stage and the surviving crabs were weighed and dried to determine ash weights and carbon, hydrogen, and nitrogen composition. The authors showed that exposure to 32 and 64 µg/L highly increased larval mortality (from hatch to C-1), 63% and 75%, respectively. Exposure to 2 µg/L led to a reduction in mean dry weight of 28.3%, the greatest decrease being obtained at 64 µg/L with a reduction of 35.2% in dry weights compared to the control. This effect is non-negligible because it is well-known that smaller-sized crabs are subject to increased mortality due to predation and their reproduction potential is also lower [29]. Altered patterns of carbohydrate, lipid, and protein use and storage were observed. Energy content (joules/individual) of all larval stages was reduced. Megalopa showed significant reductions for all the concentrations, but the most important decrease in mean energy content equaled 71% (compared with the control) at 32 µg/L. The effect persisted into the juvenile crab stage with energy reduced from 54% to 74%. At the population level, this can impact fecundity and hatching success. Hill [28] used an exponential decay model (Equation 7.1) to characterize the evolution of the population (N) of *E. depressus*. Z is the total mortality rate, which is composed of the instantaneous mortality rate due to fishing efforts and the instantaneous rate of mortality due to all the other factors such as predation and disease:

$$N_t = N_0 e^{-Zt} \tag{7.1}$$

An increase of annual mortality rate of 58% as observed in *E. depressus* by the author is equivalent to a drop in Z of 0.55. Using Equation 7.1, a decrease of 40% in the population is expected after one time interval and a decrease of 78% after three time intervals. Consequences on shrimp fishery production in Cuba were extrapolated from simulation results [28].

7.2.2 Matrix Model

The effects of methoprene on the survival and reproduction of *Americamysis bahia* were determined [30] by exposing the organisms to nominal concentrations of 0, 4, 8, 16, 32, 62, and 125 μg/L, which were selected to test mysid responses over a range of sublethal and lethal concentrations. Fifteen juveniles <24 h old were placed into each of the three replicates of each treatment concentration. Methoprene was added into the vessels with a time step of 3 min. Survival was observed daily throughout the life cycle. Ovigerous females were isolated and paired with mature males. The number of young released by each female was recorded daily. Stage-specific survival and reproduction data were used as inputs in the matrix models. Only data for concentrations ≤62 μg methoprene/L were used because at 125 μg/L, total mortality occurred within 4 days of exposure. The life cycle of *A. bahia* was divided into the seven following stages: (1) early juveniles; <24 h old; (2) juveniles of 1–4 days old; (3) advanced juveniles of 5–10 days old; (4) young adults of 11–16 days old; (5) early breeders of 17–20 days old; (6) intermediate breeders of 21–24 days old; and (7) late breeders of 25–28 days old. The three last stages allowed to better identify the potential effects of delayed brood release induced by methoprene on the population dynamics.

For each methoprene concentration (x_i), the following density-independent, deterministic population projection matrix ($\boldsymbol{A}_{(x)}$) was developed [30]:

$$\boldsymbol{A}_{(x)} = \begin{bmatrix} 0 & 0 & 0 & 0 & F_{20(x)} & F_{24(x)} & F_{28(x)} \\ G_{1(x)} & P_{4(x)} & 0 & 0 & 0 & 0 & 0 \\ 0 & G_{4(x)} & P_{10(x)} & 0 & 0 & 0 & 0 \\ 0 & 0 & G_{10(x)} & P_{16(x)} & 0 & 0 & 0 \\ 0 & 0 & 0 & G_{16(x)} & P_{20(x)} & 0 & 0 \\ 0 & 0 & 0 & 0 & G_{20(x)} & P_{24(x)} & 0 \\ 0 & 0 & 0 & 0 & 0 & G_{24(x)} & P_{28(x)} \end{bmatrix}$$

In $\boldsymbol{A}_{(x)}$, $P_{i(x)}$ is the probability of surviving and remaining in stage i, $G_{i(x)}$ is the probability of an individual growing from stage i to stage $i+1$, and $F_{i(x)}$ is the reproductive output of female in stage i. $P_{i(x)}$, $G_{i(x)}$, and $F_{i(x)}$ values, later generically denoted a_{ij}, were calculated from the experimental data using Equations 7.2 to 7.4:

$$P_{i(x)} = \sigma_{i(x)}[1-\gamma_{i(x)}] \tag{7.2}$$

$$G_{i(x)} = \sigma_{i(x)}\gamma_{i(x)} \tag{7.3}$$

$$F_{i(x)} = \sigma_{i(x)}^{1/2}\frac{(1+P_i)m_i + G_i m_{i+1}}{2} \tag{7.4}$$

where

$\sigma_{i(x)}$ is the stage-specific survival probability
$\gamma_{i(x)}$ is the transition probability
m_i is the average number of offspring per female in stage i

The lower order vital rates, $\sigma_{i(x)}$ and $\gamma_{i(x)}$, were calculated as follows:

$$\sigma_{i(x)} = 1 - \frac{\text{Number of deaths in stage } i}{\text{Total individual days in stage } i} \tag{7.5}$$

$$\gamma_{i(x)} = \frac{\left[\sigma_{i(x)}/\lambda_{(x)}\right]^{Ti} - \left[\sigma_{i(x)}/\lambda_{(x)}\right]^{Ti-1}}{\left[\sigma_{i(x)}/\lambda_{(x)}\right]^{Ti} - 1} \tag{7.6}$$

In Equation 7.6, T is the duration of the stage interval and $\lambda_{(x)}$ is the population growth rate calculated as the dominant eigenvalue of the matrix $\boldsymbol{A}_{(x)}$. An iterative approach was used to calculate $\gamma_{i(x)}$ values by setting an initial value of $\lambda_{(x)}$ to 1.1. Generally a $\lambda_{(x)}$ value of 1 is used and a population with a $\lambda_{(x)} < 1$ will decline, whereas those with $\lambda_{(x)} > 1$ will increase [31]. Here, the value of 1.1 was selected to avoid mathematical complication where the denominator of Equation 7.6 was reduced to zero. The resulting values of $\gamma_{i(x)}$ were used to estimate the entries of a second matrix used to produce a second value of $\lambda_{(x)}$, and this was repeated till stabilization of $\lambda_{(x)}$ values.

To evaluate the contribution of each vital rate, a_{ij}, to λ for each concentration x, the sensitivity (s_{ij}) of λ to changes in a_{ij} was calculated for each concentration from Equation 7.7 [30]:

$$s_{ij} = \frac{\partial\lambda}{\partial a_{ij}} = \frac{v_i w_j}{\langle vw\rangle} \tag{7.7}$$

In Equation 7.7, w, the stable age distribution, is the right eigenvector of the matrix, v is the corresponding left eigenvector representing the reproductive value of each stage, and ⟨vw⟩ is the scalar product. An elasticity analysis was used to determine which stage-specific individual-level response showed the largest influence on population growth rate λ.

A sensitivity matrix derived from Equation 7.7 determined the change in λ if a change occurred to a_{ij}. An elasticity matrix for each methoprene concentration established the relative contribution of each vital rate to λ and was related to sensitivity (Equation 7.8):

$$e_{ij} = \left(\frac{a_{ij}}{\lambda}\right)\left(\frac{\partial\lambda}{\partial a_{ij}}\right) \tag{7.8}$$

A decomposition method was used to estimate how each vital rate changed with the concentrations of methoprene and the relative contribution of these impacts on the changes in λ. The decomposition analysis was based on the sensitivity of λ to the vital rates $\partial\lambda/\partial a_{ij}$ and the change of vital rate with methoprene concentration, $(\partial a_{ij(x)}/\partial x)$. The latter expression was obtained as the slope of a nonparametric regression using loess smoother. The smoothing parameter was selected by cross validation to minimize the prediction error. The decomposition of λ change with methoprene concentration, computed with Equation 7.9, is a function of the sensitivity value of each vital rate and magnitude of its change:

$$\frac{\Delta\lambda}{\Delta x} = \Sigma\left(\frac{\partial\lambda_{(x)}}{\partial a_{ij(x)}}\right)\left(\frac{\partial a_{ij(x)}}{\partial x}\right) \tag{7.9}$$

The stage-structured model showed a population decline of about 5% with an increase in the concentration of methoprene from 0 to 62 μg/L, the population growth rate (λ) always remaining >1.

The nonparametric regression of the effects of methoprene concentrations on age-specific survival (P_i), growth (G_i), and fecundity (F_i) was affected at different concentrations. At $\leq$ 16 μg/L, P_i of early juveniles through early breeders were less than control populations, the largest decrease being observed in young adults of 11–16 days old. P_i of breeders of 17–28 days old were higher than controls at <16 μg/L but decreased between 16 and 31 μg/L and then stabilized till 62 μg/L. Limited effects were observed on G_i at the highest concentrations. The most important effects were observed with F_i. A delay in reproduction was not observed; but rather, a decrease in the reproduction appeared in all adult stages at 4 and 8 μg/L. Young production then increased at 16 μg/L in all stages. At the highest concentration, the greatest impact was observed on the early breeders of 17–20 days old [30].

Despite the high value of P_i, measured by the elasticity analysis, the decomposition analysis showed that neither P_i nor G_i contributed significantly to the change in λ at the opposite of F_i.

Thus, the modeling study showed that population growth rate (λ) of *A. bahia* generally declined with the increase in tested methoprene concentrations. This leads to anticipate a population-level effect with a risk of population extinction if the concentrations are too high [32]. This might be a realistic scenario as McKenney and Celestial [33] showed that 125 μg/L of methoprene led to 100% of mortality of mysids within 4 days of exposure.

It is noteworthy that in another study, the modeling results obtained with methoprene were compared to those recorded with endosulfan, tribufos, fenthion, thiobencarb, and silver nitrate [34].

7.3 TERRESTRIAL COMPARTMENT

7.3.1 Compartment Model

Honey bees are beneficial insects playing a key role in pollinating wild and crop plants. Unfortunately, during their foraging activity they can be exposed to insecticides and other xenobiotics. The members of the colony can also be poisoned indirectly by contaminated food brought back to the hive by the foragers. As a result, there is a need to estimate the adverse effects of chemicals against all the hive inhabitants [35].

Thompson et al. [17] studied the effects of fenoxycarb on the honey bee colony development, queen rearing, and drone sperm production. Test solution was prepared by diluting fenoxycarb with 50% w/v sucrose to a rate equivalent to its maximum application rate on crops (0.6 kg Insegar/200 L = 750 mg AI/L). On day zero, a control dosage was performed. Cells were marked to follow the evolution of the eggs as well as the other developmental stages. Colonies were fed with fenoxycarb in sucrose, while another one was only fed with sucrose to be used as control. Doses were administered via a glass beaker containing 500 mL treated or control sucrose placed within the brood chamber. After 1 week, the beaker was removed and the remaining volume was recorded. A mean of 400 mL was consumed by the five colonies treated with fenoxycarb. The calculated dose per brood cell was estimated to be 50 μg fenoxycarb/brood cell. An overall mean replacement/removal of 24% for the eggs marked over the 5 week period was observed in the control, while it was of 46% in the treated colonies with fenoxycarb. The maximum replacement was 30% in week 1 for the controls versus 60% for fenoxycarb. Replacement of all brood stages was assessed in the colonies weekly after treatment. Control replacement of all stages of brood was 5% versus 21% one week after fenoxycarb treatment. Fenoxycarb did not affect sperm production. Colonies treated with fenoxycarb declined during the season earlier than the control and started the season slower. One colony did not survive over the winter. The fenoxycarb treatment resulted in significantly lower numbers of bees in the month after treatment and in the following year with significantly lower numbers of bees in May than in the control. The same trend was observed with the brood. The number of queens, which successfully mated and laid eggs, was 74% in the control and 0% (0/23) in the fenoxycarb-treated colonies. In addition, 17/23 queens were present but showed virgin queen characteristics (e.g., small abdomen) suggesting they had not been mated. Studies in the second year included placing the mated queens into small colonies to record their rate of egg laying. There were no eggs laid by any queens treated with fenoxycarb [17]. These experimental data were used to run a population dynamics model [17, 36–38] simulating egg laying in the colony, worker bee brood (eggs/larvae/pupae), drone bee brood, and an adult honey bee population consisting of younger nurse bees and older forager bees. Adult honey bee workers normally change over from brood-rearing work to foraging work at about 21 days old.

The number of adult bees, A, at the beginning of Julian day $t+1$ was described by Equation 7.10 [36]:

$$A_{t+1} = A_t + e_{t-21} - A_t Q_t \tag{7.10}$$

where

e_{t-21} is the number of eggs laid 21 days previously
Q_t is the mortality parameter at time t

The rate at which the queen lays eggs is dependent on season and various factors such as pollen and nectar availability. A simplified representation was produced with a curve drawn to represent the seasonal factors and hence the egg-laying rate. This seasonal curve was initially based on the proportion of days in each month considered as suitable for bee foraging. The number of eggs laid on a particular day was defined by Equation 7.11 [36]:

$$e_t = \mu S_t \quad \text{with} \quad 0 \le S_t \le 1 \tag{7.11}$$

In Equation 7.11, μ is the maximum worker egg-laying rate of the queen. In the initial model [36], it was fixed to 1500. S_t is the seasonal parameter value on day t.

A feedback mechanism was introduced in the model so that the amount of reared brood was limited by the number of nurse bees available as this occurs several weeks after a period of high brood mortality. The mortality rate of adult bees increases in summer when they forage extensively. Consequently, a mortality curve was created, which applied higher mortality rates during summer. Like this, adult bee populations did not grow unrealistically high. The autumn and winter mortality rates were set to levels that resulted in population reaching the same size as 1 year before.

Because in field experiments fenoxycarb induced about 50% brood mortality over the first 2 weeks after treatment, Thompson et al. [17,38,39] ran their model by reducing the egg-laying rate by 50% for 2 weeks following an application, keeping the larvae/nurse (*L*/*N*) ratio at 1.5. This resulted in approximately 20% loss of productivity (number of adult bees) when applied in June. Egg mortality of 100% for 2 weeks led to a maximum loss of productivity of 40%–45% when applied in May/June.

Heylen et al. [40] showed that fenoxycarb (tested at 100 ppm) highly impacted the bee hypopharyngeal glands. Indeed, after only 7 days, fenoxycarb-treated glands displayed features typical of the onset of foraging in older bees. Thus, it appears likely that exposure to fenoxycarb may induce precocious foraging in the exposed bees and reduce the number of nurses available to rear brood. To simulate the impact of a reduction in the number of nurse bees, the model was adapted to reduce the number of larvae that could be reared by each nurse from 1.5 to 0.5. Action on the hypopharyngeal glands also reduces the lifespan of the contaminated bees.

The decreased lifespan was simulated by increasing the mortality rate of adult bees emerging 2.5 months after treatments because fenoxycarb can persist for 2.5 months [41] and its effects have been reported for 2–3 months after treatment [42].

With a 50% brood loss applied for 14 days and an *L/N* ratio of 1 for bees that emerged for 2.5 months after treatment, a 50% reduction in mean colony size was observed when fenoxycarb was applied in April. An *L/N* ratio of 0.5 resulted in a 75% reduction in mean colony size.

Decreased lifespan was simulated by increasing the mortality rate of the emerging bees. 1.3-, 1.5-, and 5-fold increases in mortality during 2.5 months after treatment were used with, in addition, 50% brood loss for 14 days after treatment and an *L/N* ratio of 1.5. The 1.3- and 1.5-fold increases in mortality decreased the mean colony size by a maximum of about 10% in May when overlaid on the 50% brood loss, and the fivefold increase decreased the colony size by an additional maximum of 40% in May [38].

Fenoxycarb applied between April and October reduced the winter size of the colony, the greater effects were observed following a June or an August application. Indeed, assuming an *L/N* ratio of 1.5 for brood-rearing capability, the reduction in winter size of a colony following a fenoxycarb application in June or August, with a brood mortality of 50% for 2 weeks, was about 8%. Fenoxycarb had little or no effect when the application was made between November and March.

When the sublethal effect of early switch to foraging was simulated by using an *L/N* ratio of 0.5, the winter colony size was reduced with an 80% loss when the application was made in May or June. Obviously, an *L/N* ratio of 1.0 induced a smaller effect, but the winter colony size was still reduced by over 50% following an application in April.

7.3.2 Agent-Based Model

Even if the compartment model designed by Wilkinson and Smith [36,37] and Thompson et al. [17,38,39] presents some interest, it is important to stress that the functioning of a bee colony is much more complex than the one encoded by the few equations computed by these authors. Permanently, thousands of bees simultaneously add behavioral elements that interact to form the collective behavior of the colony, which behaves like a superorganism. In a more formal way, we define a superorganism as a set of agents that interact each other to produce collective phenomena in which the individuals with limited possibilities are able to pool their resources to reach a goal beyond their individual capacities. The total resultant is greater than the sum of the possibilities of each agent acting individually. This definition is similar conceptually to agent-based modeling (ABM), which is a stochastic computational technique used to simulate the spatiotemporal actions and interactions of real-world entities, called agents, in an effort to extract their combined effect on the system as a whole. Both space and time are discretized in an agent-based model, giving these autonomous agents the ability to move and interact with other agents and their environment at each time step over a given duration. Simple behavioral rules govern the movement and interaction of each agent in an effort to re-create or predict more complex behaviors. Such a model attempts to simulate the emergence of complex phenomena that may not be apparent when simply considering individual entities [43–46]. In this context, an agent-based model called SimBeePop (version 1.04) [47] was designed for predicting the effects of xenobiotics on the hive inhabitants.

Briefly, SimBeePop simulates the population dynamics of a complete domestic honey bee colony, during a 2 year period or more, considering each category of bees, the structure of the hive, and seasonal variations. This agent-based model, written in Java, follows the conventional rules of ABMs. It is a discrete-time model that includes an environment (i.e., hive), agents (i.e., bees), and objects (i.e., food).

The simulation time step used in SimBeePop corresponds to one Julian day. One of the greatest advantages of ABM is the ease of applying variability over time. This variability includes seasonal effects, stochasticity, and potential perturbations that can be introduced by the user.

The environment of the model is the hive. It consists in a hive body and honey supers. Hive body is divided into frames including hexagonal cells. In SimBeePop, the honey supers are directly divided into cells. By default, SimBeePop considers a Dadant hive of 10 frames, with a capability of 80,000 cells [48]. Size and number of honey supers can be set by the user.

Agents of SimBeePop are the different bees of the colony. These agents can be grouped into five main categories, which are the brood, the queen, the worker bees, the males (drones), and the winter bees. Each brood agent is assigned in a cell of the environment and cannot move. A brood agent is successively an egg, a larva, and a pupa. There are two kinds of brood, worker brood and male brood, whose development lasts 21 and 24 days, respectively. The queen agent is a unique agent that is movable and responsible for egg laying. Worker bee agents are also movable; they perform the different tasks for the functioning of the hive. SimBeePop considers four distinct activities assigned to worker agents according to their age. These are the cleaning (≤3 days), nursing (<13 days), receiving/processing (≤21 days), and foraging (rest of its life) bees. A worker bee agent, whatever its task, can be active or inactive; its recruitment depends on the needs of the colony. Male bees are passive agents. Last, winter bee agents are a particular class of worker bee agents whose task division does not depend on age but only on the needs of the colony at the beginning and the end of the season. All these different kinds of agents are characterized by specific parameters including mortality rates, food needs, and labor efficiency. In SimBeePop, a particular attention has been paid for the setting of these parameters, based on a critical analysis of the literature (see [49] for details).

One of the great advantages of an ABM is the possibility to define explicitly life rules and interaction patterns for the different agents. This is called the model paradigm. In SimBeePop, the different rules that govern the global functioning of the bee colony have been integrated. It concerns, for example, the division of labor that depends both on the age of bees (polyethism) but also on the needs of the colony at each simulation step. The laying activity of the queen simultaneously depends on the season, nurse availability, and food stock. The foraging forces are linked to the need of the colony, the filling rate of the hive, and obviously the season.

Objects in SimBeePop concern the food. Three kinds of food objects are considered, namely, the honey, the pollen, and the nectar. Nectar and pollen are brought by forager bee agents. Nectar is consumed directly by all the agents of the colony or transformed into honey and stored by receiver/processor agents. This is the same process for the pollen, unlike it can be directly stored into cells. Honey and pollen stored in the hive can then be used to feed the colony if the foraging activity does not satisfy the needs in food of the population such as during winter. SimBeePop paradigm has been more thoroughly described in Devillers et al. [50].

SimBeePop allows users to apply a wide range of perturbations and variations on the normal population dynamics. Thanks to the properties of ABMs, these perturbations can affect either selected individuals, a specific group of agents, or the whole population. They can be applied punctually (e.g., 1 day), during a given period (e.g., weeks, months), or the whole simulation duration. This offers invaluable opportunities to test and extrapolate the effects of various chemicals on bee population dynamics both on short- and long-term periods.

The ABM was run from the data and modeling hypotheses selected by Thompson et al. [17,38,39]. Thus, for example, 50% and 100% brood mortality were applied 2 weeks after a treatment on the 1st of June and on the 1st of August, keeping a larvae/nurse ratio equivalent to the 1.5 ratio adopted by Thompson et al. [17,38,39] in this experiment. The obtained simulation results for 50 consecutive runs are displayed in Figure 7.1. Inspection of Figure 7.1a and b, dealing with June application, shows that 50% and 100% of brood mortality during 14 consecutive days lead to a rapid loss of productivity (number of adult bees) of about 40% and 70%, respectively. This is about 20% higher than the percentages obtained by Thompson et al. [17,38,39], but we assume that the difference of modeling approach, paradigm, and parameter setting could explain this variation in the percentages. However, while these authors did not mention an impact the following months, Figure 7.1a and b show that the adult bee population remains perturbed the months following fenoxycarb application. This is particularly true when 100% of brood mortality is applied (Figure 7.1b). In that case, a second important peak, even slightly higher than the first one, is observed in November. More importantly, Figure 7.1a shows that the adult population seems to rapidly recover its normal dynamics, while this is not the case with 100% of brood mortality (Figure 7.1b). The August application also leads to about the same percentages of adult losses, and the population is impacted the following months of the year (Figure 7.1c and d). A second peak, corresponding to a loss of 80%, is observed in November with 100% of brood mortality (Figure 7.1d). Thompson et al. [17,38,39] showed that an August application impacted less the adult bee population than an earlier application. Indeed, analysis of their results reveals in fact that an August application only induces about 5% of loss in the adult population with the scenario of 100% mortality for 14 days and even significantly less with 50% of morality [38]. This definitively does not correspond to the reality because August is always a critical period in the honey bee life. Even more surprising, their simulation results show that 50% of mortality for 2 weeks in May/June or in August impact similarly the winter bee population. In both situations, a decrease of less than 10% in the winter bee population was observed [39]. Again, this is not logical because if we consider that the winter bees generally start to appear at the end of August, beginning of September, a fenoxycarb application in August should impact more importantly this population rather than an application in May or June. This is what

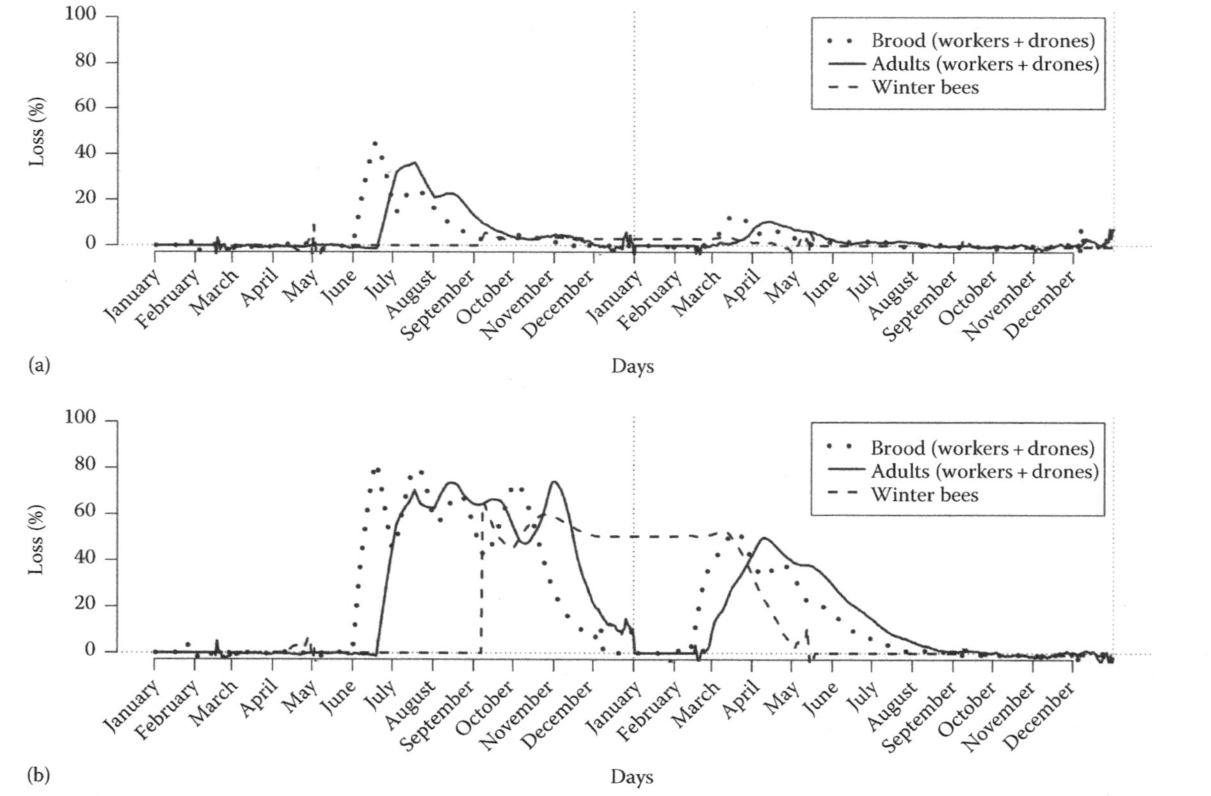

FIGURE 7.1 Effects of a fenoxycarb treatment applied on the 1st of June (a and b) and on the 1st of August (c and d) on brood, adult bee, and winter bee losses (in %) when 50% (a and c) and 100% (b and d) brood mortality are applied for 2 weeks after treatment.

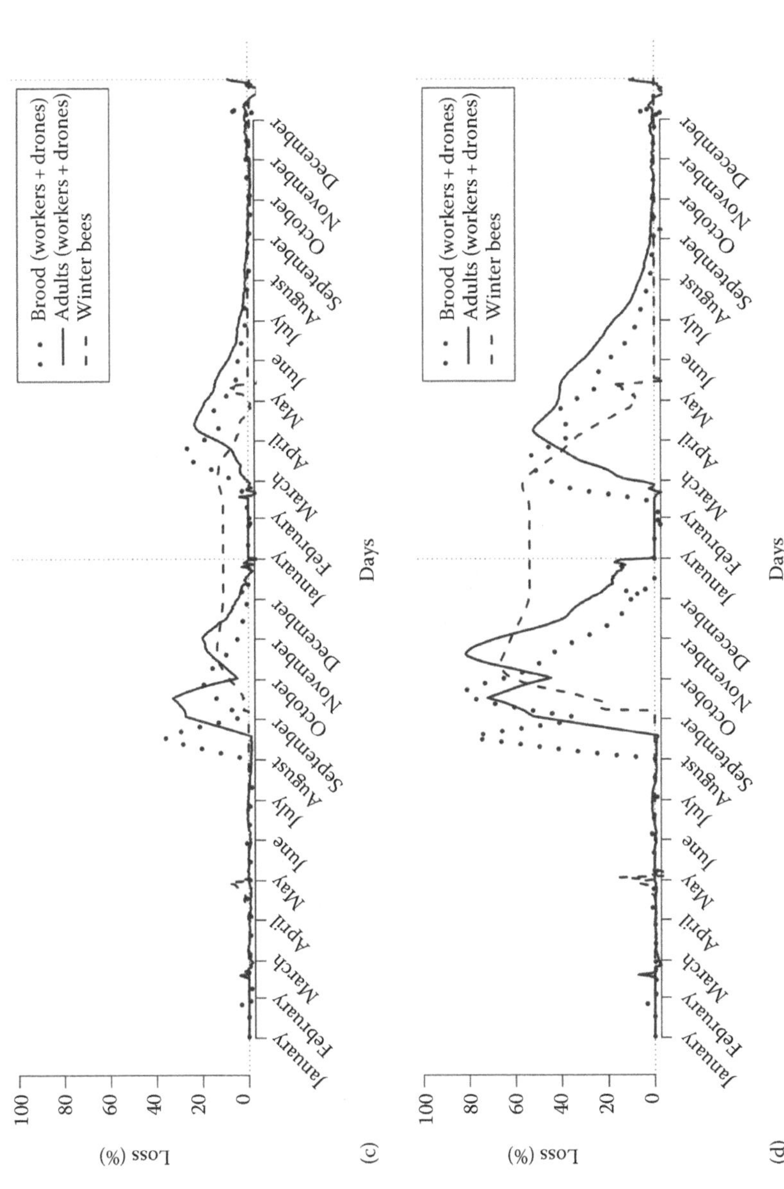

FIGURE 7.1 (continued) Effects of a fenoxycarb treatment applied on the 1st of June (a and b) and on the 1st of August (c and d) on brood, adult bee, and winter bee losses (in %) when 50% (a and c) and 100% (b and d) brood mortality are applied for 2 weeks after treatment.

we can be observed on Figure 7.1. In the same way, Thompson et al. [17,38,39] did not simulate the impacts on the hive inhabitants 1 year after the treatment, while this is of first importance. Figure 7.1 reveals that whatever the selected scenario, the hive population will be impacted 1 year later.

In another simulation exercise, Thompson et al. [17,38,39] studied the impact of an increase in the mortality rate of adult bees emerging for 2.5 months after a fenoxycarb treatment. The 1.3-, 1.5-, and 5-fold increases in mortality were tested always using an *L/N* ratio of 1.5 and 50% brood loss for 14 days after exposure. Such a scenario is of crucial interest with the juvenile hormone analogues. Unfortunately, the normal mortality rate of the emerging bees in their model being unknown, it is impossible to estimate at what percentage corresponds 1.3-, 1.5-, and 5-fold increases in mortality. Nevertheless, with the ABM model we have tested the effects of a survival rate of 78% at emergence during 2.5 months after a treatment on the 1st of May or August. The simulation results are displayed on Figure 7.2. Again, a strong divergence exists between our results and those of Thompson et al. [38] as regards the August application. Indeed, whatever their scenario, an August application induces very limited effects on the adult population, while it is definitive not the case with the ABM (Figure 7.2b). As expected, the brood and winter bees are also affected inducing a strong impact the next year (Figure 7.2b). Adverse effects on the second year are also observed with the May application (Figure 7.2a).

Our ABM model provides more realistic results than the compartment model proposed by Wilkinson and Smith [36,37] and Thompson et al. [17,38,39]. The reason mainly relies on the radically different modeling paradigm.

A compartment model is rooted on a top-down approach. To build the model, the modeler groups individuals with similar characteristics into compartments and specifies rules of how individuals flow from one compartment to another. This modeling approach approximates the average behavior of a system as a whole. Such a model fails to correctly capture the different local relationships between the system components that can be the cause of high-level behaviors.

At the opposite, an agent-based model uses a bottom-up or individual-level approach by specifying the rules that govern the behavior of individuals and that allow the overall behavior of the system to emerge from the interactions of those individuals. Like compartment models, the ABMs group individuals with similar characteristics. However, the transmission or flow between states is determined by the behavior of the individual, not the group as a whole. Such a modeling approach is particularly suited for encoding the characteristics of a honey bee population.

7.4 CONCLUSION

The complexity of the relationships between the adverse effects of a chemical observed at the individual level with the modifications of its life history traits and the subsequent possible impacts at the population level makes necessary the use of modeling approaches especially when the ecotoxicological outcomes are expected to be predictive instead of only descriptive. Different modeling approaches, varying in complexity, exist to extrapolate from the individuals to the population in an ecotoxicological context. Thus, in this chapter, different types of population

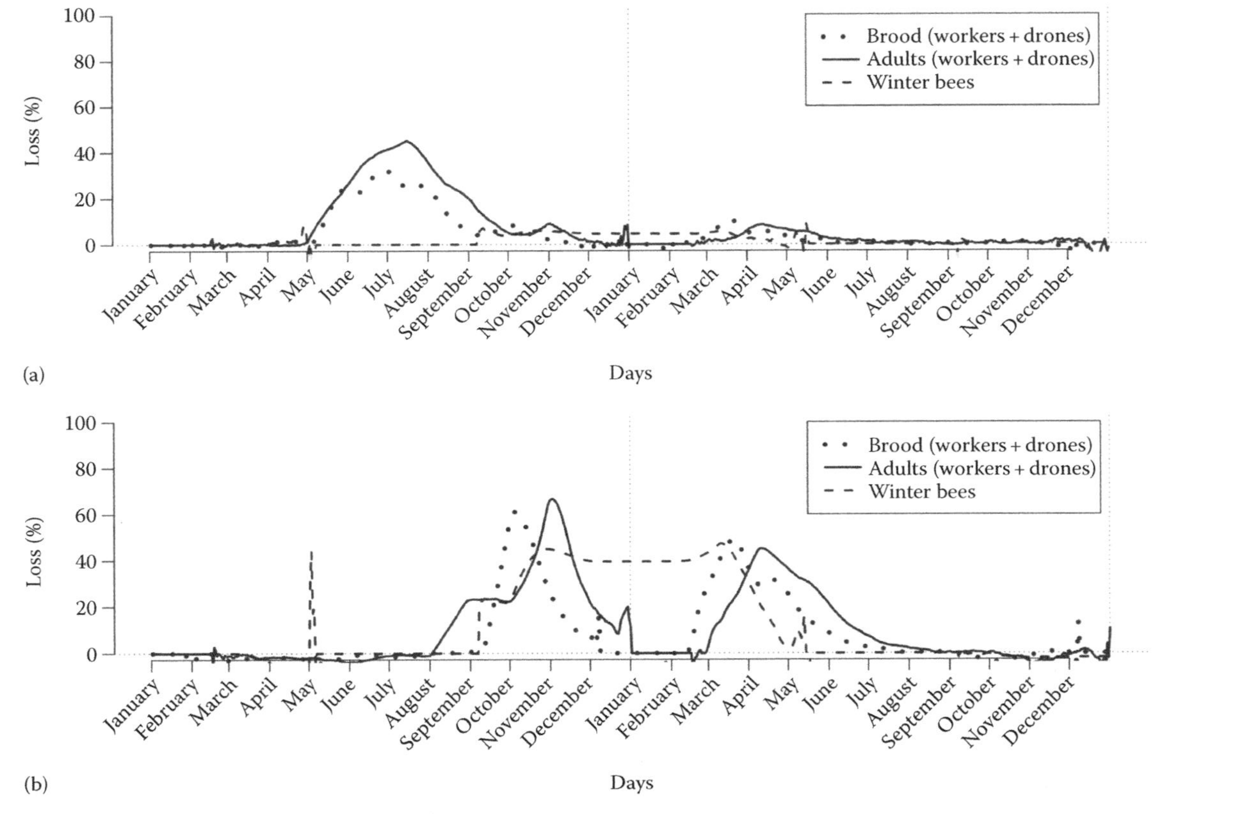

FIGURE 7.2 Effect on brood, adult bee, and winter bee losses (in %) of a survival rate of 78% at emergence during 2.5 months after a treatment on the 1st of May (a) or August (b).

dynamics models aiming at predicting the endocrine disruption potential of juvenile hormone analogues on different aquatic and terrestrial organisms have been discussed. The question to answer is what is the most interesting modeling approach? Even if programming skill is not a limitation, the answer cannot be definitive. To work around the problem, the deterministic and stochastic modeling approaches can be conveniently opposed to summarize the pros and cons of each of them. The former category, including the matrix models, can be used to simulate incremental changes in survivorship, fecundity, or other life history traits of specific individuals from their birth to death and ultimately can be used to extrapolate effects at the population level. They are relatively easy to parameterize and use, but due to their relative simplicity, it is preferable to use them in an exploratory perspective to obtain general trends.

Stochastic models, such as agent-based models, can account for the continuous development and interaction of individuals throughout their lifetime and within their population that can be related to environmental parameters. They give more realistic and robust outcomes but need much more input data than the deterministic models. Endocrine disruption being by essence a very complex phenomenon, we assume that the agent-based models are the most suited to simulate the ecotoxicological impacts of juvenoids on the nontarget species found in the aquatic and terrestrial ecosystems.

ACKNOWLEDGMENT

The financial support from the French Ministry of Ecology, Sustainable Development, Transport and Housing (MEDDTL) is gratefully acknowledged (PNRPE program).

REFERENCES

1. J. Lintelmann, A. Katayama, N. Kurihara, L. Shore, and A. Wenzel, *Endocrine disruptors in the environment*, Pure Appl. Chem. 75 (2003), pp. 631–681.
2. J.A. McLachlan, *Environmental signaling: What embryos and evolution teach us about endocrine disrupting chemicals*, Endocr. Rev. 22 (2001), pp. 319–341.
3. J. Oehlmann and U. Schulte-Oehlmann, *Endocrine disruption in invertebrates*, Pure Appl. Chem. 75 (2003), pp. 2207–2218.
4. I. Newton and I. Wyllie, *Recovery of a sparrowhawk population in relation to declining pesticide contamination*, J. Appl. Ecol. 29 (1992), pp. 476–484.
5. P. Olsen, P. Fuller, and T.G. Marples, *Pesticide-related eggshell thinning in Australian raptors*, Emu 93 (1993), pp. 1–11.
6. C.M. Markey, B.S. Rubin, A.M. Soto, and C. Sonnenschein, *Endocrine disruptors: From Wingspread to environmental developmental biology*, J. Steroid Biochem. Mol. Biol. 83 (2003), pp. 235–244.
7. S.J.M. Blaber, *The occurrence of a penis-like outgrowth behind the right tentacle in spent females of* Nucella lapillus (L.), Proc. Malacol. Soc. Lond. 39 (1970), pp. 231–233.
8. B.S. Smith, *Sexuality in the American mud snail*, Nassarius obsoletus *Say*, Proc. Malacol. Soc. Lond. 39 (1971), pp. 377–378.
9. P. Matthiessen and P.E. Gibbs, *Critical appraisal of the evidence for tributyltin-mediated endocrine disruption in mollusks*, Environ. Toxicol. Chem. 17 (1998), pp. 37–43.

10. B.S. Smith, *Reproductive anomalies in stenoglossan snails related to pollution from marinas*, J. Appl. Toxicol. 1 (1981), pp. 15–21.
11. B.S. Smith, *Male characteristics on female mud snails caused by antifouling bottom paints*, J. Appl. Toxicol. 1 (1981), pp. 22–25.
12. IPCS, *Global assessment of the state-of-the-science of endocrine disruptors*, in International Program on Chemical Safety, IPCS, 2002.
13. T. Soin and G. Smagghe, *Endocrine disruption in aquatic insects: A review*, Ecotoxicology 16 (2007), pp. 83–93.
14. L. Lagadic, M.A. Coutellec, and T. Caquet, *Endocrine disruption in aquatic pulmonate molluscs: Few evidences, many challenges*, Ecotoxicology 16 (2007), pp. 45–59.
15. S. Höss and L. Weltje, *Endocrine disruption in nematodes: Effects and mechanisms*, Ecotoxicology 16 (2007), pp. 15–28.
16. S. Campiche, K. Becker-van Slooten, C. Ridreau, and J. Tarradellas, *Effects of insect growth regulators on the nontarget soil arthropod* Folsomia candida (Collembola), Ecotoxicol. Environ. Safety 63 (2006), pp. 216–225.
17. H.M. Thompson, S. Wilkins, A.H. Battersby, R.J. Waite, and D. Wilkinson, *The Effects of four insect growth-regulating (IGR) insecticides on honeybee (*Apis mellifera *L.) colony development, queen rearing and drone sperm production*, Ecotoxicology 14 (2005), pp. 757–769.
18. H.Y. Wang, A.W. Olmstead, H. Li, and G.A. LeBlanc, *The screening of chemicals for juvenoid-related endocrine activity using the water flea* Daphnia magna, Aquat. Toxicol. 74 (2005), pp. 193–204.
19. A.E. Hershey, L. Shannon, R. Axler, C. Ernst, and P. Mickelson, *Effects of methoprene and* Bti (Bacillus thuringiensis var. israelensis) *on non-target insects*, Hydrobiologia 308 (1995), pp. 219–227.
20. J.K. Peterson, D.R. Kashian, and S.I. Dodson, *Methoprene and 20-OH-ecdysone affect male production in* Daphnia pulex, Environ. Toxicol. Chem. 20 (2001), pp. 582–588.
21. R.B. Forward and J.D. Costlow, *Sublethal effects of insect growth regulators upon crab larval behavior*, Water Air Soil Pollut. 9 (1978), pp. 227–238.
22. J. Devillers, *In silico methods for modeling endocrine disruption*, in *Endocrine Disruption Modeling*, J. Devillers, ed., CRC Press, Boca Raton, FL, 2009, pp. 1–9.
23. W.S.C. Gurney, *Modeling the demographic effects of endocrine disruptors*, Environ. Health Perspect. 114 (2006), pp. 122–126.
24. A.R. Brown, P.F. Robinson, A.M. Riddle, and G.H. Panter, *Population dynamics modeling. A tool for environmental risk assessment of endocrine disrupting chemicals*, in *Endocrine Disruption Modeling*, J. Devillers, ed., CRC Press, Boca Raton, FL, 2009, pp. 47–82.
25. S. Charles, E. Billoir, C. Lopes, and A. Chaumot, *Matrix population models as relevant modeling tools in ecotoxicology*, in *Ecotoxicology Modeling*, J. Devillers, ed., Springer, New York, 2009, pp. 261–298.
26. J. Ridgway and G. Shimmield, *Estuaries as repositories of historical contamination and their impact on shelf seas*, Est. Coast. Shelf Sci. 55 (2002), pp. 903–928.
27. C. Ribeiro, M.A. Pardal, M.E. Tiritan, E. Rocha, R.M. Margalho, and M.J. Rocha, *Spatial distribution and quantification of endocrine-disrupting chemicals in Sado River estuary, Portugal*, Environ. Monit. Assess. 159 (2009), pp. 415–427.
28. R.L. Hill, *Incorporating toxic disturbance effects into a population model of a crustacean fishery*, Proceedings of the 49th Gulf and Caribbean Fisheries Institute, Bridgetown, Barbados, GCFI:49 (1997), pp. 139–155.
29. D.B. Tyler-Schroeder, *Use of the grass shrimp* (Palaemonetes pugio) *in a life-cycle toxicity test*, in *Aquatic Toxicology*, L.L. Marking and R.A. Kimerle, eds., American Society for Testing and Materials, ASTM STP 667, Philadelphia, PA, 1979, pp. 159–170.
30. S. Raimondo and C.L. McKenney, *Projecting population-level responses of mysids exposed to an endocrine disrupting chemical*, Integr. Comp. Biol. 45 (2005), pp. 151–157.

31. A. Kuhn, W.R. Munns, S. Poucher, D. Champlin, and S. Lussier, *Prediction of population-level response from mysid toxicity test data using population modeling techniques*, Environ. Toxicol. Chem. 19 (2000), pp. 2364–2371.
32. Y. Tanaka, *Ecological risk assessment of pollutant chemicals: Extinction risk based on population-level effects*, Chemosphere 53 (2003), pp. 421–425.
33. C. L. McKenney and D. M. Celestial, *Modified survival, growth and reproduction in an estuarine mysid* (Mysidopsis bahia) *exposed to a juvenile hormone analogue through a complete life cycle*, Aquat. Toxicol. 35 (1996), pp. 11–20.
34. S. Raimondo and C.L. McKenney, *From organisms to populations: Modeling aquatic toxicity data across two levels of biological organization*, Environ. Toxicol. Chem. 25 (2006), pp. 589–596.
35. J. Devillers and M.H. Pham-Delègue, *Honey Bees: Estimating the Environmental Impact of Chemicals*, Taylor & Francis Group, London, U.K., 2002.
36. D. Wilkinson and G.C. Smith, *A model of the mite parasite*, Varroa destructor, *on honeybees* (Apis mellifera) *to investigate parameters important to mite population growth*, Ecol. Model. 148 (2002), pp. 263–275.
37. D. Wilkinson and G.C. Smith, *Modeling the efficiency of sampling and trapping* Varroa *destructor in the drone brood of honey bees* (Apis mellifera), Am. Bee J. 142 (2002), pp. 209–212.
38. H.M. Thompson, S. Wilkins, A.H. Battersby, R.J. Waite, and D. Wilkinson, *Modelling long-term effects of IGRs on honey bee colonies*, Pest Manag. Sci. 63 (2007), pp. 1081–1084.
39. H.M. Thompson, S. Wilkins, A.H. Battersby, R.J. Waite, and D. Wilkinson, *The Effects of Endocrine Disrupters on Honeybee Populations*, Final Project Report DEFRA PN0936, York, U.K., 2004.
40. K. Heylen, B. Gobin, L. Arckens, R. Huybrechts, and J. Billen, *The effects of four crop protection products on the morphology and ultrastructure of the hypopharyngeal gland of the European honeybee*, Apis mellifera, Apidologie 42 (2011), pp. 103–116.
41. C. Tomlin, *The Pesticide Manual*, 10th ed., The Royal Society of Chemistry and The British Crop Protection Council, Cambridge, U.K., 1994.
42. J.N. Tasei, *Effects of insect growth regulators on honey bees and non-*Apis *bees. A review*, Apidologie 32 (2001), pp. 527–545.
43. E. Bonabeau, *Agent-based modeling: Methods and techniques for simulating human systems*, Proc. Natl. Acad. Sci. 99 (2002), pp. 7280–7287.
44. H. Devillers, J.R. Lobry, and F. Menu, *An agent-based model for predicting the prevalence of* Trypanosoma cruzi *I and II in their host and vector populations*, J. Theor. Biol. 255 (2008), pp. 307–315.
45. J. Devillers, H. Devillers, A. Decourtye, and P. Aupinel, *Internet resources for agent-based modelling*, SAR QSAR Environ. Res. 21 (2010), pp. 337–350.
46. M. Azimi, Y. Jamali, and M.R.K. Mofrad, *Accounting for diffusion in agent based models of reaction-diffusion systems with application to cytoskeletal diffusion*, PLoS One 6(9), 2011, p. e25306. doi:10.1371/journal.pone.0025306
47. SimBeePop, version 1.04, CTIS, Rillieux La Pape, France.
48. F. Anchling, *Avril, c'est deux saisons dans un même mois*, Abeille de France, 1995, pp. 913.
49. J. Devillers, A. Decourtye, and P. Aupinel, *Réponses Individuelles et Populationnelles des Abeilles aux Perturbateurs Endocriniens Xénobiotiques*, Programme National de Recherche "Perturbateurs Endocriniens" (PNRPE)-APR 2008, 2011.
50. J. Devillers, H. Devillers, A. Decourtye, J. Fourrier, P. Aupinel, and D. Fortini, *Agent-based modeling of the long term effects of pyriproxyfen on honey bee population*, in *In Silico Bees*, J. Devillers, eds., CRC Press, Boca Raton, FL, 2013 (in press).

8 SAR and QSAR Modeling of Juvenile Hormone Mimics

James Devillers

CONTENTS

ABSTRACT

Juvenoids are chemicals that mimic the activity of the juvenile hormone (JH). They selectively target and disrupt the endocrine system of insects. When they are applied to juvenile insects, during the period when the JH titer is normally low, abnormal developments occur that can lead to the appearance of malformations and/or the death of the insects at the emergence. Juvenoids are particularly suited as larvicides for the control of pest and disease vectoring insects such as mosquitoes. Consequently, QSAR models have been derived from structurally diverse sets of molecules for estimating their JH activity against various insect species. In this context, an attempt was made to review about forty years of QSAR research on juvenoids. From the available data, new models were also computed from nonlinear methods for comparison purposes. Our study clearly shows that the steric parameters outperform the other molecular descriptors in the design of QSARs, which are species dependent. The nonlinear methods allow us to obtain more powerful models than the linear regression techniques. Our study also shows the interest of such modeling approaches for selecting candidate molecules for further evaluation, for identifying the structural features and/or the physicochemical properties explaining an activity, and for comparing the difference of sensibility among the targeted species.

KEYWORDS

SAR, QSAR, Juvenile hormone mimics, Linear regression, Nonlinear techniques

8.1 INTRODUCTION

By the end of World War II, drastic changes in farming practice were observed in different parts of the world but especially in North America, Australia, and western Europe. Tractors have replaced horses, and chemical fertilizers were increasingly used in place of organic manure. Use of synthetic pesticides became widespread. Farmers were encouraged to stop crop rotation and to devote large acreages of their farms to the cultivation of a single crop, which undoubtedly required the use of increasing quantities of synthetic fertilizers and pesticides.

Both were touted as miraculous cure that would eradicate hunger in the world. Even if their use did not eradicate hunger and famine in the world, undoubtedly they have boosted agricultural productivity, increasing crop yield and reducing postharvest losses.

Unfortunately, these successes have been bought at a price. The surge in fertilizer use has led to greater contamination of ground and surface waters. The massive use of pesticides in agriculture but also for vector control has induced resistances in target organisms, contamination of the ecosystems, and adverse effects on nontarget species including humans.

Mass awareness regarding the adverse effects of the excessive use of pesticides was generated only after the publication of the Rachel Carson's book, "*Silent Spring*" [1], considered by many as instrumental in spurring the environmental movements in the 1960s, first in the United States and then in the rest of the world. It became crucial to find safer alternatives to control pest animals.

In this context, juvenile hormone mimics (JHMs) have opened new perspectives in agriculture as well as in vector control.

The six major juvenile hormone (JH) homologues identified to date are sesquiterpenes that regulate a wide spectrum of critical events in holometabolous insects including development, metamorphosis, reproduction, diapause, polyethism, and migration [2–4]. Synchronization of the activity of JH and 20-hydroxyecdysone (20E), the molting hormone, is essential for the normal development and metamorphosis of insects. During larval development the presence of JH in the hemolymph maintains the insect in a juvenile status, and when the JH titer goes down, it signals the insect to undergo metamorphosis that is under the control of 20E titer.

As their name implies, JHMs, also termed juvenoids, are man-made chemicals that behave similarly to the JHs. Thus, applied to critical periods, they will disrupt metamorphosis and lead to various deleterious effects in the insects. These chemicals are much more target specific than conventional synthetic pesticides, their environmental fate profile is also better, and they are safe to vertebrates. JHMs are divided roughly into two classes, terpenoid and non-terpenoid types.

Qualitative structure–activity relationship (SAR) and quantitative structure–activity relationship (QSAR) modeling has become an indispensable tool in drug design to better understand mechanisms of action of molecules and find potential leads in a reduced time and with an optimized cost [5,6]. These *in silico* approaches have found applications in the design of JHMs active against target insects damaging crops or that are the vectors of diseases. Consequently, the aim of this study was to review the SAR and QSAR models dedicated to JHMs. For comparison purposes, an attempt was also made to derive new models from biological data and molecular descriptors used by others. It is noteworthy that only the 2D models were analyzed, the specific case of the 3D QSAR models being addressed in Chapter 9.

8.2 STRUCTURAL FEATURES ASSOCIATED WITH JUVENILE HORMONE ACTIVITY

Numerous studies have tried to identify structural features associated with JH activity (JHA) of molecules tested on different insect species. Historically, molecules with JHA were found in the wood of evergreen trees, especially in the balsam fir *Abies balsamea* [7]. Then, they were identified as juvabione [8] and dehydrojuvabione [9] (Figure 8.1). Ten analogues of these two molecules in which the alicyclic ring was replaced by an aromatic ring (Figure 8.1) were tested by topical application on the cuticle of freshly molted last instar larvae of *Pyrrhocoris apterus* and *Neodysdercus intermedius* (Hemiptera, Pyrrhocoridae) [10]. The JHA was scored according to the degree of preservation of the larval epidermal structures after the final molt. The activity was expressed according to a scale ranged from 0 to 5. A score of 0 indicated the formation of perfect adults, while at the opposite, a score of 5 revealed a maximum activity with the formation of perfect extralarval instars. Figure 8.1 shows the JHA of the molecules tested by Sláma et al. [10] at 1 μg in 1 μL of acetone on *P. apterus* and *N. intermedius*, respectively. The obtained results show that the JHA of these molecules is dependent on the chemical nature of the ester radical as well as the number of double bonds and the nature of the substituents in the aliphatic side chain. In addition, *N. intermedius* appears a little bit more sensitive to these molecules than *P. apterus*. It is noteworthy that topical tests were also performed on last instar larvae of *Graphosoma italicum* (Hemiptera), *Acheta domesticus*, and *Locusta migratoria* (Orthoptera) as well as on freshly molted pupae of *Tenebrio molitor (T.m.)* and *Leptinotarsa decemlineata* (Coleoptera) and *Galleria mellonella* (*G.m.*) (Lepidoptera). All these insects were inactive except *G. italicum* for which chemicals A4, A6, A8, and A10 were slightly active at the dose of 100 μg in 1 μL of acetone. Sláma et al. [11] showed that the JHA of farnesenic acid esters (Figure 8.2) increased from the methyl to the ethyl ester and then decreased for the cyclohexyl and benzyl esters when tested topically on *P. apterus* and *Dysdercus cingulatus* (Hemiptera, Pyrrhocoridae) and *G. italicum* (Hemiptera, Pentatomidae) and by injection in the pupae of *T.m.* (Coleoptera, Tenebrionidae). The decrease was less obvious with *P. apterus* and it was only significant for the benzyl ester. The same trend was observed with the dihydrodichloro farnesenic acid esters (Figure 8.2). JHA increased from the methyl to the ethyl ester and then decreased for the propyl, *n*-butyl, *t*-butyl, amyl, hexyl, and cyclohexyl esters. This was only observed with the three hemipteran species. The results obtained on *P. apterus* and *D. cingulatus* showed that an important increase was obtained after the addition of two chlorine

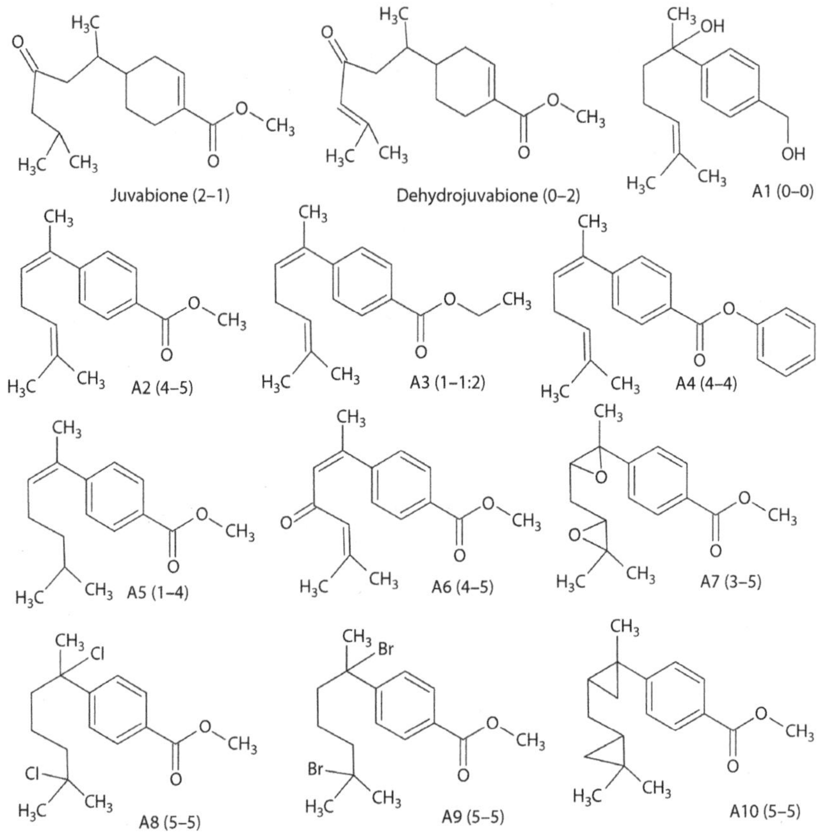

FIGURE 8.1 Structures of juvabione, dehydrojuvabione, and their aromatic derivatives (A1–A10) with, in parentheses, their JHA on *P. apterus* and *N. intermedius*, respectively (see text for explanation).

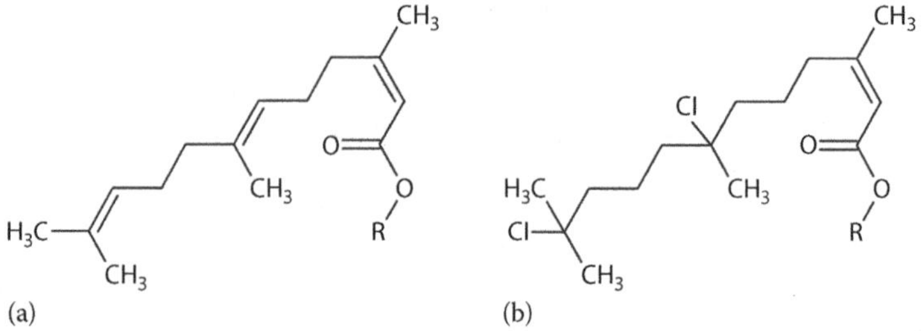

FIGURE 8.2 General structure of farnesenic acid esters (a) and dihydrodichloro farnesenic acid esters (b).

atoms on the double bonds of the farnesenic acid esters. However, the same change in chemical structure considerably decreased the activity on *Tenebrio* pupae [11]. Fifteen methyl and ethyl esters of 3,7,11-trimethyl dodecanoic acid derivatives differing in the amount and number of double bonds as well as the presence of chlorine and epoxy substituents were tested topically on the following Hemiptera: *P. apterus* and *D. cingulatus* (family Pyrrhocoridae), *Lygaeus equestris* (family Lygaeidae), and *G. italicum*, *Aelia acuminata*, and *Eurygaster integriceps* (family Pentatomidae). The 15 chemicals were also tested topically and via injection on the following Coleoptera: *T.m.* (family Tenebrionidae) and *Dermestes vulpinus* (family Dermestidae) [12]. The number of double bonds differently influences the tested taxa. In the Pyrrhocoridae and Lygaeidae, the JHA does not significantly change with an increase in the degree of unsaturation, while a continuous increase in the JHA in the Pentatomidae in relation to an increase in the degree of unsaturation of the tested molecules is observed. The reverse was observed with the coleopterans, where the esters with only one double bond are the most active and those with four double bonds are 100–1000 times less active or completely inactive. When the 10,11 double bond is transformed into an oxirane ring, the JHA is substantially and slightly increased in the hemipterans and coleopterans, respectively. The increase of JHA is less important when the epoxidation is made on more unsaturated chemicals. Addition of a chlorine atom on the 10,11 double bond leads to similar effects on JHA. However, when a chlorine atom is added on the 10,11 and 6,7 double bonds of the methyl and ethyl esters of the 3,7,11-trimethyl-2,6,10-dodecatrienoic acid, a very important increase in the JHA is observed with the pyrrhocorid and lygaeid bugs, whereas only a small change is noted with the pentatomids. A large decrease of JHA is observed with the beetles. Sláma et al. [12] also tested the influence of stereoisomery on JHA. They showed that the *trans* isomers were more active than the *cis* isomers. The higher JHA of the stereochemical *trans* isomers was also found in other studies [13,14]. This is also the case for the enhanced JHA induced by the addition of an epoxy group [15–19] and by the addition of chlorine atoms [19–21]. Walker and Bowers [21] showed the effects of the addition of halogens on aromatic terpenoid ethers. They compared the JHA of these chemicals, applied topically to *T.m.*, *L. decemlineata*, and *Epilachna varivestis*, which are species of Coleoptera belonging to the families of Tenebrionidae, Chrysomelidae, and Coccinellidae, respectively. The chemicals were applied at 10 times dilution intervals in 1 μL of acetone to the ventral abdomen of *T.m.* pupae and to the middorsal abdomen of *L. decemlineata* pupae and *E. varivestis* prepupae. The JHA was expressed as the lowest dose that reduced the emergence of normal-appearing adults of *T.m.* by 80% or more and of *L. decemlineata* and *E. varivestis* by 90% or more. The obtained results are listed in Table 8.1. Besides the difference of sensibility among the species that is a constant feature found with the JHMs, Table 8.1 shows that the number and position of the chlorine atoms on the molecules influence their JHA. In addition, a substitution by a fluorine atom is less effective than that with chlorine and bromine atoms. Table 8.1 shows that the para-chlorinated and para-brominated derivatives show the same JHA on the three Coleoptera species. This is in agreement with the results found by Sláma et al. [10].

TABLE 8.1
Lowest Doses (in μg) of Halogenated Aromatic Terpenoid Ethers That Prevent Adult Emergence

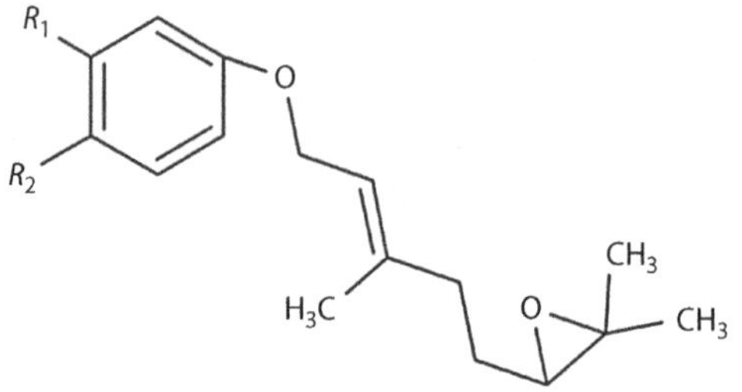

R_1	R_2	*T.m.*	*L. decemlineata*	*E. varivestis*
H	H	10	1	10
H	Cl	0.1	0.1	0.01
Cl	H	10	10	1
Cl	Cl	0.1	10	1
H	Br	0.1	0.1	0.01
H	F	1	1	0.1

8.3 SAR MODELS FOR DETECTING JUVENILE HORMONE MIMICS

Iwamura and colleagues developed different series of molecules for estimating their JHA potential against *Culex pipiens* (Diptera, Culicidae). These chemicals were (4-phenoxyphenoxy)- and (4-benzylphenoxy)alkanaldoxime *O*-ethers, their ether and hydroxylamine congeners [22,23], and (4-alkoxyphenoxy)- and (4-alkylphenoxy) alkanaldoxime *O*-ethers and their congeners in which the oxime group was replaced by the function ether, ester, amide, carbamate, urea, or aromatic benzene or pyridine [24,25]. Attempts were made to relate the structure of these molecules to their JHA against the common mosquito. The results showed that the most important features for JHA were the overall length of the chain molecule, the dimensions of the molecular ends, and the position of the functional group incorporated at one end of the molecule [24,26]. Other authors already pinpointed the importance of these kinds of structural characteristics for explaining the JHA of molecules (see, e.g., [14,15,27,28]), but to our knowledge, Iwamura and colleagues were the first to have proposed molecular descriptors for encoding these structural features in their studied series of molecules. For *C. pipiens*, the optimum length *D* is about 21–22 Å, and the position-specific interaction site of the functional groups in the molecules with optimum activity is about 4.6 Å distant from one end of the molecule [29]. Electrostatic potentials were calculated for a variety of functional groups found in active insect JHMs studied by Iwamura's group. Quantum chemical calculations showed that the contours of the electrostatic potentials of these functions have negative peak in the plane that perpendicularly bisects their skeletal plane [26]. Hayashi et al. [29]

showed that the conditions for high potent JHA, namely, an optimum molecular length and the position of the functional group, were the same for *Musca domestica* (*M.d.*) (Diptera, Muscidae) and *Spodoptera litura* (Lepidoptera, Noctuidae).

8.4 2D QSAR MODELS FOR PREDICTING JUVENILE HORMONE ACTIVITY

A series of 28 aryl terpene ethers, selected on the basis of variations on the structure of 6,7-epoxygeranyl 4-ethylphenyl ether, were tested by Nilles et al. [30] for their JHA by induction of oviposition in the adult prediapause cereal leaf beetle (*Oulema melanopus*), a Coleoptera belonging to the family of the Chrysomelidae. A compound showed maximum JHA if initiation of oviposition and number of eggs laid in a given time were the same as would be expected from a postdiapause beetle after emerging overwintering or laboratory storage [31]. The Hansch lipophilicity parameter (π) and an empirical steric parameter (ω) were used to derive QSAR regression equations on subsets of eight chemicals [30]. The equations were of very limited interest being overparameterized and derived from only eight chemicals.

Niwa et al. [32] used a data set of 48 (4-phenoxyphenoxy)- and (4-benzylphenoxy) alkanaldoxime *O*-ethers for which JHA data on *C. pipiens* were available. JHA was expressed in terms of pI50 (*M*), the logarithm of the reciprocal of the concentration at which 50% inhibition of mosquito metamorphosis was observed (Table 8.2). To express the steric features of the molecules, the authors defined the D axis that passes through the alkoxy oxygen atom and the β-oxygen atom to the oxime carbon atom in the fully extended conformation of the acetaldoxime *O*-ethers. By definition, the angle between the D axis and the bond that connects the β-oxygen atom to the oxime moiety is 40.02°C. For the propionaldoxime *O*-ethers, the D axis was drawn so as to pass on the β-oxygen atom to the oxime carbon and satisfying this angle condition. D_1, the length along the D axis of the terminal benzyl and phenoxy moieties, and D_2, the length of the alkoxy end, were also defined. To express the steric dimensions of the two terminal moieties in more detail, the authors further defined other steric parameters. W_1 was defined as the width of the benzene end in the direction perpendicular to its connecting axis to the rest of the molecule. W_1° corresponded to the minimum width of the substituents. T_2 was the thickness of the alkoxy moiety in vertical direction to the zigzag backbone plane. All the steric parameters were expressed in angstroms [32]. The authors also experienced the usefulness of the 1-octanol/water partition coefficient (log P) in the design of the QSAR models. Their best model was the following:

$$\text{pI50} = 2.91D - 0.12D^2 + 11.47W_1 - 0.78W_1^2 - 1.92W_i^{\circ} + 4.67T_2 - 0.48T_2^2 + 1.30Ip - 59.81 \quad (8.1)$$

$$n=48,\ s=0.34,\ r=0.95,\ F=43.14$$

The parameters included in Equation 8.1 (excluding their squared value) are listed in Table 8.2. Ip is an indicator variable having a value of 1 for the propionaldoxime series and 0 for acetaldoximes. In Equation 8.1, n is the number of chemicals, s is the

TABLE 8.2
JHA and Physicochemical Properties of (4-Phenoxyphenoxy)- and (4-Benzylphenoxy)alkanaldoxime *O*-Ethers

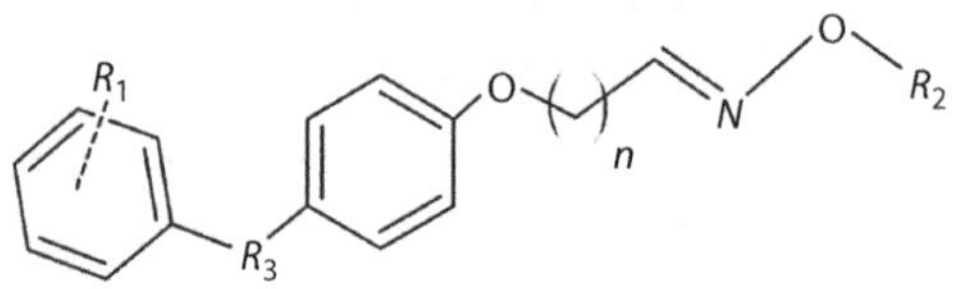

IN	R_1	R_2	R_3	*n*	pI50	*D*	W_1	W_1°	T_2	I_p
1	H	Et	CH_2	1	7.09	11.06	6.23	1	3.8	0
2	H	Pr	CH_2	1	7.97	12.34	6.23	1	3.8	0
3[a]	H	*i*-Pr	CH_2	1	7.72	11.06	6.23	1	5.05	0
4[b]	H	Allyl	CH_2	1	8.14	12.31	6.23	1	3.8	0
5[a]	H	Propargyl	CH_2	1	7.46	12.06	6.23	1	3.8	0
6	H	Bu	CH_2	1	7.41	13.57	6.23	1	3.8	0
7	H	*i*-Bu	CH_2	1	8.48	12.34	6.23	1	5.05	0
8	H	*s*-Bu	CH_2	1	7.42	12.34	6.23	1	5.05	0
9	H	Pentyl	CH_2	1	6.72	14.84	6.23	1	3.8	0
10[b]	H	*i*-Pentyl	CH_2	1	8.10	13.57	6.23	1	5.05	0
11	H	*c*-Pentyl	CH_2	1	8.10	11.96	6.23	1	5.95	0
12	H	*c*-Hexyl	CH_2	1	8.12	12.34	6.23	1	5.05	0
13[b]	H	Benzyl	CH_2	1	6.75	13.44	6.23	1	6.23	0
14	H	Et	CH_2	2	8.76	12.27	6.23	1	3.8	1
15	H	Pr	CH_2	2	8.37	13.5	6.23	1	3.8	1
16	H	*i*-Pr	CH_2	2	9.68	12.27	6.23	1	5.05	1
17	H	*c*-Pentyl	CH_2	2	8.42	13.19	6.23	1	5.9	1
18[a]	H	H	O	2	6.58	9.6	6.23	1	2.7	1
19	H	Et	O	2	9.09	12.1	6.23	1	3.8	1
20	H	*i*-Pr	O	2	9.76	12.1	6.23	1	5.05	1
21[b]	3-Me	Et	O	2	9.46	13.21	7.18	1	3.8	1
22[a]	3-Me	*i*-Pr	O	2	10.0	13.21	7.18	1	5.05	1
23	H	Pr	O	1	7.54	12.17	6.23	1	3.8	0
24	2-Me	Pr	O	1	7.36	12.17	7.44	1.52	3.8	0
25	2-MeO	Pr	O	1	6.81	12.22	8.55	1.35	3.8	0
26	2-F	Pr	O	1	7.35	12.17	6.78	1.35	3.8	0
27	2-Cl	Pr	O	1	6.89	12.17	7.6	1.8	3.8	0
28	2-CF_3	Pr	O	1	6.18	12.17	7.97	1.98	3.8	0
29	3-Me	Pr	O	1	8.71	13.28	7.18	1	3.8	0
30	3-Et	Pr	O	1	7.85	14.59	7.27	1	3.8	0
31	3-MeO	Pr	O	1	7.41	14.43	7.22	1	3.8	0
32	3-F	Pr	O	1	8.27	12.74	6.78	1	3.8	0
33[a]	3-Cl	Pr	O	1	8.12	13.59	7.6	1	3.8	0
34	3-CF_3	Pr	O	1	7.47	13.78	7.66	1	3.8	0
35[b]	3-NO_2	Pr	O	1	7.75	13.86	8.01	1	3.8	0

TABLE 8.2 (continued)
JHA and Physicochemical Properties of (4-Phenoxyphenoxy)- and (4-Benzylphenoxy)alkanaldoxime *O*-Ethers

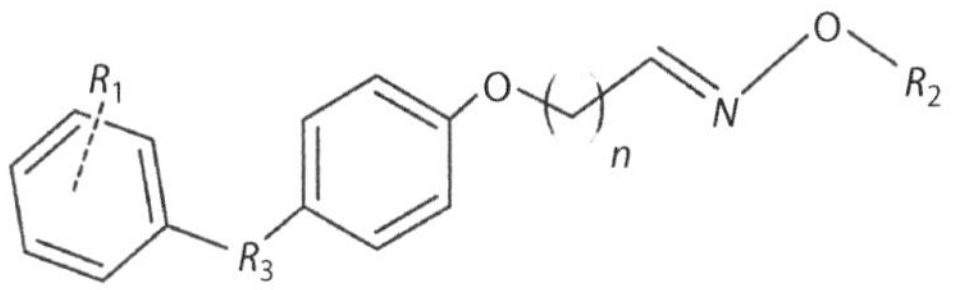

IN	R_1	R_2	R_3	*n*	pI50	*D*	W_1	W_1°	T_2	I_p
36[a]	4-Me	Pr	O	1	6.84	13.03	6.23	1	3.8	0
37	4-Et	Pr	O	1	6.42	14.25	6.23	1	3.8	0
38	4-Pr	Pr	O	1	6.36	15.54	6.6	1	3.8	0
39	4-MeO	Pr	O	1	7.34	14.08	6.23	1	3.8	0
40	4-EtO	Pr	O	1	6.86	15.36	6.47	1	3.8	0
41	4-F	Pr	O	1	7.58	12.28	6.23	1	3.8	0
42	4-Cl	Pr	O	1	7.26	13.05	6.23	1	3.8	0
43	4-NO_2	Pr	O	1	7.08	13.59	6.23	1	3.8	0
44[b]	2,3-Me_2	Pr	O	1	7.90	13.28	7.44	1.52	3.8	0
45[b]	2,5-Me_2	Pr	O	1	6.61	13.28	8.4	1.52	3.8	0
46	3,5-Me_2	Pr	O	1	7.98	13.28	8.14	1	3.8	0
47	2,3,5-Me_3	Pr	O	1	6.22	13.28	8.4	1.52	3.8	0
48[a]	2,3,6-Me_3	Pr	O	1	5.68	13.28	8.66	1.52	3.8	0

[a] TS chemicals.
[b] VS chemicals.

standard deviation, *r* is the correlation coefficient, and *F* is the Fisher's test. Figure 8.3 shows the plot of the observed versus calculated pI50 values obtained with Equation 8.1.

The squared descriptor parameters in Equation 8.1 let to suppose that some degrees of nonlinearity exist between the structure of the studied molecules and their JHA. Consequently, an attempt was made to use a purely nonlinear statistical tool for modeling the data. Thus, a three-layer perceptron (TLP) [33], with the five initial descriptors (Table 8.2) as input layer neurons, was experienced. Three neurons were selected for constituting the hidden layer and one neuron, the pI50 values, constituted the output layer. Calculations were made with the data mining module of Statistica™ software (StatSoft, Fr). Niwa et al. [32] only tried to find a model that fitted at best the whole data set without verifying the prediction performances of Equation 8.1, while according to the OECD principles for validation of the (Q)SAR models [34], the applicability domain of the models must always be estimated, preferentially from an external test set (TS). This is even more important for the TLPs because they are well known to suffer from problems of overfitting if they are not correctly used [33]. To avoid this problem, the data set was randomly split into a learning set (LS) to train the TLP, a TS for monitoring the learning phase via the estimation of the performances of the network while it is under training, and a validation set (VS)

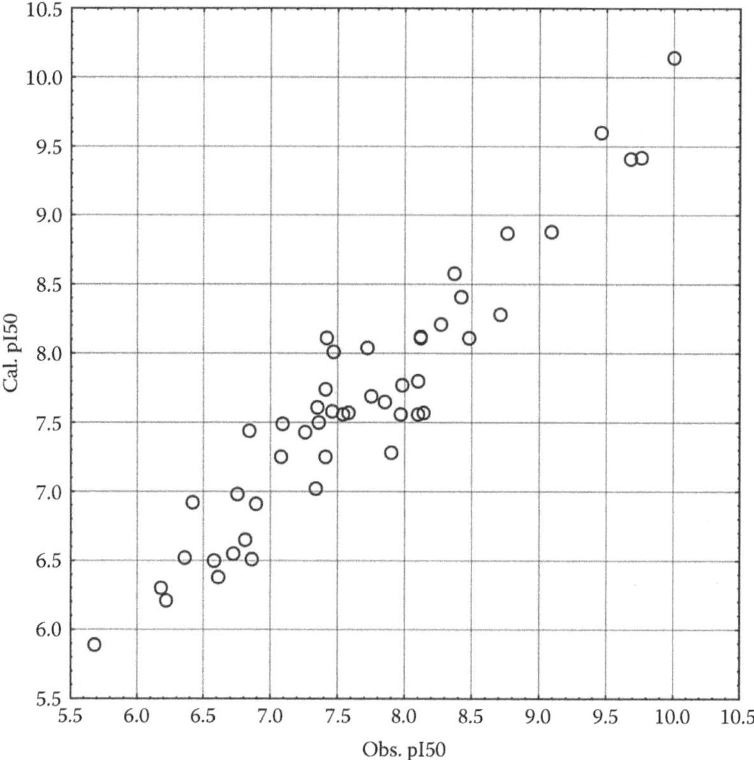

FIGURE 8.3 Observed versus calculated pI50 values obtained with Equation 8.1.

to determine how well the network predicts "new" data that were used neither to train the model nor to test its performance when being trained. The LS, TS, and VS included 34, 7, and 7 chemicals, respectively. TS included chemicals #3, 5, 18, 22, 33, 36, and 48, while VS included chemicals #4, 10, 13, 21, 35, 44, and 45. A 5/3/1 TLP with a bias connected to the hidden and output layers and using the BFGS (Broyden–Fletcher–Goldfarb–Shanno) second-order training algorithm allowed us to obtain the best prediction results. The hidden and output activation functions were both a negative exponential function. The convergence was obtained after only 40 cycles. With such a configuration, the correlation coefficients for the LS, TS, and VS were equal to 0.92, 0.94, and 0.93, respectively. An overall correlation coefficient of 0.92 was calculated from the prediction results obtained with the three sets. Figure 8.4 shows the plot of the observed versus calculated pI50 values obtained with the TLP model. Comparison of Figures 8.3 and 8.4 shows that the residual values (difference between observed and calculated data) are lower when using Equation 8.1 in place of the TLP model. However, the predictive performances of the neural network model have been estimated. In addition, these performances are satisfying, while nothing was done for Equation 8.1 [32]. It is noteworthy that other random selections of the three sets, only keeping the same proportions between them, led to broadly the same prediction performances. Because, the aforementioned selected

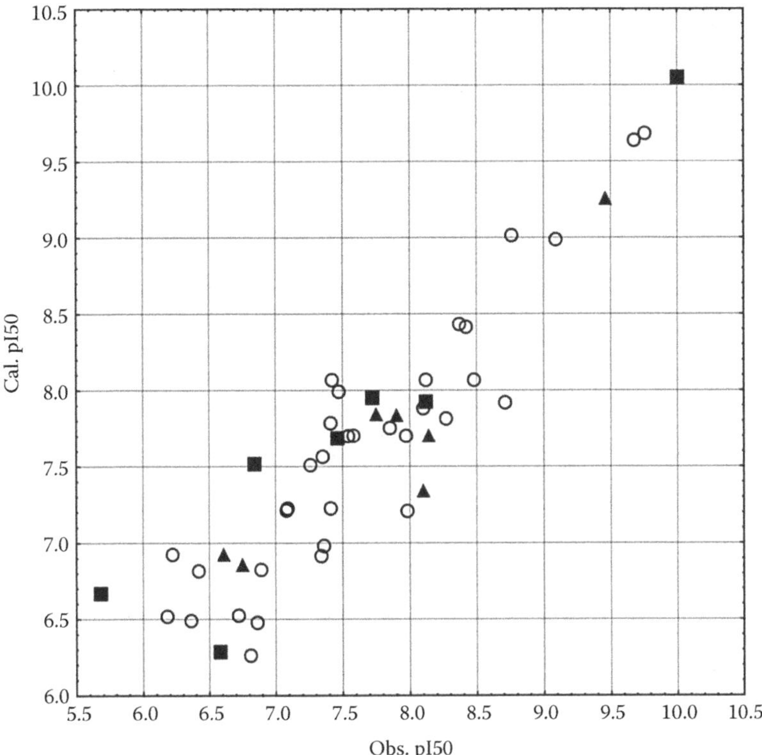

FIGURE 8.4 Observed versus calculated pI50 values obtained with the TLP model. The empty circles, black squares, and black triangles are LS, TS, and VS, respectively.

TLP model included a rather high number of connections, attempts were made to reduce the number of connections in the network. A 5/2/1 TLP with the same configuration and after 26 cycles led to correlation coefficients of 0.90, 0.95, and 0.92 for the LS, TS, and VS, respectively. An overall correlation coefficient of 0.90 was calculated. We also tried to use a support vector regression (SVR) analysis [35,36] for modeling the data from the five initial descriptors (Table 8.2). The data set was randomly split into an LS of 36 chemicals and an external TS of 12 chemicals (i.e., #7, 14, 18, 25, 27, 28, 32, 33, 37, 40, 42, 44). An SVR type 1 with a radial basis function as kernel provided the best prediction results with C, ε, and γ set to 27, 0.1, and 0.60, respectively. With such a configuration, the correlation coefficients for the LS and external TS were equal to 0.96 and 0.92, respectively. An overall correlation coefficient of 0.95 was calculated from the prediction results obtained with the two sets. Figure 8.5 shows the plot of the observed versus calculated pI50 values obtained with the SVR model. The simulation performances of the SVR model are quite satisfying and we assume that they outperform those obtained with Equation 8.1 despite a same correlation coefficient of 0.95 because the selection of this configuration has been made on the basis of its performances on a randomly selected LS and external TS at the opposite of Equation 8.1 where attempts were only made to fit at best

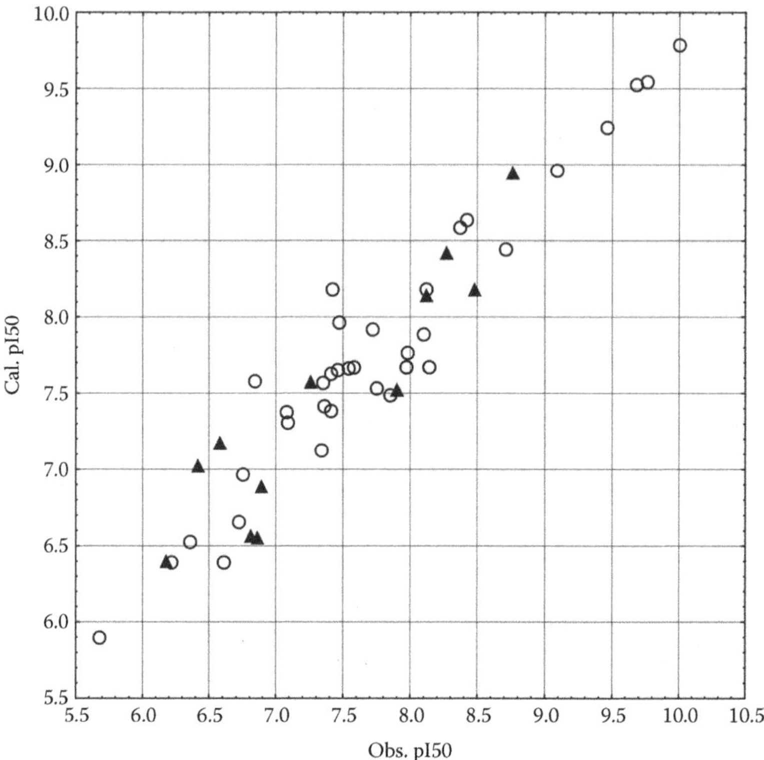

FIGURE 8.5 Observed versus calculated pI50 values obtained with the SVR model. The empty circles and black triangles are the LS and external TS, respectively.

the whole data set (Table 8.2). Notice that other selections of LSs and external TSs showed similar performances.

The same kind of Hansch analysis was performed by Hayashi et al. [37] on 30 (4-alkoxyphenoxy)alkanaldoxime *O*-ethers tested on fourth larval instars of *C. pipiens*. This led to the design of a four-parameter regression model (Equation 8.2):

$$\text{pI50} = 3.40D - 0.08D^2 + 1.12Ip + 1.59I_{br}(OR) - 31.06 \tag{8.2}$$

$$n = 30,\ s = 0.46,\ r = 0.93,\ F = 41.65$$

In Equation 8.2, *D* is the length in the fully extended conformation of the whole molecule along the axis that passes on the phenoxy oxygen atom and has an angle of 40.02° to its connecting bond to the oxime moiety. *Ip* is an indicator variable that takes 1 for propionaldoximes and 0 for acetaldoximes. $I_{br}(OR)$ is an indicator variable that takes 1 for compounds having a branch at β of the 4-alkoxy substituents and 0 otherwise. Equation 8.2 was extended to (4-alkylphenoxy)alkanaldoxime *O*-ethers by mixing descriptors in an unusual way to a six-parameter regression equation derived from 44 molecules ($r = 0.96$, $s = 0.46$).

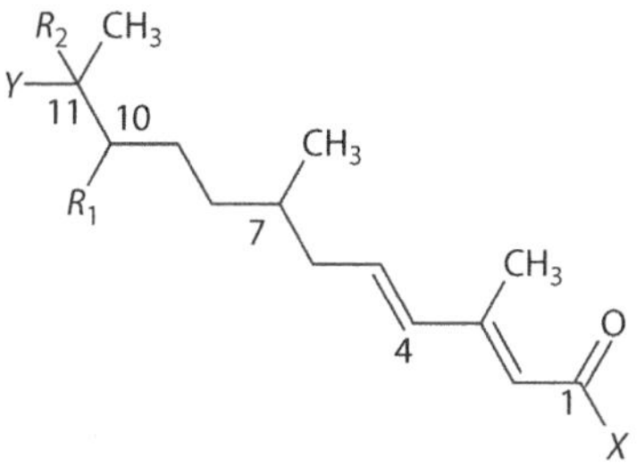

FIGURE 8.6 Structure of JHM 2,4-dodecadienone derivatives with X=OR, SR, NHR, NR2, Alkyl; R_1=H, OR, SEt, 10-ene, 11-ene, 10-epoxy, oxo; Y=OR, SR, OCOR, Me, Et; and R_2=H, Me, Cl.

JHA values on *Aedes aegypti (A.a.)* and *T.m.*, originally reported by Henrick et al. [18] for about 100 structurally diverse (2*E*,4*E*)-3,7,11-trimethyl-2,4-dodecadienoates, were used by Nakayama et al. [38] for computing different QSAR models. The JHA originally reported by Henrick et al. [18] on *A.a.* in ppm and *T.m.* in μg/pupa was converted into mole units (IC50) and converted to pI50 by taking the logarithm of its reciprocal. Figure 8.6 shows the common skeleton of the 2,4-dodecadienones in which X expresses the substituents at the carbonyl C_1 atom of the dodecadienone skeleton and Y is the longest one of the C_{11} substituents in terms of length L_y along the bond axis. The molecules were described by different steric and hydrophobic parameters and used to compute step by step various QSAR models based on different functional groups. Equation 8.3 represents the final QSAR model on *A.a.* selected by Nakayama et al. [38]:

$$\begin{aligned} \text{pI50} = {} & 3.65(\pm 1.26)L_x - 0.35(\pm 0.11)L_x^2 + 1.08(\pm 1.12)D - 0.06(\pm 0.06)D^2 \\ & + 1.90(\pm 1.13)\log P - 0.14(\pm 0.09)\log P^2 + 0.57(\pm 0.25)B_x - 0.71(\pm 0.41)I_N \\ & + 0.86(\pm 0.35)I_{OR} + 1.39(\pm 0.65)I_{br} - 0.65(\pm 0.37)I_{(-)} - 16.35(\pm 5.21) \end{aligned} \quad (8.3)$$

$$n = 85,\ s = 0.53,\ r = 0.89$$

In Equation 8.3, L_x is the length of the X end along the bond axis (C_1–X). D is the maximum length of the whole molecule. B_x expresses the bulkiness toward the carbonyl group of α-substituents in the alcohol moiety of ester derivatives and/or thiol ester derivatives. I_N is an indicator variable that takes the value of 1 for the amides and 0 for the other chemicals. I_{OR} is an indicator variable that takes the value of 1 for the compounds whose Y moiety is an alkoxy or hydroxy group and otherwise zero. I_{br} is an indicator variable for the chemicals having a branch at any position in the X moiety of ketone derivatives. $I(-)$ is an indicator variable coding enantiomeric effects. It takes the value of 1 for the (R)-(−)-isomer and zero for the (±) mixture. The hydrophobicity of the whole molecule is estimated by log P [38,39].

The pI50 values calculated with Equation 8.3 as well as the corresponding residual values are listed in Table 8.3.

TABLE 8.3
Observed (Obs.) pI50 Values (mM) of 2,4-Dodecadienones Tested on *A.a.* and the Calculated (Cal.) Activity Values Computed from Equation 8.3 and the 10/4/1 TLP with the Corresponding Residual Values (Res.)

						Equation 8.3		TLP	
IN[a]	X[b]	*Y*	R_1	R_2	Obs.	Cal.	Res.	Cal.	Res.
1**	OMe	Me	H	H	3.23	3.51	−0.28	3.64	−0.41
2	OEt	Me	H	H	4.53	4.32	0.21	4.17	0.36
3*	O-*n*-Pr	Me	H	H	4.24	4.18	0.06	3.90	0.34
4	O-*i*-Pr	Me	H	H	5.17	5.04	0.13	5.18	−0.01
5**	O-*i*-Bu	Me	H	H	4.07	4.13	−0.06	3.81	0.26
6	O-*sec*-Bu	Me	H	H	4.21	4.84	−0.63	4.40	−0.19
8	$OCH_2C{\equiv}CH$	Me	H	H	3.08	3.60	−0.52	3.54	−0.46
9	$OCH_2C{\equiv}CCH_3$	Me	H	H	2.99	2.65	0.34	2.25	0.74
10	$OCH_2CH{=}CH_2$	Me	H	H	4.14	4.07	0.07	4.01	0.13
12	OCH_2Ph	Me	H	H	1.52	1.14	0.38	2.20	−0.68
13	OH	Me	H	H	1.26	1.41	−0.15	1.27	−0.01
14	OMe	OMe	H	Me	4.15	4.18	−0.03	4.61	−0.46
15*	OEt	OMe	H	Me	5.17	4.94	0.23	4.47	0.70
16*	O-*n*-Pr	OMe	H	Me	5.00	4.77	0.23	4.40	0.60
17	O-*i*-Pr	OMe	H	Me	6.26	5.80	0.46	5.93	0.33
18	O-*n*-Bu	OMe	H	Me	4.02	3.80	0.22	3.69	0.33
19	O-*sec*-Bu	OMe	H	Me	5.15	5.56	−0.41	4.79	0.36
20	O-*t*-Bu	OMe	H	Me	5.33	5.89	−0.56	5.44	−0.11
21	O-*i*-amyl	OMe	H	Me	2.96	3.82	−0.85	3.54	−0.58
22	$OCH_2C{\equiv}CH$	OMe	H	Me	3.18	3.82	−0.64	3.23	−0.05
23**	$O(CH_2)_2C{\equiv}CH$	OMe	H	Me	4.51	4.24	0.27	4.69	−0.18

24	$OCH_2CH{=}CH_2$	OMe	H	Me	4.26	4.50	−0.24	4.16	0.10
25	$OCH_2CH{=}CHMe$	OMe	H	Me	3.96	3.51	0.45	3.61	0.35
26	$OCH(Me)CH{=}CH_2$	OMe	H	Me	4.99	5.35	−0.37	5.11	−0.12
27	O-*c*-Pr	OMe	H	Me	5.11	5.18	−0.07	4.70	0.41
28	O-*c*-Bu	OMe	H	Me	5.09	5.10	−0.01	5.31	−0.22
30**	OH	OMe	H	Me	1.22	2.23	−1.01	1.29	−0.07
31	OEt	Me	H	Cl	4.55	4.29	0.26	4.23	0.32
32	OEt	OH	H	Me	5.02	4.44	0.58	5.31	−0.29
33	OEt	OAc	H	Me	3.84	3.31	0.53	3.38	0.46
34	OEt	OEt	H	Me	5.71	4.96	0.75	5.75	−0.04
35*	OEt	SMe	H	Me	3.19	3.63	−0.44	3.39	−0.20
36	OEt	OCONHEt	H	Me	2.25	2.48	−0.23	3.07	−0.82
37	OEt	Me	H	Me	3.88	3.68	0.20	3.64	0.24
38**	OEt	Et	H	H	3.99	4.28	−0.29	3.65	0.34
39	OEt	Me	OH	H	3.45	3.58	−0.13	3.52	−0.07
41	OEt	Me	SEt	H	3.15	3.72	−0.57	3.38	−0.23
43	OEt	Me	10-ene		3.42	4.24	−0.82	4.27	−0.85
44*	OEt	Me	11-ene		4.10	4.24	−0.14	4.27	−0.17
45*	OEt	Me	10-epoxy		3.41	3.65	−0.24	3.71	−0.30
46	OEt	Me	oxo	H	2.76	3.34	−0.58	2.71	0.05
47	O-*i*-Pr	Me	H	Cl	5.20	4.39	0.81	4.74	0.46
48	O-*i*-Pr	OH	H	Me	6.06	5.40	0.66	6.04	0.02
49	O-*i*-Pr	OAc	H	Me	4.30	4.16	0.14	4.02	0.28
50	O-*i*-Pr	OCHO	H	Me	5.06	4.10	0.96	4.63	0.43
51	O-*i*-Pr	OEt	H	Me	6.15	5.76	0.39	6.36	−0.21
52*	O-*i*-Pr	O-*i*-Pr	H	Me	5.97	5.79	0.18	5.42	0.55
54	O-*i*-Pr	OCONHEt	H	Me	3.18	3.32	−0.14	3.43	−0.25

(*continued*)

TABLE 8.3 (continued)
Observed (Obs.) pI50 Values (mM) of 2,4-Dodecadienones Tested on *A.a.* and the Calculated (Cal.) Activity Values Computed from Equation 8.3 and the 10/4/1 TLP with the Corresponding Residual Values (Res.)

						Equation 8.3		TLP	
IN[a]	X[b]	Y	R_1	R_2	Obs.	Cal.	Res.	Cal.	Res.
55	O-*i*-Pr	Me	H	Me	3.55	4.35	−0.80	4.50	−0.95
56	O-*i*-Pr	Me	OMe	H	4.97	4.89	0.08	4.88	0.09
59*	O-*i*-Pr	Me	10-ene		4.03	5.02	−0.99	4.12	−0.09
60	O-*i*-Pr	Me	11-ene		4.44	4.37	0.07	4.83	−0.39
61	SMe	Me	H	H	4.65	3.96	0.70	4.14	0.51
62	SEt	Me	H	H	5.10	4.30	0.80	3.86	1.24
63	S-*i*-Pr	Me	H	H	4.07	4.17	−0.10	4.30	−0.23
64	$SCH_2C{\equiv}CH$	Me	H	H	3.47	2.93	0.54	3.46	0.01
65	$SCH_2CH{=}CH_2$	Me	H	H	3.39	3.76	−0.37	3.12	0.27
66	SMe	OMe	H	Me	5.04	4.93	0.11	4.44	0.60
67	SEt	OMe	H	Me	5.89	5.24	0.65	5.79	0.10
68	S-*i*-Pr	OMe	H	Me	4.66	5.24	−0.58	4.66	0.00
70	$SCH_2CH{=}CH_2$	OMe	H	Me	5.10	4.49	0.61	5.13	−0.03
71	NHEt	Me	H	H	3.65	3.21	0.44	3.31	0.34
72	NH-*i*-Pr	Me	H	H	4.00	3.38	0.62	3.31	0.69
74	$NHCH_2CH{=}CH_2$	Me	H	H	2.85	2.92	−0.07	2.87	−0.02

75**	$N(Me)_2$	Me	H	H	3.31	2.36	0.95	3.31	0.00
76	$N(Et)_2$	Me	H	H	3.24	3.49	−0.25	3.31	−0.07
77	NHEt	OMe	H	Me	3.43	3.33	0.10	3.31	0.12
78	NH-*i*-Pr	OMe	H	Me	4.02	3.64	0.38	3.31	0.71
79	NH-*c*-Pr	OMe	H	Me	2.64	3.37	−0.73	3.31	−0.67
80	$N(Me)_2$	OMe	H	Me	3.05	2.61	0.44	3.31	−0.26
81	$N(Et)_2$	OMe	H	Me	3.08	3.88	−0.80	3.31	−0.23
83	Et	Me	H	H	4.02	3.68	0.34	4.02	0.00
84**	*n*-Pr	Me	H	H	3.84	4.37	−0.53	4.13	−0.29
85	*i*-Bu	Me	H	H	3.10	2.98	0.12	4.05	−0.95
88	Et	OMe	H	Me	5.27	4.34	0.93	4.72	0.55
89	*n*-Pr	OMe	H	Me	4.02	4.99	−0.97	4.41	−0.39
90	*i*-Bu	OMe	H	Me	3.85	3.73	0.12	4.32	−0.47
91	*sec*-Bu	OMe	H	Me	3.49	3.73	−0.24	4.32	−0.83
93	SEt	OH	H	Me	4.17	4.48	−0.31	4.41	0.03
94**	SEt	OAc	H	Me	3.75	3.51	0.24	3.20	0.55
95	SEt	Me	H	Me	3.90	3.52	0.38	3.39	0.51
97	NHEt	Me	H	Me	1.88	2.73	−0.85	2.10	−0.22
98	NHEt	Me	10-ene		2.70	2.91	−0.21	3.31	−0.61
100	$N(Et)_2$	Me	10-epoxy		1.78	1.79	−0.01	1.98	−0.20
102	Et	OH	H	Me	3.62	3.55	0.07	3.55	0.07

[a] The indentification numbers are those of Nakayama et al. [38]. Missing figures correspond to missing or approximate values in their paper [38]. *TS chemicals and **VS chemicals.

[b] X, Y, R_1, and R_2 substituents of 2,4-dodecadienone in Figure 8.6.

Equation 8.3 is quite complex including a lot of parameters with their square terms while leading to rather modest results. Hansch and Leo [40] have tried to obtain a simpler model than Equation 8.3 by deriving a bilinear model [41] from the descriptors calculated by Nakayama et al. [38]. To do so, the L_x, L_x^2, D, and D^2 descriptors were omitted and 10 chemicals were removed without justification. This led to Equation 8.4:

$$\begin{aligned} \text{pI50} = {} & 1.16(\pm 0.71)\ \log P - 1.04(\pm 0.81)\ \log\left(\beta\ 10^{\log P} + 1\right) \\ & + 0.42(\pm 0.25)\ B_x - 0.57(\pm 0.41)\ I_N + 0.96(\pm 0.27)\ I_{OR} \\ & - 1.18(\pm 0.63)\ I_{br} - 0.91(\pm 0.31)\ I_{(-)} - 1.78(\pm 3.20) \end{aligned} \tag{8.4}$$

$$n = 75,\ s = 0.52,\ r = 0.86$$

According to Hansch and Leo [40], the relatively high standard deviations and low correlation coefficients of these equations clearly called for further analyses.

Consequently, an attempt was made to compute a model presenting better performances while including a fewer number of descriptors. To do so, a TLP was used as statistical engine to optimize the search for complex relationships between the pI50 values recorded on *A.a.* and the molecular descriptors selected by Nakayama et al. [38] without accounting for their squared term. The original data set (Table 8.3) was randomly split into an LS, a TS, and a VS of 69, 8, and 8 chemicals, respectively. TS included chemicals #3, 15, 16, 35, 44, 45, 52, and 59, while VS included chemicals #1, 5, 23, 30, 38, 75, 84, and 94. Use of L_x, D, B_x, log P, I_N, I_{OR}, and $I(-)$ as input neurons in the TLP allowed us to obtain a model with good performances. Because with the Statistica™ software a Boolean descriptor is encoded as two input neurons, a 10/4/1 TLP was designed. Such a TLP with a bias connected to the hidden and output layers and the use of the BFGS second-order training algorithm led to satisfying prediction results. The hidden and output activation functions were a hyperbolic tangent (Tanh) function and a negative exponential function, respectively. The convergence was obtained after 138 cycles. With such a configuration, the correlation coefficients for the LS, TS, and VS were equal to 0.92, 0.94, and 0.95, respectively. An overall correlation coefficient of 0.92 was calculated from the prediction results obtained with the three sets. The TLP model has been more robustly derived than Equation 8.3 and it shows better predictive performances. Thus, for example, inspection of Table 8.3 shows that 13 chemicals (15%) have residuals ≥ 0.8 (in absolute values) when Equation 8.3 is used to estimate the pI50 values of the 85 studied 2,4-dodecadienones, while the use of the TLP model leads to more than twice less chemicals with a residual value ≥ 0.8 (7%). Other trials with random selections of 80% (LS), 10% (TS), and 10% (VS) of the data set and with the same TLP architecture led to broadly the same prediction results. In the same way, use of five or six neurons on the hidden layer increased the performances of the models. It is noteworthy that with such TLP architecture, most of the time, good performances were obtained on the TS and VS despite the high number of connections within the networks.

Equation 8.5 represents the final QSAR model on *T.m.* selected by Nakayama et al. [38]:

$$\text{pI50} = 4.63(\pm1.69)W_x - 0.63(\pm0.21)W_x^{\,2} + 2.58(\pm1.14)D - 0.15(\pm0.06)D^2$$
$$+ 1.76(\pm1.13)\log P - 0.13(\pm0.09)\log P^2 + 0.61(\pm0.30)\pi_{x'} + 0.87(\pm0.26)B_x$$
$$+ 2.89(\pm0.47)I_{NR} - 2.51(\pm0.71)I_{br} - 23.32(\pm5.96) \tag{8.5}$$

$$n=84,\ s=0.54,\ r=0.90$$

In Equation 8.5, W_x is the width of the X end perpendicular to the L axis and in the direction in which the longest chain of the substituent extends in the staggered conformation (Figure 8.6). I_{NR} is an indicator variable for the amides and ketones. It is equal to 1 when X=NHR, NR2, or R and to zero otherwise. The variation in the hydrophobicity of the X end is expressed by the π value of the C(Me)=CH–CO–X moiety. D, B_x, log P, and I_{br} have been described earlier (see Equation 8.3).

Again, an attempt was made to derive a TLP model including fewer descriptors as input neurons than in Equation 8.5 as well as with a minimum of neurons on the hidden layer. The data set was randomly partitioned into an LS, TS, and VS, including 80%, 10%, and 10% of the chemicals, respectively. Rather good results were obtained by selecting W_x, D, B_x, log P, $\pi_{x'}$, and I_{NR} as input neurons and five neurons on the hidden layers. The 7/5/1 TLP also included a bias connected to the hidden and output layers. The BFGS algorithm was used to train the network. The hidden and output activation functions were a hyperbolic tangent (Tanh) function and a negative exponential function, respectively. The convergence was obtained after only 66 cycles. With such a configuration, the correlation coefficients for the LS, TS, and VS were equal to 0.94, 0.93, and 0.92, respectively. An overall correlation coefficient of 0.94 was calculated from the prediction results obtained with the three sets. Inspection of Table 8.4 shows that the TLP models significantly outperform Equation 8.5. Different trials with random selections of 80% (LS), 10% (TS), and 10% (VS) of the data set and with the same TLP architecture allowed us to obtain similar prediction results, but the number of satisfying configurations was lower than in the case of runs performed for selecting the TLP model on the *A.a.* pI50 data.

Basak et al. [42] also used the biological data reported by Henrick et al. [18] to derive QSAR models, focusing on the selection of the most informative descriptors as well as the most efficient statistical method. They increased the number of species and chemicals used in the modeling processes. Thus, they used last larval instars of yellow fever mosquitoes (*A.a.*), fresh pupae of greater wax moths (*G.m.*) and of yellow mealworms (*T.m.*), full grown larvae of house flies (*M.d.*), 2nd- and 3rd instar nymphs of pea aphids (*Acyrthosiphon pisum*, *A.p.*), and larvae of tobacco budworms (*Heliothis virescens*, *H.v.*). The numbers of 2,4-dienoates (+ the censored data) used to compute the QSAR models on *A.a.*, *G.m.*, *T.m.*, *M.d.*, *A.p.*, and *H.v.* were 143(+35), 125(+52), 152(+26), 121(+56), 55(+32), and 109(+51), respectively.

TABLE 8.4

Observed (Obs.) pI50 Values (μM/Pupa) of 2,4-Dodecadienones Tested on *T.m.* and the Calculated (Cal.) Activity Values Computed from Equation 8.5 and the 7/5/1 TLP with the Corresponding Residual Values (Res.)

						Equation 8.5		TLP	
IN[a]	*X*[b]	*Y*	R_1	R_2	Obs.	Cal.	Res.	Cal.	Res.
1	OMe	Me	H	H	2.20	1.43	0.77	2.09	0.11
2	OEt	Me	H	H	3.03	2.81	0.22	2.84	0.19
3*	O-*n*-Pr	Me	H	H	2.71	3.11	−0.40	3.09	−0.38
4	O-*i*-Pr	Me	H	H	4.03	4.11	−0.08	3.66	0.37
5	O-*i*-Bu	Me	H	H	3.80	3.30	0.50	3.26	0.54
6	O-*sec*-Bu	Me	H	H	3.87	4.38	−0.51	4.06	−0.19
8	$OCH_2C{\equiv}CH$	Me	H	H	2.33	1.90	0.43	2.28	0.05
9	$OCH_2C{\equiv}CCH_3$	Me	H	H	2.12	2.55	−0.43	2.67	−0.55
10**	$OCH_2CH{=}CH_2$	Me	H	H	3.51	2.69	0.82	2.87	0.64
11	OPh	Me	H	H	1.50	1.88	−0.38	1.26	0.24
14	OMe	OMe	H	Me	0.89	1.67	−0.78	0.88	0.01
15	OEt	OMe	H	Me	1.92	2.69	−0.77	2.80	−0.88
16	O-*n*-Pr	OMe	H	Me	1.65	2.66	−1.01	2.19	−0.54
17*	O-*i*-Pr	OMe	H	Me	4.89	4.11	−0.78	4.01	0.88
18**	O-*n*-Bu	OMe	H	Me	0.93	1.68	−0.75	1.38	−0.45
19	O-*sec*-Bu	OMe	H	Me	4.67	4.05	0.62	4.38	0.29
20**	O-*t*-Bu	OMe	H	Me	3.81	4.44	−0.63	4.32	−0.51
21	O-*i*-amyl	OMe	H	Me	0.81	0.56	0.25	1.39	−0.58
22	$OCH_2C{\equiv}CH$	OMe	H	Me	1.34	1.23	0.11	1.13	0.21
24	$OCH_2CH{=}CH_2$	OMe	H	Me	1.41	2.11	−0.70	1.54	−0.13
25	$OCH_2CH{=}CHMe$	OMe	H	Me	2.25	1.74	0.51	1.23	1.02

26	OCH(Me)CH=CH_2	OMe	H	Me	4.39	3.57	0.82	3.87	0.52
27	O-*c*-Pr	OMe	H	Me	2.89	3.10	−0.21	3.08	−0.19
28	O-*c*-Bu	OMe	H	Me	2.79	2.73	0.06	2.61	0.18
29	O-*c*-C_5H_4	OMe	H	Me	4.32	3.76	0.56	4.41	−0.09
31*	OEt	Me	H	Cl	3.50	2.79	0.71	2.86	0.64
33	OEt	OAc	H	Me	2.28	2.37	−0.09	2.30	−0.02
34	OEt	OEt	H	Me	2.77	2.48	0.29	2.56	0.21
35*	OEt	SMe	H	Me	2.24	2.79	−0.55	2.80	−0.56
37	OEt	Me	H	Me	2.50	2.81	−0.31	2.79	−0.29
38	OEt	Et	H	H	3.47	2.81	0.66	2.71	0.76
40	OEt	Me	OMe	H	2.54	2.57	−0.03	2.81	−0.27
42	OEt	OMe	OMe	Me	1.31	2.09	−0.78	1.66	−0.35
43	OEt	Me	10-ene		2.38	2.75	−0.37	2.88	−0.50
44	OEt	Me	11-ene		2.49	2.75	−0.26	2.88	−0.39
45	OEt	Me	10-epoxy		2.22	2.22	0.00	2.05	0.17
47	O-*i*-Pr	Me	H	Cl	3.75	4.11	−0.36	3.73	0.02
48	O-*i*-Pr	OH	H	Me	4.33	3.70	0.63	3.96	0.37
49	O-*i*-Pr	OAc	H	Me	3.81	3.78	0.03	3.93	−0.12
50**	O-*i*-Pr	OCHO	H	Me	4.43	3.95	0.48	4.11	0.32
51	O-*i*-Pr	OEt	H	Me	4.51	3.84	0.67	3.87	0.64
52	O-*i*-Pr	O-*i*-Pr	H	Me	3.01	3.87	−0.86	3.80	−0.79
53	O-*i*-Pr	SMe	H	Me	3.77	4.12	−0.35	3.83	−0.06
54	O-*i*-Pr	OCONHEt	H	Me	1.79	1.76	0.03	2.03	−0.24
55	O-*i*-Pr	Me	H	Me	3.53	4.06	−0.53	3.56	−0.03
56*	O-*i*-Pr	Me	OMe	H	3.82	3.99	−0.17	3.94	−0.12
57	O-*i*-Pr	OMe	OMe	Me	3.88	3.64	0.24	4.05	−0.17
58	O-*i*-Pr	OMe	OH	Me	2.61	3.04	−0.43	2.81	−0.20

(continued)

TABLE 8.4 (continued)
Observed (Obs.) pI50 Values (μM/Pupa) of 2,4-Dodecadienones Tested on *T.m.* and the Calculated (Cal.) Activity Values Computed from Equation 8.5 and the 7/5/1 TLP with the Corresponding Residual Values (Res.)

						Equation 8.5		TLP	
IN[a]	*X*[b]	*Y*	R_1	R_2	Obs.	Cal.	Res.	Cal.	Res.
59	O-*i*-Pr	Me	10-ene		3.67	4.10	–0.43	3.79	–0.12
60**	O-*i*-Pr	Me	11-ene		4.10	4.10	0.00	3.79	0.31
61	SMe	Me	H	H	3.25	2.92	0.33	3.09	0.16
62	SEt	Me	H	H	4.05	3.57	0.48	3.37	0.68
63	S-*i*-Pr	Me	H	H	4.22	4.43	–0.21	4.54	–0.32
64	$SCH_2C{\equiv}CH$	Me	H	H	3.47	2.70	0.77	2.76	0.71
65	$SCH_2CH{=}CH_2$	Me	H	H	3.26	2.45	0.81	3.41	–0.15
66	SMe	OMe	H	Me	3.47	3.30	0.17	3.25	0.22
67	SEt	OMe	H	Me	3.49	3.60	–0.11	3.45	0.04
68	S-*i*-Pr	OMe	H	Me	5.04	4.59	0.45	5.06	–0.02
69*	$SCH_2C{\equiv}CH$	OMe	H	Me	2.15	2.54	–0.39	2.47	–0.32
70	$SCH_2CH{=}CH_2$	OMe	H	Me	2.06	1.99	0.07	1.88	0.18
71	NHEt	Me	H	H	4.76	4.47	0.29	4.49	0.27
72	NH-*i*-Pr	Me	H	H	4.33	4.80	–0.47	4.05	0.28
74	$NHCH_2CH{=}CH_2$	Me	H	H	4.33	4.25	0.08	4.07	0.26

75	$N(Me)_2$	Me	H	H	3.70	3.20	0.50	4.27	−0.57
76	$N(Et)_2$	Me	H	H	4.63	5.19	−0.56	4.52	0.11
78	NH-*i*-Pr	OMe	H	Me	4.21	4.32	−0.11	4.28	−0.07
79	NH-*c*-Pr	OMe	H	Me	3.74	3.89	−0.15	4.43	−0.69
80	$N(Me)_2$	OMe	H	Me	3.80	3.04	0.76	3.59	0.21
81	$N(Et)_2$	OMe	H	Me	5.51	4.85	0.66	5.42	0.09
82	NH_2	OMe	H	Me	0.05	0.74	−0.69	0.07	−0.02
83*	Et	Me	H	H	4.89	4.59	0.30	4.35	0.54
84	*n*-Pr	Me	H	H	4.70	5.84	−1.14	4.00	0.70
85**	*i*-Bu	Me	H	H	3.14	3.43	−0.29	2.72	0.42
86**	Ph	Me	H	H	3.52	3.36	0.16	3.47	0.05
88*	Et	OMe	H	Me	5.78	4.79	0.99	5.54	0.24
89	*n*-Pr	OMe	H	Me	5.73	5.67	0.06	5.09	0.64
90	*i*-Bu	OMe	H	Me	3.49	3.38	0.11	3.15	0.34
91	*sec*-Bu	OMe	H	Me	3.54	3.52	0.02	4.27	−0.73
93	SEt	OH	H	Me	3.43	3.59	−0.16	3.72	−0.29
94**	SEt	OAc	H	Me	2.83	3.08	−0.25	2.88	−0.05
97	NHEt	Me	H	Me	4.45	4.63	−0.18	4.48	−0.03
98	NHEt	Me	10-ene		4.78	4.20	0.58	4.48	0.30
100	$N(Et)_2$	Me	10-epoxy		3.85	4.24	−0.39	4.20	−0.35
101	Et	Me	H	H	4.10	4.64	−0.54	4.44	−0.34

[a] The indentification numbers are those of Nakayama et al. [38]. Missing figures correspond to missing or approximate values in their paper [38]. *TS chemicals and **VS chemicals.

[b] X, Y, R_1, and R_2 substituents of 2,4-dodecadienone in Figure 8.6.

The molecules were described by means of topostructural indices (e.g., Wiener number, Randic indices), topochemical indices (e.g., valence and bond connectivity indices, E-state indices), triplet indices and Balaban's J indices, geometric or 3D indices (Kappa shape indices) [43], and atom pairs. Briefly, an atom pair was defined as a substructure consisting of two non-hydrogen atoms i and j and their interatomic separation: {atom descriptor$_i$} – {separation} – {atom descriptor$_j$}, where {atom descriptor} contains the information regarding atom type, the number of non-hydrogen neighbors, and the number of π electrons. Thus, for example, C1X2-6-O0X2 represents a carbon atom with one π electron and two non-hydrogen neighbors (atom #1) and an oxygen atom with no π electrons and two non-hydrogen neighbors (atom #2), with an interatomic separation (including both the atoms) of 6. Initially, 1173 descriptors were computed, including 915 atom pairs. Ridge regression (RR), principal component regression (PCR), and partial least squares (PLS) regression were used as statistical tools, and the modified Gram–Schmidt orthogonalization was employed to trim the 258 global molecular descriptors to a size of 100. The leave-one-out cross-validated q^2 values for the RR [44], PCR [45], and PLS [46] methods were ranged from 0.323 to 0.386, −0.054 to 0.224, and 0.104 to 0.327, respectively. RR giving the best results, it was used to compute the models from atom pairs and from the whole set of descriptors after their trimming by means of the soft threshold method. Twenty-one models were derived with q^2 values ranging from 0.030 to 0.586. Analysis of the best model obtained for each species showed that the top ten descriptors, based on t statistics, mainly included atom pairs. Most of the atom pairs included a heteroatom, such as oxygen or sulfur, and five or more vertices. From this, Basak et al. [42] deducted that the interaction between the ligand and the critical target depended not only on the size of the ligand but also on the chemical nature such as electronegativity, polarity, and polarizability. This inference was supported by the presence of a number of topochemical indices such as neighborhood complexity indices, bond and valence connectivity indices, and internal hydrogen bonders. Because analysis of the residuals obtained with the different models suggested some nonlinearity among the data, recursive partitioning [47] was used to capture any nonlinear relation between the JHA measured on the six species and the global descriptors. The r^2 values obtained for the dendrograms and predictors that lead to significant splits in recursive partitioning equaled 0.54, 0.55, 0.50, 0.61, 0.54, and 0.59 for *A.a.*, *G.m.*, *T.m.*, *M.d.*, *A.p.*, and *H.v.*, respectively. Inspection of the descriptors involved in the partitioning revealed that some of them were common to different species and especially to a specific development stage. Nevertheless, this does not mean that the species behave in a same way against the 2,4-dienoates. This is clearly shown in Figures 8.7 through 8.9 when pairwise comparisons are made between pI50 values. While the different results obtained by Basak et al. [42] are interesting, the performances of their numerous models remain rather modest and the models obtained on *A.a.* do not outperform those obtained by Nakayama et al. [38] (Equation 8.3), by Hansch and Leo [40] (Equation 8.4), and by us in the present study.

In another study, Basak et al. [48] used the same types of descriptors and statistical methods (i.e., RR, PCR, PLS) for modeling a large set of JHMs tested

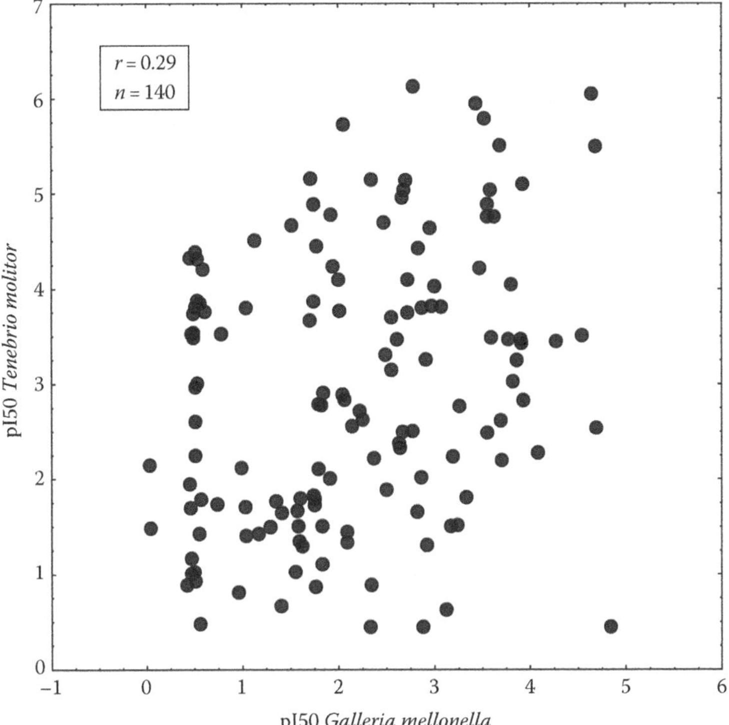

FIGURE 8.7 *Galleria mellonella* versus *Tenebrio molitor* pI50 values for 140 2,4-dienoates.

on *C. pipiens*. Three different predictor-thinning methods, namely, a modified Gram–Schmidt algorithm, a marginal soft thresholding algorithm, and a least absolute shrinkage and selection operator (LASSO), were utilized to reduce the number of descriptors prior to developing QSAR models. The models were derived from 304 pI50 (logarithm of the reciprocal of the concentration (M) at which 50% of metamorphosis was observed) obtained on *C. pipiens* larvae. Molecules were described by a large number of diverse descriptors including 268 global molecular descriptors (topostructural, topochemical, and geometrical), 13 quantum chemical descriptors, and 915 atom pairs (substructural counts). The initial data set was split into five calibration data sets of random samples of sizes 60/110/160/210/260 and the remaining 244/194/144/94/44 chemicals were used for validations. For each of these five divisions of data, the three predictor-thinning methods were used. Subsets of 50, 100, 150, 200, and 250 predictors were chosen by each method (thought this was not always possible, depending on the size of the calibration data). RR was used to compute the models that were evaluated for the calculation of the 10-fold cross-validated q^2 values. LASSO was not found to be a very effective method in handling a large set of molecular descriptors because the number of predictors retained could not exceed the number of observations. The results revealed that the modified Gram–Schmidt algorithm was suited to trim

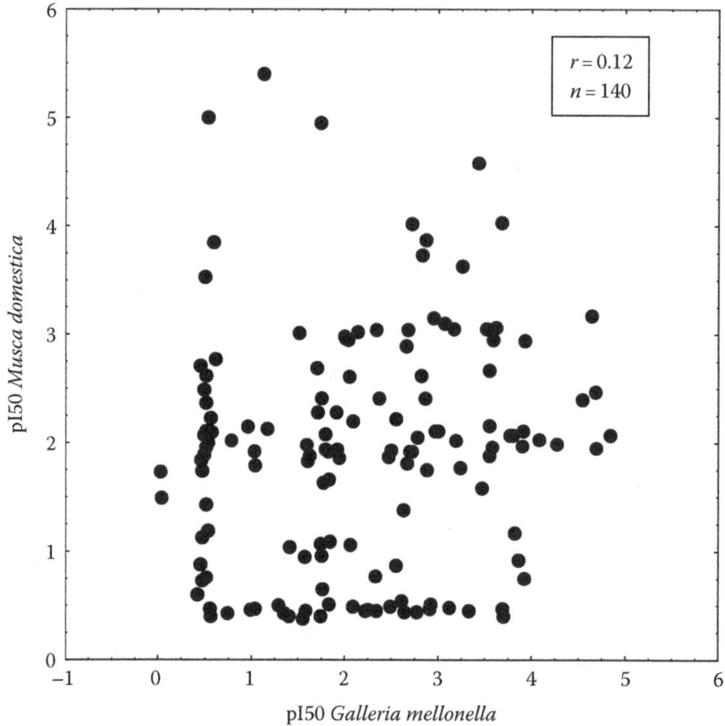

FIGURE 8.8 *Galleria mellonella* versus *Musca domestica* pI50 values for 140 2,4-dienoates.

the number of predictors in the global molecular descriptor set where collinearity of the descriptors was the major concern. On the contrary, the soft thresholding approach was found to be an effective tool in subset selection from a diverse set of descriptors having both sparsity and multicollinearity, as in the case of the combined set of atom pairs and global molecular descriptors. After the evaluation of the applicability of descriptor trimming, Basak et al. [48] decided to fit the JHA of all 304 chemicals using 250 predictors selected from the combined set of predictors by the marginal soft threshold method and to use the RR as statistical tool for deriving the model. The final model was derived on 244 descriptors and leads to a 10-fold q^2 equals to 0.60. Among the 16 predictors with the highest t values, a majority of them were atom pairs and a few were triplet indices. The electrotopological state of oxygen (–O–) was found to be the most important factor that affects JHA. The E-state index of an atom is a measure of the intrinsic state that is perturbed by every other atom in the molecule, accounting for the valence electronegativity of the atom and its local chemical environment. Four of the atom pairs that contain oxygen also appeared as important moieties affecting the JHA of the studied chemicals. The presence in the final model of triplet indices derived from both adjacency and distance matrices also indicated the importance of shape of the ligand.

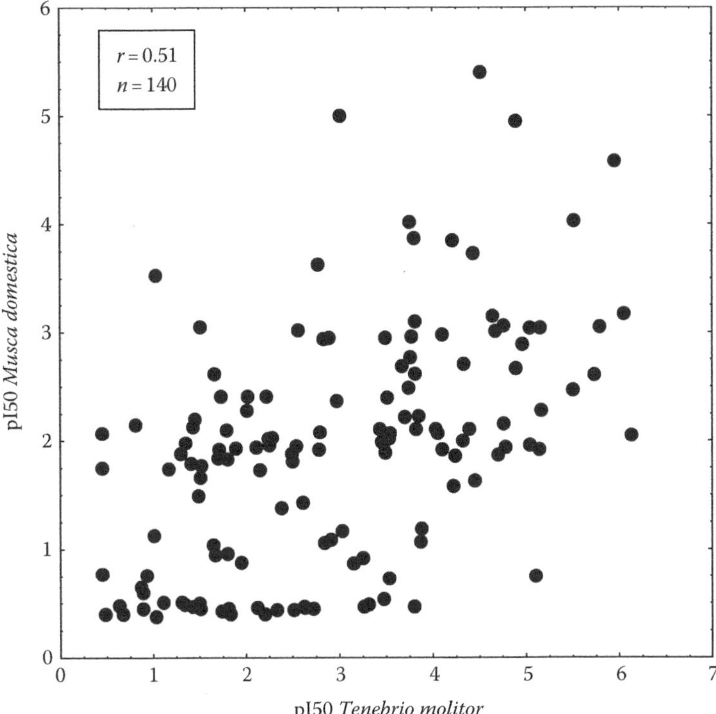

FIGURE 8.9 *Tenebrio molitor* versus *Musca domestica* pI50 values for 140 2,4-dienoates.

8.5 CONCLUSIONS

SAR and QSAR modeling has become a crucial tool in drug design because its implementation in the drug discovery and development process allows us to reduce the time and cost required in lead optimization. The SAR and QSAR modeling techniques play an equally important role in toxicology and ecotoxicology for predicting the adverse effects of chemicals on humans as well as on the invertebrates and vertebrates occupying different trophic levels in the aquatic and terrestrial ecosystems. These methods are also particularly suited to better understand the mechanism of action of the chemicals. Consequently, it is not surprising that these tools have found applications in the discovery of potential new JH analogues. The survey of the published literature and the new models derived in the frame of this study allow us to pinpoint the particularities of the QSAR models aiming at predicting the JHA of chemicals.

A small change in the structure of a chemical can lead to a drastic change in its JHA. This can be observed on one species, while another one, even if belonging to the same taxonomical group, will be only marginally or not affected. This explains the poor interspecies correlations independently of the species as well as their stage of development. Consequently, this is not surprising that for a same chemical data set tested on different insect species for their JHA, different molecular descriptors will be necessary for computing the corresponding QSAR models. In practice, this means that a model derived for a species cannot be used for simulating the same endpoint on another species.

Analysis of the different molecular descriptors included in the models shows that the most important are those encoding steric effects. The 1-octanol/water partition coefficient appears only marginally significant.

Lastly our study reveals that the supervised nonlinear methods outperform the linear regression techniques in the design of powerful predictive QSAR models. This finding is not surprising because these nonlinear tools have shown their great interest to find complex relationships between the activity and the property of structurally diverse sets of molecules and their structure [49,50; see also Chapter 10].

ACKNOWLEDGMENT

The financial support from the French Ministry of Ecology, Sustainable Development, Transport and Housing (MEDDTL) is gratefully acknowledged (PNRPE program).

REFERENCES

1. R. Carson, *Silent Spring*, Houghton Mifflin, Boston, MA, 1962.
2. K. Hartfelder, *Insect juvenile hormone: From "status quo" to high society*, Braz. J. Med. Biol. Res. 33 (2000), pp. 157–177.
3. L.I. Gilbert, N.A. Granger, and R.M. Roe, *The juvenile hormones: Historical facts and speculations on future research directions*, Insect Biochem. Mol. Biol. 30 (2000), pp. 617–644.
4. S.R. Palli, *Recent advances in the mode of action of juvenile hormones and their analogs*, in *Biorational Control of Arthropod Pests*, I. Ishaaya and A.R. Horowitz, eds., Springer, New York, 2009, pp. 111–129.
5. W. Karcher and J. Devillers, *Practical Applications of Quantitative Structure–Activity Relationships (QSAR) in Environmental Chemistry and Toxicology*, Kluwer Academic Publishers, Dordrecht, The Netherlands, 1990, p. 475.
6. J. Devillers and W. Karcher, *Applied Multivariate Analysis in SAR and Environmental Studies*, Kluwer Academic Publishers, Dordrecht, The Netherlands, 1991, p. 530.
7. K. Sláma and C.M. Williams, *Juvenile hormone activity for the bug* Pyrrhocoris apterus, Proc. Nat. Acad. Sci. 54 (1965), pp. 411–414.
8. W.S. Bowers, H.M. Fales, M.J. Thompson, and E.C. Uebel, *Juvenile hormone: Identification of an active compound from balsam fir*, Science 154 (1966), pp. 1020–1021.
9. V.L. Cerný, L. Dolejš, L. Lábler, F. Šorm, and K. Sláma, *Dehydrojuvabione-a new compound with juvenile hormone activity from balsam fir*, Tetrahedron Lett. 12 (1967), pp. 1053–1057.
10. K. Sláma, M. Suchý, and F. Šorm, *Natural and synthetic materials with insect hormone activity. 3. Juvenile hormone activity of derivatives of p-(1,5-dimethyl-hexyl) benzoic acid*, Biol. Bull. 134 (1968), pp. 154–159.
11. K. Sláma, M. Romaňuk, and F. Šorm, *Natural and synthetic materials with insect hormone activity. 2. Juvenile hormone activity of some derivatives of farnesenic acid*, Biol. Bull. 136 (1969), pp. 91–95.
12. K. Sláma, K. Hejno, V. Jarolim, and F. Šorm, *Natural and synthetic materials with insect hormone activity. 5. Specific juvenile hormone effects in aliphatic sesquiterpenes*, Biol. Bull. 139 (1970), pp. 222–228.

13. R.T. Yamamoto and M. Jacobson, *Juvenile hormone activity of isomers of farnesol*, Nature 196 (1962), pp. 908–909.
14. V.B. Wigglesworth, *Chemical structure and juvenile hormone activity: Comparative tests on* Rhodnius prolixus, J. Insect Physiol. 15 (1969), pp. 73–94.
15. J.W. Patterson and M. Schwarz, *Chemical structure, juvenile hormone activity and persistence within the insect of juvenile hormone mimics for* Rhodnius prolixus, J. Insect Physiol. 23 (1977), pp. 121–129.
16. J.W. Patterson and M. Schwarz, *The activity of juvenile hormone mimics for the eggs of* Rhodnius prolixus, J. Insect Physiol. 25 (1979), pp. 399–404.
17. V. Jarolím, K. Hejno, F. Sehnal, and F. Šorm, *Natural and synthetic materials with insect hormones activity. 8. Juvenile activity of the farnesane-type compounds on* Galleria mellonella, Life Sci. 8 (1969), pp. 831–841.
18. C.A. Henrick, W.E. Willy, and G.B. Staal, *Insect juvenile hormone activity of alkyl (2E,4E)-3,7,11-trimethyl-2,4-dodecadienoates. Variations in the ester function and in the carbon chain*, J. Agric. Food Chem. 24 (1976), pp. 207–218.
19. M.M. Metwally and F. Sehnal, *Effects of juvenile hormone analogues on the metamorphosis of beetles* Trogoderma granarium (Dermestidae) *and* Caryedon gonagra (Bruchidae), Biol. Bull. 144 (1973), pp. 368–382.
20. M. Romaňuk, K. Sláma, and F. Šorm, *Contribution of a compound with a pronounced juvenile hormone activity*, Proc. Natl. Acad. Sci. 57 (1967), pp. 349–352.
21. W.F. Walker and W.S. Bowers, *Comparative juvenile hormone activity of some terpenoid ethers and esters on selected Coleoptera*, J. Agric. Food Chem. 21 (1973), pp. 145–148.
22. A. Niwa, H. Iwamura, Y. Nakagawa, and T. Fujita, *Development of (phenoxyphenoxy)- and (benzylphenoxy)propyl ethers as potent insect juvenile hormone mimetics*, J. Agric. Food Chem. 37 (1989), pp. 462–467.
23. A. Niwa, H. Iwamura, Y. Nakagawa, and T. Fujita, *Development of N,O-disubstituted hydroxylamines and N,N-disubstituted amines as insect juvenile hormone mimetics and the role of the nitrogenous function for activity*, J. Agric. Food Chem. 38 (1990), pp. 514–520.
24. T. Hayashi, H. Iwamura, and T. Fujita, *Insect juvenile hormone mimetic activity of (4-substituted)phenoxyalkyl compounds with various nitrogenous and oxygenous functions and its relationship to their electrostatic and stereochemical properties*, J. Agric. Food Chem. 39 (1991), pp. 2029–2038.
25. T. Hayashi, H. Iwamura, and T. Fujita, *Development of 4-alkylphenyl aralkyl ethers and related compounds as potent insect juvenile hormone mimetics and structural aspects of their activity*, J. Agric. Food Chem. 38 (1990), pp. 1965–1971.
26. T. Hayashi, H. Iwamura, and T. Fujita, *Electrostatic and stereochemical aspects of insect juvenile hormone active compounds: A clue for high activity*, J. Agric. Food Chem. 38 (1990), pp. 1972–1977.
27. H.A. Schneiderman, A. Krishnakumaran, V.G. Kulkarni, and L. Friedman, *Juvenile hormone activity of structurally unrelated compounds*, J. Insect Physiol. 11 (1965), pp. 1641–1649.
28. G. Brieger, *Juvenile hormone mimics: Structure–activity relationships for* Oncopeltus fasciatus, J. Insect Physiol. 17 (1971), pp. 2085–2093.
29. T. Hayashi, H. Iwamura, T. Fujita, N. Takakusa, and T. Yamada, *Structural requirements for activity of juvenile hormone mimetic compounds against various insects*, J. Agric. Food Chem. 39 (1991), pp. 2039–2045.
30. G.P. Nilles, M.J. Zabik, R.V. Connin, and R.D. Schuetz, *Synthesis of bioactive compounds. A structure-activity study of aryl terpenes as juvenile hormone mimics*, J. Agric. Food Chem. 24 (1976), pp. 699–708.
31. G.P. Nilles, M.J. Zabik, R.V. Connin, and R.D. Schuetz, *Synthesis of bioactive compounds: Juvenile hormone mimetics affecting insect diapause*, J. Agric. Food Chem. 21 (1973), pp. 342–347.

32. A. Niwa, H. Iwamura, Y. Nakagawa, and T. Fujita, *Development of (phenoxyphenoxy)- and (benzylphenoxy)alkanaldoxime O-ethers as potent insect juvenile hormone mimics and their quantitative structure-activity relationship*, J. Agric. Food Chem. 36 (1988), pp. 378–384.
33. J. Devillers, *Neural Networks in QSAR and Drug Design*, Academic Press, London, U.K., 1996.
34. Anonymous, *The Principles for Establishing the Status of Development and Validation of (Quantitative) Structure-Activity Relationships (Q)SARs, OECD document*, ENV/JM/TG(2004), 27.
35. V.N. Vapnik, *The Nature of Statistical Learning Theory*, Springer, New York, 1995.
36. N. Cristianini and J. Shawe-Taylor, *An Introduction to Support Vector Machines and other Kernel-Based Learning Methods*, Cambridge University Press, Cambridge, U.K., 2000.
37. T. Hayashi, H. Iwamura, Y. Nakagawa, and T. Fujita, *Development of (4-alkoxyphenoxy)- and (4-alkylphenoxy)alkanaldoxime O-ethers as potent insect juvenile hormone mimics and their structure-activity relationships*, J. Agric. Food Chem. 37 (1989), pp. 467–472.
38. A. Nakayama, H. Iwamura, and T. Fujita, *Quantitative structure-activity relationship of insect juvenile hormone mimetic compounds*, J. Med. Chem. 27 (1984), pp. 1493–1502.
39. H. Iwamura, K. Nishimura, and T. Fujita, *Quantitative structure-activity relationships of insecticides and plant growth regulators: Comparative studies toward understanding the molecular mechanism of action*, Environ. Health Perspect. 61 (1985), pp. 307–320.
40. C. Hansch and A. Leo, *Exploring QSAR. Fundamentals and Applications in Chemistry and Biology*, ACS Professional Reference Book Series, Washington, DC, 1995.
41. S. Bintein, J. Devillers, and W. Karcher, *Nonlinear dependence of fish bioconcentration on n-octanol/water partition coefficient*, SAR QSAR Environ. Res. 1 (1993), pp. 29–39.
42. S.C. Basak, R. Natarajan, D. Mills, D.M. Hawkins, and J.J. Kraker, *Quantitative structure-activity relationship modeling of insect juvenile hormone activity of 2,4-dienoates using computed molecular descriptors*, SAR QSAR Environ. Res. 16 (2005), pp. 581–606.
43. J. Devillers and A.T. Balaban, *Topological Indices and Related Descriptors in QSAR and QSPR*, Gordon and Breach Science Publishers, Amsterdam, the Netherlands, 1999.
44. A.E. Hoerl and R.W. Kennard, *Ridge regression: Applications to nonorthogonal problems*, Technometrics 12 (1970), pp. 69–82.
45. J. Devillers, D. Zakarya, M. Chastrette, and J.C. Doré, *The stochastic regression analysis as a tool in ecotoxicological QSAR studies*, Biomed. Environ. Sci. 2 (1989), pp. 385–393.
46. H. Abdi, *Partial least squares regression and projection on latent structure regression (PLS Regression)*, WIREs Comput. Stat. 2010, DOI: 10.1002/wics.051.
47. S.S. Young and D.M. Hawkins, *Using recursive partitioning to analyze a large SAR data set*, SAR QSAR Environ. Res. 8 (1998), pp. 183–193.
48. S.C. Basak, R. Natarajan, D. Mills, D.M. Hawkins, and J.J. Kraker, *Quantitative structure-activity relationship modeling of juvenile hormone mimetic compounds for* Culex pipiens *larvae, with a discussion of descriptor-thinning methods*, J. Chem. Inform. Model. 46 (2006), pp. 65–77.
49. J. Devillers, J.P. Doucet, A. Panaye, N. Marchand-Geneste, and J.M. Porcher, *Structure-activity of a diverse set of androgen receptor ligands*, in *Endocrine Disruption Modeling*, J. Devillers, ed., CRC Press, Boca Raton, 2009, pp. 335–355.
50. J. Devillers, J.P. Doucet, A. Doucet-Panaye, A. Decourtye, and P. Aupinel, *Linear and non-linear QSAR modelling of juvenile hormone esterase inhibitors*, SAR QSAR Environ. Res. 23 (2012), pp. 357–369.

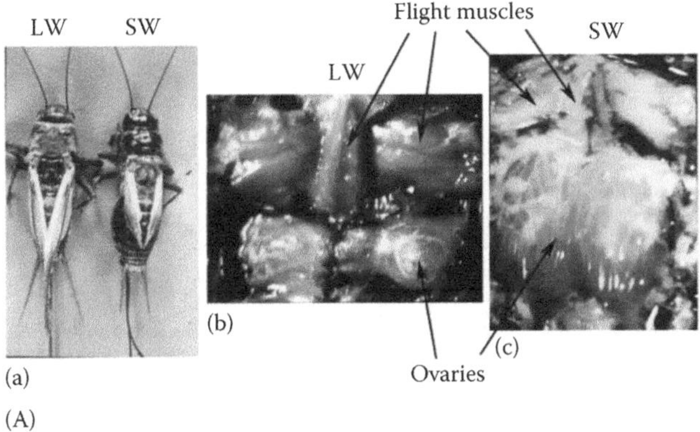

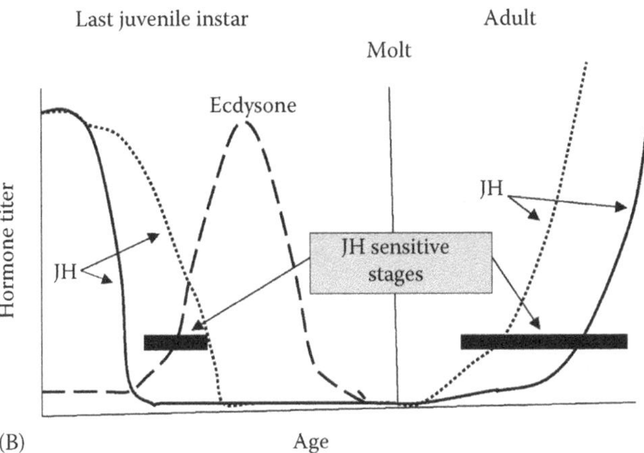

FIGURE 3.1 (A) Flight-capable, LW and flightless, SW female morphs of *G. firmus* of the same age (day 5 of adulthood). In the panel (a), the forewings have been removed to show variation in the hind wings. The panels (b and c) illustrate dissections of the same-aged morphs showing much larger, functional flight muscles but much smaller ovaries in the LW females relative to SW females. (B) The "classic model" of JH-regulated complex polymorphism. The panel illustrates hypothetical variation in the JH titer during restricted periods (denoted by solid bars) of the juvenile and adult stages that regulates morph-specific development and reproduction. Note: "LW" is the same as "LW(f)" in some figures as follows. (From Zera, A.J., *Evol. Dev.*, 9, 499, 2007; Zera, A.J., *Integr. Comp. Biol.*, 45, 511, 2005, by permission of Oxford University Press.)

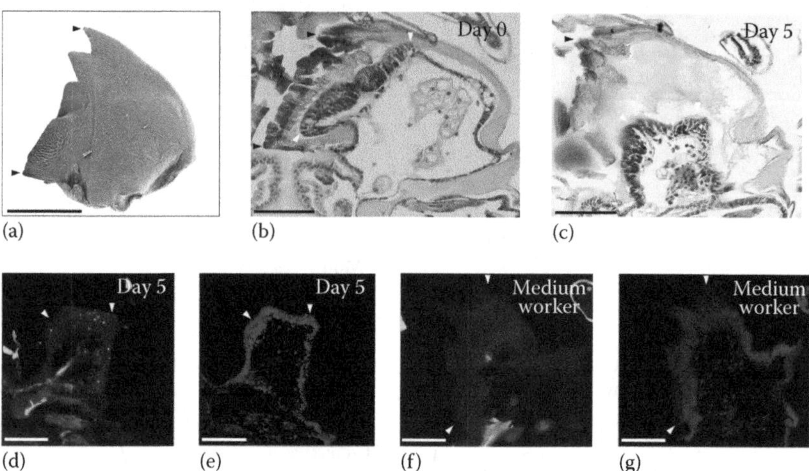

FIGURE 4.8 Scanning electron micrographs of right mandible of minor workers (a). Sections of day 0 (b) and day 5 (c) minor workers. TUNEL and DAPI staining of sectioned mandibles from day 5 minor workers (d, e) and a gut-purged medium worker prior to the molting into major workers (f, g). Black and white arrowheads show corresponding regions. Scale bars indicate 100 μm. (With kind permission from Springer Science and Business Media: *Naturwissenschaften*, The TUNEL assay suggests mandibular regression by programmed cell death during presoldier differentiation in the nasute termite *Nasutitermes takasagoensis*, 98, 801–806, 2011, Toga, K., Yoda, S., and Maekawa, K.)

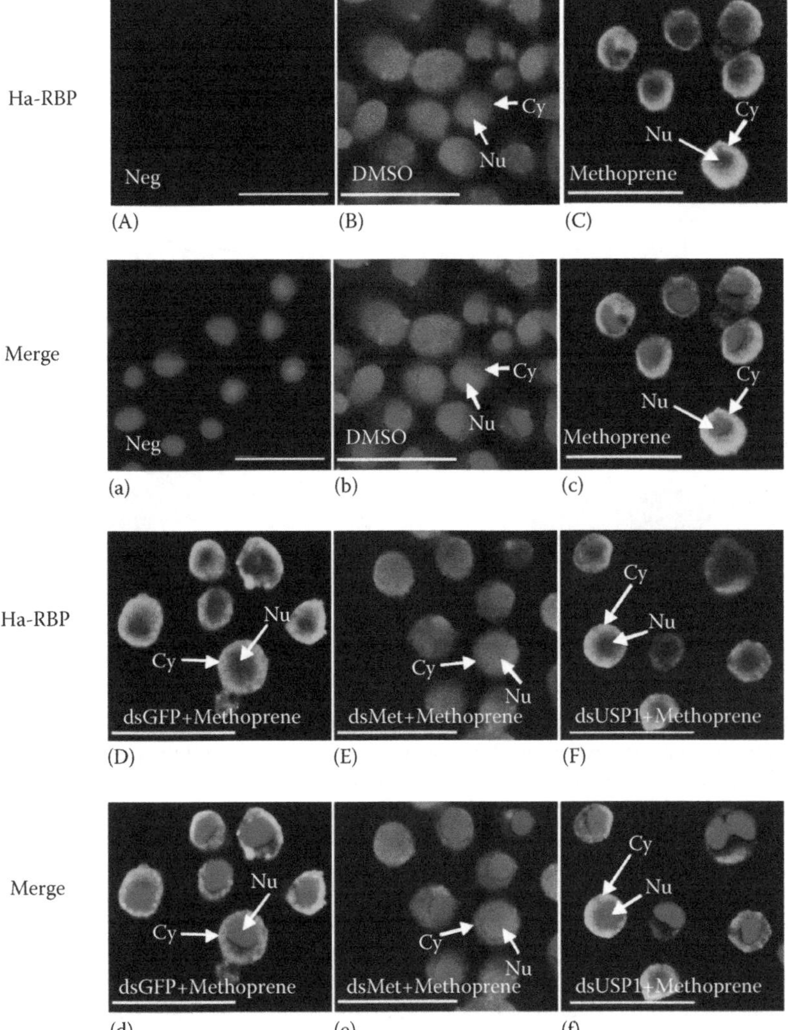

FIGURE 5.3 Methoprene regulation on the subcellular translocation of Ha-RBP in HaEpi cells. Panels (A) and (a), negative control that cells treated with pre-immune serum; panels (B) and (b), cells treated with DMSO; panels (C) and (c), cells treated with methoprene for 6 h at a final concentration of 1 μM; panels (D) and (d), cells incubated with dsGFP followed by methoprene incubation; panels E and e, cells incubated with dsMet followed by methoprene incubation; panels F and f, cells incubated with dsUSP1 followed by methoprene incubation. Green portions (Alexa 488) indicate Ha-RBP detected with anti-Ha-RBP. Blue portions (DAPI) indicate nuclei. Nu, nucleus; Cy, cytoplasm. Size bars = 50 μm. At least three independent experiments were performed.

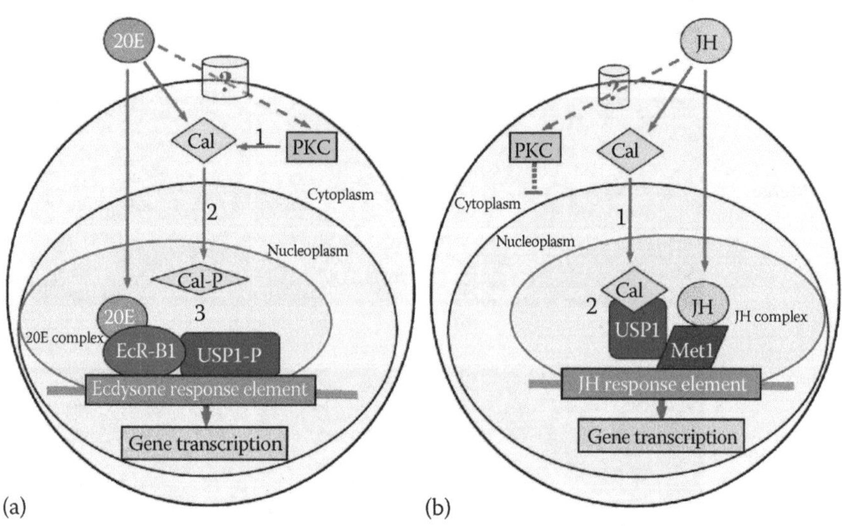

FIGURE 5.5 Chart explaining the function of HaCal in the cross talk between 20E and JH pathways. (a) 20E pathway: 20E signaling leads HaCal protein phosphorylation by PKC via an unknown membrane pathway (1), HaCal is translocated into the nuclei (2), and phosphorylated HaCal does not bind with phosphorylated USP1 in 20E pathway (3). EcR binds 20E and USP and other chaperone protein to form transcription complex, combining the 20E response element initiating the gene transcription. (b) JH pathway: Methoprene maintains HaCal non-phosphorylation and translocates HaCal into the nuclei (1) and non-phosphorylated HaCal binds with non-phosphorylated USP1 (2). Met binds JH and interacts with USP and other chaperone proteins, and then this complex binds JH response element via Met to initiate JH signaling pathway. (From Li, M. et al., *Proc. Natl. Acad. Sci. USA*, 108, 638, 2011; Miura, K. et al., *FEBS J.*, 272, 1169, 2005; Li, Y. et al., *J. Biol. Chem.*, 282, 37605, 2007; Riddiford, L.M. et al., *Insect Biochem. Mol.*, 33, 1327, 2003; Antoniewski, C. et al., *Mol. Cell. Biol.*, 16, 2977, 1996; Hiruma, K. and Riddiford, L.M., *Dev. Biol.*, 272, 510, 2004; Lan, Q. et al., *Mol. Cell Biol.*, 19, 4897, 1999; Stone, B.L. and Thummel, C.S., *Cell*, 75, 307, 1993; Yao, T.P. et al., *Nature*, 366, 476, 1993; Yao, T.P. et al., *Cell*, 71, 63, 1992.)

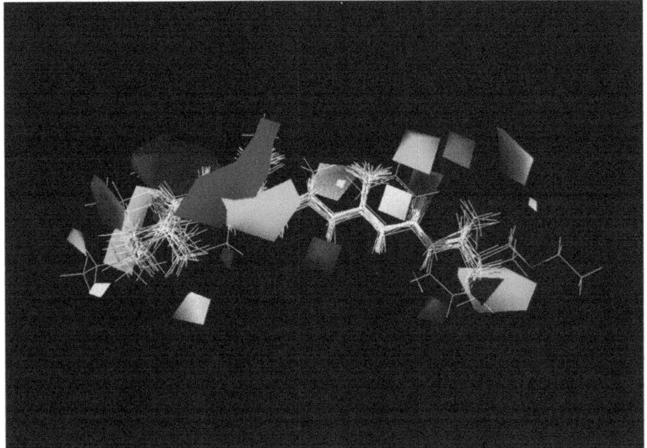

(a)

(b)

FIGURE 9.8 Graphic representation of CoMFA steric and electrostatic field contour plot for Class I (a) and Class II (b) juvenoids. Yellow-polyhedron-surrounded regions indicate sites where less bulky substituents are appreciated for increasing biological activity, whereas green polyhedra represent sterically favored regions where more bulky substituents increase biological activity. The red contours represent regions where negative charge is favorable, while blue polyhedra represent electrostatic regions where positively charged groups will be favorable and will enhance biological activity. The importance of the electronegative oxygens or nitrogens on the ends of the aligned structures is indicated by red polyhedra near the positions of these atoms. The presence of smaller red polyhedra in the middle of the JH agonist structure indicates an additional site where an electronegative atom or group can enhance biological activity importantly. The contour map of Class II (b) differs from that of Class I (a) mainly in the steric fields. The large green polyhedra of Class II (b) near the part of the phenoxyphenyl group (left side) indicate that the presence of steric bulk substituents in this part of the molecule enhances biological activity.

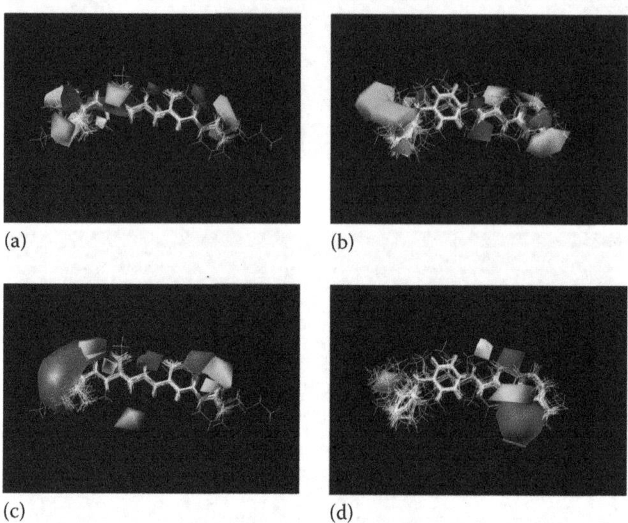

FIGURE 9.9 Contour maps of Class I and Class II JH agonists as revealed by CoMSIA analyses. Green and yellow polyhedra for sterical fields and blue plus red polyhedra for electrostatic fields have the same representative meaning as for CoMFA maps in Figure 9.8 for both Class I (a) and Class II (b) JH agonists. Contour maps for the hydrogen bond acceptor and donor fields are illustrated in (c) (for Class I agonists) and (d) (for Class II agonists). Magenta areas indicate regions where hydrogen bond acceptors are favorable for increasing biological activity (oxygens and nitrogens in the ligand); cyan areas indicate fields where hydrogen bond donors are favorable (NH and OH groups in ligand). Orange polyhedra surround area where H-bond acceptors are unfavorable and white polyhedral areas where H-bond donors are unfavorable.

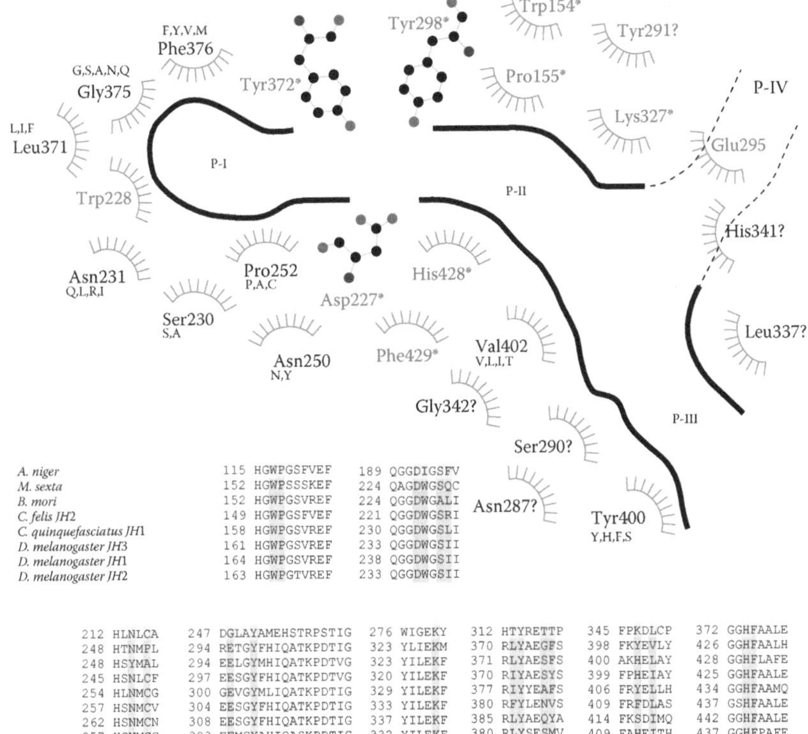

A. niger	115 HGWPGSFVEF	189 QGGDIGSFV
M. sexta	152 HGWPSSSKEF	224 QAGDWGSQC
B. mori	152 HGWPGSVREF	224 QGGDWGALI
*C. felis JH*2	149 HGWPGSFVEF	221 QGGDWGSRI
*C. quinquefasciatus JH*1	158 HGWPGSVREF	230 QGGDWGSLI
*D. melanogaster JH*3	161 HGWPGSVREF	233 QGGDWGSII
*D. melanogaster JH*1	164 HGWPGSVREF	238 QGGDWGSII
*D. melanogaster JH*2	163 HGWPGTVREF	233 QGGDWGSII

212 HLNLCA	247 DGLAYAMEHSTRPSTIG	276 WIGEKY	312 HTYRETTP	345 FPKDLCP	372 GGHFAALE
248 HTNMPL	294 RETGYFHIQATKPDTIG	323 YLIEKM	370 RLYAEGFS	398 FKYEVLY	426 GGHFAALH
248 HSYMAL	294 EELGYMHIQATKPDTVG	323 YILEKF	371 RLYAESFS	400 AKHELAY	428 GGHFLAFE
245 HSNLCF	297 EESGYFHIQATKPDTVG	320 YILEKF	370 RIYAESYS	399 FPHEIAY	425 GGHFAALE
254 HLNMCG	300 GEVGYMLIQATKPDTIG	329 YILEKF	377 RIYYEAFS	406 FRYELLH	434 GGHFAAMQ
257 HSNMCV	304 EESGYFHIQATKPDTIG	333 YILEKF	380 RFYLENVS	409 FRFDLAS	437 GSHFAALE
262 HSNMCN	308 EESGYFHIQATKPDTIG	337 YILEKF	385 RLYAEQYA	414 FKSDIMQ	442 GGHFAALE
257 HSNMCG	303 EEMGYAHIQASKPDTIG	332 YILEKF	380 RLYSESMV	409 FAHEITH	437 GGHFPAFE

FIGURE 11.2 (Top) Binding site of JHEHs; a number of residues correspond to *M. sexta* JHEH, pockets P-I-P-IV are labeled, and the meaning of colored letters, asterisks, and question marks is explained in the text. (Bottom) Alignment of relevant motifs between several JHEHs and the *A. niger* EH; residues in the top part of the figure are highlighted in cyan in the sequences.

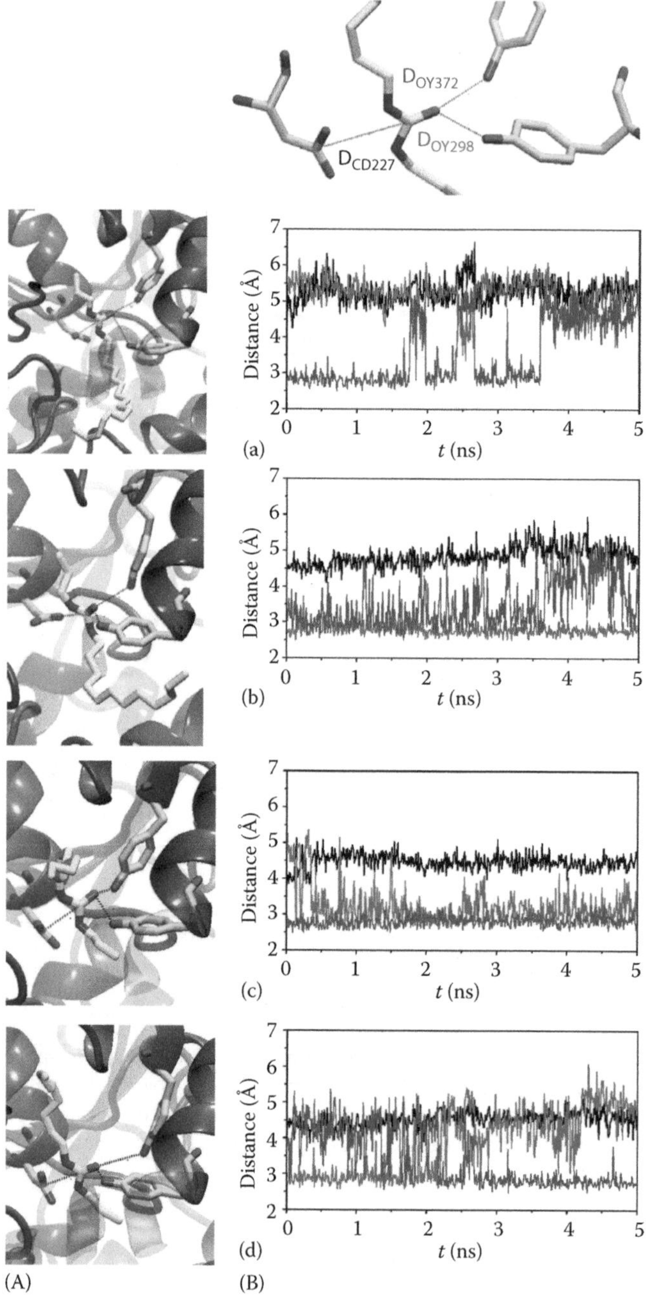

FIGURE 11.5 MD of complexes containing compounds **1–4** (a–d). (A) Orientation of the inhibitor (yellow) within JHEH's active site; Asp227, Tyr298, and Tyr372 are in cyan and stick representation. (B) Distances between groups of inhibitor and residues of JHEH for the complexes extracted from MD simulations. D_{CD227}, D_{OY298}, and D_{CD372} are represented in black, red, and blue, respectively.

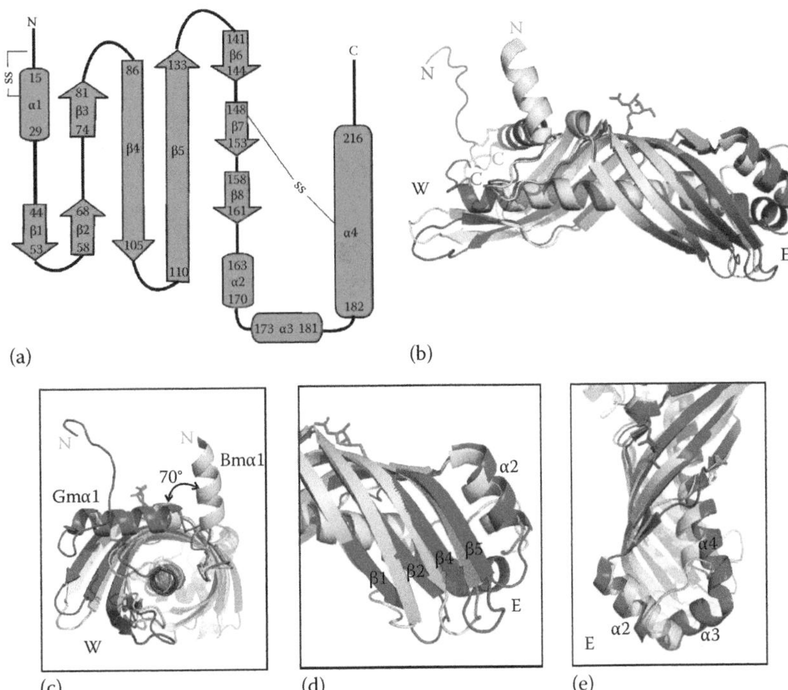

FIGURE 12.2 Overall structure of JHBP and structural comparison of GmJHBP (PDB code: 2rck) and BmJHBP (3aot). (a) Topology diagram of GmJHBP. Cylinders and arrows represent α-helices and β-strands, respectively. The secondary structure elements are numbered from α1 to α4 and from β1 to β8. All numbers correspond to amino acid residues in the GmJHBP sequence. (b) A cartoon representation of JHBP fold. The GmJHBP structure shown in red is superposed on the apo-BmJHBP structure (3aot) shown in green. The Cys10–Cys17 disulfide bridge (yellow) is visible near the N-terminus. At the top of the GmJHBP molecule, one *N*-acetylglucosamine residue is attached to Asn94. The axial poles of the molecule are defined as its W and E ends. (c) The W pole of the JHBP structures is shown illustrating different conformations of helix α1 in GmJHBP (Gmα1) and apo-BmJHBP (Bmα1). (d) The strands β1–β2 and β4–β5 are curved differently in the two structures. (e) The BmJHBP structure has an extended loop instead of the helix α3 present in GmJHBP. This and all other structural illustrations were prepared in PyMol. (From W. L. DeLano, The PyMOL Molecular Graphics System, DeLano Scientific, San Carlos, CA, (2002), http://www.pymol.org.)

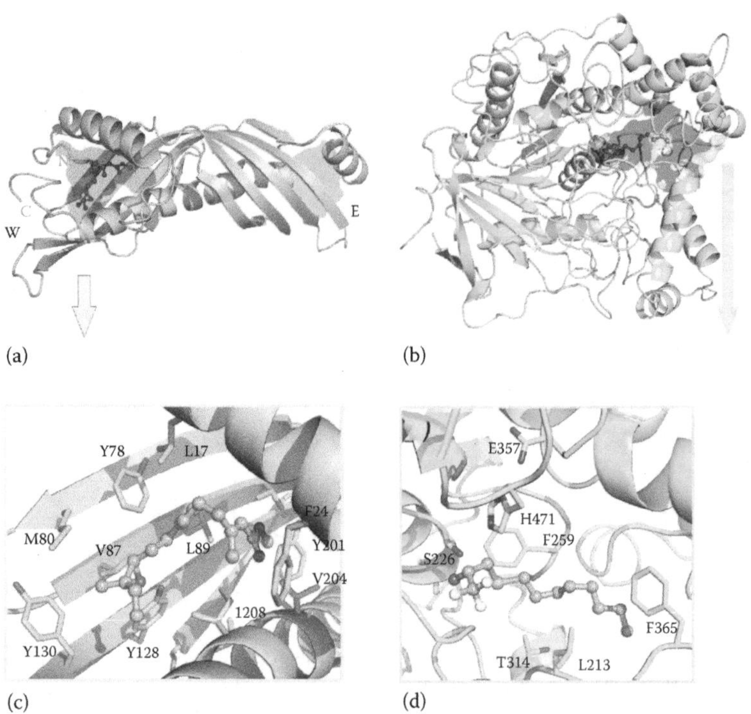

FIGURE 12.5 The JH binding sites of BmJHBP and MsJHE. (a) The BmJHBP–JH II complex (PDB code: 3aos), with the JH II molecule bound in the W cavity. (b) The overall fold of the MsJHE protein (PDB code: 2fj0) and the location of JH binding site, with a JH analog (OTFP) bound. The ligands are shown as orange sticks and balls. The binding pockets W and E in BmJHBP and the JH binding site in MsJHE are in gray. (c) The W cavity of BmJHBP. The majority of residues forming the pocket are shown in green in stick representation. (d) The JH binding site in the MsJHE structure, the catalytic triad (Ser226, E357, and His471), and the residues interacting with an JH analog (orange) are shown in green in stick representation.

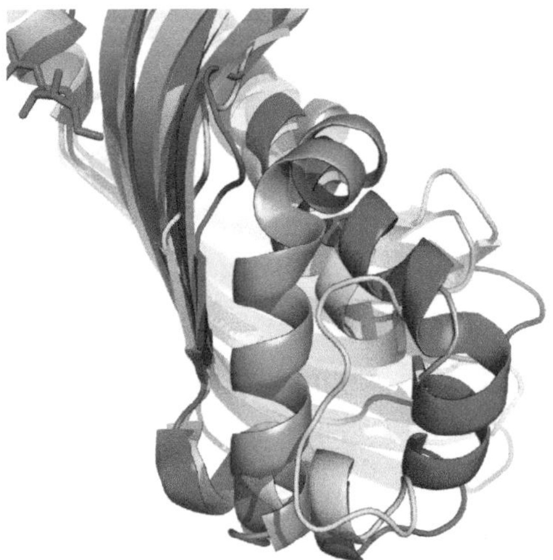

FIGURE 12.6 Structural comparison of GmJHBP (red), BmJHBP (green), and EpTol (blue) in the area of the E pole, where the molecules show the highest degree of structural divergence.

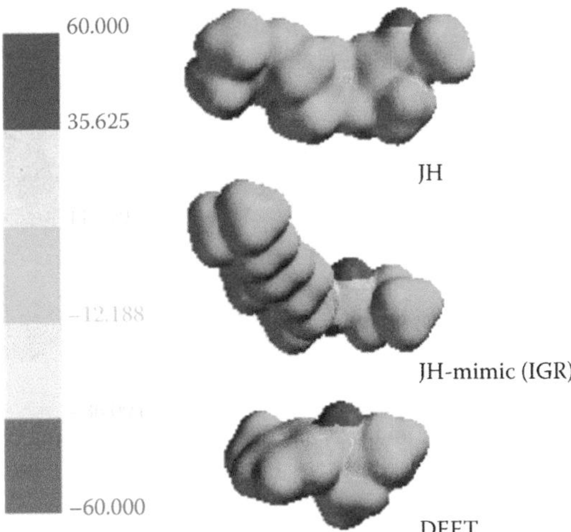

FIGURE 13.3 MEP maps of JH, JH mimic, and DEET on the van der Waals surface showing similarity of the single nucleophilic site and large area of hydrophobic similarity.

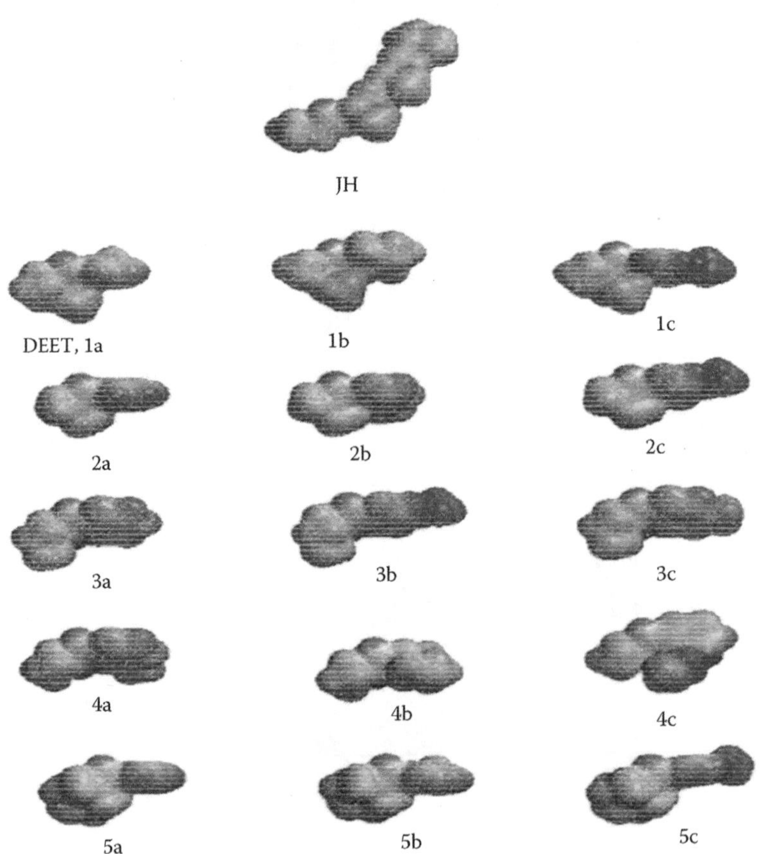

FIGURE 13.4 MEP maps of JH and DEET-like compounds showing electronic similarity of the carbonyl group. (From Bhattacharjee, A.K. et al., *J. Mol. Recognit.*, 13, 213, 2000.)

9 Using CoMFA and CoMSIA as Tools in a 3D QSAR Analysis of Juvenile Hormone Agonist Action in *Drosophila*

Robert Farkaš and Maja Polakovičová

CONTENTS

ABSTRACT

Using 3D QSAR CoMFA and CoMSIA methods, we built predictive models for a structurally diverse set of 86 JH analogs. The combined bioassay- and *in silico*-generated pharmacophore model of juvenoids has revealed important structural features necessary for morphogenetic JH activity in *Drosophila* and has specified the dominant interactions that these juvenoids exhibit with putative JH receptor. By dividing the training set of molecules in two subsets, Class I and Class II, we gained models with better statistical parameters. This may indicate that these two classes of molecules exhibit a different mode of binding to the JH receptor. By means of this approach we have obtained both computer-aided and species-specific pharmacophore fingerprints of JH and its agonists, which revealed that the most active compounds must possess an electronegative atom (oxygen or nitrogen) at both ends of the molecule. When either of these electronegative atoms are replaced by carbon or the distance between them is shorter than 11.5 Å or longer than 13.5 Å, their biological activity is widely decreased. The presence of an electron-deficient moiety in the middle of the farnesene-type JH agonist is also essential for high activity. The information from 3D QSAR provides guidelines and a mechanistic scope for the identification of steric and electrostatic properties as well as donor and acceptor hydrogen-bonding that are important features of the ligand-binding cavity of a JH target protein. In order to refine the pharmacophore analysis and evaluate the outcomes of the CoMFA and CoMSIA study, we have used the pseudoreceptor modeling software PrGen to generate a putative binding site surrogate that is composed of eight amino acid residues corresponding to the defined molecular interactions for both classes of juvenoids. We have shown that computational CoMFA and CoMSIA models, with their graphical interpretation, can also serve to guide the design of novel JH-compounds and predict their biological activity. The low activity or inactivity of several juvenoids in *Drosophila* with relatively high or very high activity in other species already suggests that 3D QSAR pharmacophore models based on numerous bioassays can reveal biologically meaningful, distinct species-specific differences, as well as help uncover more of the atomistic details of the JH mode of action.

KEYWORDS

Juvenile hormone agonists, 3-D QSAR, Pharmacophore analysis, Atomistic pseudoreceptor model, *Drosophila*

9.1 INTRODUCTION

Juvenile hormones (JHs) are a unique group of simple sesquiterpenic compounds produced by the specialized endocrine gland corpora allata whose major function is to prevent precocious metamorphosis in immature larval stages [1–4]. In addition, JH plays critical roles in a rich array of other processes, including reproduction, behavior, pheromone production, adult diapause, morph and caste determination, and polyphenism [5,6]. Besides JH, insect development is primarily regulated by the steroid hormone ecdysone (Ec), pulses of which drive the transition between each developmental stage as well as contribute to the regulation of reproduction [7–9]. Compared to our understanding of the mode of action of JH, the mode of Ec action is relatively well understood, in part due to extensive research on vertebrate steroid endocrinology that supported research on Ec action in *Drosophila* and other insects (for reviews, see [10–12]). In general, it is thought that while Ec induces every molt, JH regulates its outcome; by another words, JH modulates action of Ec [8,13,14]. Ec, like all other steroids, acts by binding to its cognate receptor, EcR. This requires heterodimerization with another nuclear receptor partner, Usp, encoded by the *ultraspiracle* (*usp*) gene. The ligand-bound heterodimer then recognizes Ec-responsive elements (EcREs) within the regulatory regions of a particular gene(s) to regulate its expression [15–20]. During metamorphosis, the EcR/Usp complex triggers the expression of a small series of early (regulatory) genes, which in turn orchestrate a complex cellular response through the coordinated expression of the cascade of late (effector) genes [9,11,12,21,22].

Our incomplete understanding of JH action is not due to a lack of effort [23–26] but rather originates from the unique properties of JH as a hormone [27]. Therefore, Gilbert et al. [28] in one of his reviews correctly wrote "in all of endocrinology there is no more wondrous name for a hormone than the insect juvenile hormone." Most insects have JH-III (epoxy farnesoic acid methyl ester) as a natural JH [29]. A few insect groups have also less frequent JH-I, JH-II, and JH-0 hormones. Williams [3] was the first to demonstrate exogenous JH effects without needing to transplant active corpora allata. He was able to develop a simple assay for JH activity using lipid extracts, which later motivated Law to prepare a synthetic mixture of farnesoates [30] and the discovery of so-called paper factors (juvabione and dehydrojuvabione) as the active JH compounds from balsam fir [31]. Since then, large numbers of JH analogs, called juvenoids (see in the following), have been synthesized by different research laboratories and agrochemical companies. Very early on, Williams [32] suggested that these juvenoids could serve as a new generation of insecticides. A comprehensive survey of available juvenoid structures tested prior to 1985 revealed more than 4000 compounds for which JH activity has been documented in at least one insect species [33–38]. There are also potential juvenoids with hypothesized biological activity (mainly from the patent literature), so the total number of juvenoid compounds is close to 6000.

The elementary structural type of many juvenoids is related to natural JH compounds and derived from a common sesquiterpenoid backbone. This is the case for farnesol, the first chemical compound discovered to have JH activity. Farnesol was identified by Schmialek [39] long before it was known that 10,11-epoxyfarnesoates and their homologues were endogenous JH molecules. However, the chemical structure of many juvenoid molecules is totally unrelated to that of endogenous JH or sesquiterpenoids. Currently, a large series of the juvenoids have chemical structures that categorically exclude them from having a common biogenetic origin with endogenous JH. It is for this reason that the term juvenoid was proposed as a generic name for all the hormonomimetic compounds exhibiting pharmacobiological properties of JH in insects.

Juvenoid structures are one of the most structurally divergent groups of compounds to share common biological activities. This may, in part, explain why it has been so difficult to identify the JH receptor. At the same time, the great structural diversity of JH analogs or agonists has provided a valuable set of pharmacological tools to explore JH action at the molecular level.

9.2 RATIONALE OF PHARMACOPHORE ANALYSIS *IN SILICO* EXPLORING *DROSOPHILA*

Although studies in *Drosophila* have contributed greatly to establishing molecular mechanism of action of Ec and steroid hormones in general—it is the best known representative of cyclorrhaphous diptera and an ideal genetic model organism—it was mostly omitted from the juvenoid research, perhaps because it is not a pest. Only a limited number of juvenoids had been tested in *Drosophila* [40–46] and their structure–activity relationships were never systemically evaluated. Even though it was known that the application of JH to *Drosophila* and other related cyclorrhaphous dipterans causes a complex set of morphogenetic effects (perturbation of abdominal bristle differentiation including shortened or completely missing bristles, defects in abdominal cuticle pigmentation, misrotation of male genitalia, disruption of metamorphosis of the nervous and muscular systems, inhibition of eclosion), most tests of compounds for JH activity in different insect species, including *Drosophila*, have relied on using inhibition of eclosion as a simple parameter to evaluate their biological effectiveness. Since the eclosion of adults in many insect species can be inhibited by numerous nonhormonal chemicals due to nonspecific and generally toxic effects, we chose to evaluate the JH activity of each compound by taking into account a specific, unambiguous morphogenetic effect of JH action in *Drosophila*. To exclude such nonspecific effects, we initially evaluated each compound for its ability to cause a minimum of two independent morphogenetic effects, i.e., the misrotation of male genitalia and the perturbation of abdominal bristle differentiation (Figure 9.1). Then, to quantify the effects of different juvenoids in a structure–activity study, the ED_{50} of a compound was expressed as the number of micrograms of the compound able to cause macrochaeteless and microchaeteless tergites in the abdomens of 50% of the animals.

In the genomic era, studies utilizing *Drosophila* offer considerable promise for understanding the molecular mechanism of JH action and to identify the JH receptor and thus to explain the plethora of observations on JH accumulated in the past decades. As a prerequisite to this goal, we performed a pharmacophore analysis to understand the 3D structure of the site in the JH receptor able to bind natural JH. We evaluated the hormonomimetic properties of 86 juvenile hormone agonists/analogues (JHAs) and used data on their biological activities and physical structures to feed software able to compute and model 3D quantitative structure–activity relationships (QSAR). Current *in silico* 3D QSAR and related procedures have proven to be very useful tools for elucidating the molecular action of many compounds, including drugs, and predicting new pharmaceutically successful compounds. Examples include steroid agonists and antagonists [47], acetylcholinesterase inhibitors for Alzheimer symptoms treatment [48], dopamine receptor agonists [49], antimalarial drugs [50] or multidrug

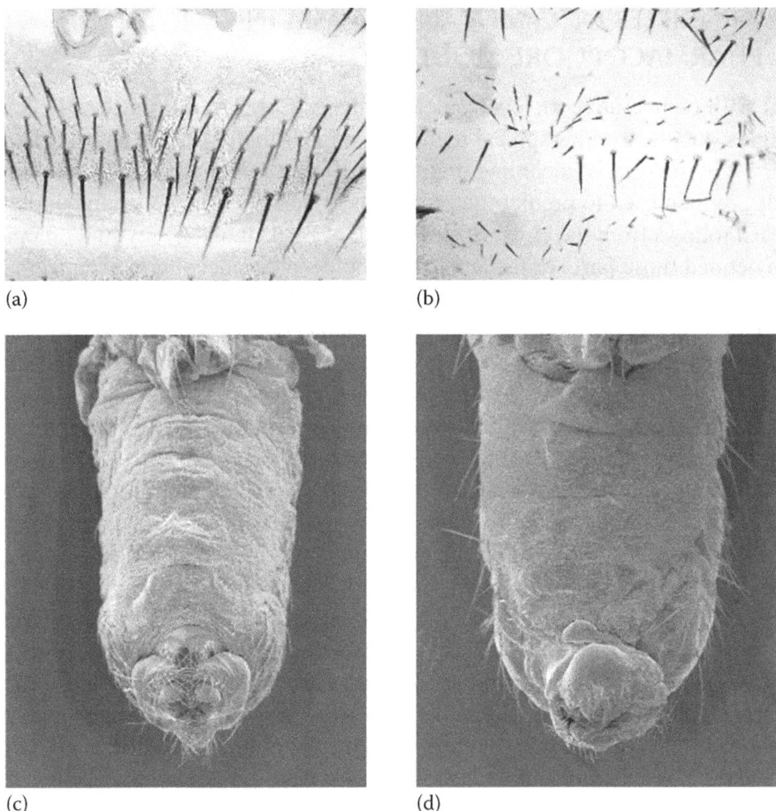

FIGURE 9.1 Morphogenetic effects evoked by exogenous administration of juvenoids to last instar larvae of *D. melanogaster* that were used to evaluate hormonomimetic activity of tested ligands. (a) designates light microscopic view of control wild-type fourth abdominal segment (tergite) with normal pattern of bristles (microchaetae and macrochaetae). (b) shows perturbation of abdominal bristle differentiation and segment underpigmentation, as seen in pharate adults incapable of eclosion. (c) depicts scanning electron microscopic view of ventral side of the abdomen with normally rotated genital arch in control wild-type male. In contrast, (d) illustrates incompletely rotated genital arch of uneclosed pharate adult male as a consequence of juvenoid administration to wandering last instar larva.

resistance modulators [51], *N*-methyl-D-aspartate (NMDA)–glycine B antagonists [52], histamine H4 receptor ligands [53], inhibitors of HIV-1 integrase [54], carbonic anhydrase inhibitors [55], Aurora A kinase inhibitors [56], B-Raf serine/threonine kinase inhibitors [57], and, of course, ecdysteroid receptor agonists [58–63]. These and notable commercial successes using 3D QSAR technology suggest that applying it will provide for further improvement to existing juvenoid-based insecticides and the development of new juvenoid-based insecticides in the near future. As will be documented in the following, these goals will be more easily reached since many of the parameters and the steps utilized during our JH pharmacophore analysis in *Drosophila* can be transformed or extrapolated to already existing bioassays in various insect species, including pests.

9.3 PRINCIPLES OF CoMFA AND CoMSIA IN PHARMACOPHORE MODELING

QSAR with comparative molecular field analysis (CoMFA) within the Sybyl package builds statistical and graphical regression models that relate the physicochemical properties of molecules (including their biological activity) to their structures. These models are then used to predict the properties or activity of novel compounds. This approach follows from the early seminal work of Backett and Casey [64] on opioids, which defined those parts of the active molecules (pharmacophoric groups) that are essential for activity. Analyses of the trimethylammonium receptor site by Kier and Aldrich [65] further developed the pharmacophore concept and applied it to rationalize the SAR of several systems. The application of molecular modeling has dichotomized to develop methods dealing with biological systems having an unknown receptor where there is no available structural information for the receptor at the atomic level and methods dealing with situations where a 3D structure is known from crystallography or NMR spectroscopy [66]. Significant progress in characterizing the molecular properties of a ligand-binding site has been made using 3D QSAR methods by considering the 3D structures and binding modes of ligands and comparing these compounds by mapping their molecular fields in a 3D grid. The initial success of QSAR methods led to efforts to extend the analysis to noncongeneric series where the structural similarity between active compounds in the same bioassay was not obvious, so that the parts of active molecules essential for activity as pharmacophoric groups could be defined. Knowledge of the pharmacophore can be used to infer the important elements within the binding site of the protein and elucidate the possible molecular mechanism of ligand–protein interaction.

A standard CoMFA procedure, as implemented in the Sybyl software from Tripos Inc., consists of few sequential steps: First, the 3D structures of all of the molecules in the dataset are generated in their presumed bioactive conformations. Second, all of the molecules are superimposed or aligned in a 3D spatial lattice/grid using either manual or automated methods. Third, steric and electrostatic fields are calculated around the aligned molecules with different probe groups positioned at all intersections of the grid. Fourth, the interaction energy or field values are correlated with the biological activity data using partial least squares (PLS) and related statistical techniques to identify and extract the quantitative influence of specific chemical features of molecules on their biological activity. Fifth, the results are articulated as correlation equations with a number of latent variable terms, each of which is a linear combination of original independent lattice descriptors. CoMFA has been used as the method of choice in hundreds of published QSAR studies. Statistical tools in QSAR with CoMFA include principal component analysis (PCA or factor analysis) for uncovering relationships between descriptors, previously mentioned PLS regression for analyzing continuous response data (IC_{50}/ID_{50}), and soft independent modeling of class analogy (SIMCA) for analyzing data that are categorical rather than continuous (i.e., active vs. inactive). A hierarchical clustering tool groups compound into classes having similar properties. Bootstrapping and cross-validation techniques are provided to test a model's predictive power, diagnose chance correlation, and insure model robustness.

Data and results of statistical analyses can be displayed as scatter plots, distributions, or histograms. Graphs, structures, and spreadsheets interact with each other to facilitate exploration of the data. The results of CoMFA analyses are displayed as color-coded contours around molecules, allowing visual identification of regions responsible for favorable or unfavorable interactions with the receptor. QSAR with CoMFA stores project details, making it possible to regraph, compare analyses, and predict the properties of new compounds. All of the data can be easily reevaluated if the underlying molecular structures are modified.

An extensive set of physicochemical descriptors (structural, conformational, geometric, electrostatic, and thermodynamic) is built into QSAR with CoMFA. Molecular structures, descriptors, and properties can be organized and managed within a friendly molecular spreadsheet. 3D QSAR methods such as CoMFA require a set of aligned molecules. For example, Field Fit optimizes the alignment of molecules to a previously calculated steric or electrostatic field, if it is available. An alternative, more frequent approach is to align molecules in a database to a template molecule based on a common substructure.

Once a set of molecules is aligned, CoMFA calculates the steric and electrostatic interaction energy of a probe atom with each molecule at points on a grid surrounding the molecules. CoMFA descriptors can be used alone or in conjunction with other descriptors. Comparative molecular similarity indices analysis (CoMSIA) is similar to CoMFA but uses a Gaussian function rather than Coulombic and Lennard-Jones potentials to assess steric, electrostatic, hydrophobic, and hydrogen bond donor/acceptor fields. In principle, this does not necessarily increase the accuracy or predictive ability of the model, but it makes the contour maps easier to interpret. If the bioactive conformation of a molecule is not known, multiple conformers can be stored in the molecular spreadsheet. Though this allows alternative conformers and alignments to be considered in a CoMFA or CoMSIA analysis, it can increase the computational load (workload) significantly.

In general, at the present 3D QSAR methods are primarily employed in so-called ligand-based drug/molecular design tasks to design new pharmacologically active compounds (drugs, inhibitors, etc.) or analyze effects of toxicants and pollutants. As shown in the following, selecting such a complex approach can provide a computer-aided and species-specific pharmacophore model with contour maps able to reveal the steric and electrostatic properties of important features of the ligand-binding cavity of the JH target protein. In addition, the outcomes of the CoMFA and CoMSIA studies can be used in the pseudoreceptor modeling to generate a putative atomistic binding site model.

9.4 METHODOLOGY OF JUVENOID 3D QSAR ANALYSIS

9.4.1 Obtaining Data on Biological Activity of JH Agonists

All tests of juvenoid activity were performed on the *Oregon R* wild-type *Drosophila melanogaster* (from Bloomington Stock Center, Indiana University, Bloomington, IN). Flies were cultured using standard method as described elsewhere [67].

JHs and their analogs (listed in Table 9.1) were prepared freshly by dissolution in acetone (and never stored) and applied topically in a 0.5 μL volume on the surface of

TABLE 9.1
List of JH Agonists Used in This Study and Their Biological Activities

No.	Compound[a]	ED_{50} (μg)	ED_{50} (nM)	$-\log ED_{50}$ (nM)
1	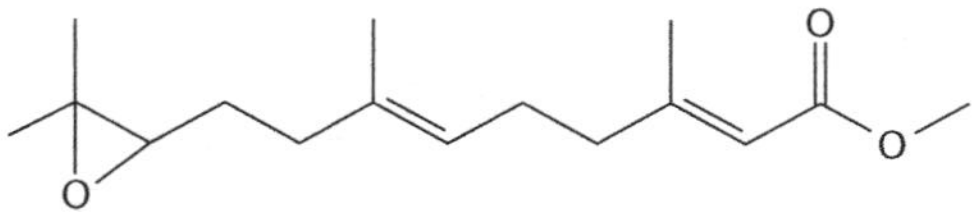	0.0100	0.0375	1.4259
2	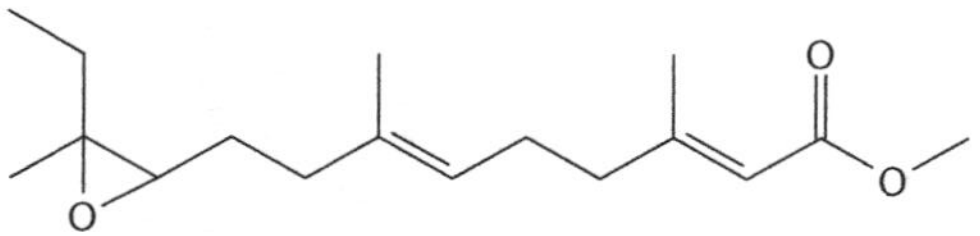	1.0000	3.5695	−0.5526
3	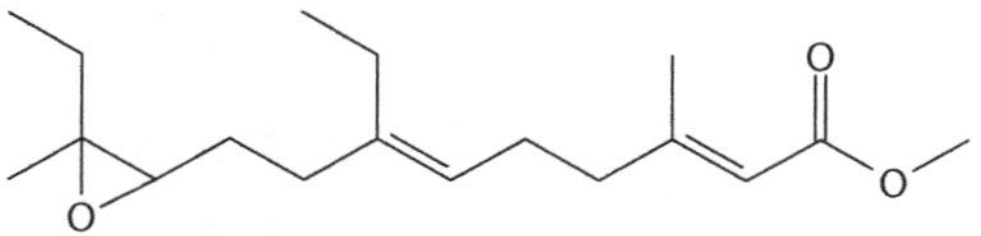	0.0500	0.1699	0.7698
4	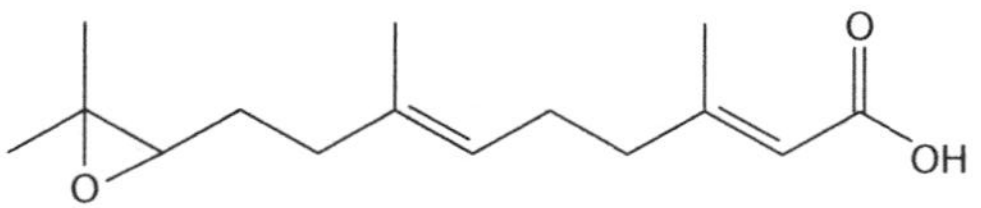	0.5000	1.8922	−0.2769
5	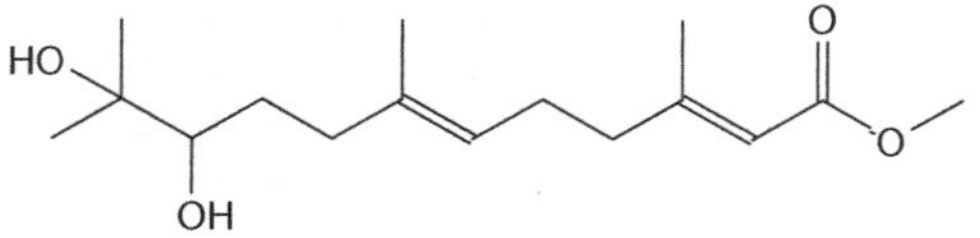	1.5000	5.2418	−0.7194
6	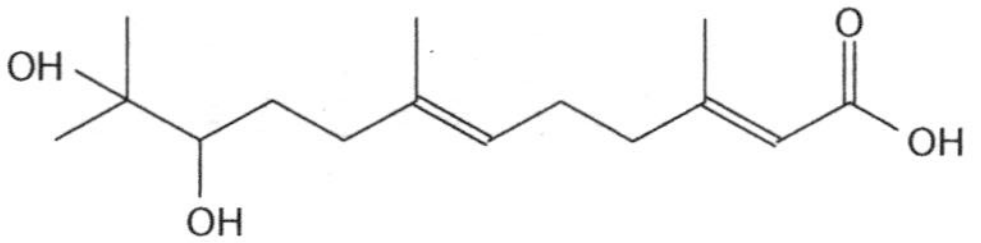	2.5000	8.9482	−0.9517

7	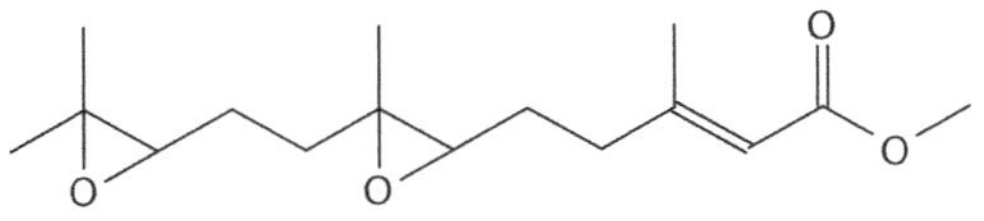	0.1150	0.4070	0.3904
8	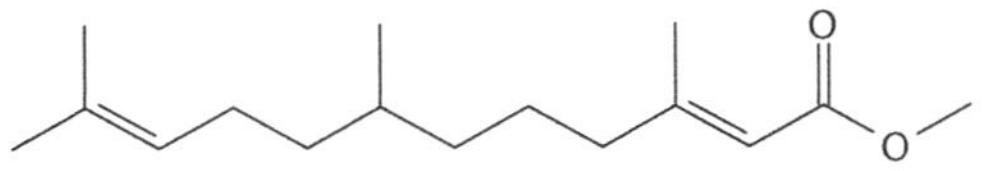	0.0080	0.0317	1.4989
9	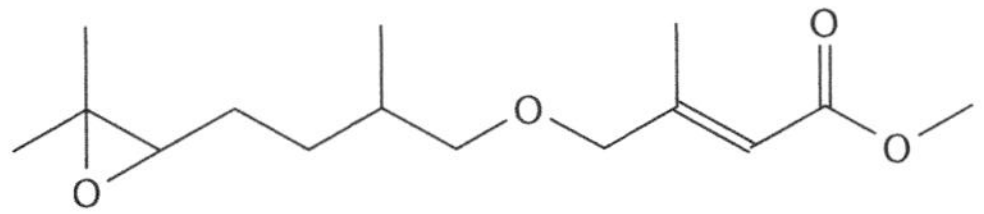	0.00005	0.0002	3.6989
10	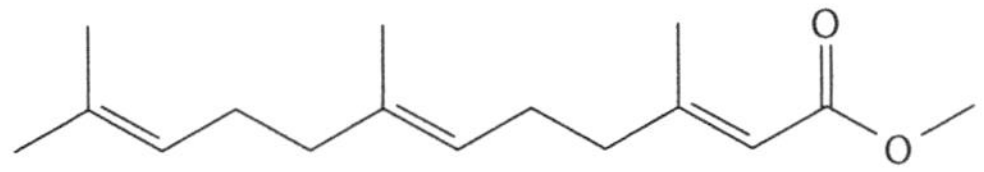	5.0000	20.1499	–1.3042
11	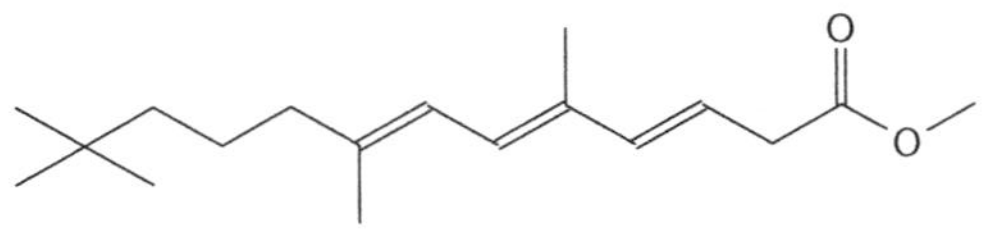	0.0250	0.0873	1.0589
12	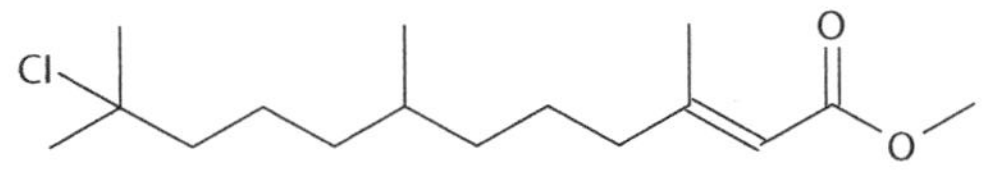	2.0000	6.9302	–0.8407
13	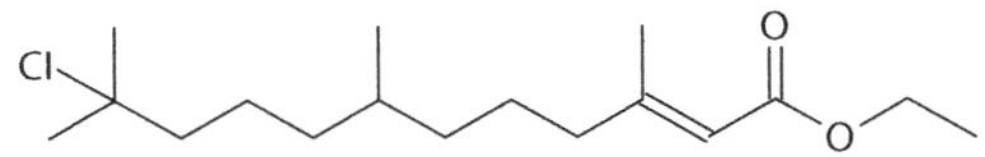	2.5000	8.3166	–0.9199
14	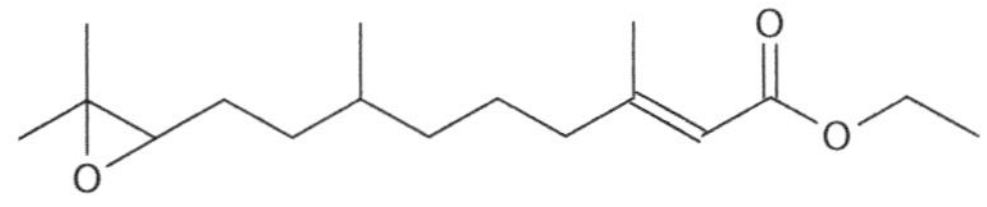	0.0050	0.0177	1.7520

(continued)

TABLE 9.1 (continued)
List of JH Agonists Used in This Study and Their Biological Activities

No.	Compound[a]	ED_{50} (μg)	ED_{50} (nM)	$-\log ED_{50}$ (nM)
15	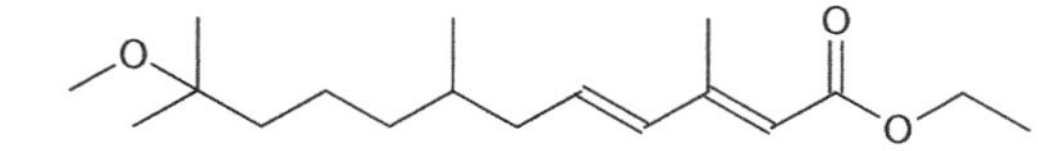	0.0010	0.0034	2.4685
16	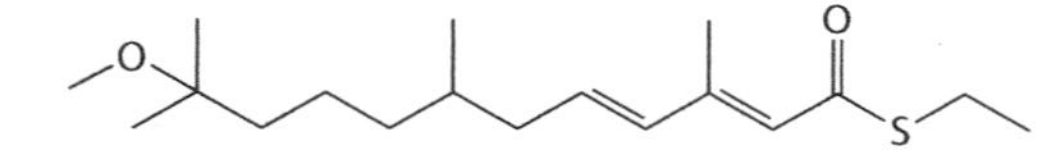	0.1250	0.5266	0.2785
17	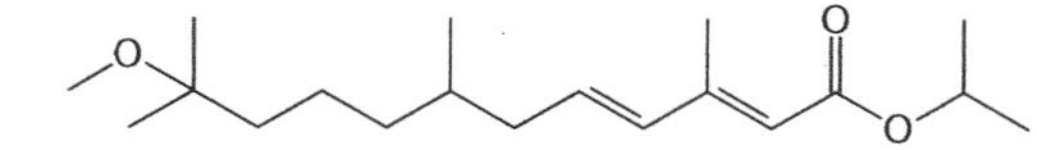	0.00125	0.0040	2.3979
18	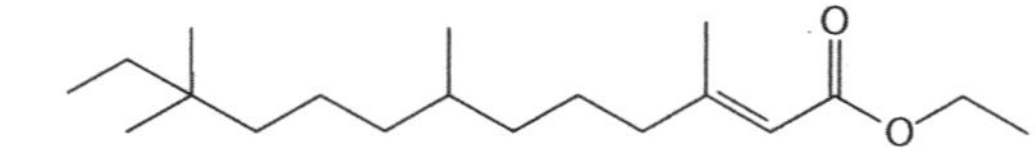	4.0000	13.5057	−1.1305
19	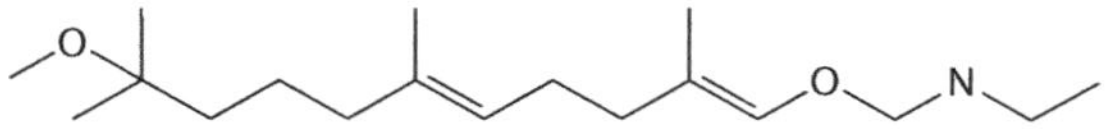	5.0000	16.8827	−1.2274
20	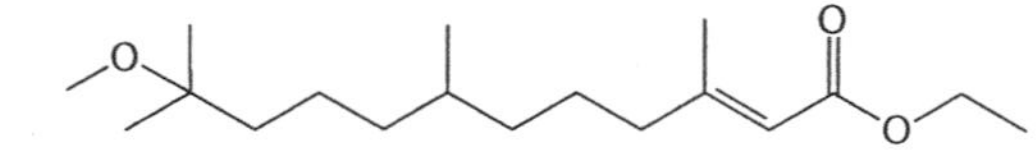	0.0050	0.0167	1.7773
21	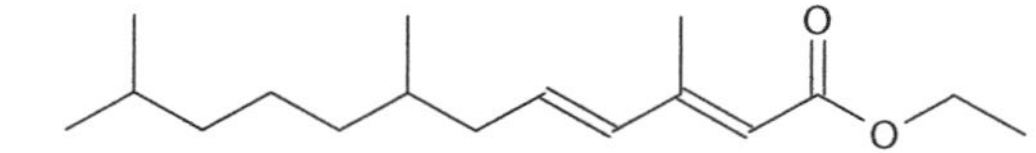	0.2000	0.7515	0.1207
22	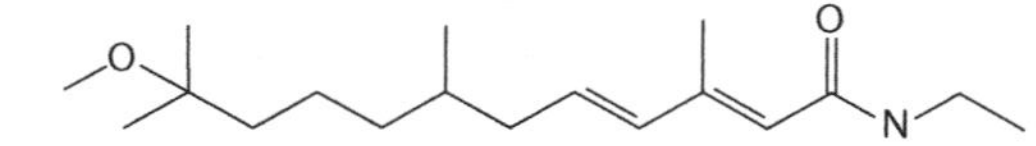	0.4000	1.3506	−0.1322

23	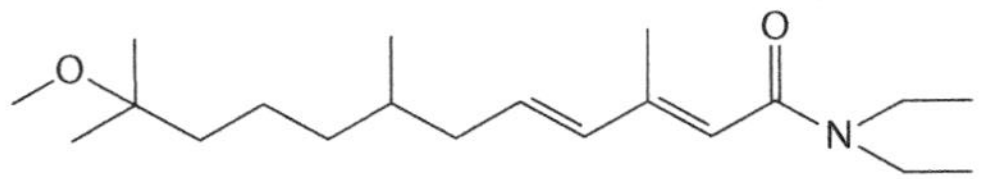	0.5000	1.5471	−0.1895
24	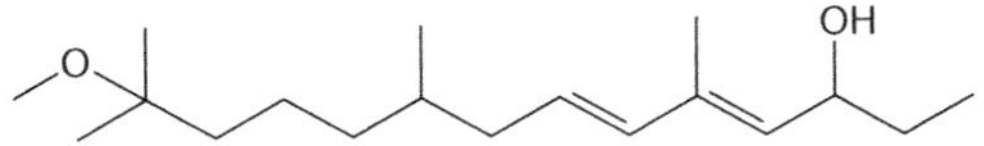	0.5000	1.7783	−0.2500
25	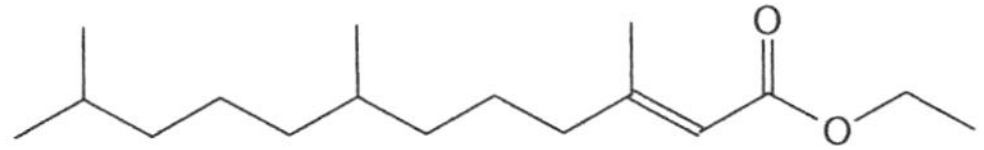	1.0000	16.8827	−1.2274
26	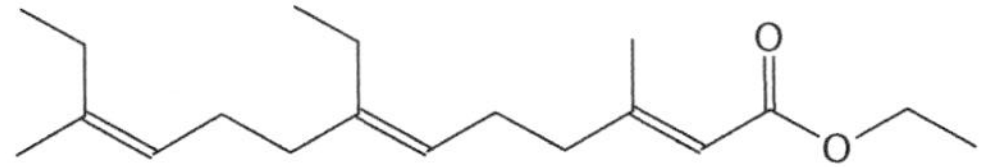	1.5000	5.1339	−0.7104
27	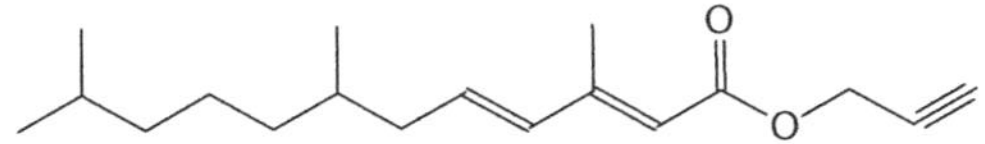	5.0000	17.8475	−1.2515
28	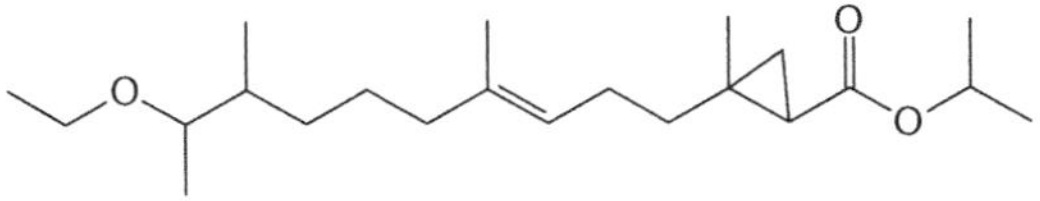	1.0000	2.8411	−0.4535
29	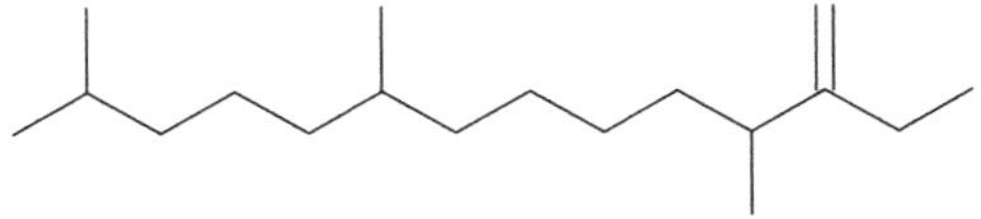	10.0000	39.9712	−1.6017
30	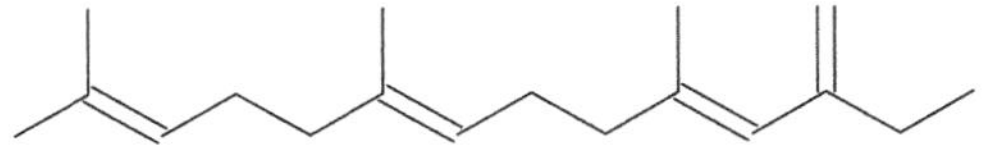	10.0000	40.8163	−1.6108
31	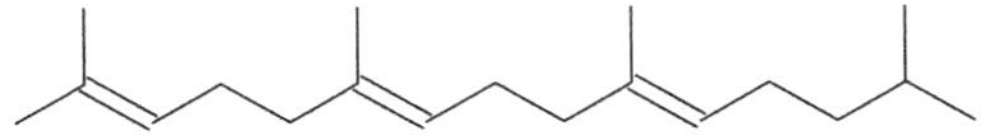	10.0000	38.1402	−1.5814

(continued)

TABLE 9.1 (continued)
List of JH Agonists Used in This Study and Their Biological Activities

No.	Compound[a]	ED_{50} (μg)	ED_{50} (nM)	$-\log ED_{50}$ (nM)
32	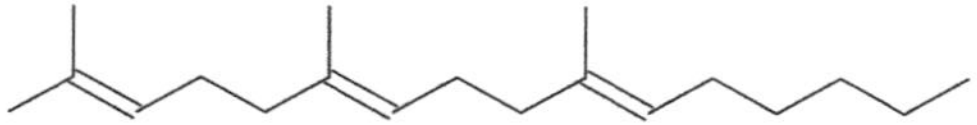	10.0000	38.1402	−1.5814
33	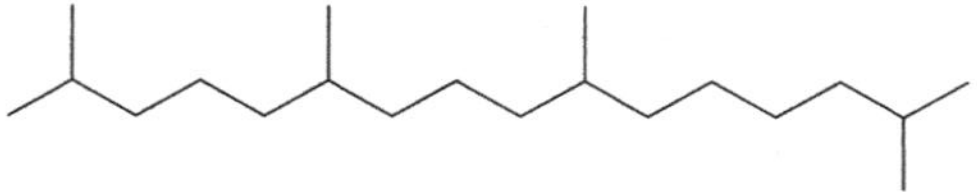	10.0000	35.4358	−1.5494
34	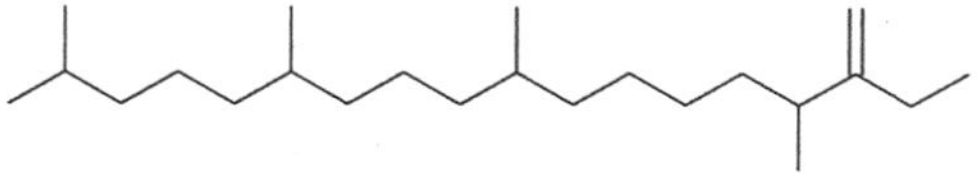	10.0000	31.2275	−1.4945
35	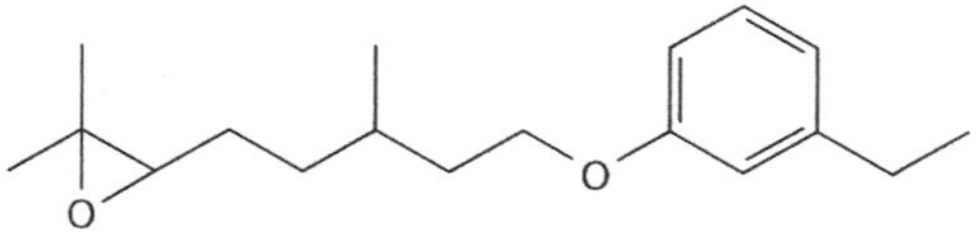 O O	0.0040	0.0144	1.8416
36	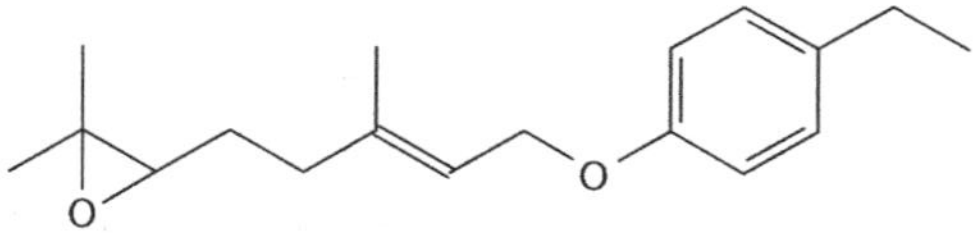 O O	0.0100	0.0364	1.4380
37	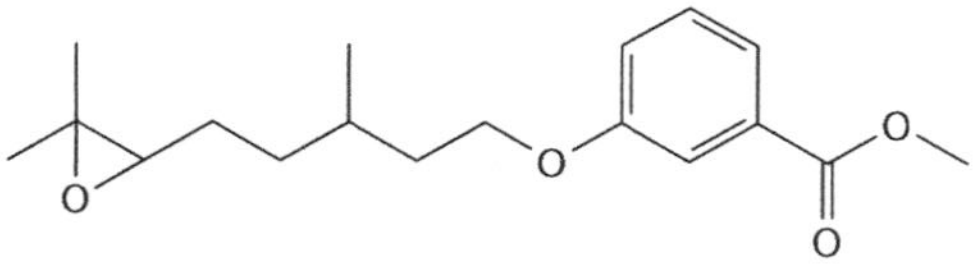 O O O O	5.0000	16.3323	−1.2130

38	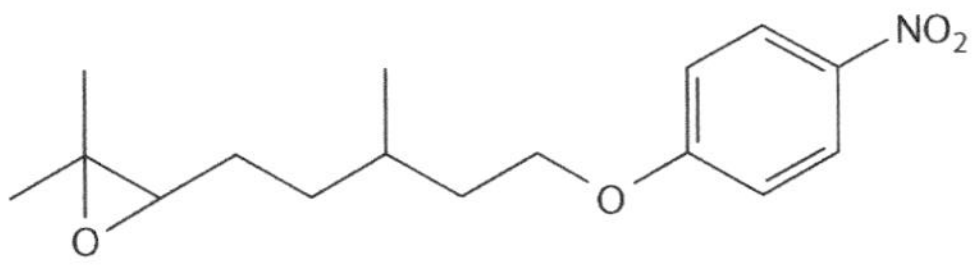	10.0000	34.1157	−1.5329
39	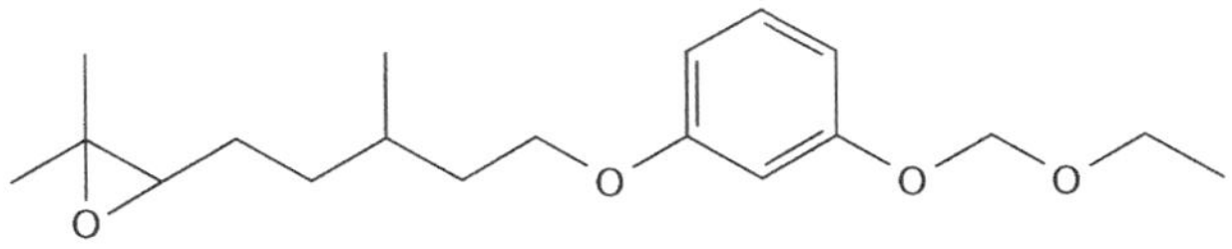	10.0000	31.0414	−1.4919
40	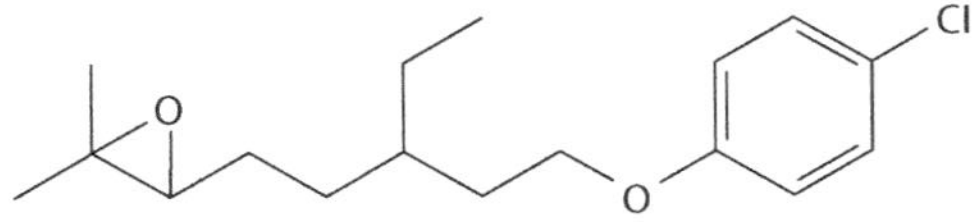	4.0000	14.1548	−1.1509
41	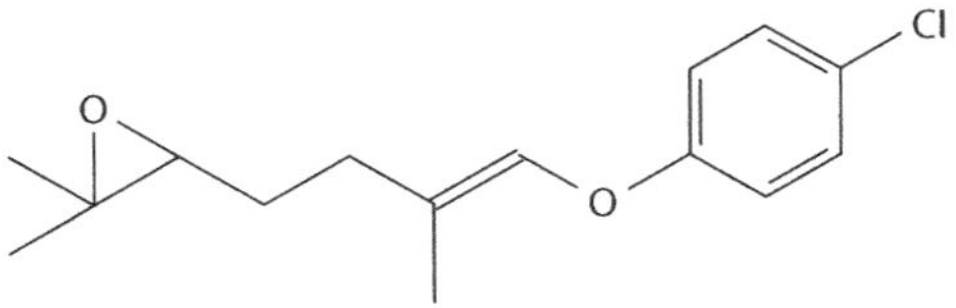	1.0000	3.7512	−0.5742
42	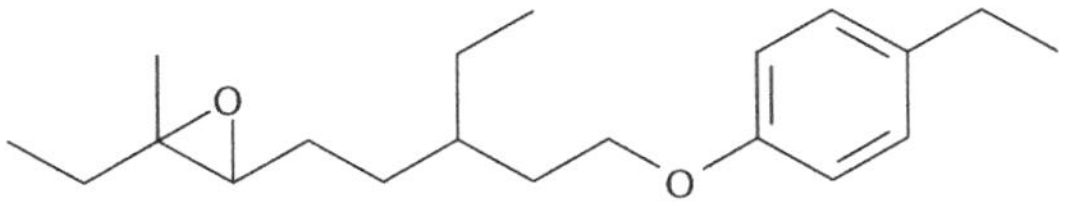	4.0000	13.1501	−1.1189
43	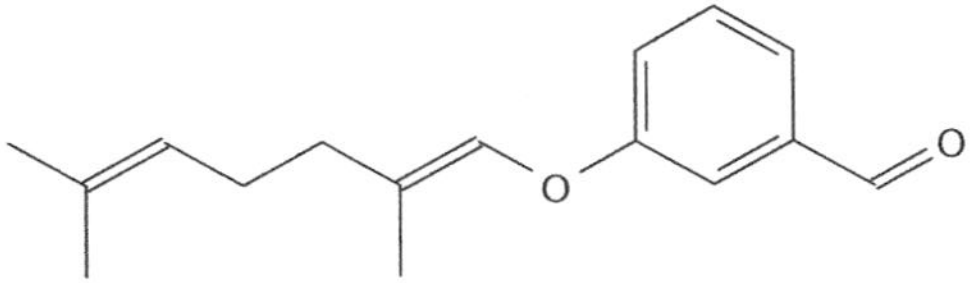	1.2500	5.1200	−0.7092

(continued)

TABLE 9.1 (continued)
List of JH Agonists Used in This Study and Their Biological Activities

No.	Compound[a]	ED_{50} (µg)	ED_{50} (nM)	$-\log ED_{50}$ (nM)
44	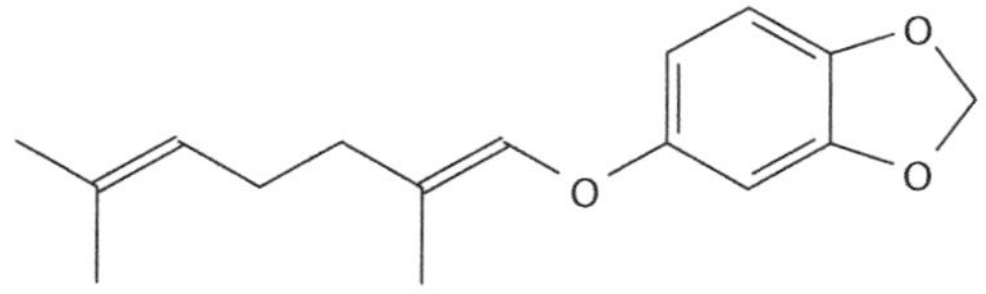	1.2500	4.8053	−0.6817
45	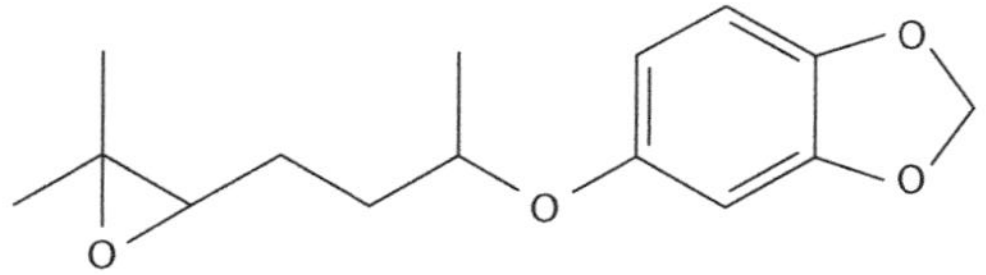	5.0000	18.9315	−1.2771
46	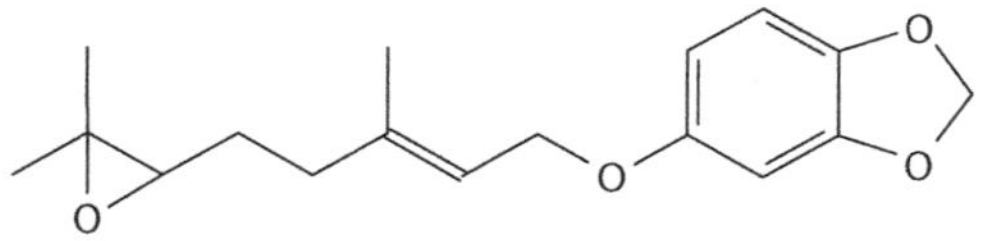	0.0050	0.0172	1.7645
47	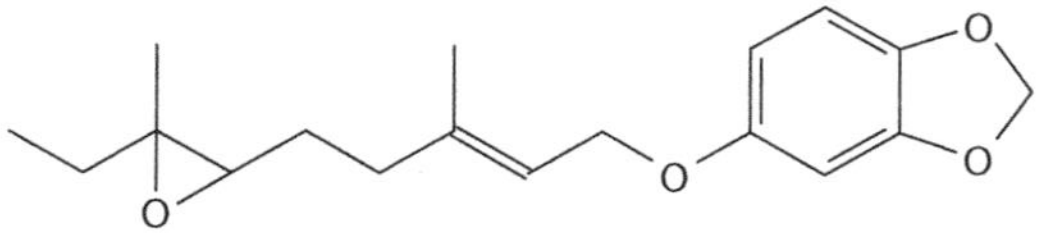	0.5000	1.6439	−0.2158
48	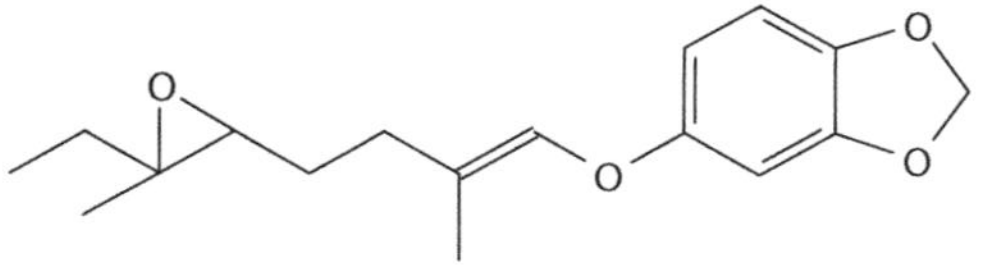	0.3500	1.2063	−0.0814

49	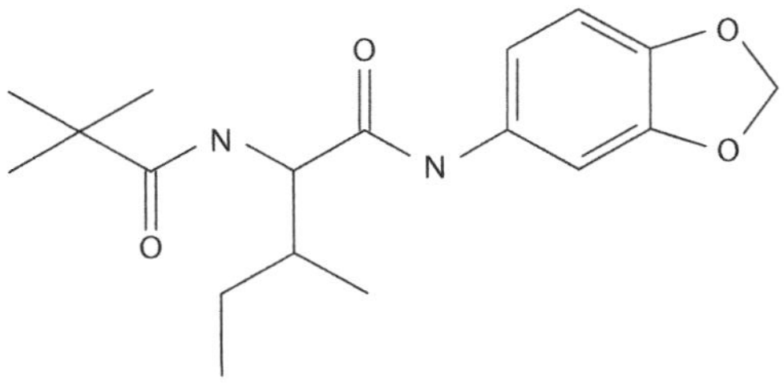	7.5000	22.3751	−1.3497
50	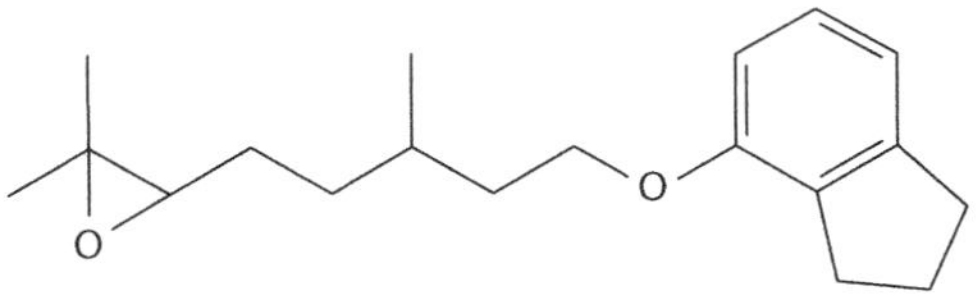	0.4000	1.3880	−0.1423
51	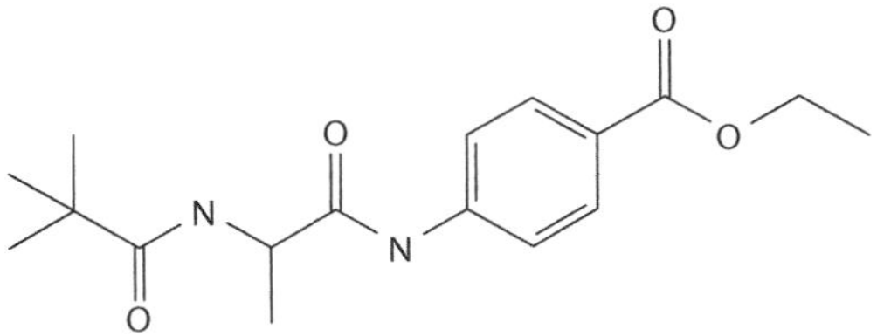	5.0000	14.9701	−1.1752
52	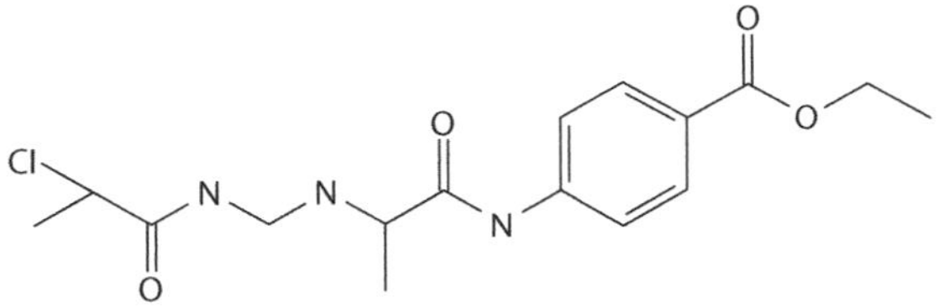	5.0000	14.0619	−1.1480

(continued)

TABLE 9.1 (continued)
List of JH Agonists Used in This Study and Their Biological Activities

No.	Compound[a]	ED_{50} (μg)	ED_{50} (nM)	$-\log ED_{50}$ (nM)
53		5.0000	11.7799	−1.0711
54		0.0100	0.0356	1.4485
55		0.0400	0.1423	0.8467
56		0.0013	0.0041	2.3872

57	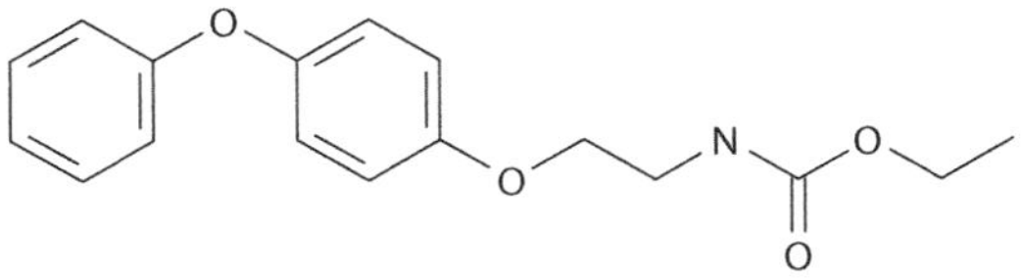	0.0500	0.1663	0.7791
58	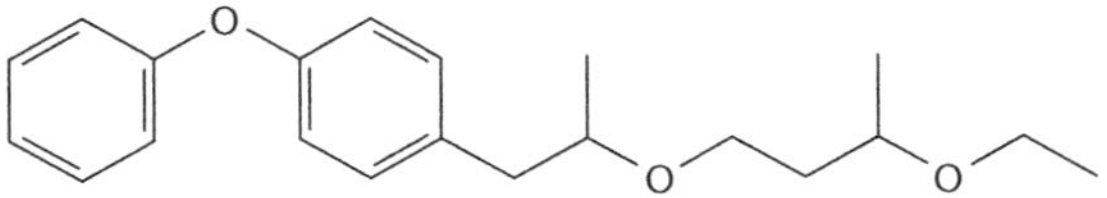	0.0500	0.1523	0.8171
59	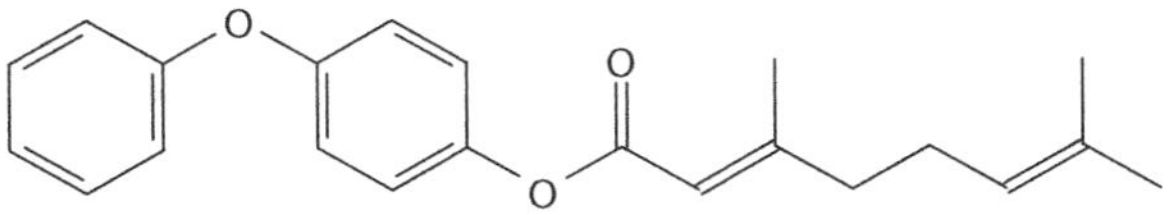	2.0000	5.9667	−0.7757
60	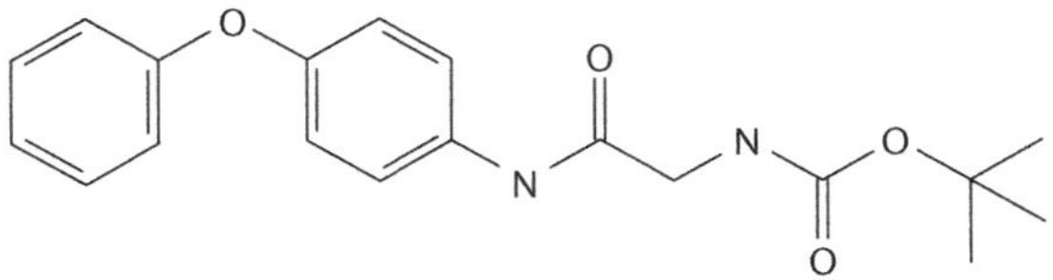	0.8500	2.6061	−0.4159
61	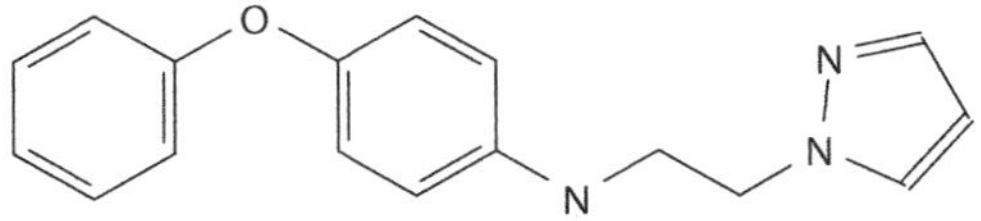	5.0000	17.8475	−1.2515
62	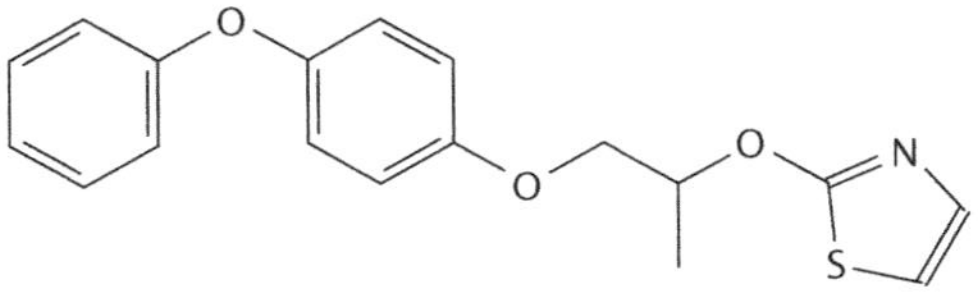	0.8000	2.4487	−0.3889

(continued)

TABLE 9.1 (continued)
List of JH Agonists Used in This Study and Their Biological Activities

No.	Compound[a]	ED_{50} (μg)	ED_{50} (nM)	−log ED_{50} (nM)
63	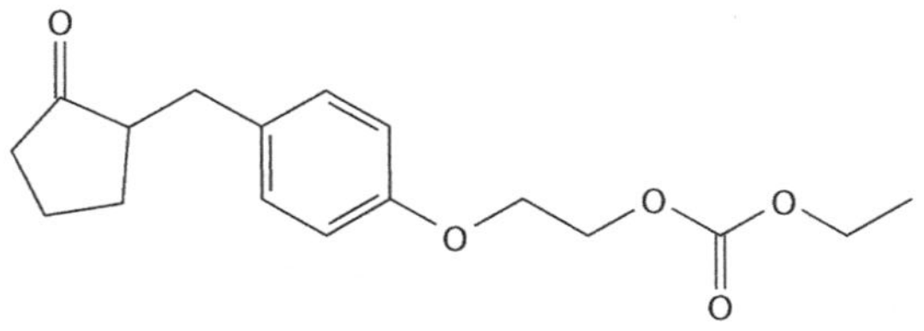	1.2000	3.9328	−0.5947
64	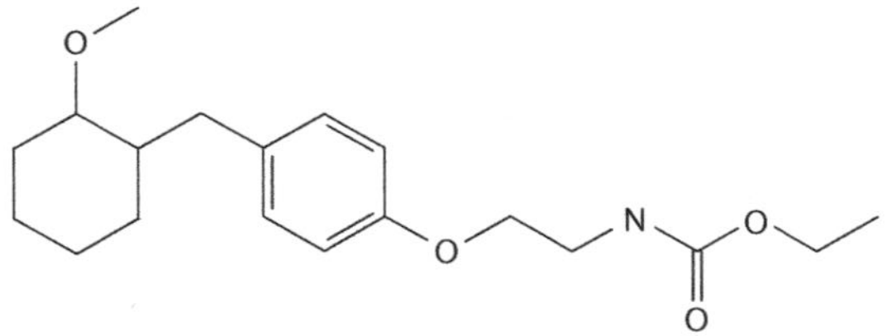	1.0000	2.9837	−0.4747
65	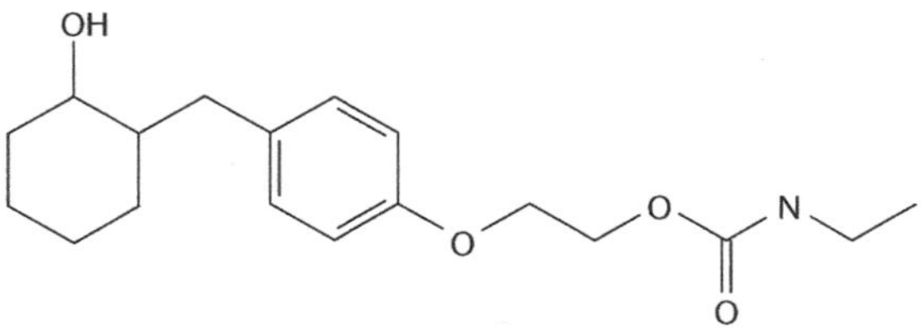	3.5000	10.9669	−1.0401

66	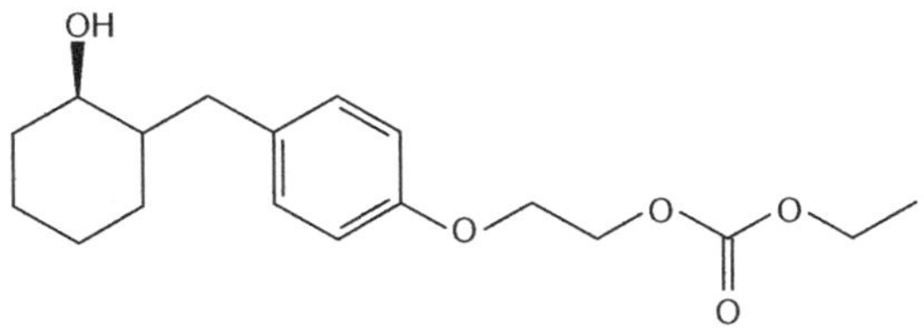	1.4000	4.3455	−0.6380
67	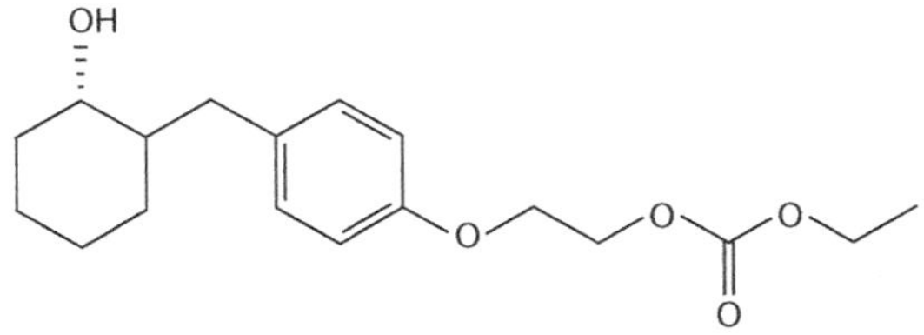	1.0000	3.1043	−0.4919
68	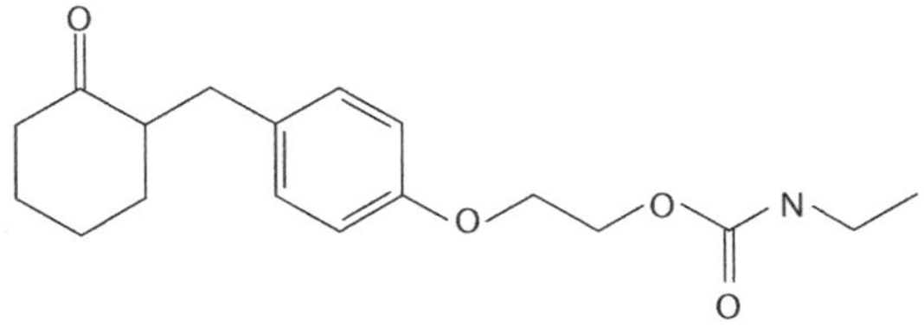	1.4000	4.3867	−0.6421
69	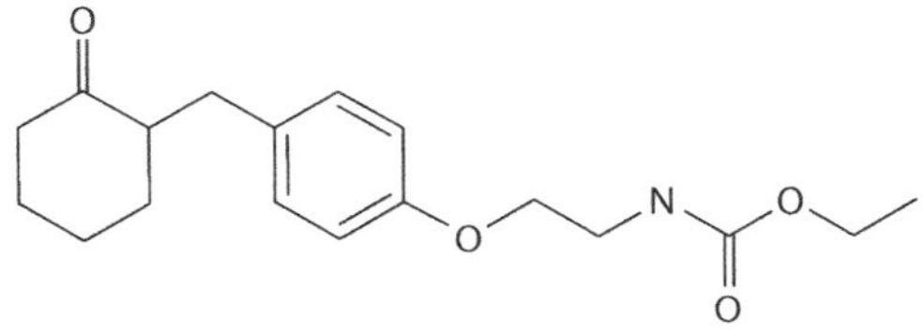	0.3000	0.9895	0.0045

(continued)

TABLE 9.1 (continued)
List of JH Agonists Used in This Study and Their Biological Activities

No.	Compound[a]	ED_{50} (μg)	ED_{50} (nM)	$-\log ED_{50}$ (nM)
70	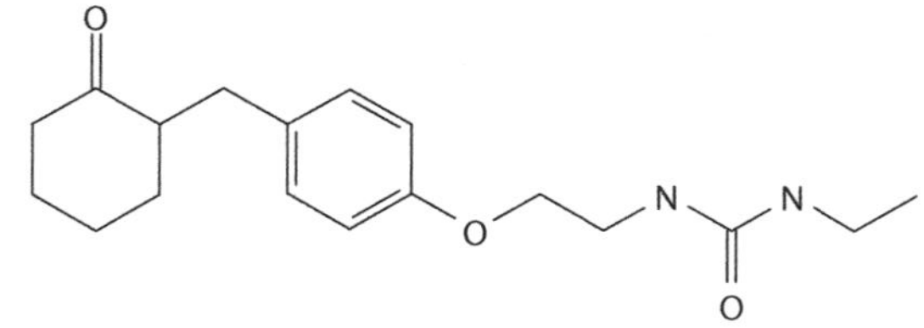	1.1500	3.6146	–0.5580
71	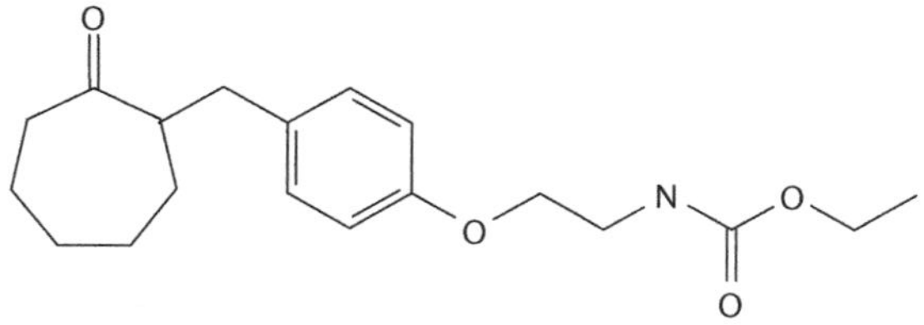	2.3000	6.9038	–0.8391
72	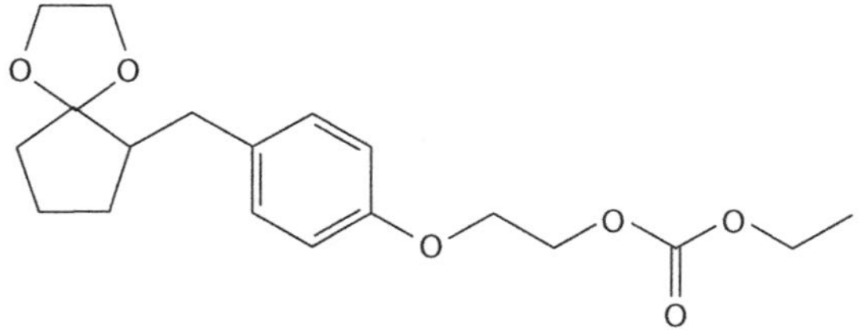	1.8000	5.1409	–0.7110

73	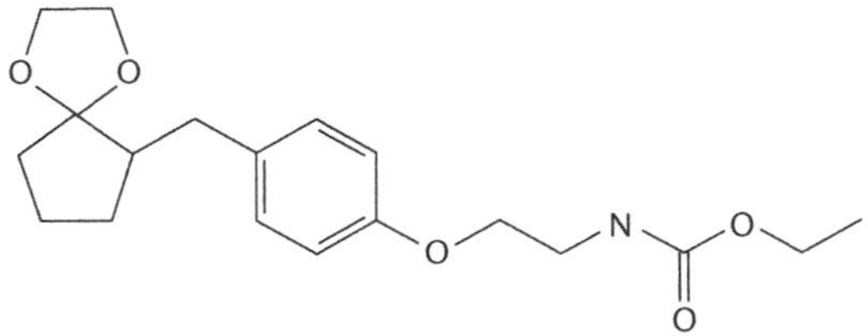	0.8500	2.4345	–0.3864
74	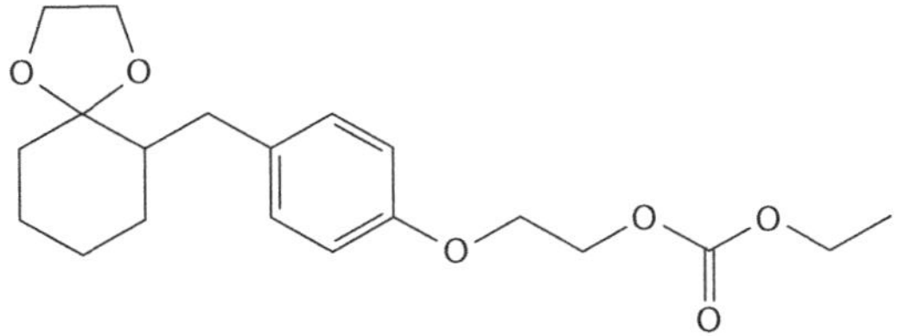	1.5000	4.1192	–0.6148
75	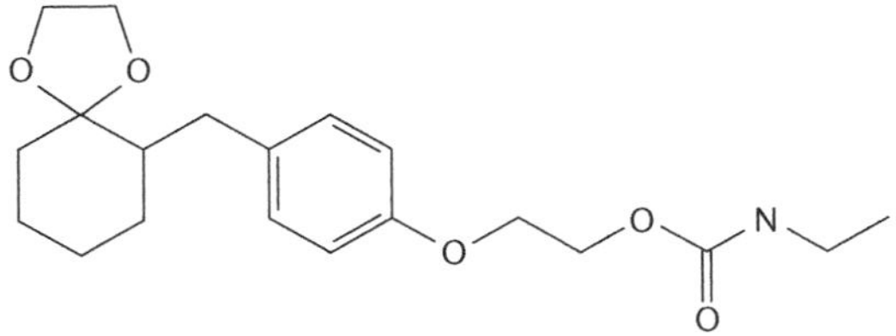	1.9000	5.2319	–0.7186
76	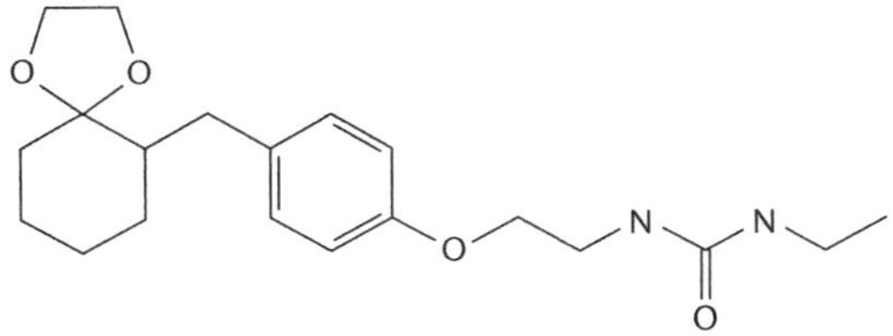	1.0000	2.6583	–0.4246

(continued)

TABLE 9.1 (continued)
List of JH Agonists Used in This Study and Their Biological Activities

No.	Compound[a]	ED_{50} (µg)	ED_{50} (nM)	$-\log ED_{50}$ (nM)
77	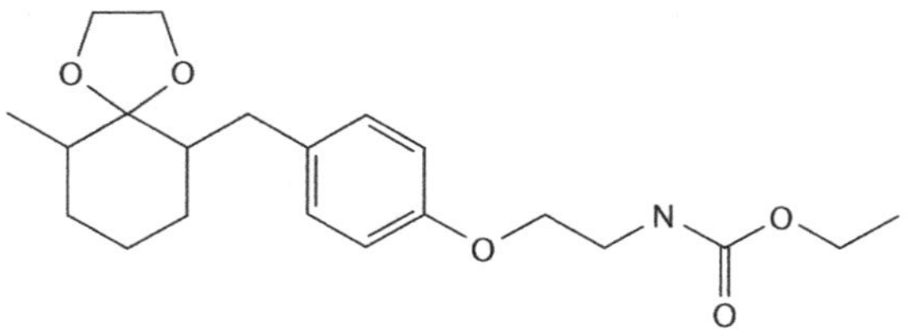	2.9000	7.6890	−0.8858
78	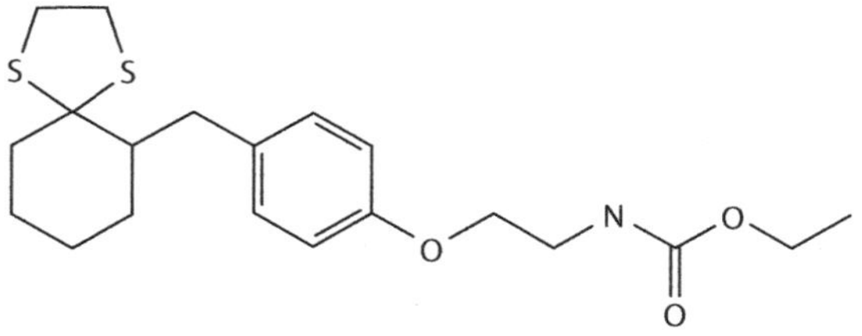	3.2700	0.2426	0.6151
79	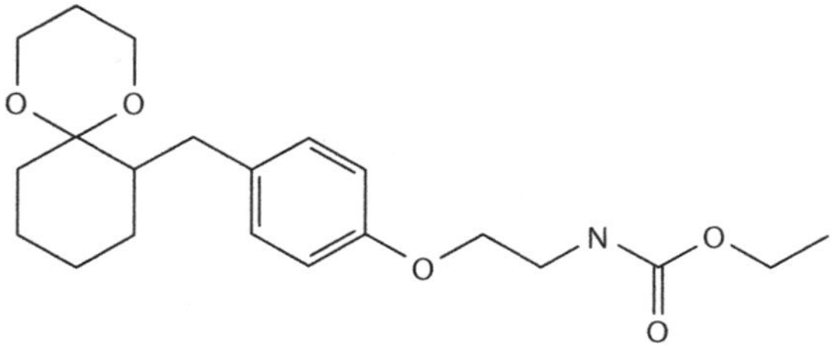	0.7500	1.9885	−0.2985

80	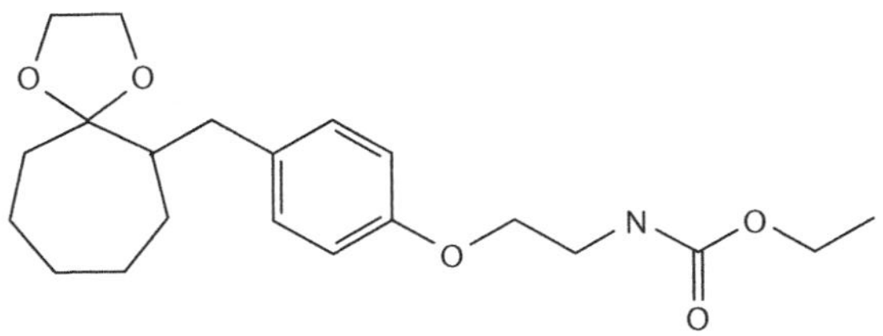	1.6000	4.2422	−0.6275
81	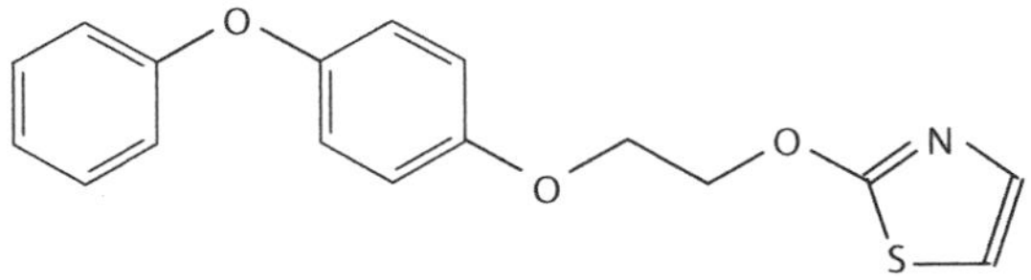	0.0020	0.0063	2.2006
82	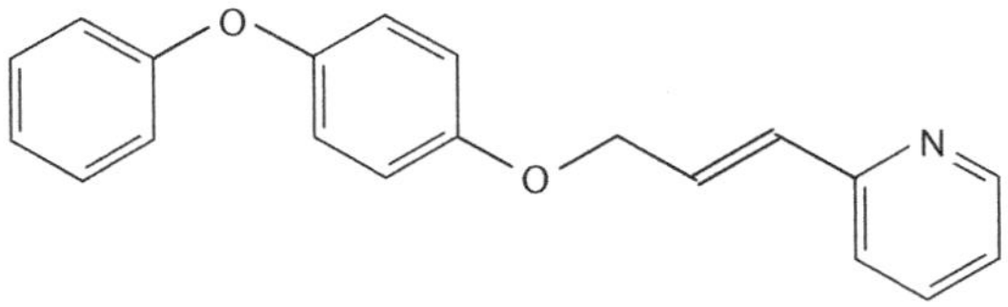	0.0060	0.0198	1.7033
83	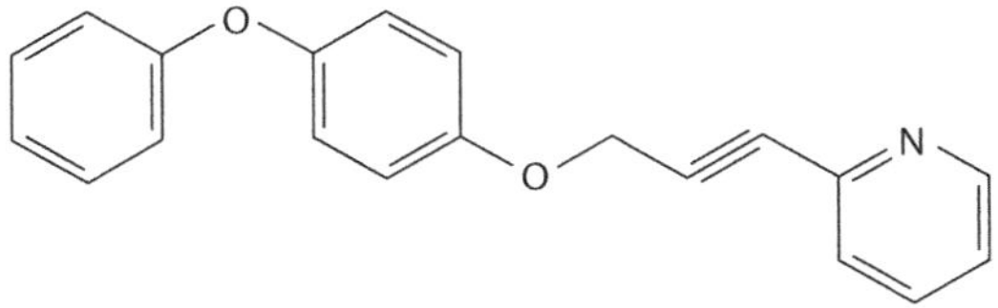	0.0060	0.0199	1.7011

(continued)

TABLE 9.1 (continued)
List of JH Agonists Used in This Study and Their Biological Activities

No.	Compound[a]	ED_{50} (µg)	ED_{50} (nM)	$-\log ED_{50}$ (nM)
84	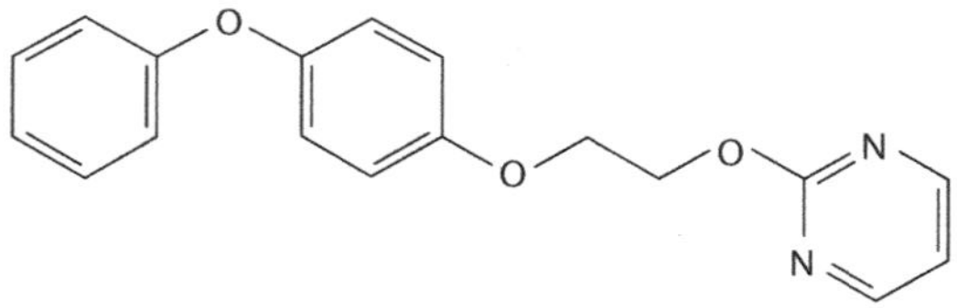	0.0050	0.0162	1.7904
85	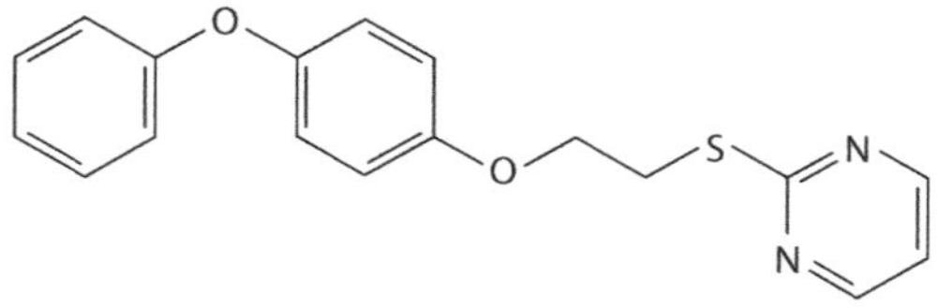	0.0055	0.0169	1.7721
86	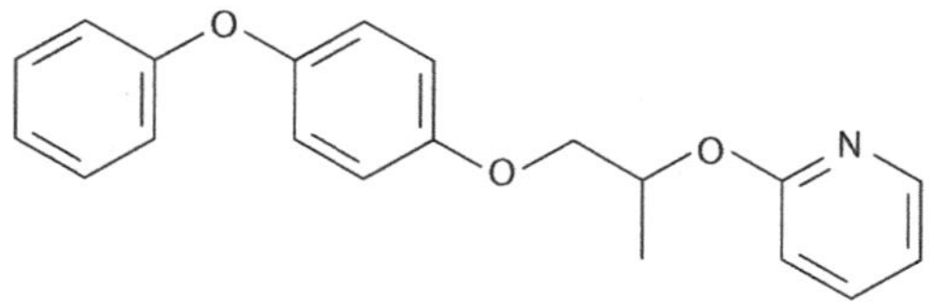	0.0005	0.00156	2.8068

The biological activity (ED_{50}) is expressed in µg of the compound per animal, then it is converted to nmol per animal and finally to −log of nmol values that are used for CoMFA and CoMSIA computations.

[a] Chemical names are given in [67].

late wandering third instar larvae (cca 8–12 h prior to pupariation), according to Ashburner [40]. All compounds were initially tested at a dose of 1 μg/animal and afterward tested at more diluted concentrations. If a compound displayed low or no activity at 1 μg/animal, doses were gradually increased to 10 μg/animal. Each concentration of every evaluated compound was tested on 200 larvae and in triplicate. The morphogenetic effects of a compound on animals were evaluated and scored after control siblings completed eclosion (120 h after pupariation) (for more details, see [67]).

Samples were processed as follows: *Drosophila* cuticles were excised, fixed in 4% formaldehyde + 50% acetic acid, rinsed in ethanol, and delipidated in 1N KOH at 90°C for 5 min. After washing in saline, cuticles were dehydrated in ethanol, cleared in xylene, and mounted in Euparal (Chroma GmbH.) or Entellan (E. Merck GmbH.). Alternatively, cuticles were fixed in glycerol–acetic acid (1:4), mounted, and cleared overnight in Hoyer's medium at 60°C [68]. These two techniques gave comparable results.

Preparation of adult flies for scanning electron microscopy was done as we described previously [69] with modifications for pupae and pharate adults as follows [70]. Briefly, glutaraldehyde-fixed and osmium tetroxide-postfixed specimens were enzymatically cleared of dirt contaminants that derived from exuvial fluid and then washed in 15% Triton X-100 plus 5% Tween 20. After dehydration through an ascending ethanol series, animals were critical point dried using hexamethyldisilazane (Sigma GmbH.) as described by Nation (1983), sputter-coated with platinum or gold–palladium in a Balzers SCD-030, and viewed in a field emission electron source Hitachi S-800 scanning electron microscope operating at 10 kV. More details can be found in [67].

Since biological data are generally found to be skewed, we used a logarithmic (log) transformation to move the data to a nearly normal distribution. Thus, to generate models using measurements of responses under equilibrium conditions, we usually transformed the data to express doses or concentration values as negative log [IC_{50} or EC_{50} or ED_{50}].

9.4.2 Computational Analysis

All the computational tasks were performed on Silicon Graphics Origin (SGI) 2000 (R10000) and O_2 (R10000) servers running under the IRIX (ver. 6.5.8 and 6.5.19) operating system.

9.4.2.1 Structure Building

The entire set of the studied JH compounds in 3D was created by the sketch function of Sybyl 6.8 molecular modeling software (Tripos-Certara Co., St. Louis, MO) on a SGI workstation and thereupon fully energetically minimized using the standard Tripos force field, with a 0.05 kcal/mol energy gradient convergence criterion and a distance-dependent dielectric constant. To identify local energy-minimum conformations for all of the studied juvenoids, the complete systemic conformational search with a 30.0 angle increment was performed. In addition, for finding the low-energy conformations of juvenoid structures with more than 3 rotatable bonds, a random search routine needs to be employed. Then, all these low-energy conformations were fully reoptimized using the AM1 semiempirical quantum chemical method; the AM1 was also used for a partial charges calculation. The resulting congener structures were refined against the x-ray or NMR structural data of highly related compounds, the coordinates of which were obtained from Cambridge Crystallographic Database. JH and some of its agonists are known to form chiral mixtures [71]. In general, many enantiomer-forming compounds appear to affect the biological activity or play an important role in enzymatic reactions, stereoselective degradation, and other mechanisms [72–74]. Several x-ray structural studies have shown that biologically less active enantiomers of ligands for nuclear hormone receptors are also able to bind a receptor pocket and adopt bioactive conformation similar to more active enantiomers, but this conformation is energetically less favored [75–77]. Therefore, when appropriate the enantioselective isomers and chiral centers of the juvenoids were considered according to the descriptors for 3D QSAR analysis as described by Golbraikh et al. [78], Paier et al. [79], and Kovatcheva et al. [80]. To avoid some of the inherent deficiencies arising from the functional form of the Lennard-Jones and Coulomb potentials, in CoMSIA computations we used Gaussian function for the distance dependence between the probe atom and the molecule atoms [81].

9.4.2.2 Molecular Alignment

Structural alignment is known to be the most critical step in CoMFA and CoMSIA studies and the resulting model is often sensitive to a particular atom-group arrangement. It became recognized and widely accepted that the global energy-minimum conformation may not necessarily be adopted in the compound–receptor complexes. As a practical starting point for statistical comparisons of flexible structures within both the CoMFA and the CoMSIA models to minimize this problem, reasonably low-energy conformations in the alignment can be used. Empirically, it became evident that in our study of juvenoids, taking the lowest energy conformation of the most rigid and highly active molecule (No. **86**) as the template structure for the alignment provided the best result. Subsequently, JH molecules were superimposed by minimizing the root-mean-square (RMS) distance between atom pairs that belong to the fitting molecule and the template molecule, respectively. The alignment of all 86 compounds including the test set is shown in Figure 9.2a. Then, as a basis, the superimposition of JH ligands was based on manually selected overlap of electronegative atoms of oxygens or nitrogen at the ends and quaternary or sp^2 hybridized carbon in the middle of the structures (Figure 9.2b).

9.4.2.3 CoMFA Protocol

A 3D cubic lattice with a 2 Å grid spacing was generated automatically around the aligned JH molecules with the grid extending molecular dimensions up to 4 Å in all directions. The steric and electrostatic fields were calculated using the sp^3 hybridized carbon atom with a charge of +1 separately for each molecule. An energy cutoff at 30 kcal/mol was applied, which means that calculated energies greater than 30 kcal/mol are truncated to fit this value and to avoid infinite energy values inside molecule. To analyze the relationship between the calculated steric and electrostatic energies and −log ED values, the method of PLS was used. The leave-one-out (LOO)

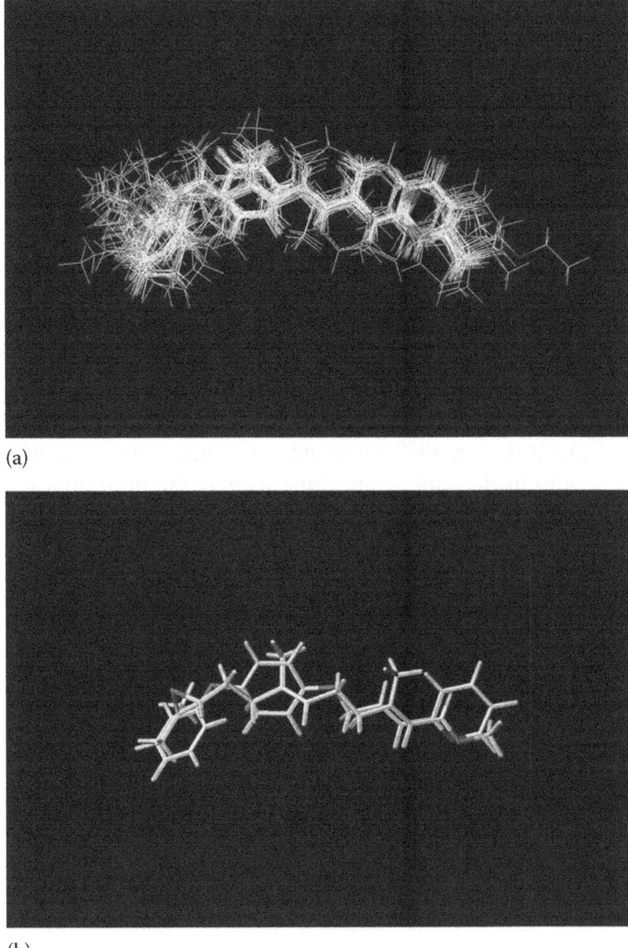

(a)

(b)

FIGURE 9.2 The superpositional alignment of juvenoids (congeners of JH agonists) analyzed in this study. Complete set of all 86 JH compounds is shown in (a), whereas (b) shows alignment of two selected agonists, natural JH-III (**1**) as representative of Class I and the most rigid structure of ZR-10852 (**82**) as representative of Class II.

procedure was performed to cross validate PLS calculation. Column filtering was set at 1.0 kcal/mol, so that only steric and electrostatic energies with values greater than 1.0 kcal/mol could be considered, which helped to speed up and reduce noise in the PLS analysis. The number of analyzed components corresponding to the lowest predictive error of estimate (PRESS) value was selected for obtaining the final PLS regression models and to maintain the optimum number of PLS components as well as minimize the tendency to overfit the data. The estimated models were based on the cross-validated LOO r^2 (expressed as q^2), standard error of prediction (SEP), the non-cross-validated conventional correlation coefficient r^2, and the standard error of estimate (SEE).

The overall predictive ability of CoMFA analysis was evaluated by the q^2 term, which was calculated according to following equation:

$$q^2 = 1 - \frac{\sum_Y (Y_{pred} - Y_{act})^2}{\sum_Y (Y_{act} - Y_{mean})^2}$$

Some of the earlier indices change and others are omitted as meaningless if cross validation is used in conjunction with PLS. Then, the key difference lies in the definition of the standard error value (s). When the analysis is run without cross validation, the standard error is the uncertainty remaining after the least-squares fit has been performed. If cross validation is applied, the standard error becomes the expected uncertainty in prediction for the individual compound based on the data available from other compounds. In this context, s becomes the root mean PRESS. It should be emphasized that it is more difficult to predict values that are not used in deriving a model than it is to fit the same values that include the minimum model. In most CoMFA studies, the cross-validated correlation coefficient q^2 is usually lower than the conventional correlation coefficient r^2. The uncertainty of prediction is characterized by the following equation:

$$\text{SPRESS} = \frac{\text{PRESS}}{(n-k-1)}$$

For the reliability of predictions, it is widely accepted that PRESS and q^2 are generally proving to be better indicators than standard error and r^2. Although in time as knowledge becomes accumulated, this can change; currently, any model with a q^2 value greater than 0.4 or 0.5 is usually considered to be significant. The steric and electrostatic fields of the CoMFA analysis are illustrated as contour maps (see later Figure 9.8). For their interpretation, it should be noted that the color polyhedra surround those lattice points in which the QSAR strongly associates changes of compound field values with changes in biological potency. By default setting in the program, green polyhedra represent sterically favored regions where more bulky substituents are expected to increase biological activity, while yellow polyhedra surround regions where less bulky substituents are predicted to increase biological activity. Blue polyhedra paint electrostatic regions in which positively charged moieties are favorable and will enhance biological activity, whereas the red contours represent regions for which negatively charged groups are favorable. The optimal test for the predictive power of the model is its ability to predict the activity of ligands in test set, which was not used in generating the model and generating a calculation of $r^2_{predictive}$, which is defined similarly to q^2, where the predicted activities are activities calculated for members of this test set.

9.4.2.4 CoMSIA Setup

CoMSIA was developed to overcome some of the limitations of CoMFA. It provides a different way for the calculation of grid point energies and uses different similarity probes. This enables it to explicitly include the hydrophobic and hydrogen-bonding parameters of molecules into a correlation model. In CoMSIA, molecular similarity indices were employed as descriptors to simultaneously consider steric, electrostatic, hydrophobic, and hydrogen-bonding properties. These indices are estimated indirectly by comparing the similarity of each molecule in the dataset with a common probe atom usually having a radius of 1 Å, charge of +1 and hydrophobicity of +1 positioned at the intersections of a surrounding grid/lattice. For computing similarity at all of the grid points, the mutual distances between the probe atom and the atoms of the molecules in the aligned dataset are also taken into account. To describe this distance dependence and calculate the molecular properties, Gaussian-type functions are employed. Since the underlying Gaussian-type functional forms are smooth with no singularities, their slopes are not as steep as the Coulombic and Lennard-Jones potentials in CoMFA; therefore, no arbitrary cutoff limits are required to be defined [81]. In evaluation of JH biological activities, all five physicochemical descriptors, steric (S), electrostatic (E), hydrophobic (H), hydrogen bond donor (D), and acceptor (A) properties, were taken. In comparison to CoMFA, involvement of these five different descriptor fields substantially help to (1) increase the model's significance and its predictive power as well as help to (2) partition the ligand-binding properties into the spatial locations where they play a decisive role in determining biological activity. Similarity indices (AFk) between the compounds and a probe atom are calculated according to the following equation:

$$A^{q}_{\mathrm{F,k}}(j) = -\sum_{i} w_{\mathrm{probe},k} w_{ik} e^{-\alpha r_{iq}^{2}}$$

where q is the grid point for molecule j, with the summation index over all atoms of the molecule j; w_{ik} represents the actual value of physicochemical property k of atom i; $w_{\mathrm{probe},k}$ indicates probe atom with charge +1, radius 1 Å, hydrophobicity +1, and H-bond donor and acceptor property +1; α is the attenuation factor (default is 0.3), and r_{iq} is the mutual distance between probe atom at grid point q and atom i of the test molecule [82].

Both LOO run and a nonvalidation PLS analysis were performed by utilizing all five CoMSIA descriptors for the explanatory variables. The series of obtained models were ordinarily estimated on the cross-validated LOO q^2, the PRESS, the non-cross-validated conventional correlation coefficient r^2, the SEP, and the SEE. As an outcome, the colored polyhedra surround those lattice points where the QSAR strongly associates changes in compound field values with changes in biological activity.

For more detailed information on CoMFA and CoMSIA protocols, readers are recommended to consult Sybyl manuals (Tripos-Certara Co., St. Louis, MO).

9.4.2.5 Pharmacophore Development and Pseudoreceptor Model Generation

Pharmacophore model represents description of features, which are necessary for molecular recognition of a ligand by a target. By other words, information contents of the pharmacophore are an ensemble of steric and electronic features and atomistic distances that are necessary to ensure the optimal supramolecular interactions with a specific biological target. In our specific case, the pharmacophore model explains how structurally diverse ligands, juvenoids, can bind to a common receptor site. Summarizing the results of CoMFA and CoMSIA, we proposed a putative pharmacophore model that explains the key structural requirements for the activity of JH and its agonists in *Drosophila*.

To extend and efficiently refine the pharmacophore model, the program PrGen (3R Biographics Laboratory Foundation, Basel, Switzerland) was used to generate and calibrate a pseudoreceptor. This method serves to create spatial atomistic surrogate and aims to predict the relative free energies of binding based solely on the structures of ligand molecules. The pseudoreceptor model is validated by its ability to reproduce the data experimentally obtained for the set of ligands. For pseudoreceptor purposes, the 3D coordinates of JH-III and five most active JH agonists from Class I and six most active JH agonists from Class II superimposed in the conformations from the CoMFA study were used; this served as starting point for the alignment in pseudoreceptor modeling. PrGen program generates vectors for each functional group from the overlapped ligands indicating steric, electrostatic, hydrogen-bonding, and lipophilic interactions as apparent from both CoMFA and CoMSIA analyses. A pseudoreceptor is then created from individually selected residues that are positioned at the tips of the vectors.

In the first step in pseudoreceptor, the force field types are automatically assigned to imported ligand atom coordinates in Protein Data Base (PDB) format, followed by calculations of atomic partial charges, strain-free reference energies, conformational entropy corrections, and solvation energies. Intermediate step is addition of amino acid residues to properly aligned ligands selected from training set. To do so, it is necessary to position vectors reflecting orientation of CoMFA and CoMSIA fields with particular interactions, followed by geometry optimization. Based on the results from CoMFA and CoMSIA, every type of interaction has several, albeit limited variables that need to be efficiently minimized by retrospective comparison of each biologically active ligand from the distinct agonist class. To generate pseudoreceptor model, individual vectors must correspond to optimal orientation of potential amino acid groups involved in the interaction and fit to the limits of its length (in Å).

In the last step, the amino acid residues were chosen specifically to fit the type of interaction of each vector as characterized by overlapped molecules and CoMFA/CoMSIA-based pharmacophore. The experimental free energies of ligand binding were calculated by a linear regression according to methodology reported by Vedani et al. [83], Bassoli et al. [84], and Zbinden et al. [85] using the following equation:

$$\Delta G_{\text{exp}} = \text{RT} \ln K_d$$

where K_d is experimental dissociation constant. Because dissociation constant (K_d) of JH agonists is unknown, it was approximated by $\ln \text{ED}_{50}$.

In PrGen, free energies of binding, ΔG_0, are estimated from the following equation:

$$E_{\text{binding}} = E_{\text{ligand-receptor}} - T\Delta S - \Delta G_{\text{ligand solvation}} + \Delta E_{\text{internal ligand}}$$

Algorithms to calculate these quantities are included in PrGen package. The predicted free energies of ligand binding, ΔG_{pred}, are obtained by means of a linear regression between ΔG_{exp} and E_{binding}:

$$\Delta G_{\text{pred}} = aE_{\text{binding}} + b$$

The resulting complex of superimposed ligand molecules and amino acid residues of the pseudoreceptor was optimized by a conformational search protocol combined with energy minimization. This step is repeated until the functional groups of the ligand(s) interact with a pseudoreceptor residue. An interactive algorithm, equilibration protocol, was used to obtain the best correlation between the experimental and predicted free energies of binding. The predictive power of the resulting model was assessed using test set.

9.4.2.6 Molecular Volume Computations

The van der Waals volumes of JH and juvenoids were calculated using Sybyl–Base software. For computation of the cavity volume, distances between individual amino acids of the pseudoreceptor model were measured and then artificially connected using the MidasPlus package [86,87] or Biopolymer module of Sybyl to create continuous cavity, and the energy of the resulting structure was iteratively minimized. The final cavity volume was calculated using the Woidoo program [88] and SiteID module within the Sybyl package (Tripos-Certara Inc., St. Louis, MO).

9.5 KEY OUTCOMES OF JUVENOID 3D QSAR ANALYSIS

9.5.1 Structural Diversity in the Light of a Biological Activity of JH Agonists

In our pilot study, we tested the biological activity of over 200 JH agonists and ex post facto evaluated set of 86 compounds whose members spanned the range of wide structural diversity as illustrated in Figure 9.2a and Table 9.1. Even though JH agonists belong to various chemical entities from a structural point of view (farnesol and geraniol derivatives, trimethyl or tetramethyl-dodecenoate or undecenoate derivatives, juvabiones, various derivatives of benzoic acid, acetophenone, aniline, nitrophenol, halophenol, benzenesulfonic acid or carbamate, then ω-alkoxy-ω, ω-dimethyl derivatives, oxime ethers, phenoxyphenoxy, and other oligocyclic derivatives, and peptidic juvenoids), the entire set of compounds under study can be divided into two large classes. This classification is substantiated in detail in the description of how we optimized the CoMFA and CoMSIA models. The first class is composed of linear flexible terpenoid or terpenoid-related molecules with several freely rotatable bonds that include also natural JH-I, JH-II, and JH-III (Class I structures **1–3**). The second class has more rigid compounds (one or few rotatable bonds) containing phenoxy or other cyclic groups on both ends of the molecule exemplified by ZR-10183, ZR-10131, and pyriproxyfen (Class II structures **81**, **85**, **86**). Several independent 3D QSAR reports have documented that it is difficult to analyze very flexible molecules to generate reliable CoMFA model. As obvious, many of the agonist molecules used in this study have linear, farnesene-derived flexible structures; the CoMFA model that is presented here is based on the superposition of both structurally diverse compound classes and shows unexpectedly acceptable predictive ability. Table 9.1 presents the biological activity of 86 previously mentioned agonists, which is expressed as ED_{50} and ranges from 0.00005 to 10 μg per animal where picogram amounts reflect most active compounds. Those JH agonists that show biological activity above 1 μg/animal are considered practically as nonactive.

9.5.2 Optimization Steps for the CoMFA Model

We could easily conclude that highly active compounds of both classes (I and II) display an electronegative oxygen at both ends of the molecule or a nitrogen replacing the oxygen at one end (**1–3**, **14–17**, **19**, **81–86**). As apparent from the summary table of the biological activities, it is dramatically decreased when these oxygens or the nitrogen is replaced by uncharged carbon (e.g., **25**, **26**, **29**, **31**, **32**, **43**, **44**). Therefore, it becomes interesting to understand how these atoms contribute to different longitudinal shifts of juvenoid structures in the alignments.

In most cases, the nitrogen present in JH agonists is part of an unsaturated heterocycle, carbamate, or amide, whereas the oxygen is found mostly within ester, etheric, epoxide, or phenoxyphenol moieties (see Table 9.1). The oxygen in the phenoxyphenol group of the Class II compounds is found to be sterically hindered by benzene rings, and its free electron pairs participate in the conjugation system with these unsaturated benzene rings. This circumstance causes the phenoxyphenol oxygen poorly reactive for intermolecular interactions including hydrogen bonding. Conversely, the oxygen within the epoxy moiety of Class I compounds can provide electron pairs very easily for H-bonding or for other electrostatic interaction. On the opposite side of these compounds, the oxygen atom is part of an ester group, while nitrogen is part of an amide, carbamate, or heterocycle group where all these atoms can provide relatively weak interaction potential. This indicates that oxygen and nitrogen in the different compounds provide substantially different chemical reactivities, atom charges, and abilities to form hydrogen bonds or electrostatic interactions. The combination of these variable properties poses severe challenges for generating an acceptable alignment of JH agonists. Because the structural alignment is critical in CoMFA, we sequentially constructed multiple CoMFA models considering all of these aspects of the properties of JH agonists to find an appropriate alignment.

To obtain optimal alignment, initially we generated several of its variants containing the entire set of tested JH compounds (Figure 9.3). A CoMFA model was optimized by creating a training set from 76 compounds and a test set containing 10 structures (**13**, **14**, **18**, **21**, **31**, **59**, **63**, **69**, **76**, **83**) in both cases representing two classes

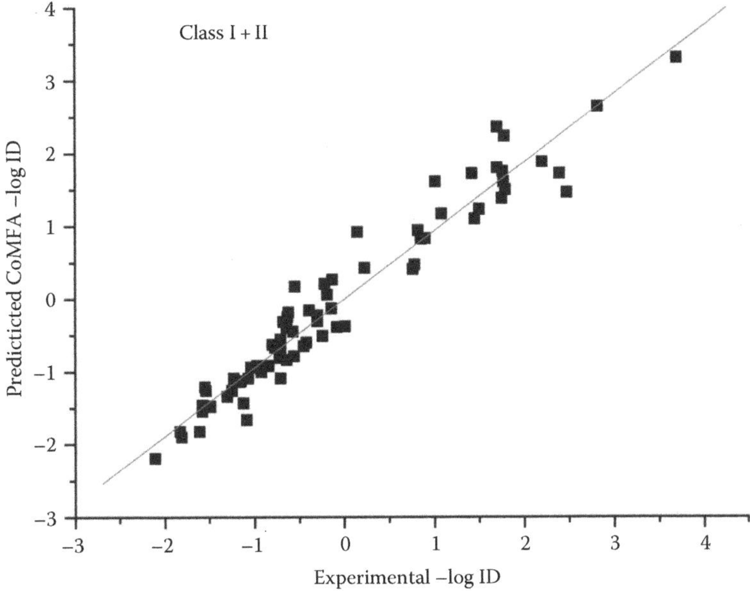

FIGURE 9.3 Basic comparative plot of experimental versus CoMFA-predicted biological activities (–log ED_{50}) of common training set (Class I+II) of 76 JH agonists. Despite being structurally diverse, most active compounds in *Drosophila* share some common features, i.e., an electronegative atom (oxygen or nitrogen) at one end of the molecule and electronegative atom (epoxy oxygen) or electron-rich moiety (oxyphenyl group) on the molecule's opposite end (see compounds **1–3**, **14–17**, **19**, **81–86**). Nonetheless, the terpenoid and rigid phenoxy structures have very different chemical reactivities, atom charges, and abilities in forming hydrogen bonds or electrostatic interactions. Indeed, this was one major reason to divide the complete training set into two classes. The oxygen in phenoxyphenol group of Class II compounds is sterically hindered by benzene rings that make the phenoxyphenol oxygen poorly reactive for intermolecular hydrogen bonding, while the oxygen within an epoxy moiety of Class I compounds can easily provide electron pairs for H-bonding or for other electrostatic interactions. The difference between Class I and II analogs is reflected also in their negative charge distribution. In the Class I structures, it is concentrated near electronegative, ether, or epoxy oxygen, whereas in Class II structures, it is localized to the phenyl rings.

of compounds (linear and cyclic). Every alignment was produced with respect to the electronegative oxygens or nitrogens located at one end of the molecules. After all tasks, the alignment with the best statistical parameters was selected for the final CoMFA model.

A large number of flexible linear compounds (e.g., **14**, **15**, **17**, **19**, **22**, **23**, **26**) are 1.5-fold longer in their extended conformation than typical rigid cyclic compounds (**56**, **66**, **72**, **74**, **80–86**). To this end, bent and therefore shorter conformations with low energy were selected because sterically they fit better to the cyclic compounds and seemingly improve q^2. As anticipated, varying the energy cutoff from 10 to 30 kcal/mol did not show a significant effect on the predictive ability of the model. However, the best q^2 was possible to achieve with a column filtering of 1 kcal/mol. Then, a non-cross-validated PLS analyses were performed. Their final parameters

TABLE 9.2
Summary of Results from CoMFA Analyses for the Common, Unseparated Training Set of 76 JH Agonists as well as for This Set Split into Class I and Class II Compounds

			Cross-Validated		Non-Cross-Validated			Fraction	
Class	n	Comp	q^2	SPRESS	r^2	SEE	F Test	S	E
I+II	76	6	0.508	0.969	0.948	0.306	196,93	0.426	0.574
I	45	6	0.576	1.007	0.983	0.202	325.47	0.403	0.597
II	31	6	0.686	0.767	0.987	0.155	308.77	0.524	0.466

n, number of compounds; Comp, number of PLS components in analysis; q^2, squared correlation coefficient of a cross-validated analysis; SPRESS, standard deviation of error of prediction; r^2, standard correlation coefficient of a non-cross-validated analysis; SEE, standard deviation of a non-cross-validated analysis (standard error of estimate); Fraction, field contribution from CoMFA; S, steric; E, electrostatic.

TABLE 9.3
Test Set Class I+II (CoMFA: Predictive $r^2=0.49$, CoMSIA: Predictive $r^2=0.51$)

Compound	Experimental	Calculated	Residual
13	−0.91	−1.38	−1.23
14	1.75	1.37	1.48
18	−1.13	−0.76	−0.65
21	0.12	0.31	0.24
31	−1.58	−1.33	−1.21
59	−1.34	−0.78	−0.89
63	−0.59	−0.32	−0.86
69	0.01	−0.25	−0.21
76	−0.42	−0.72	−0.64
83	1.70	2.12	1.93

and the statistics ($q^2=0.508$, $r^2=0.948$) for the common training set (designated Class I+II) are documented in the first row of Table 9.2. The predictive ability was externally evaluated through the probing of a test set consisting of 10 previously listed ligands representing both Class I and II compounds. These tasks resulted in CoMFA predictive coefficient $r^2=0.49$ and CoMSIA predictive coefficient $r^2=0.51$ (Table 9.3) for all 10 compounds (unseparated Class I+II test set) or in CoMFA $r^2=0.54$ and CoMSIA $r^2=0.59$ for a test set of 5 Class I ligands (Table 9.4), plus CoMFA $r^2=0.60$ and CoMSIA $r^2=0.63$ for 5 Class II ligands (Table 9.5).

TABLE 9.4
Test Set I (CoMFA: Predictive $r^2 = 0.54$, CoMSIA: Predictive $r^2 = 0.59$)

Compound	Experimental	Calculated	Residual
13	−0.91	−1.24	−1.12
14	1.75	1.81	1.69
18	−1.13	−0.89	−0.96
21	0.12	0.24	0.19
31	−1.58	−1.83	−1.36

TABLE 9.5
Test Set II (CoMFA: Predictive $r^2 = 0.60$, CoMSIA: Predictive $r^2 = 0.63$)

Compound	Experimental	Calculated	Residual
59	−1.34	−1.56	−1.49
63	−0.59	−0.47	−0.68
69	0.01	−0.10	−0.05
76	−0.42	−0.65	−0.59
83	1.70	1.89	1.83

In a next encounter, we reconsidered the previously mentioned diversity of JH agonist structures, and as a consequence, their common merger was split into two separated alignments. Out of several empirical trials based on various structural aspects, we concluded that one alignment covering each of the two structurally already mentioned Class I and Class II compounds resulted in remarkably optimal outcomes. Henceforth, this kind of split led to separate CoMFA calculations and allowed us to exhaustively explore how JH agonist structural diversity can affect the CoMFA result. These two independent training sets showed a nearly ideal alignment and markedly better statistical parameters (Figures 9.4 and 9.5) than the initial, common training set. After CoMFA computations of separated Class I and Class II subsets were performed, the final statistical parameters for each individual set (Class I: $q^2 = 0.576$, $r^2 = 0.983$; Class II: $q^2 = 0.686$, $r^2 = 0.987$) produced significantly better results than the entire common set (I + II) (see rows two and three of Table 9.2). For clarity and validity, the calculated versus actual −log ID values of compounds in the test sets of Class I and Class II are shown in Tables 9.4 and 9.5, and the calculated versus actual −log ID values for all tested agonists are provided in Table 9.6.

9.5.3 Process of the CoMSIA Model Optimization

The CoMSIA analysis serves to test standard steric plus electrostatic interactions, donor and acceptor hydrogen-bonding fields and hydrophobic fields, and their

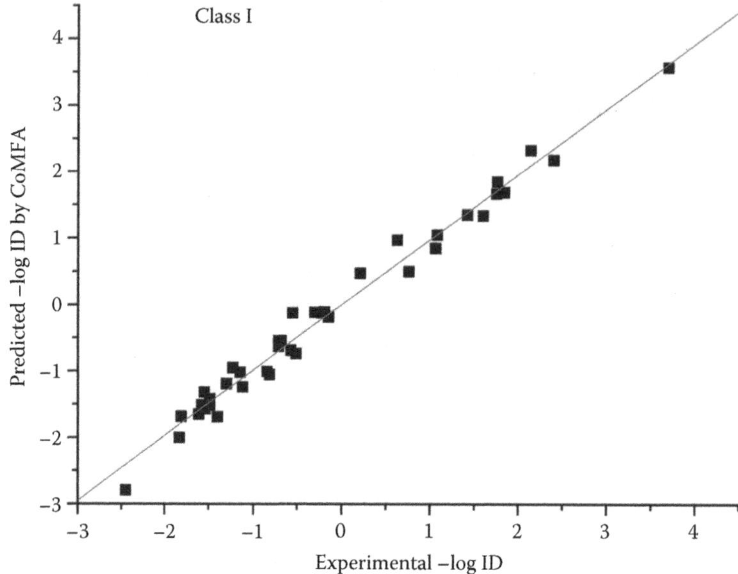

FIGURE 9.4 Plot representing observed versus CoMFA-predicted biological activities (–log ED_{50}) for training set of Class I JH agonists.

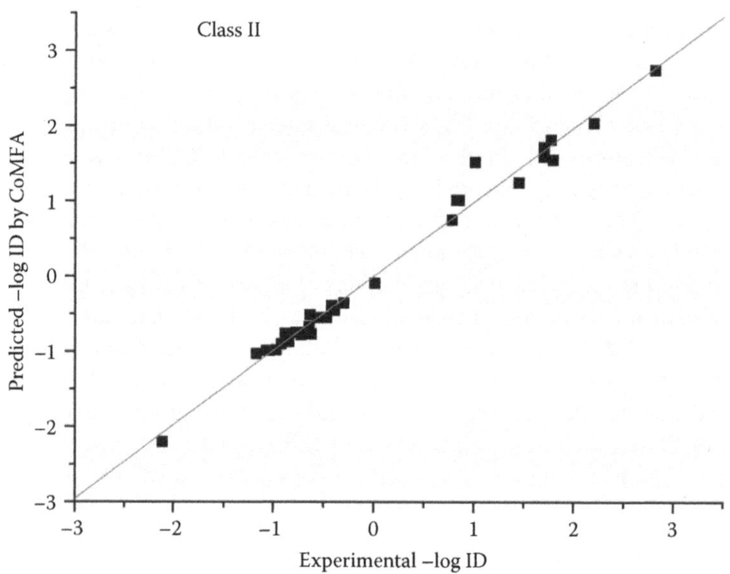

FIGURE 9.5 Correlation between experimental and CoMFA-predicted biological activities (–log ED_{50}) for training set of Class II JH agonists.

TABLE 9.6
Experimental versus CoMFA-Calculated Biological Activities (–log ED_{50}) of JH Agonists

Compound	Experimental	Calculated	Residual
1	1.42	1.72	–0.30
2	–0.55	0.17	–0.72
3	–0.76	0.41	–0.34
4	–0.27	–0.49	0.22
5	–0.72	–0.66	–0.06
6	–0.95	–0.81	–0.14
7	0.39	0.54	–0.15
8	1.50	1.23	0.27
9	3.69	3.31	0.38
10	–1.30	–1.34	0.04
11	1.06	1.17	–0.11
12	–0.84	–0.92	0.08
13	–0.92	–0.68	–0.24
14	1.75	1.38	0.37
15	2.46	1.72	0.74
16	0.28	0.22	0.06
17	2.39	1.46	0.93
18	–1.13	–0.95	–0.18
19	–1.23	–1.22	–0.01
20	1.78	2.24	–0.46
21	0.12	0.92	–0.80
22	–0.13	0.27	–0.40
23	–0.19	0.06	–0.25
24	–0.25	–0.51	0.26
25	–1.23	–1.09	–0.14
26	–0.71	–0.56	–0.15
27	–1.25	–1.19	–0.06
28	–0.45	–0.65	0.20
29	–1.60	–1.87	0.27
30	–1.61	–1.82	0.21
31	–1.58	–1.54	–0.04
32	–1.58	–1.46	–0.12
33	–1.55	–1.21	–0.34
34	–1.49	–1.48	–0.01
35	1.84	1.77	0.06
36	1.43	1.19	0.24
37	–1.21	–1.30	0.09
38	–1.53	–1.26	–0.27
39	–1.49	–1.47	–0.02
40	–1.15	–1.13	–0.02
41	–0.57	–0.45	–0.12

(continued)

TABLE 9.6 (continued)
Experimental versus CoMFA-Calculated Biological Activities (–log ED_{50}) of JH Agonists

Compound	Experimental	Calculated	Residual
42	−1.12	−1.43	0.31
43	−0.71	−0.70	−0.01
44	−0.68	−0.32	−0.36
45	−1.27	−1.44	0.17
46	1.76	1.75	0.01
47	−0.22	0.21	−0.43
48	−0.08	−0.39	0.31
49	−1.35	−1.30	−0.05
50	−0.14	−0.12	−0.02
51	−1.17	−1.14	−0.03
52	−1.15	−1.23	0.08
53	−1.07	−1.09	0.02
54	1.45	1.10	0.35
55	0.85	0.82	0.03
56	2.38	2.50	−0.12
57	0.78	0.47	0.31
58	0.82	0.94	−0.12
59	−0.77	−0.74	−0.03
60	−0.42	−0.60	0.18
61	−1.25	−1.26	0.01
62	−0.39	−0.35	−0.04
63	−0.59	−0.62	0.03
64	−0.47	−0.43	−0.04
65	−1.04	−0.99	−0.05
66	−0.64	−0.40	−0.24
67	−0.49	−0.51	0.02
68	−0.64	−0.54	−0.10
69	0.01	−0.38	0.39
70	−0.56	−0.79	0.23
71	−0.84	−0.84	0.00
72	−0.71	−1.09	0.38
73	−0.38	−0.16	−0.22
74	−0.61	−0.19	−0.42
75	−0.72	−0.81	0.09
76	−0.42	−0.37	−0.05
77	−0.88	−0.71	−0.17
78	0.62	0.91	−0.29
79	−0.30	−0.31	0.01
80	−0.63	−0.25	−0.38
81	2.20	1.88	0.32
82	1.70	1.80	−0.10
83	1.70	2.36	−0.66

TABLE 9.6 (continued)
Experimental versus CoMFA-Calculated Biological Activities (−log ED_{50}) of JH Agonists

Compound	Experimental	Calculated	Residual
84	1.79	1.50	0.29
85	1.77	1.61	0.16
86	2.80	2.64	0.16

continuous optimization as a function of energy cutoff and column filtering. Different versions of the dataset obtained during optimization of the CoMFA model are applied in the statistical evaluation for the CoMSIA analyses. From this point of view, CoMSIA procedure is faster, since it can utilize datasets already produced during CoMFA analysis.

Similar to what was experienced during initial CoMFA, also when we analyzed the whole set of structures (both Class I and II) by CoMSIA, the indicator fields yielded promising albeit not ideal statistical results, as documented in Table 9.7, which summarizes the optimized parameters and q^2 values for the indicator fields. Using CoMSIA parameters and version we operated, a model with a q^2 value greater than 0.3 is usually regarded to be significant, so models with field indicators that led to $q^2<0.3$ were not considered. The cross-validated PLS analysis with best outcome gave the following values at $n=76$: $q^2=0.534$, $r^2=0.901$. Clearly the electrostatic factors are playing a major role because in the best CoMSIA model, they have shown a high q^2 value of 0.740 versus 0.260 for steric factors (Table 9.7). As would be now anticipated, another significant descriptor identified was a field of acceptor hydrogen bonding, although its q^2 was only 0.391. The graphical contour maps of the sterical and electrostatic fields are similar to the corresponding CoMFA plots. The significant q^2 values of steric, electrostatic, and hydrogen-bonding descriptors validate that these variables are critical to describe the interaction of JH compound with its target.

If this complex set of agonists was divided into Class I and Class II structures as described earlier, and separate CoMSIA calculations performed, the differences between these two classes became obvious again and even more remarkable than under CoMFA procedure. This is mirrored in the different types of indices producing the best models of separated agonist classes. Table 9.7 summarizes detailed final statistical parameters for separated classes (Class I: $n=45$, $q^2=0.637$, $r^2=0.960$ for steric and electrostatic indices; Class II: $n=31$, $q^2=0.755$, $r^2=0.956$ for the steric index), and Figures 9.6 and 9.7 interpret them graphically.

9.5.4 Resultants of 3D QSAR

Correct interpretation of results is essential for identification of individual types of interactions that play significant role in determining biological activity of the compounds. Thus, in order to visualize the information content of the CoMFA model produced, 3D electrostatic and steric contour maps were generated (Figure 9.8a and b).

TABLE 9.7
Summary of Results from CoMSIA Analyses for a Training Set of 76 Juvenoids

			Cross-Validated		Non-Cross-Validated			Fraction		
Class	**n**	**Comp**	q^2	**SPRESS**	r^2	**SEE**	**F Test**	**S**	**E**	**DA**
I+II										
S	76	2	0.371	1.065	0.744	0.805	31.07			
E	76	3	0.507	0.882	0.884	0.651	61.77			
SE	76	6	0.534	0.877	0.901	0.408	96.76	0.26	0.74	
A	76	3	0.391	1.072	0.667	0.725	45.42			
SEDA	76	3	0.485	0.901	0.736	0.643	63.86	0.08	0.36	0.56
I										
S	45	3	0.260	1.009	0.786	0.721	29.91			
E	45	6	0.740	0.934	0.926	0.396	70.80			
SE	45	6	0.637	0.875	0.960	0.290	136.18	0.29	0.71	
A	45	2	0.329	1.192	0.625	0.915	29.94			
SEDA	45	3	0.493	1.050	0.860	0.623	49.58	0.05	0.23	0.72
II										
S	31	6	0.755	0.678	0.956	0.286	87.37			
E	31	2	0.675	0.722	0.841	0.505	74.23			
SE	31	2	0.695	0.700	0.927	0.396	78.38	0.20	0.80	
A	31	3	0.537	0.878	0.836	0.522	45.91			
H	31	6	0.660	0.798	0.978	0.205	174.16			
DA	31	3	0.481	0.930	0.843	0.511	48.28			
SEDAH	31	3	0.579	0.837	0.887	0.435	70.31	0.07	0.25	0.51

n, number of compounds; Comp, number of PLS components in analysis; q^2, squared correlation coefficient of a cross-validated analysis; SPRESS, standard deviation of error of prediction; r^2, standard correlation coefficient of a non-cross-validated analysis; SEE, standard deviation of a non-cross-validated analysis; Fraction, field contribution from CoMSIA like; S, steric; E, electrostatic; A, hydrogen bond acceptor type; D, hydrogen bond donor type; H, hydrophobic.

The importance of the electronegative oxygens or nitrogens at the ends of the aligned structures is indicated by red polyhedra near the positions of these atoms. Interpretation of the steric fields causes main difference between the contour maps of Class I (Figure 9.8a) and Class II (Figure 9.8b) agonists. The large green polyhedra positioned on "left side" of the Class II map (Figure 9.8b) are situated near the phenoxyphenyl group and indicate that the presence of steric bulk substituents in this part of the molecule is expected to enhance JH biological activity.

The CoMSIA contour maps are easier to interpret than CoMFA plots because they partition variance into the higher number of different field types. The steric and electrostatic CoMSIA fields for these maps are shown in Figure 9.9a (Class I) and B (Class II). Like in the CoMFA contour maps, the green polyhedra in CoMSIA also represent sterically favored regions in which more bulky substituent increases

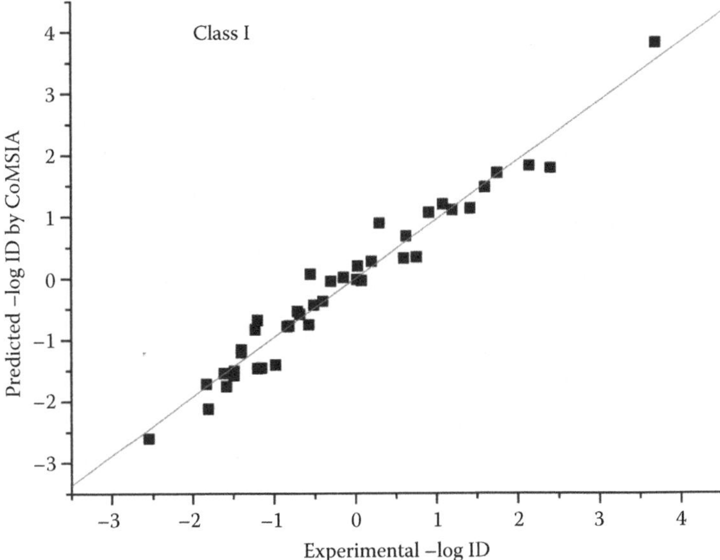

FIGURE 9.6 Graphical representation of experimentally observed versus CoMSIA-predicted biological activities (–log ED_{50}) for training set of Class I JH agonists displaying acceptably good correlation.

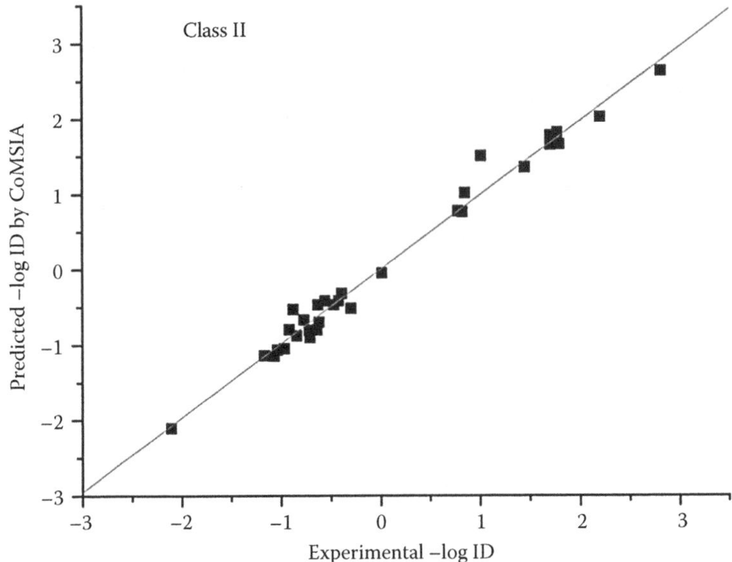

FIGURE 9.7 Experimental versus CoMSIA-predicted biological activities (–log ED_{50}) for training set of Class II JH agonists. For illustration, see Figures 9.8 and 9.9 where the significant difference between Class I and Class II molecules (e.g., in the large green area) in the steric contour maps of both CoMFA and the CoMSIA indicates that more bulky substituents in these regions will enhance the biological activity in Class II compounds.

(a)

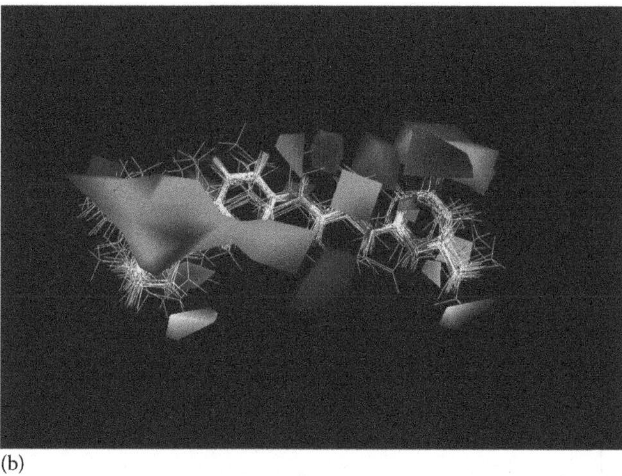

(b)

FIGURE 9.8 (See color insert.) Graphic representation of CoMFA steric and electrostatic field contour plot for Class I (a) and Class II (b) juvenoids. Yellow-polyhedron-surrounded regions indicate sites where less bulky substituents are appreciated for increasing biological activity, whereas green polyhedra represent sterically favored regions where more bulky substituents increase biological activity. The red contours represent regions where negative charge is favorable, while blue polyhedra represent electrostatic regions where positively charged groups will be favorable and will enhance biological activity. The importance of the electronegative oxygens or nitrogens on the ends of the aligned structures is indicated by red polyhedra near the positions of these atoms. The presence of smaller red polyhedra in the middle of the JH agonist structure indicates an additional site where an electronegative atom or group can enhance biological activity importantly. The contour map of Class II (b) differs from that of Class I (a) mainly in the steric fields. The large green polyhedra of Class II (b) near the part of the phenoxyphenyl group (left side) indicate that the presence of steric bulk substituents in this part of the molecule enhances biological activity.

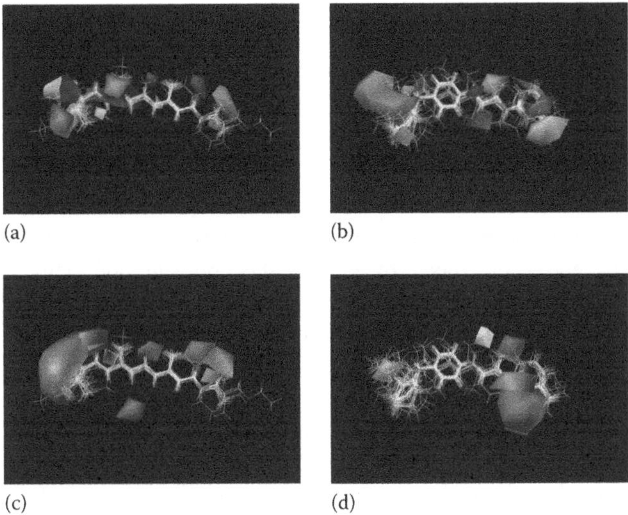

FIGURE 9.9 (See color insert.) Contour maps of Class I and Class II JH agonists as revealed by CoMSIA analyses. Green and yellow polyhedra for sterical fields and blue plus red polyhedra for electrostatic fields have the same representative meaning as for CoMFA maps in Figure 9.8 for both Class I (a) and Class II (b) JH agonists. Contour maps for the hydrogen bond acceptor and donor fields are illustrated in (c) (for Class I agonists) and (d) (for Class II agonists). Magenta areas indicate regions where hydrogen bond acceptors are favorable for increasing biological activity (oxygens and nitrogens in the ligand); cyan areas indicate fields where hydrogen bond donors are favorable (NH and OH groups in ligand). Orange polyhedra surround area where H-bond acceptors are unfavorable and white polyhedral areas where H-bond donors are unfavorable.

biological activity, whereas yellow polyhedra reflect sterically disfavored regions where a smaller substituent can increase hormonal activity. In the CoMSIA electrostatic contour plot, red-colored polyhedra represent favorable regions where negatively charged groups enhance activity, while blue polyhedra indicate disfavored regions of the molecule where positively charged groups may enhance activity. The presence of smaller red polyhedra in the middle of a JH agonist structure indicates an additional site where an electronegative atom or group enhances the biological activity, as exemplified by the oxygen in the middle of SJ-68 epoxide (**9**) or double bond in JH-III (**1**). This conclusion was further confirmed by inclusion of activity data from another 30 oxa-juvenoids (but not yet included in the current text) that are "siblings" of SJ-68 epoxide. The steric and electrostatic fields of CoMSIA maps are generally in good accordance with the field distribution of CoMFA contour maps. They do, however, indicate more sterical freedom for Class II compounds on their phenoxyphenol side than would be possible to read from CoMFA results.

Substantially dramatic difference between Class I and Class II compounds can be seen in the CoMSIA hydrogen bond acceptor and donor fields (Figure 9.9c and d). These interaction fields highlight the areas beyond ligands where putative hydrogen bond partners (amino acid residues) in the putative receptor could form hydrogen bonds that are so important for biological activity via ligand binding to target pocket.

Areas in magenta color indicate where hydrogen bond acceptors are favorable for increasing biological activity (oxygens and nitrogens in ligand), while cyan areas designate those fields in which hydrogen bond donors are favorable (e.g., NH and OH groups in the ligand). Conversely, orange polyhedra surround the area where H-bond acceptor seems to be unfavorable and white polyhedra surround the area for which H-bond donor is not favorable. The importance of the hydrogen bond acceptor interaction for the Class I contour map (Figure 9.9c) is accented at both ends of the molecules in the positions of the oxygen within the ester group and in the position of the epoxy or etheric oxygen on the opposite side. This can be documented by comparison of the ED_{50} values between structures **15** and **19**, since the only difference between these two compounds is that an esteric oxygen (a hydrogen bond acceptor) is replaced by an amidic nitrogen (a hydrogen bond donor), which resulted in 100-fold decrease in the biological activity of **19**. In fact, these conclusions become also apparent from both the CoMFA and the CoMSIA electrostatic contour maps in which a negative charge enhances activity (red color). It needs to be stressed that when these electronegative atoms in Class I structures are replaced by carbon (e.g., compounds **21**, **25**, **26**, **29**, **31**, **32**) or distance between these electronegative end points was shorter than 11.5 Å or longer than 13.5 Å (compounds **28**, **40**, **41**, **48**, **49**), biological activity drops down dramatically. Moreover, the presence of an electronegative oxygen or a double bond in the middle of the molecule (e.g., **9**, **15**, **17**), which results in a site with an increased concentration of negative charge and less electron charge at the ends, appears to have additional importance for increased biological activity. In the CoMFA and CoMSIA contour maps, these areas are represented by blue polyhedra. The contour map of Class II compounds specifically lacks polyhedra for a favorable H-bond donor or acceptor field on phenoxyphenol side of structures, and this result produces one of the most remarkable differences between two presented pharmacophore models.

Although size of juvenoid compounds is very important feature predetermining biological activity, the CoMFA as well as CoMSIA computations unambiguously indicated that the electrostatic requirements are more important than the steric ones because the electrostatic fraction had higher values than the steric fraction for both Class I and Class II (see Table 9.2). In CoMFA, the electrostatic fraction is 0.574 versus 0.426 for a steric fraction. This is even more remarkable in the CoMSIA calculation where the electrostatic fraction is 0.740, while the steric fraction is 0.260. Thus, the presence of negative charge-rich atoms or groups at the ends of the molecules represented by electronegative atoms like an oxygen or nitrogen or an unsaturated cycle is the crucial element that contributes to high-affinity binding. This is additionally supported by CoMSIA field indices and hydrogen bond acceptor fractions (Table 9.7), which specify that these electronegative atoms are most probably involved in hydrogen bonding with the putative receptor cavity. In other words, the "extraction" of data from analysis of CoMFA and CoMSIA models for Class I and Class II juvenoids revealed the points and regions that are highly correlated to the activity of tested compounds. Altogether these data from 3D QSAR analyses are summarized in the form of two pharmacophore models that depict the key structural requirements for the biological activity of JH agonists in *Drosophila* (Figure 9.10a and b).

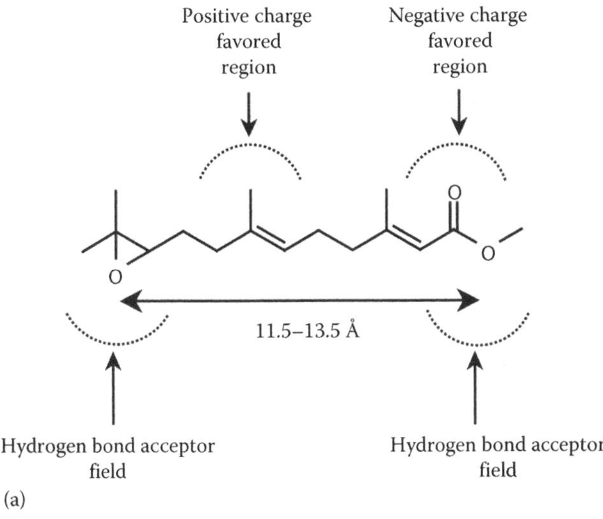

FIGURE 9.10 The pharmacophore models of Class I and Class II juvenoids based on CoMFA and CoMSIA results. Models show key structural elements responsible for the hormonal activity of JH agonists. Both classes of JH compounds share regions that favor negative charge and a field that requires hydrogen bond acceptor. In both cases, these elements are located at the sides of JH agonists. However, Class I compounds (a) have also additional hydrogen bond acceptor element located on the esteratic side and a region that favors positive charges centrally. A specific feature of Class II compounds (b) is a hydrophobic region that is favored around the outside edge of a phenoxyphenol moiety. Due to the high flexibility of Class I compounds, the distance between hydrogen bond acceptor atoms (e.g., esteratic and epoxy oxygens at two sides) must be between 11.5 and 13.5 Å to possess agonist biological activity in *Drosophila*. Highly rigid Class II compounds that show JH biological activity in *Drosophila* automatically fit to this requirement.

9.5.5 Pseudoreceptor Model

Pseudoreceptor modeling is one of the receptor mapping approaches, in which a paucity of information concerning protein structure has spawned techniques that project the properties of the bioactive ligands into three dimensions around their appropriately superimposed molecular framework. Based on the information summarized in the pharmacophore model, we used specifically the PrGen program to produce an atomistic pseudoreceptor model. This pseudoreceptor construction comprises the essential features of an active site by assuming complementarity between the bioactive conformations of the set of JH compounds and the shape and properties of the receptor site and thus mimics the real receptor-binding surface. To generate corresponding pseudoreceptors, the alignments of natural JH-III plus the five most active ligands of Class I and the six most active ligands of Class II were used. The appropriate amino acid residues for pseudoreceptor construction were selected by considering two important factors. First, we used data about molecular interactions derived from the CoMFA and CoMSIA structure–activity relationships. Second, we applied information about amino acids and their interactions, which are most frequently involved in forming binding pocket in members of the nuclear receptor superfamily recognizing small lipophilic ligands. Due to these criteria, polar amino acids (e.g., Arg, Asp) were placed complementarily to the vectors of atoms with local electron deficit, and residues acting as H-bond donors (e.g., Tyr, Asp) were placed complementarily to the vectors of ester or etheric oxygens. Hydrophobic amino acids (e.g., Met, Leu, Iso, Tyr) were gradually located around the rest of the molecular alignment. For acceptor-type hydrogen bond interactions, we used bonding with Ser, Thr, Tyr, His, Arg, or Gln, whereas Ser, Thr, Tyr, His, Arg, Gln, Lys, or Asp was fully appropriate to mediate electrostatic interactions. An intermediate step is the addition of individual amino acids to properly aligned ligands selected from training set. To do so, it is necessary to place vectors reflecting orientation of CoMFA and CoMSIA fields with particular interactions for Class I (**a**) and Class II (**b**) ligands (Figure 9.11a and b), followed by geometry optimization. After the whole complex of a pharmacophore surrounded by a putative active site was generated, the energy equilibration protocol was applied to produce an energetically relaxed model until the best possible correlation between calculated and experimental free energy of ligand binding could be achieved. This procedure was repeated iteratively with each amino acid combination selected for their appropriate positions at the tips of vectors in order to obtain the optimal energy equilibrium. From several trials, for an atomistic pseudoreceptor model of Class I ligands, a combination of Trp, Thr, Leu, Thr, Leu, Ile, Val, and Tyr amino acid residues yielded the best correlation coefficient ($r^2=0.91$) with predictive cross-validated coefficient of $q^2=0.53$, when the experimental free energies of binding were confronted with predicted free energies of binding (Figure 9.12a). In this model, the hydrogen bond donors Tyr and Thr are positioned in the vicinity of the esteric or amidic group of the aligned ligands, and Tyr is positioned near the epoxy group on the opposite site of the molecular alignment. Trp has optimal position near the electron-deficient middle part of the Class I alignment. Application of the same procedure to data gained for the Class II pharmacophore model resulted in the combination of Tyr, Met, Val, Val, Thr, Leu, Phe, and Ile amino acid residues

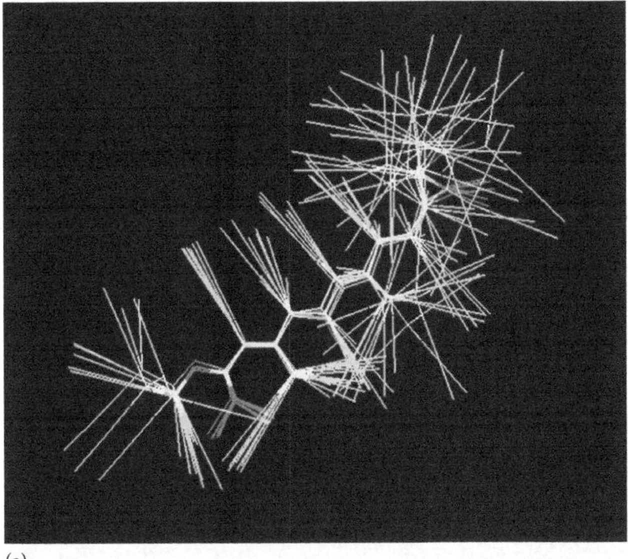

(a)

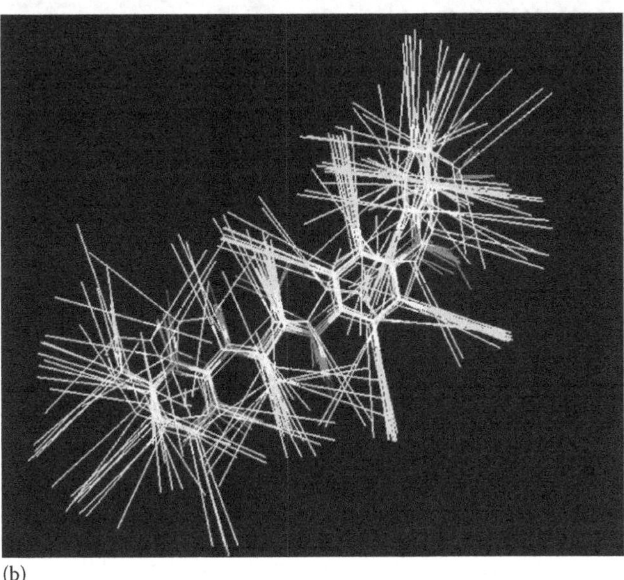

(b)

FIGURE 9.11 Intermediate step in the generating of pseudoreceptor models for Class I (a) and Class II (b) ligands. For the addition of individual amino acids to properly aligned ligands selected from training set, it is necessary to place vectors reflecting orientation of CoMFA and CoMSIA fields with particular interactions, followed by geometry optimization. Based on CoMFA and CoMSIA results, each type of interaction has several, albeit limited variables that need to be efficiently minimized by retrospective comparison of each biologically active ligand of the distinct agonist class. To generate pseudoreceptor model, individual vectors must correspond to optimal orientation of potential amino acid groups involved in the interaction and fit to the limits of its length (in Å).

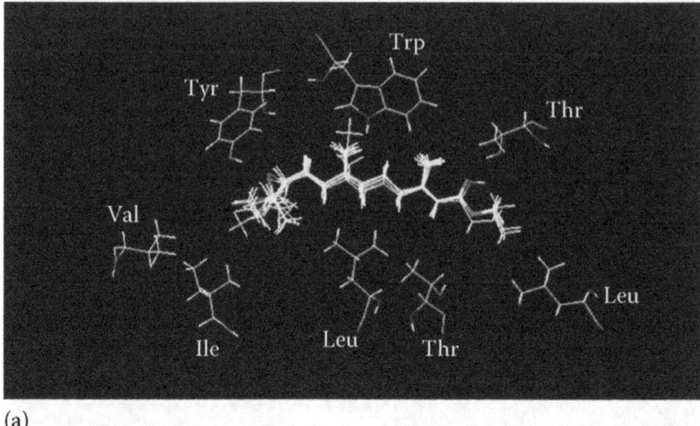

(a)

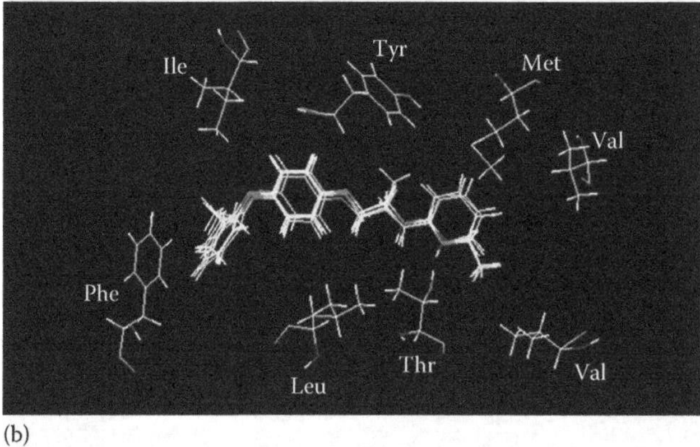

(b)

FIGURE 9.12 The 3D structure of the pseudoreceptor models for Class I (a) and Class II (b) JH agonists. Selected juvenoids are aligned in the middle with the surrogate of eight amino acids surrounding them. Both models are composed of eight amino acid residues reflecting all interactions that are predicted to be required from pharmacophore analysis. For the sake of clarity, bonding interactions and vectors have not been displayed.

and yielded the best statistical correlation coefficient $r^2=0.92$ and cross-validated predictive coefficient $q^2=0.43$. Similar to the pseudoreceptor model created for the Class I structures, in the model reflecting data of Class II structures (Figure 9.12b), Thr is placed as a hydrogen bond donor in the vicinity of the esteric or amidic group of the agonist alignment, and Tyr is positioned near the middle electron-deficient region. Considering weak interactions described in the pharmacophore model, the hydrophobic residues (Ile, Phe, Leu) are placed against the phenoxyphenyl group. The theoretical relative Gibbs free energies computed for the binding of both series of agonists versus experimental relative Gibbs energies are shown in Tables 9.8 and 9.9.

9.5.6 Design of Novel *Drosophila* Juvenoids

Our understanding of the pharmacophore properties of the ligands we have analyzed can be used to design a set of lead or related compounds with predicted activities

TABLE 9.8
ΔG Experimental versus ΔG PrGen-Predicted Activities for Juvenoids of Class I

Comp. No	ΔG_{exp}	ΔG_{pred}
1	2.02	1.44
2	−1.09	−1.42
8	−2.17	−2.72
9	−5.25	−4.70
15	3.40	3.86
19	2.54	1.73

TABLE 9.9
ΔG Experimental versus ΔG PrGen-Predicted Activities for Juvenoids of Class II

Comp. No	ΔG_{exp}	ΔG_{pred}
53	−2.05	−2.08
54	−1.90	−2.16
56	−1.50	−1.80
82	−2.41	−1.98
84	−2.26	−1.08
86	−3.98	−3.94

that can be evaluated *in silico*; and after synthesis, also *in vivo*. Such compounds can be designed using the contour maps generated during the CoMFA and CoMSIA procedures and focusing attention on the key aspects of compounds, e.g., where to add or remove steric bulk, where to strive for negative or positive substituents, and where such changes are unlikely to strongly affect activity. The basis for confidently designing new compounds is rationalized by rederiving the model with the aim of predicting the "omitted QSAR observations." The most frequent goal of this pharmacophore approach is to design compounds with novel properties or higher activity. Thus, we used the color-coded contour maps to identify the steric, electrostatic, and additional properties that would be challenged if we changed the structural details of a molecule. For example, we can predict the effects of adding specific atoms or chemical groups within the concentrated electron density depicted in red-colored fields or adding bulky substituents within green fields. In contrast, these bulky substituents would be removed in the yellow-colored fields, and in the case of blue fields, the present atoms would be exchanged for atoms with lower electron density. Subsequently, energy minimization is performed, charges are calculated for these changes, the newly designed molecules are realigned by the program, and a set of regression equations is applied to compute the potential activity of the models. This is found in the new rows of common spreadsheet. In addition to several low but active or nonactive compounds, we obtained four novel highly active juvenoids (**87–90**) presented in the Table 9.10, the predicted activity of which reaches values comparable to the most active agonists observed in our bioassays (**8**, **9**, **15**, **46**, **56**, **81**, **85**, **86**).

9.6 CONSIDERATIONS AND SOME IMPLICATIONS OF *DROSOPHILA* JUVENOID PHARMACOPHORE ANALYSIS

The primary aim of this effort was to establish an initial juvenoid 3D QSAR by using CoMFA and CoMSIA models generated from an extensive set of juvenoid molecules and to find out whether this kind of pharmacophore analysis can be applied to so divergent group of compounds as JH agonists are. The earliest trials in the past to look at action of juvenoids via structure–activity relationships [35,89–91] faced several problems: (1) The potentially largest obstacle could be in combination of testing, collecting, and comparing data simultaneously among several even unrelated species. Reliable approach could be hampered by the inability of obtaining by computer interpretable data specific for a single given insect species. It can be anticipated that the molecular mode of JH action relies on an interaction between a small ligand and its protein target; there are likely species-specific differences in amino acid sequence and tertiary protein structure. Then, heterophilic data would impede development of a pharmacophore model useful for predicting the molecular properties of a JH receptor. (2) A second and frequently neglected problem was associated with the selection of the characteristics used to measure ID_{50}, IC_{50}, or ED_{50}. To simplify the work and for efficient data collection, researchers were often satisfied with counting the inhibition of metamorphosis or the prevention of eclosion. Using these parameters, such data, at least in some cases, easily could reflect a compound's insecticidal toxicity rather than its morphogenetic action, producing somewhat misleading data. (3) A third problem can be associated with the method of juvenoid administration. Its choice can make a conceptual difference in the outcome of tested JH biological activity. For instance, very different dose–response curves and ID_{50} or IC_{50} values will be obtained when animals are given a tested compound via a single direct topical application than via its continuous presence in food or the living environment (e.g., in the water for aquatic insects [92]). Under conditions where a compound is repeatedly applied or continuously present, there is an increased chance that toxic effects will have prevalence over real hormonomimetic biological activity. (4) Although logical and historically understandable, until now the structure–activity studies focused on testing JH analogs in agricultural pests or disease vectors that, in contrast to *Drosophila*, could not provide suitable genetic or molecular tools to dissect the mechanism of JH action. We would like to offer a realistic hope that findings from the presented approach can be combined with the power of *Drosophila* genetics to identify candidates for the JH receptor. (5) Finally, versatile bioinformatics-aided and drug-design-oriented Unix- or Linux-operating high-performance computer programs to evaluate data were unavailable at the time the most active research on

TABLE 9.10
***De Novo* Designed Juvenoids with High JH Activity as Predicted from CoMFA and CoMSIA Calculations, Providing Most Efficient Interactions within Putative Binding Protein Cavity**

No.	Compound[a]	ED_{50} (µg)	ED_{50} (nM)	$-\log ED_{50}$ (nM)
87	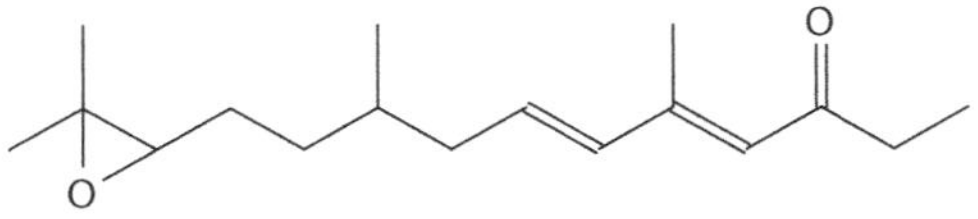	0.000085	0.000321	3.4935
88	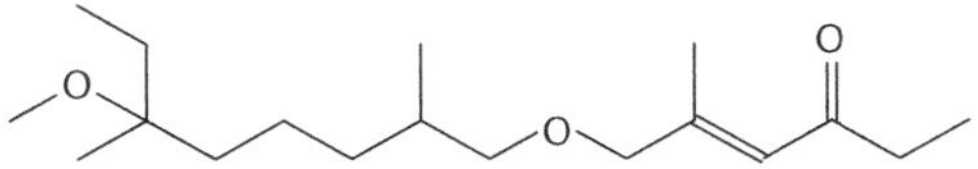	0.000048	0.0001608	3.7937
89	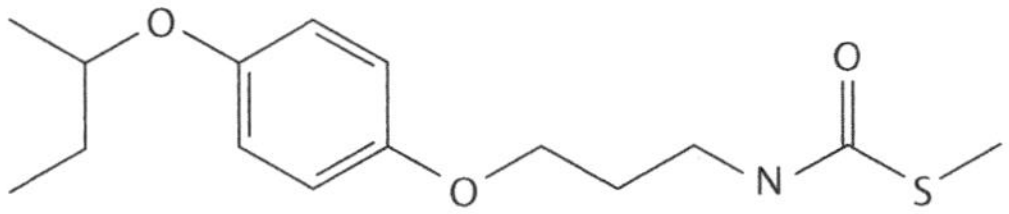	0.00033	0.001113	2.9535
90	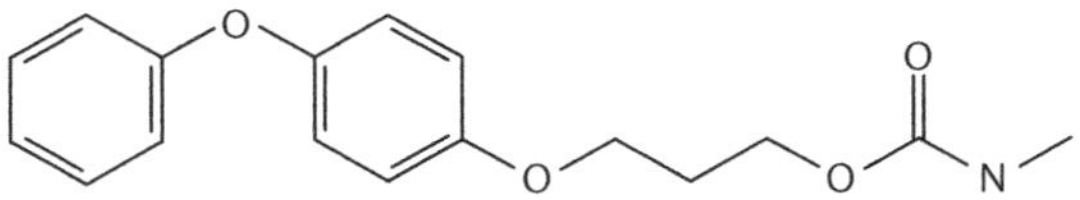	0.0018	0.00599	2.2226

[a] Chemical names are given in [67].

development and production of JH analogs was being undertaken. To avoid problems and circumvent previously listed challenges, we decided to use only one insect species, in this case, *Drosophila*, since it is a well-characterized genetic model. To ascertain evaluation of *bona fide* hormonal activity of juvenoid compounds and distinguish it from any potential toxic effects, we used a strictly defined and specific morphogenetic effect [40,43,46,93] as a criterion to record ED_{50} data. Finally, to avoid any misunderstanding and misinterpretation, we designed assays, generated, collected, processed, and evaluated all data using a well-established 3D QSAR computational approach, all exclusively within our research group.

In majority of recent CoMFA and CoMSIA studies, datasets that are being analyzed originate from *in vitro* assays in which a recombinant receptor is transiently expressed in host cells and where a series of compounds is tested for reporter activity. Even though this approach produces the predictions that are much easier and straightforward to test and is widely used in drug development, the lipophilic compounds like JH agonists almost always pose a unique challenge. In contrast to this, (1) we used more complex *in vivo* assay model, and (2) we employed a reverse approach where quantitative structure–activity data and results from CoMFA/CoMSIA analyses are expected to shed light on the properties of JH receptor, which is not known yet.

It became generally accepted that the biological activity of JH analogs in *in vivo* tests is always to some extent affected by their solubility, penetration through cuticle, transport via hemolymph, and delivery to target tissues (for reviews, see [34,35,94–96]). It should be noted that these conclusions were based on the lipophilic properties of JH and its analogs. There is no doubt that these factors cannot be eliminated completely. Notwithstanding this potential complication, presented CoMFA and CoMSIA analyses demonstrate that the hydrophobic interactions along with other weak interactions of Class I JH agonists have very poor q^2 and r^2 values. These types of interactions, under certain circumstances, reflect a complex of less-specific properties of biological molecules such as their solubility, penetration, and transport [97–102]. Indeed, the highly significant q^2 and r^2 values for the steric and electrostatic fields document that the QSAR data obtained reflect very closely the real binding of a ligand to its protein target, possibly a receptor, rather than nonspecific interactions [103]. Also, two additional important considerations can shed more light on explanation of this result. First, JH compounds that in our bioassay show biological activity are active at relatively low concentrations, which are below the threshold that would produce additive or nonspecific effects associated with penetrability, solubility, etc. And this seems to be fully consistent with the negligible role that hydrophobic interactions analyzed by 3D QSAR protocols are observed to play in biological activities of numerous small hormonal ligands. Secondly, as already discussed elsewhere [67], the composition of the *Drosophila* exoskeleton with its cuticular layers [104,105] appears to be relatively favorable for JH penetration and in contrast to many other insect groups offers little if any resistance to hormonal effects. The previously mentioned view is supported by the observation that the administration of JH compounds topically or in drinking water renders a very different biological activity response in different insects (e.g., the linden bugs, *Pyrrhocoris apterus* and *Dysdercus cingulatus* [34,94]). Last but not least, the juvenoids we tested in *Drosophila* were applied prior to pupariation, and their morphogenetic effect was evaluated more than 120 h later. Even in the case that some morphogenetic processes could be influenced by JH agonists 16 or 48 h after administration, this time is more than sufficient to minimize, if not eliminate, nonspecific factors like visceral solubility or transport. Although excretory activity of *Drosophila* is not turned off during the prepupal, pupal, and pharate adult stages, it cannot be as fast and efficient as in feeding larva, and the excretory system needs to be reconstructed during the larval-to-adult metamorphic transformation. Thus, detoxification and excretory mechanisms during this period are likely to be at least slightly less efficient and should not significantly contribute to nonspecific factors affecting biological activity of JH and its agonists in tests performed.

Important aspect that deserves more explanation is the fact that we have come out with two separate classes of juvenoids. In spite of their structural diversity, most active juvenoids in *Drosophila* share some common features, i.e., an electronegative atom (oxygen or nitrogen) at one end of the molecule and electronegative atom (epoxy oxygen) or electron-rich moiety (oxyphenyl group) on the molecule's opposite end (compounds **1–3**, **11–14**, **16**, **78–83**). Nonetheless, the terpenoid and rigid phenoxy structures possess very different chemical reactivities, atom charges, and abilities to form hydrogen bonds or electrostatic interactions. These differences represent some of the major reasons for dividing the complete training set into two classes. In detail, the phenoxyphenol group oxygen of Class II compounds is sterically hindered by benzene rings that make this oxygen poorly reactive for intermolecular hydrogen bonding, whereas the oxygen of the epoxy moiety of Class I compounds can easily provide electron pairs for H-bonding or for other electrostatic interactions. Another substantial difference between Class I and II analogs is reflected in the distribution of their negative charge. In the Class I structures, negative charge is concentrated near electronegative, ether, or epoxy oxygen, while in Class II structures, it is localized to the phenyl rings. Subdivision of compounds into two chemotypes for QSAR analyses was helpful also in the case of COX-2 inhibitors [106] and was published for steroid hormones to reflect the unusual conformational adaptation of nuclear receptor ligand-binding domains to agonist variety [60,107]. It can be stressed here that for very high biological activity of some synthetic JH agonists and not observed in natural JH (blue and cyan polyhedral regions in CoMFA and CoMSIA contour maps, respectively; see Figures 9.8 and 9.9a and b) is essential the presence of an electron-deficient moiety in the middle of the JH agonist molecule. On the other hand, requirements for the presence of more bulky substituent in Class I compounds (green polyhedra) to enhance biological activity are indicated by CoMFA and CoMSIA steric contour maps. It is equally important that larger substituent (yellow polyhedra) in the vicinity of the phenoxyphenyl or epoxy groups of Class II juvenoids would decrease their biological activity (Figures 9.8b and 9.9b). Collectively, our data suggest that each of the two independent training sets has produced a nearly ideal alignment and markedly better statistical parameters than the original common set. These conclusions actually corroborate the original idea of Humber et al. [108–111] for the receptor-bound conformation of ligands as inferred from the study of diverse dopamine receptor agonists (apomorphine, chlorpromazine, and butaclamol) that are difficult to align while maintaining the coplanarity of their only centroid, the aromatic ring.

Another detail regarding difference between Class I and Class II agonists that should be brought to the reader's attention is in their hydrogen-bonding availability, which is markedly visible in the CoMSIA hydrogen bond contour maps illustrated in Figure 9.9c and d. The significant difference between Class I and Class II molecules is in the CoMFA and CoMSIA steric contour maps (depicted as the large green area), indicating that more bulky substituents in these regions should enhance the biological activity in Class II compounds. Given these data, we reasoned that there is a bigger or more flexible binding site in the cavity surrounding this region of the agonists. On the other hand, the green area in this part of the structures in Class I compounds is much smaller, and so it could signify a tighter contact with the receptor-binding site. It is assumed that steric and electrostatic contour maps generated by CoMFA and CoMSIA hold the potential to indicate the shape and surface requirements of the ligand-binding protein cavity, the putative JH receptor. From this it can be inferred that the receptor cavity is supposed to display charged residues lengthwise along its borders and negatively charged or neutral residues in its middle.

The approach we presented delineates several important moments that allow us to ask the question, what do these structurally divergent JH compounds have in common that enable them to retain identical biological activity as naturally occurring JH? This question was addressed and answered in large part by defining five key elements of the most active Class I agonists in the first pharmacophore model. These key elements are (1) an acceptor type of hydrogen bond related to esteratic or equivalent oxygen (as shown from studies with **17**, **23,** and **24**); (2) a second acceptor hydrogen bond coming from the epoxy oxygen on the opposite side of the molecule (as apparent in, e.g., **35**); (3) an electrostatic interaction originating from a carbonyl (keto) oxygen on the same side of the molecule (see **1–12**, **15**, **17**, **20**, **22**, **25**, **26**), which potentially may serve as hydrogen-bonding element; (4) a strictly determined distance (11.5–13.5 Å) between these hydrogen acceptor oxygens or nitrogens; and (5) a highly favorable positively charged group(s) in the middle of the molecule. JH-III as a natural ligand has only four of these elements, which provides explanation why several JH agonists, in addition to their commonly known resistance to metabolism [6,34,94], can be more potent than natural JH. While keeping orientation as in Figure 9.10, the pharmacophore model of Class II agonists has three elements: (1) The first element is a hydrogen bond acceptor on the right side of the ligand, as in the Class I agonists; (2) a negatively charged group on the left side of the pharmacophore; and (3) a sterically large hydrophobic substituent on the same side where the phenoxyphenol moiety is located. In addition, juvenoids belonging to Class I have an average molecular volume of 291.47 Å^3, whereas Class II compounds have an average molecular volume of 253.65 Å^3, which is partially reflecting space requirements for these agonists to fit within the putative binding cavity of the JH receptor.

The models described herein have relevance to two of the most frequently cited proposals that try to explain the molecular mode of JH action. The first proposal is that of the Usp model, based on the proposed ability of the receptor to bind with low affinity ($K_d \sim 4$ μM) to JH and JH-like ligands. In transfected lepidopteran cells, JH treatment slightly activates expression of a reporter driven by the juvenile hormone esterase (JHE) gene core promoter [23,112,113]. One criticism of the Usp model has been that this retinoid X receptor (RXR) homologue cannot be the JH receptor since the K_d of JH:Usp binding is ~600 nmol, and a K_d of 10^{-9} M is viewed as an upper limit for a true hormone receptor [28]. By measuring a conformational change in Usp, Wozniak et al. [114] found that JH agonists lacking an epoxide group bind more strongly than those containing an epoxide group. This is in striking contrast to the CoMFA and CoMSIA models presented here, where compounds lacking an epoxide group or oxygen in this position have no biological activity. Interestingly, most recent conclusions of Jones and Jones [115] indicate that JH is unlikely to be the ligand for Usp and that another farnesoid product of the ring gland, most probably methyl farnesoate, may be the naturally occurring ligand for Usp. Therefore, the data and analysis presented here suggest that the binding of JH and other agonists to Usp does not reflect biological activity and an interaction with a *bona fide* JH receptor. Contributing support to this view is the finding that two phospholipids, phosphatidylethanolamine and phosphatidylglycerol, exhibit very strong binding to bacterially expressed Usp [116,117]. That this binding is mediated by only one hydrogen bond (to polar Gln338 within helix H6) close to the entrance of the ligand-binding pocket argues that numerous hydrophobic rather than electrostatic interactions are involved in ligand recognition and binding by Usp. Moreover, evidence from mass spectrometric studies suggests that the phospholipid ligand in the Usp ligand binding domain (LBD) cannot be displaced by competition with JH ester or with methoprene [116]. This too is in direct contrast to the JH pharmacophore model developed here, which favors at least two hydrogen bonds and three electrostatic interactions, but no dominant hydrophobic one. These types of JH interactions and their number are characteristic of steroids, thyroids, retinoids, or vitamin D_3 with their nuclear receptors and are known from crystallographic studies [118–124]. The very recent structural and transactivation studies of Iwema et al. [125] show that LBD of *Tribolium* Usp expressed in transgenic *Drosophila* is not activated by ligands that include RXR agonists and JH analogs. This suggests that the JH receptor, albeit not Usp, most probably belongs to the family of classical nuclear receptors, possibly one of numerous nuclear *Drosophila* hormone receptors (DHRs) (for recent reviews, see [11,126,127]).

A second insight about the molecular mode of JH action from our JH pharmacophore model is related to genetic evidence showing that the *Methoprene-tolerant* (*Met*) gene, when mutated, confers resistance to JH [27,128,129] and the finding that loss-of-function *Met*$^-$ mutants display reduced intracellular JH-binding activity [130,131]. The Met protein was found in the nuclei; it has been detected in several tissues, some of which are known JH targets [129]. In addition, Miura et al. [25] claimed that recombinant Met binds JH and can transactivate a JH-dependent reporter. The possibility that Met binds JH can be theoretically explored using our model, but the results do not favor Met as the protein target of JH binding. Met is the member of the family of basic and helix-loop-helix (bHLH) and period–ARNT–single-minded (PAS) proteins to which belongs also the dioxin or aryl hydrocarbon receptors (AHR) that bind various pollutants and toxic chemicals including dioxins, polychlorinated biphenyls, and other environmentally hazardous agents. Recognition and binding of pollutants to AHRs are mediated by a PAS-B domain [132–135] and all known interactions of aryl hydrocarbons are predominantly hydrophobic [136–141]. The pharmacophore model we present precludes JH from interactions with Met. Although this may appear at first glimpse to contradict genetic data, since *Met*⁻ mutations are resistant to JH, it does not necessarily mean that there are no other possibilities. It is possible that the genetic data indicate that the Met protein plays another and more intriguing role in the molecular action of JH that does not involve direct ligand binding. However, a recent paper reports that JH binds to the Met protein potentially by forming a single hydrogen bond with Tyr252 based on an AutoDock procedure in a model of the PAS-B domain of flour beetle *Tribolium*. All of the best scoring outcomes of docking JH-III with Met produced this single interaction but left the other types of pharmacophore-derived interactions unoccupied [142]. Although docking of pyriproxyfen into the PAS-B domain produced an additional theoretical contact of the juvenoid ether groups to Lys311, this interaction does not correspond to the pharmacophore model of Class II molecules. Thus, further studies on juvenoid binding in Met proteins from other insect species or other potential target proteins will be required to clarify these potential discrepancies.

The pseudoreceptor ligand-binding site was constructed using information about the amino acids most frequently involved in forming interactions with small lipophilic ligands in nuclear receptors [143–148] and by considering molecular nature of natural JH and its agonists (similar to retinoids, free fatty acids, and steroids). Thus, these considerations had to be expanded for the type of interactions predicted from pharmacophore models arguing strongly that the ligand-binding pocket of the receptor for JH holds properties, which are similar to members of nuclear receptor superfamily. Both pseudoreceptor models of a 3D receptor surrogate have been validated, leading to high correlation and predictive power as well as an agreement with the pharmacophore models. Furthermore, as mentioned earlier, juvenoids belonging to Class I have an average molecular volume 291.47 Å^3, whereas Class II compounds have average molecular volume 253.65 Å^3. The volume of the predicted binding cavity as revealed from our models is between 680 and 820 Å^3, which corresponds to the cavity to ligand ratio of 2.3 and 2.8, respectively. Although this ratio varies from 1.4 to 4.7 for known receptors and ligands, the majority of the receptors (including AHRs) and their ligands show distribution of their volume ratios between 1.8 and 2.9, thus further suggesting that putative JH receptor should not deviate too much from the families of known intracellular receptors [149,150]. Therefore, the data that are presented here have allowed for the first time the rationalization of JH agonist SAR at a qualitative level and also have provided quantitative relationships between the structure of JH compounds and their biological activity as summarized and reflected in likelihood models of the docking pocket of a putative JH receptor. We believe that this approach, in combination with 3D-object recognition based on scanning, pair-wise comparison, or similar protocols to identify the spatial coordinates of individual atoms [151,152], will be useful to find potential siblings in structural databases.

From the practical point of view, we have demonstrated that 3D QSAR analysis of highly divergent juvenoid compounds is a useful analytical tool to evaluate *in vivo* biological activities in *Drosophila*. Due to its effects on morphogenesis, the molecular mode of JH action is expected to be conserved across insect species. Consequently, the CoMFA/CoMSIA analysis we have presented here holds promise for addressing the similar questions in non-drosophilid species that are important pests and disease vectors. Support for the optimistic view that this approach promises to make significant inroads into identifying the JH receptor, and more effective juvenoids also comes from the work of Basak et al. [153] who generated a 2D model of some hundred and eighty 2,4-dienoates that were tested in six different insect species, using topological and geometrical descriptors, and for the yellow fever mosquito *Aedes aegypti*. They reached data on electronegativity of the ligand–target interaction, similar to what we present here for our Class I pharmacophore model. For many of the pests, there already exist abundant data on the activity of hundreds or even thousands of juvenoids. After a careful survey and selection following uniform, standardized tests, these data can be used to develop realistic models of the pharmacophore, as we currently are doing in our laboratory, to design new JH compounds that are species specific and serve as powerful insecticides.

9.7 CONCLUSIONS

Currently, there are very few, if any, small number of hormones in the animal kingdom without identified receptor, except insect JH. Tremendous effort to find JH receptor during last several decades did not lead to its unambiguous identification. Therefore, there is a need to employ various even untraditional approaches that can help to shed more light on JH receptor identity. Along these lines, we have applied 3D QSAR CoMFA and CoMSIA high-performance computational protocols to obtain reliable pharmacophore model for the analysis of molecular interactions that are critical for JH biological activity. Using these methods, we built predictive models for an 86 structurally divergent JH agonists. The combined bioassay- and *in silico*-generated pharmacophore model of juvenoids has revealed important structural features necessary for morphogenetic JH activity in *Drosophila* and has specified the dominant interactions that these juvenoids exhibit with putative JH receptor. By dividing the training set of molecule in two subsets, Class I and Class II, we gained models with better statistical parameters. This may indicate that these two classes of molecules exhibit a different mode of binding to the JH receptor. By means of this approach, we have obtained both computer-aided and species-specific pharmacophore fingerprints of JH and its agonists, which revealed that compounds with highest activity require an electronegative atom (oxygen or nitrogen) at both ends of the molecule. If either of these electronegative atoms are replaced by uncharged carbon or the distance between them is shorter than 11.5 Å or longer than 13.5 Å, the JH biological activity is dramatically decreased. The presence of an electron-deficient moiety in the middle of the farnesene-type JH agonist boosts high activity even more. These data from 3D QSAR provide guidelines and a mechanistic scope for the identification of steric and electrostatic properties as well as donor and acceptor hydrogen bonding that are important features of the ligand-binding cavity of a JH target protein. With the aim to meliorate the pharmacophore analysis and evaluate the outcomes of the CoMFA and CoMSIA study, we have selected one of the pseudoreceptor modeling software tools, PrGen from Biographics Laboratory 3R, to obtain a putative binding site surrogate that is composed of a defined number of amino acid residues (8), which corresponds to the molecular interactions revealed for both classes of juvenoids. We have shown that computational CoMFA and CoMSIA models, with their graphical interpretation, can also serve to guide the design of novel JH compounds and predict their biological activity. Several juvenoids with low activity or inactivity in *Drosophila* show relatively high or even very high activity in non-dipteran species, and this suggests that 3D QSAR pharmacophore analysis based on numerous bioassays can reveal biologically meaningful, distinct species-specific differences and significantly can help to gain deeper insight on the atomistic details of the JH mode of action.

ACKNOWLEDGMENTS

Authors would like to thank Denisa Beňová-Liszeková and Milan Beňo for their skillful help in preparing this manuscript. Bruce A. Chase is greatly acknowledged for critical reading of the manuscript and his comments. We thank Karel Sláma, František Sehnal, Clive A. Henrick, William S. Bowers, Hans Laufer, Lawrence I. Gilbert, Zdeněk Wimmer, and Martin Rejzek for providing many of JH compounds for testing. The JH work in our lab was gradually supported by VEGA grants 2/999533, 2/7194/20 and 2/3025/23, and 2/0170/10; APVT-51-027402 grant; NATO grants CRG-972173 and LST.CLG-977559; and EEA-Norwegian FM SK-0086 grant to R.F.

REFERENCES

1. V.B. Wigglesworth, *The physiology of ecdysis in* Rhodnius prolixus, *II. Factors controlling moulting and metamorphosis*, Q. J. Microsc. Sci. 77 (1934), pp. 191–222.
2. V.B. Wigglesworth, *The function of the corpus allatum in the growth and reproduction of* Rhodnius prolixus *(Hemiptera)*, Q. J. Microsc. Sci. 79 (1936), pp. 91–121.
3. C.M. Williams, *The juvenile hormone of insects*, Nature 178 (1956), pp. 212–213.
4. V.J.A. Novák, *Insect Hormones*, 2nd ed., Chapman and Hall, London, U.K., 1975.
5. M. Bownes, *The roles of juvenile-hormone, ecdysone and the ovary in the control of* Drosophila vitellogenesis, J. Insect. Physiol. 35 (1989), pp. 409–413.
6. T. Flatt, M.P. Tu, and M. Tatar, *Hormonal pleiotropy and the juvenile hormone regulation of* Drosophila *development and life history*, Bioessays 27 (2005), pp. 999–1010.
7. L.M. Riddiford, *Hormones and* Drosophila *development*, in *The Development of* Drosophila melanogaster, Vol. 2, M. Bate and A. Marinez-Arias, eds., Cold Spring Harbor Press, New York, 1993, pp. 899–939.
8. V.C. Henrich, R. Rybczynski, and L.I. Gilbert, *Peptide hormones, steroid hormones, and puffs: Mechanisms and models in insect development*, Vitam. Horm. 55 (1999), pp. 73–125.
9. C.S. Thummel, *Ecdysone-regulated puff genes 2000,* Insect Biochem. Mol. Biol. 32 (2002), pp. 113–120.
10. L.M. Riddiford, P. Cherbas, and J.W. Truman, *Ecdysone receptors and their biological actions,* Vitam. Horm. 60 (2000), pp. 1–73.
11. K. King-Jones and C.S. Thummel, *Nuclear receptors–A perspective from* Drosophila, Nat. Rev. Genet. 6 (2005), pp. 311–323.
12. Y. Nakagawa and V.C. Henrich, *Arthropod nuclear receptors and their role in molting,* FEBS J. 276 (2009), pp. 6128–6157.
13. G. Jones, *Molecular mechanisms of action of juvenile hormone*, Annu. Rev. Entomol. 40 (1995), pp. 147–169.
14. L.M. Riddiford, *Juvenile hormone: The status of its "status quo" action*, Arch. Insect Biochem. Physiol. 32 (1996), pp. 271–286.

15. A.E. Oro, M. McKeown, and R.M. Evans, *Relationship between the product of the* Drosophila *ultraspiracle locus and the vertebrate retinoid X receptor*, Nature 347 (1990), pp. 298–301.
16. L. Cherbas, K. Lee, and P. Cherbas, *Identification of ecdysone response elements by analysis of the* Drosophila *Eip28/29 gene*, Genes Dev. 5 (1991), pp. 120–131.
17. M.R. Koelle, W.S. Talbot, W.A. Segraves, M.T. Bender, P. Cherbas, and D.S. Hogness, *The* Drosophila *EcR gene encodes an ecdysone receptor, a new member of the steroid receptor superfamily*, Cell 67 (1991), pp. 59–77.
18. W.S. Talbot, E.A. Swyryd, and D.S. Hogness, Drosophila *tissues with different metamorphic responses to ecdysone express different ecdysone receptor isoforms*, Cell 73 (1993), pp. 1323–1337.
19. W.A. Segraves, *Steroid receptors and orphan receptors in* Drosophila *development*, Semin. Cell Biol. 5 (1994), pp. 105–113.
20. C.S. Thummel, *From embryogenesis to metamorphosis. The regulation and function of* Drosophila *nuclear receptor superfamily members*, Cell 83 (1995), pp. 871–877.
21. M. Ashburner, *Puffs, genes, and hormones revisited*, Cell 61 (1990), pp. 1–3.
22. S. Russell and M. Ashburner, *Ecdysone regulated chromosome puffing in* Drosophila melanogaster, in *Metamorphosis: Postembryonic Reprogramming of Gene Expression in Amphibian and Insect Cells*, L.I. Gilbert, J.R. Tata, and B.G. Atkinson, eds., Academic Press, New York, 1996, pp. 109–144.
23. G. Jones and P.A. Sharp, *Ultraspiracle: An invertebrate nuclear receptor for juvenile hormones*, Proc. Natl. Acad. Sci. USA 94 (1997), pp. 13499–13503.
24. M. Ashok, C. Turner, and T.G. Wilson, *Insect juvenile hormone resistance gene homology with the bHLH-PAS family of transcriptional regulators*, Proc. Natl. Acad. Sci. USA 95 (1998), pp. 2761–2766.
25. K. Miura, M. Oda, S. Makita, and Y. Chinzei, *Characterization of the* Drosophila *Methoprene-tolerant gene product. Juvenile hormone binding and ligand-dependent gene regulation*, FEBS J. 272 (2005), pp. 1169–1178.
26. Y. Li, Z. Zhang, G.E. Robinson, and S.R. Palli, *Identification and characterization of a juvenile hormone response element and its binding protein*, J. Biol. Chem. 28 (2007), pp. 37605–37617.
27. T.G. Wilson, *The molecular site of action of juvenile hormone and juvenile hormone insecticides during metamorphosis: How these compounds kill insects*, J. Insect Physiol. 50 (2004), pp. 111–121.
28. L.I. Gilbert, N.A. Granger, and R.M. Roe, *The juvenile hormones: Historical facts and speculations on future research directions,* Insect Biochem. Mol. Biol. 30 (2000), pp. 617–644.
29. K.J. Judy, D.A. Schooley, L.L. Dunham, M.S. Hall, B.J. Bergot, and J.B. Siddall, *Isolation, structure, and absolute configuration of a new natural insect juvenile hormone from* Manduca sexta, Proc. Natl. Acad. Sci. USA 70 (1973), pp. 1509–1513.
30. J.H. Law, C. Yuan, and C.M. Williams, *Synthesis of a material with high juvenile hormone activity*, Proc. Natl. Acad. Sci. USA 55 (1966), pp. 576–578.
31. K. Sláma and C.M. Williams, *Juvenile hormone activity for the bug* Pyrrhocoris apterus, Proc. Natl. Acad. Sci. USA 54 (1965), pp. 411–414.
32. C.M. Williams, *Third generation pesticides*, Sci. Am. 217 (1967), pp. 13–17.
33. K. Sláma, *Insect juvenile hormone analogues*, Annu. Rev. Biochem. 40 (1971), pp. 1079–1102.
34. K. Sláma, *Pharmacology of insect juvenile hormones*, in *Comprehensive Insect Physiology, Biochemistry, and Pharmacology*, Vol. 11, G.A. Kerkut and L.I. Gilbert, eds., Pergamon Press, Oxford, U.K., 1985, pp. 357–394.
35. G.B. Staal, *Insect growth regulators with juvenile hormone activity*, Annu. Rev. Entomol. 20 (1975), pp. 417–460.

36. F. Sehnal, *Action of juvenoids on different groups of insects*, in *The Juvenile Hormones*, L.I. Gilbert, ed., Plenum Press, New York, 1976, pp. 301–321.
37. F. Sehnal, *Juvenile hormone analogues*, in *Endocrinology of Insects*, R.G.H Downer and H. Laufer, eds., Alan. R. Liss, New York, 1983, pp. 657–672.
38. C.A. Henrick, *Juvenile hormone analogues: Structure-activity relationships*, in *Insecticidal Modes of Action*, J.R. Coats, ed., Academic Press, New York, 1982, pp. 315–402.
39. P. Schmialek, *Die Identifizierung zweier im Tenebriokot und in Hefe vorkommender Substanzen mit Juvenilhormonwirkung*, Z. Naturf. 16b (1961), 461–464.
40. M. Ashburner, *Effects of juvenile hormone on adult differentiation of* Drosophila melanogaster, Nature 227 (1970), pp. 187–189.
41. J. Bouletreau-Merle, *Réceptivité sexuelle et vitelogenése chez les femelles de Drosophila melanogaster: Effects d'une application d'hormone juvénile et de deux analogues hormonaux*, C. R. Acad. Sci. Paris 277 (1973), pp. 2045–2048.
42. K. Madhavan, *Morphogenetic effects of juvenile hormone and juvenile hormone mimics on adult development of* Drosophila, J. Insect Physiol. 19 (1973), pp. 441–453.
43. J.H. Postlethwait, *Juvenile hormone and the adult development of* Drosophila, Biol. Bull. 147 (1974), pp. 119–135.
44. F. Sehnal, V. Jarolím, S. Abou-Halawa, J. Žd'árek, D.S. Daoud, and F. Šorm, *Activities of oxa-analogues of farnesane-type juvenoids on cyclorrhaphous flies*, Acta Entomol. Bohemoslov. 72 (1975), pp. 353–359.
45. F. Sehnal and J. Žd'árek, *Action of juvenoids on the metamorphosis of cyclorrhaphous diptera*, J. Insect Physiol. 22 (1976), pp. 673–682.
46. L.M. Riddiford and M. Ashburner, *Effects of juvenile hormone mimics on larval development and metamorphosis of* Drosophila melanogaster, Gen. Comp. Endocrinol. 82 (1991), pp. 172–183.
47. I.M. Spitz, H.B. Croxatto, and A. Robbins, *Antiprogestins: Mechanism of action and contraceptive potential*, Annu. Rev. Pharmacol. Toxicol. 36 (1996), pp. 47–81.
48. M. Recanatini, A. Cavalli, F. Belluti, L. Piazzi, A. Rampa, A. Bisi, S. Gobbi, P. Valenti, V. Andrisano, M. Bartolini, and V. Cavrini, *SAR of 9-amino-1,2,3,4- tetrahydroacridine-based acetylcholinesterase inhibitors: Synthesis, enzyme inhibitory activity, QSAR, and structure-based CoMFA of tacrine analogues*, J. Med. Chem. 43 (2000), pp. 2007–2018.
49. R.E. Wilcox, W.H. Huang, M.Y. Brusniak, D.M. Wilcox, R.S. Pearlman, M.M. Teeter, C.J. DuRand, B.L. Wiens, and K.A. Neve, *CoMFA-based prediction of agonist affinities at recombinant wild type versus serine to alanine point mutated D2 dopamine receptors*, J. Med. Chem. 43 (2000), pp. 3005–3019.
50. D. Muthas, D. Noteberg, Y.A. Sabnis, E. Hamelink, L. Vrang, B. Samuelsson, A. Karlen, and A. Hallberg, *Synthesis, biological evaluation, and modeling studies of inhibitors aimed at the malarial proteases plasmepsins I and II*, Bioorg. Med. Chem. 13 (2005), pp. 5371–5390.
51. X.F. Zhou, Q. Shao, R.A. Coburn, and M.E. Morris, *Quantitative structure-activity relationship and quantitative structure-pharmacokinetics relationship of 1,4-dihydropyridines and pyridines as multidrug resistance modulators*, Pharm. Res. 22 (2005), pp. 1989–1996.
52. B.A. Krueger, T. Weil, and G. Schneider, *Comparative virtual screening and novelty detection for NMDA-GlycineB antagonists*, J. Comput. Aided Mol. Des. 23 (2009), pp. 869–881.
53. K. Sander, T. Kottke, E. Proschak, Y. Tanrikulu, E.H. Schneider, R. Seifert, G. Schneider, and H. Stark, *Lead identification and optimization of diaminopyrimidines as histamine H4 receptor ligands*, Inflamm. Res. 59(Suppl. 2) (2010), pp. S249–S251.
54. P. Lu, X. Wei, and R. Zhang, *CoMFA and CoMSIA 3D-QSAR studies on quinolone caroxylic acid derivatives inhibitors of HIV-1 integrase*, Eur. J. Med. Chem. 45 (2010), pp. 3413–3419.

55. K.K. Sethi, S.M. Verma, N. Prasanthi, S.K. Sahoo, R.N. Parhi, and P. Suresh, *3D-QSAR study of benzene sulfonamide analogs as carbonic anhydrase II inhibitors,* Bioorg. Med. Chem. Lett. 20 (2010), pp. 3089–3093.
56. P. Lan, W.N. Chen, and W.M. Chen, *Molecular modeling studies on imidazo[4,5-b]pyridine derivatives as Aurora A kinase inhibitors using 3D-QSAR and docking approaches*, Eur. J. Med. Chem. 46 (2011), pp. 77–94.
57. K.C. Shih, C.Y. Lin, J. Zhou, H.C. Chi, T.S. Chen, C.C. Wang, H.W. Tseng, and C.Y. Tang, *Development of novel 3D-QSAR combination approach for screening and optimizing B-Raf inhibitors* in silico, J. Chem. Inf. Model. 51 (2011), pp. 398–407.
58. L. Dinan, R.E. Hormann, and T. Fujimoto, *An extensive ecdysteroid CoMFA*, J. Comput. Aided Mol. Des. 13 (1999), pp. 185–207.
59. Y. Nakagawa, K. Takahashi, H. Kishikawa, T. Ogura, C. Minakuchi, and H. Miyagawa, *Classical and three-dimensional QSAR for the inhibition of [3H]ponasterone A binding by diacylhydrazine-type ecdysone agonists to insect Sf-9 cells*, Bioorg. Med. Chem. 13 (2005), pp. 1333–1340.
60. C.E. Wheelock, Y. Nakagawa, T. Harada, N. Oikawa, M. Akamatsu, G. Smagghe, D. Stefanou, K. Iatrou, and L. Swevers, *High-throughput screening of ecdysone agonists using a reporter gene assay followed by 3-D QSAR analysis of the molting hormonal activity*, Bioorg. Med. Chem. 14 (2006), pp. 1143–1159.
61. B. Bordas, I. Belai, A. Lopata, and Z. Szanto, *Interpretation of scoring functions using 3D molecular fields. Mapping the diacyl-hydrazine-binding pocket of an insect ecdysone receptor*, J. Chem. Inf. Model. 47 (2007), pp. 176–185.
62. G. Holmwood and M. Schindler, *Protein structure based rational design of ecdysone agonists,* Bioorg. Med. Chem. 17 (2009), pp. 4064–4070.
63. T. Harada, Y. Nakagawa, M. Akamatsu, and H. Miyagawa, *Evaluation of hydrogen bonds of ecdysteroids in the ligand-receptor interactions using a protein modeling system*, Bioorg. Med. Chem. 17 (2009), pp. 5868–5873.
64. A.H. Beckett and A.F. Casey, *Synthetic analgesics: Stereochemical considerations*, J. Pharm. Pharmacol. 6 (1954), pp. 986–999.
65. L.B. Kier and H.S. Aldrich, *A theoretical study of receptor site models for trimethylammonium group interaction*, J. Theor. Biol. 46 (1974), pp. 529–541.
66. G.R. Marshall, *Introduction to chemoinformatics in drug discovery*, in *Chemoinformatics in Drug Discovery*, Vol. 23, T.I. Oprea, ed., Wiley-VCH Verlag GmbH. and Co., Weinheim, Germany, 2005, pp. 1–22.
67. D. Liszeková, M. Polakovičová, M. Beňo, and R. Farkaš, *Molecular determinants of juvenile hormone action as revealed by 3D QSAR analysis in* Drosophila, PloS One 4 (2009), e6001, doi:10.1371/journal.pone.0006001.
68. J.M. van der Meer, *Optical clean and permanent whole mount preparation for phase-contrast microscopy of cuticular structures of insect larvae*, Drosoph. Inf. Serv. 52 (1977), pp. 160–161.
69. B. Yun, R. Farkaš, K. Lee, and L. Rabinow, *The Doa locus encodes a member of a new protein kinase family and is essential for eye and embryonic development in* Drosophila melanogaster, Genes Dev. 8 (1994), pp. 1160–1173.
70. M. Beňo, D. Liszeková, and R. Farkaš, *Processing of soft pupae and uneclosed pharate adults of* Drosophila *for scanning electron microscopy,* Microsc. Res. Tech. 70 (2007), pp. 1022–1027.
71. G.D. Prestwich and C. Wawrzenczyk, *High specific activity enantiomerically enriched juvenile hormones: Synthesis and binding assay*, Proc. Natl. Acad. Sci. USA 82 (1985), pp. 5290–5294.
72. J.T. Cody, S. Valtier, and S.L. Nelson, *Amphetamine excretion profile following multidose administration of mixed salt amphetamine preparation*, J. Anal. Toxicol. 28 (2004), pp. 563–574.

73. P. Gadler and K. Faber, *New enzymes for biotransformations: Microbial alkyl sulfatases displaying stereo- and enantioselectivity*, Trends Biotechnol. 25 (2007), pp. 83–88.
74. M.G. Nillos, G. Rodriguez-Fuentes, J. Gan, and D. Schlenk, *Enantioselective acetylcholinesterase inhibition of the organophosphorous insecticides profenofos, fonofos, and crotoxyphos*, Environ. Toxicol. Chem. 26 (2007), pp. 1949–1954.
75. B.P. Klaholz, J.P. Renaud, A. Mitschler, C. Zusi, P. Chambon, H. Gronemeyer, and D. Moras, *Conformational adaptation of agonists to the human nuclear receptor RAR gamma*, Nat. Struct. Biol. 5 (1998), pp. 199–202.
76. B.P. Klaholz, A. Mitschler, M. Belema, C. Zusi, and D. Moras, *Enantiomer discrimination illustrated by high-resolution crystal structures of the human nuclear receptor hRARgamma*, Proc. Natl. Acad. Sci. USA 97 (2000), pp. 6322–6327.
77. B.P. Klaholz, A. Mitschler, and D. Moras, *Structural basis for isotype selectivity of the human retinoic acid nuclear receptor*, J. Mol. Biol. 302 (2000), pp. 155–170.
78. A. Golbraikh, D. Bonchev, and A. Tropsha, *Novel chirality descriptors derived from molecular topology*, J. Chem. Inform. Comput. Sci. 41 (2001), pp. 147–158.
79. J. Paier, T. Stockner, A. Steinreiber, K. Faber, and W.M. Fabian, *Enantioselectivity of epoxide hydrolase catalysed oxirane ring opening: A 3D QSAR study*, J. Comput. Aided Mol. Des. 17 (2003), pp. 1–11.
80. A. Kovatcheva, A. Golbraikh, S. Oloff, J. Feng, W. Zheng, and A. Tropsha, *QSAR modeling of datasets with enantioselective compounds using chirality sensitive molecular descriptors*, SAR QSAR Environ. Res. 16 (2005), pp. 93–102.
81. R.D. Cramer, D.E. Patterson, and J.D. Bunce, *Recent advances in comparative molecular field analysis (CoMFA)*, Prog. Clin. Biol. Res. 291 (1989), pp. 161–165.
82. G. Klebe and U. Abraham, *Comparative molecular similarity index analysis (CoMSIA) to study hydrogen-bonding properties and to score combinatorial libraries*, J. Comput. Aided Mol. Des. 13 (1999), pp. 1–10.
83. A. Vedani, P. Zbinden, J.P. Snyder, and P.A. Greenidge, *Pseudoreceptor modelling: The construction of three-dimensional receptor surrogates*, J. Am. Chem. Soc. 117 (1995), pp. 4987–4994.
84. A. Bassoli, G. Merlini, and A. Vedani, *A three-dimensional receptor model for isovanillic sweet derivatives*, J. Chem. Soc. Perkin. Trans. 2 (1998), pp. 1449–1454.
85. P. Zbinden, M. Dobler, G. Folkers, and A. Vedani, *PrGen: Pseudoreceptor modeling using receptor-mediated ligand alignment and pharmacophore equilibration*, Quant. Struct.-Act. Relat. 17 (1998), pp. 122–130.
86. T.E. Ferrin, G.S. Couch, C.C. Huang, E.F. Pettersen, and R. Langridge, *An affordable approach to interactive desktop molecular modeling*, J. Mol. Graph 9 (1991), pp. 27–32.
87. G.S. Couch, E.F. Pettersen, C.C. Huang, and T.E. Ferrin, *Annotating PDB files with scene information*, J. Mol. Graph 13 (1995), pp. 153–158.
88. G.J. Kleywegt and T.A. Jones, *Detection, delineation, measurement and display of cavities in macromolecular structures*, Acta Cryst. D 50 (1994), pp. 178–185.
89. N. Punja, C.N. Ruscoe, and C. Treadgold, *Insect juvenile hormone mimics: A structural basis for activity*, Nat. New Biol. 242 (1973), pp. 94–96.
90. A. Nakayama, H. Iwamura, and T. Fujita, *Quantitative structure-activity relationship of insect juvenile hormone mimetic compounds*, J. Med. Chem. 27 (1984), pp. 1493–1502.
91. H. Iwamura, K. Nishimura, and T. Fujita, *Quantitative structure-activity relationships of insecticides and plant growth regulators: Comparative studies toward understanding the molecular mechanism of action*, Environ. Health Perspect. 61 (1985), pp. 307–320.
92. S.C. Basak, R. Natarajan, D. Mills, D.M. Hawkins, and J.J. Kraker, *Quantitative structure-activity relationship modeling of juvenile hormone mimetic compounds for Culex pipiens larvae, with a discussion of descriptor-thinning methods*, J. Chem. Inform. Model. 46 (2006), pp. 65–77.

93. X. Zhou and L.M. *Riddiford, Broad specifies pupal development and mediates the "status quo" action of juvenile hormone on the pupal-adult transformation in Drosophila and Manduca*, Development 129 (2002), pp. 2259–2269.
94. K. Sláma, M. Romaňuk, and F. Šorm, *Insect Hormones and Bioanalogues*, Springer Verlarg, Wien, Austria, 1974.
95. L.M. Riddiford, *Cellular and molecular actions of juvenile hormone. I. General considerations and premetamorphic actions*, Avd. Insect Physiol. 24 (1994), pp. 213–274.
96. C.A. Henrick, *Juvenoids*, in *Agrochemicals from Natural Products*, C.R.A. Godfrey, ed., Marcel Dekker Press, Inc., New York, 1995, pp. 147–213.
97. J. Li, J.J. Masso, and S. Rendon, *Quantitative evaluation of adhesive properties and drug-adhesive interactions for transdermal drug delivery formulations using linear solvation energy relationships*, J. Control. Release 82 (2002), pp. 1–16.
98. A.J. Tervo, T.H. Nyronen, T. Ronkko, and A.A. Poso, *Structure-activity relationship study of catechol-O-methyltransferase inhibitors combining molecular docking and 3D QSAR methods*, J. Comput. Aided Mol. Des. 17 (2003), pp. 797–810.
99. T. Ghafourian, P. Zandasrar, H. Hamishekar, and A. Nokhodchi, *The effect of penetration enhancers on drug delivery through skin: A QSAR study*, J. Control. Release 99 (2004), pp. 113–125.
100. D.T. Stanton, B.E. Mattioni, J.J. Knittel, and P.C. Jurs, *Development and use of hydrophobic surface area (HSA) descriptors for computer-assisted quantitative structure-activity and structure-property relationship studies*, J. Chem. Inform. Comput. Sci. 44 (2004), pp. 1010–1023.
101. M.D. Kelly and R.L. Mancera, *A new method for estimating the importance of hydrophobic groups in the binding site of a protein*, J. Med. Chem. 48 (2005), pp. 1069–1078.
102. E. Manivannan and S. Prasanna, *First QSAR report on FSH receptor antagonistic activity: Quantitative investigations on physico-chemical and structural features among 6-amino-4-phenyltetrahydroquinoline derivatives*, Bioorg. Med. Chem. Lett. 15 (2005), pp. 4496–4501.
103. I. Zamora, T. Oprea, G. Cruciani, M. Pastor, and A.L. Ungell, *Surface descriptors for protein-ligand affinity prediction*, J. Med. Chem. 46 (2003), pp. 25–33.
104. C.E. Kaznowski, H.A. Schneiderman, and P.J. Bryant, *Cuticle secretion during larval growth in* Drosophila melanogaster, J. Insect Physiol. 31 (1985), pp. 801–813.
105. J.M. Pechine, F. Perez, C. Antony, and J.M. Jallon, *A further characterization of* Drosophila *cuticular monoenes using a mass spectrometry method to localize double bonds in complex mixtures*, Anal. Biochem. 145 (1985), pp. 177–182.
106. H.J. Kim, C.H. Chae, K.Y. Yi, K.L. Park, and S.E. Yoo, *Computational studies of COX-2 inhibitors: 3D-QSAR and docking*, Bioorg. Med. Chem. 12 (2004), pp. 1629–1641.
107. M. Togashi, S. Borngraeber, B. Sandler, R.J. Fletterick, P. Webb, and J.D. Baxter, *Conformational adaptation of nuclear receptor ligand binding domains to agonists: Potential for novel approaches to ligand design*, J. Steroid. Biochem. Mol. Biol. 93 (2005), pp. 127–137.
108. L.G. Humber, F.T. Bruderlein, A.H. Philipp, and M. Götz, *Mapping the dopamine receptor. 1. Features derived from modifications in ring E of the neuroleptic butaclamol*, J. Med. Chem. 22 (1979), pp. 761–767.
109. A.H. Philipp, L.G. Humber, and K. Voith, *Mapping the dopamine receptor. 2. Features derived from modifications in the rings A/B region of the neuroleptic butaclamol*, J. Med. Chem. 22 (1979), pp. 768–773.
110. A.A. Asselin, L.G. Humber, K. Voith, and G. Metcalf, *Drug design via pharmacophore identification. Dopaminergic activity of ^{3}H-benz[e]indol-8-amines and their mode of interaction with the dopamine receptor*, J. Med. Chem. 29 (1986), pp. 648–654.

111. H. Wikström, B. Andersson, A. Svensson, L.G. Humber, A.A. Asselin, K. Svensson, A. Ekman, A. Carlsson, I. Nilsson, and C. Chidester, *Resolved 6,7,8,9-tetrahydro-N,N-dimethyl-^{3}H-benz[e]indol-8-amine: Central dopamine and serotonin receptor stimulating properties*, J. Med. Chem. 32 (1989), pp. 2273–2276.
112. G. Jones and D. Jones, *Considerations on the structural evidence of a ligand-binding function of ultraspiracle, an insect homolog of vertebrate RXR*, Insect Biochem. Mol. Biol. 30 (2000), pp. 671–679.
113. Y. Xu, F. Fang, Y. Chu, D. Jones, and G. Jones, *Activation of transcription through the ligand-binding pocket of the orphan nuclear receptor ultraspiracle*, Eur. J. Biochem. 269 (2002), pp. 6026–6036.
114. M. Wozniak, Y. Chu, F. Fang, Y. Xu, L.M. Riddiford, D. Jones, and G. Jones, *Alternative farnesoid structures induce different conformational outcomes upon the* Drosophila *ortholog of the retinoid X receptor, ultraspiracle*, Insect Biochem. Mol. Biol. 34 (2004), pp. 1147–1162.
115. G. Jones and D. Jones, *Farnesoid secretions of dipteran ring glands: What we do know and what we can know*, Insect Biochem. Mol. Biol. 37 (2007), pp. 771–798.
116. I.M. Billas, L. Moulinier, N. Rochel, and D. Moras, *Crystal structure of the ligand-binding domain of the ultraspiracle protein USP, the ortholog of retinoid X receptors in insects*, J. Biol. Chem. 276 (2001), pp. 7465–7474.
117. I.M. Billas, T. Iwema, J.M. Garnier, A. Mitschler, N. Rochel, and D. Moras, *Structural adaptability in the ligand-binding pocket of the ecdysone hormone receptor*, Nature 426 (2003), pp. 91–96.
118. P.F. Egea, A. Mitschler, N. Rochel, M. Ruff, P. Chambon, and D. Moras, *Crystal structure of the human RXRalpha ligand-binding domain bound to its natural ligand: 9-cis retinoic acid*, Embo J. 19 (2000), pp. 2592–2601.
119. P.F. Egea, N. Rochel, C. Birck, P. Vachette, P.A. Timmins, and D. Moras, *Effects of ligand binding on the association properties and conformation in solution of retinoic acid receptors RXR and RAR*, J. Mol. Biol. 307 (2001), pp. 557–576.
120. N. Rochel, J.M. Wurtz, A. Mitschler, B. Klaholz, and D. Moras, *The crystal structure of the nuclear receptor for vitamin D bound to its natural ligand*, Mol. Cell. 5 (2000), pp. 173–179.
121. S. Hammer, I. Spika, W. Sippl, G. Jessen, B. Kleuser, H.D. Holtje, and M. Schafer-Korting, *Glucocorticoid receptor interactions with glucocorticoids: Evaluation by molecular modeling and functional analysis of glucocorticoid receptor mutants*, Steroids 68 (2003), pp. 329–339.
122. R.K. Bledsoe, K.P. Madauss, J.A. Holt, C.J. Apolito, M.H. Lambert, K.H. Pearce, T.B. Stanley, E.L. Stewart, R.P. Trump, T.M. Willson, and S.P. Williams, *A ligand-mediated hydrogen bond network required for the activation of the mineralocorticoid receptor*, J. Biol. Chem. 280 (2005), pp. 31283–31293.
123. A. Hedfors, T. Appelqvist, B. Carlsson, L.G. Bladh, C. Litten, P. Agback, M. Grynfarb, K.F. Koehler, and J. Malm, *Thyroid receptor ligands. 3. Design and synthesis of 3,5-dihalo-4-alkoxyphenylalkanoic acids as indirect antagonists of the thyroid hormone receptor*, J. Med. Chem. 48 (2005), pp. 3114–3117.
124. Y. Li, K. Suino, J. Daugherty, and H.E. Xu, *Structural and biochemical mechanisms for the specificity of hormone binding and coactivator assembly by mineralocorticoid receptor*, Mol. Cell. 19 (2005), pp. 367–380.
125. T. Iwema, I.M. Billas, Y. Beck, F. Bonneton, H. Nierengarten, A. Chaumot, G. Richards, V. Laudet, and D. Moras, *Structural and functional characterization of a novel type of ligand-independent RXR-USP receptor*, EMBO J. 26 (2007), pp. 3770–3782.
126. C.S. Thummel, *Molecular mechanisms of developmental timing in C. elegans and* Drosophila, Dev. Cell. 1 (2001), pp. 453–465.

127. C.S. Thummel and J. Chory, *Steroid signaling in plants and insects—Common themes, different pathways*, Genes Dev. 16 (2002), pp. 3113–3129.
128. T.G. Wilson and M. Ashok, *Insecticide resistance resulting from an absence of target-site gene product*, Proc. Natl. Acad. Sci. USA 95 (1998), pp. 14040–14044.
129. S. Pursley, M. Ashok, and T.G. Wilson, *Intracellular localization and tissue specificity of the Methoprene-tolerant (Met) gene product in* Drosophila melanogaster, Insect Biochem. Mol. Biol. 30 (2000), pp. 839–845.
130. L. Shemshedini and T.G. Wilson, *Resistance to juvenile hormone and an insect growth regulator in Drosophila is associated with an altered cytosolic juvenile hormone-binding protein*, Proc. Natl. Acad. Sci. USA 87 (1990), pp. 2072–2076.
131. L. Shemshedini, M. Lanoue, and T.G. Wilson, *Evidence for a juvenile hormone receptor involved in protein synthesis in* Drosophila melanogaster, J. Biol. Chem. 265 (1990), pp. 1913–1918.
132. P. Coumailleau, L. Poellinger, J.A. Gustafsson, and M.L. Whitelaw, *Definition of a minimal domain of the dioxin receptor that is associated with Hsp90 and maintains wild type ligand binding affinity and specificity*, J. Biol. Chem. 270 (1995), pp. 25291–25300.
133. B.N. Fukunaga, M.R. Probst, S. Reisz-Porszasz, and O. Hankinson, *Identification of functional domains of the aryl hydrocarbon receptor*, J. Biol. Chem. 270 (1995), pp. 29270–29278.
134. J.C. Rowlands and J.A. Gustafsson, *Aryl hydrocarbon receptor-mediated signal transduction*, Crit. Rev. Toxicol. 27 (1997), pp. 109–134.
135. J. McGuire, K. Okamoto, M.L. Whitelaw, H. Tanaka, and L. Poellinger, *Definition of a dioxin receptor mutant that is a constitutive activator of transcription: Delineation of overlapping repression and ligand binding functions within the PAS domain*, J. Biol. Chem. 276 (2001), pp. 41841–41849.
136. E. Fraschini, L. Bonati, and D. Pitea, *Molecular polarizability as a tool for understanding the binding properties of polychlorinated dibenzo-p-dioxins: Definition of a reliable computational procedure*, J. Phys. Chem. 100 (1996), pp. 10564–10569.
137. R.D. Beger, D.A. Buzatu, and J.G. Wilkes, *Combining NMR spectral and structural data to form models of polychlorinated dibenzodioxins, dibenzofurans, and biphenyls binding to the AhR*, J. Comput. Aided Mol. Des. 16 (2002), pp. 727–740.
138. M. Procopio, A. Lahm, A. Tramontano, L. Bonati, and D. Pitea, *A model for recognition of polychlorinated dibenzo-p-dioxins by the aryl hydrocarbon receptor*, Eur. J. Biochem. 269 (2002), pp. 13–18.
139. E. Kotsikorou and E. Oldfield, *A quantitative structure-activity relationship and pharmacophore modeling investigation of aryl-X and heterocyclic bisphosphonates as bone resorption agents*, J. Med. Chem. 46 (2003), pp. 2932–2944.
140. S. Arulmozhiraja and M. Morita, *Structure-activity relationships for the toxicity of polychlorinated dibenzofurans: Approach through density functional theory-based descriptors*, Chem. Res. Toxicol. 17 (2004), pp. 348–356.
141. G. Zheng, M. Xiao, and X.H. Lu, *QSAR study on the Ah receptor-binding affinities of polyhalogenated dibenzo-p-dioxins using net atomic-charge descriptors and a radial basis neural network*, Anal. Bioanal. Chem. 383 (2005), pp. 810–816.
142. J.P. Charles, T. Iwema, V.C. Epa, K. Takaki, J. Rynes, and M. Jindra, *Ligand-binding properties of a juvenile hormone receptor, Methoprene-tolerant*, Proc. Natl. Acad. Sci. USA 108 (2011), pp. 21128–21133.
143. E. Marengo and R. Todeschini, *A new algorithm for optimal, distance-based experimental design*, Chemom. Intell. Lab. Syst. 16 (1992), pp. 37–44.
144. M. Gurrath, G. Müller, and H.D. Höltje, *Pseudoreceptor modelling in drug design*, Perspect. Drug Discov. Des. 12–14, (1998), pp. 135–157.

145. F.M. Rogerson, F.E. Brennan, and P.J. Fuller, *Mineralocorticoid receptor binding, structure and function*, Mol. Cell. Endocrinol. 217 (2004), pp. 203–212.
146. C. Carlberg and F. Molnar, *Detailed molecular understanding of agonistic and antagonistic vitamin D receptor ligands*, Curr. Top. Med. Chem. 6 (2006), pp. 1243–1253.
147. N. Rochel and D. Moras, *Ligand binding domain of vitamin D receptors*, Curr. Top. Med. Chem. 6 (2006), pp. 1229–1241.
148. D.L. Bain, A.F. Heneghan, K.D. Connaghan-Jones, and M.T. Miura, *Nuclear receptor structure: Implications for function*, Annu. Rev. Physiol. 69 (2007), pp. 201–220.
149. C.L. Waller and J.D. McKinney, *Three-dimensional quantitative structure-activity relationships of dioxins and dioxin-like compounds: Model validation and Ah receptor characterization*, Chem. Res. Toxicol. 8 (1995), pp. 847–858.
150. A.A. Bogan, F.E. Cohen, and T.S. Scanlan, *Natural ligands of nuclear receptors have conserved volumes*, Nat. Struct. Biol. 5 (1998), pp. 679–681.
151. L. Holm and J. Park, *DaliLite workbench for protein structure comparison*, Bioinformatics 16 (2000), pp. 566–567.
152. J. Reichelt, G. Dieterich, and M. Kvesic, D. Schomburg, and D.W. Heinz, *BRAGI: Linking and visualization of database information in a 3D viewer and modeling tool*, Bioinformatics 21 (2005), pp. 1291–1293.
153. S.C. Basak, R. Natarajan, D. Mills, D.M. Hawkins, and J.J. Kraker, *Quantitative structure-activity relationship modeling of insect juvenile hormone activity of 2,4-dienoates using computed molecular descriptors*, SAR QSAR Environ. Res. 16 (2005), pp. 581–606.

10 Predicting Highly Potent Juvenile Hormone Esterase Inhibitors from 2D QSAR Modeling

James Devillers, Annick Doucet-Panaye, and Jean-Pierre Doucet

CONTENTS

ABSTRACT

The juvenile hormone esterase (JHE) is a member of the carboxylesterase family that hydrolyzes the α,β-unsaturated methyl ester of juvenile hormone (JH) to the corresponding carboxylic acid. JHE regulates JH titer in insect hemolymph during its larval development. Thus, it has been suggested that this enzyme could be targeted for use in insect control. JHE can also be considered as involved in the phenomenon of endocrine disruption by xenobiotics in beneficial insects. Consequently, there is a need to know the characteristics of the molecules able to act on the JHE. Trifluoromethylketones (TFKs) being the most potent JHE inhibitors found to date, different QSARs have been derived on this group of chemicals. In this context, a set of 96 TFKs, tested on *Trichoplusia ni* for their JHE inhibition, was split into a training set ($n=77$) and a test set ($n=19$) to derive QSAR models. TFKs were initially described by 42 CODESSA descriptors but a feature selection process allowed us to consider only five descriptors encoding the structural characteristics of the molecules and their reactivity. A classical multiple regression analysis, a PLS (partial least squares) regression analysis, a spline regression analysis, CART (Classification and Regression Tree), random forest regression analysis, stochastic gradient boosting tree regression analysis, a three-layer perceptron, a radial basis function network, and a support vector regression (SVR) were experienced as statistical tools. The best results were obtained with the SVR (r^2 and $r^2_{test} = 0.91$). The model provides information on the structural features and properties responsible for the high JHE inhibitory activity of TFKs. It was used to predict TFK structures having potentially a high inhibitory activity against JHE.

KEYWORDS

Juvenile hormone esterase, Trifluoromethylketones, QSAR model, Linear methods, Nonlinear methods, Inhibitory activity

10.1 INTRODUCTION

Metamorphosis in holometabolous insects is a complex process in which physiological, morphological, and behavioral events occur in a precisely timed manner to result in the transformation of a larva into an adult called imago. These different events are under the control of the juvenile hormone (JH). Thus, for example, the decrease in JH titer during the last larval instar is necessary for the apparition of the pupation stage during which the insect is immobilized and undergoes important transformations [1–3]. Reduction in JH titer occurs through a combination of a decrease in JH biosynthesis and increase in JH metabolism. JH metabolism is made through cleavage of either its epoxide or ester moiety. The relative importance of these two pathways varies with insect species, development stage, and tissue [4,5]. Normal lepidopteran metamorphosis, for example, requires the degradation of JH via ester hydrolysis. Inhibition of ester hydrolysis leads to abnormally large larvae and delayed pupation. The enzyme responsible for this ester hydrolysis is juvenile hormone esterase (JHE), which is found in the hemolymph and other insect tissues at key moments in development [5,6]. JHE hydrolyzes ester bonds in a two-step reaction. The first step consists in a nucleophilic attack of the carbonyl carbon of JH by the catalytic serine residue that is found in the substrate-binding pocket, release of methyl alcohol, and formation of an acyl–enzyme intermediate. The second step involves release of JH acid and regeneration of JHE following nucleophilic attack of the carbonyl carbon of the acyl–enzyme by an activated water molecule. Tetrahedral transition-state intermediates appear in both steps [4].

Insects have significant effects on human health, as vectors for diseases such as malaria, dengue, chikungunya, or African and American trypanosomiasis, but also in agriculture as pests or conversely in the maintenance of the biodiversity through the crucial role of some of them in the pollination. It is, therefore, important to study the function and specificity of JHE due to its key role in insect development. This will favor the design of substances for use in insect control [7,8] as well as a better understanding of the mechanisms of endocrine disruption in the insects [9]. The results of such studies can also provide insights into the mechanisms of action of other carboxylesterase enzymes, some of which being important in the hydrolysis, detoxification, or metabolic activation of a number of drugs, pesticides, and other xenobiotics [10,11].

Inhibition of JHE activity can have important biological adverse effects on the development of the insects. Paradoxically, these effects are sought in the creation of insecticides but feared in the case of the protection of the endocrine system of the beneficial insects. A number of chemicals have been tested for their potential inhibitory activity against JHE [12–14]. Among them, the trifluoromethylketones (TFKs) include the most potent JHE inhibitors found to date. The catalytic serine of JHE attacks the carbonyl carbon of the TFK containing inhibitor and forms an acyl–enzyme complex, which has a tetrahedral geometry, consisting of the trifluoromethyl group, the protein (via the serine), a hydroxyl (from the ketone), and an alkyl chain. However, because the trifluoromethyl group is a very poor leaving group, the second step in the reaction involving nucleophilic attack of the acyl–enzyme complex does not occur and the enzyme is inhibited [4–6].

A significant amount of work has been done to understand the structural features governing the inhibitory activity of TFKs. Most of the studies are based on *in silico* methods especially 2D quantitative structure–activity relationship (QSAR) approaches, which allow to relate physicochemical and/or topological properties of the organic molecules to their biological activity from a statistical method [15]. Inspection of the literature shows that the prediction of the effects of TFKs on JHE activity is mostly performed from QSAR models computed from a simple linear regression analysis [16–21], while a lot of modeling results dealing with various endpoints reveal that the nonlinear methods are particularly powerful in QSAR for discovering complex structure–activity relationships [22–25]. Nevertheless, these QSARs on JHE studies have shown that the most potent TFK inhibitors contain a long-chain aliphatic tail that is supposed to mimic the JH backbone. The presence of a methyl group at the alpha position of the carbonyl also influences the potency and selectivity of the TFKs [5]. It was stressed out that the potency of aliphatic TFKs exhibited a strong positive correlation with lipophilicity encoded by the 1-octanol/water partition coefficient (log P), with a maximum potency corresponding to compounds showing an intermediate lipophilicity (log P ~4). However, it was also pinpointed that the inhibition potency of a TFK could be increased by specific substitutions that did not strongly affect its overall lipophilicity. This is the case, for example, for TFKs having their β-carbon substituted by a sulfur atom, and Wogulis et al. [6] nicely summarized the different hypotheses explaining this fact. Recently, we showed [26] that the inhibitory potential of such chemicals against JHE was better modeled by using descriptors encoding reactivity and steric effects, while log *P* alone provided very poor predictions and this hydrophobic parameter was never selected as significant descriptor in a robust feature selection process. A parabolic equation (log *P*, log P^2), as used, for example, by Wheelock et al. [21], also did not provide satisfying results. Perhaps this is why, unlike these authors, we were able to propose models that correctly predicted the activity of TFKs being strong inhibitors of the JHE such as 1,1,1-trifluoro-3-[(1-methyloctyl)thio]-2-propanone (CAS RN 112240-78-5). Without any doubts, these good results were also due to the use of nonlinear methods as statistical tools for computing the QSAR models. Consequently, the goal of this chapter was first to extend the panel of statistical methods to definitively determine what is the best approach for modeling this set of TFKs. Then, the best model able to predict very potent TFK inhibitors of the JHE was used to propose candidates with high potential inhibitory activity.

10.2 MATERIALS AND METHODS

10.2.1 Biological Data

Because the most active TFKs included the RSCC(= O)C(F)(F)F functional skeleton (R being an aliphatic or aromatic substituent), only the molecules having this basic structural feature were considered. Thus, 82 molecules were retrieved from Wheelock et al. [21], and in addition, 14 molecules showing the same general skeleton were gathered in Székács et al. [19] (Figure 10.1, Table 10.1). A radiometric partition method [27] was used for assaying inhibition of JHE activity. Hemolymph from fifth-instar larvae of the cabbage looper (*Trichoplusia ni*) was the source of JHE. The test allows the calculation of an IC_{50} value (concentration that results in 50% inhibition). The activities in Table 10.1 are expressed in log $1/IC_{50}$ in M.

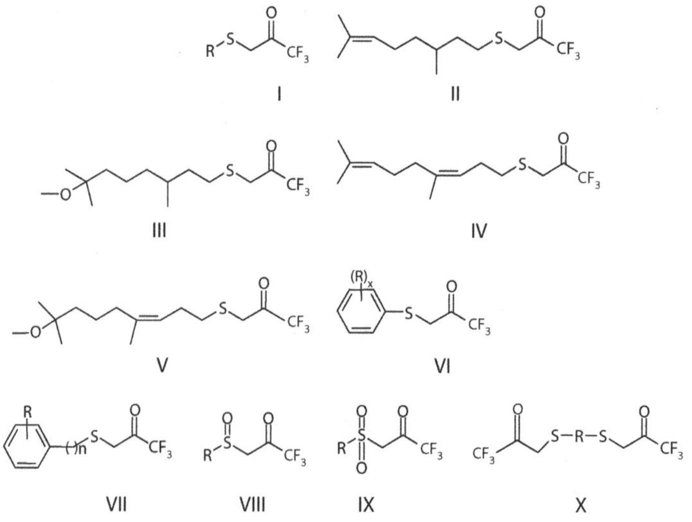

I II III IV V VI VII VIII IX X

FIGURE 10.1 Basic structural features of the TFK inhibitors.

TABLE 10.1
Structure and Inhibitory Activity of the 96 TFKs

No.	SF[a]	Substituent[b]	log (1/IC_{50})[c]	No.	SF[a]	Substituent[b]	log (1/IC_{50})[c]
1	I	n-C_4H_9	5.82	43	VI	4-F	5.08
2	I	n-C_5H_{11}	6.11	44	VI	4-Cl	5.89
3	I	$(CH_3)_3C$	5.30	45[d]	VI	4-Br	6.39
4	I	2-Me-Butyl	5.80	46	VI	4-Me	6.96
5[d]	I	3-Me-Butyl	5.60	47	VI	4-tBu	8.12
6	I	n-C_6H_{13}	7.51	48	VI	4-OH	5.21
7	I	n-C_7H_{15}	8.09	49	VI	4-OMe	5.89
8	I	n-C_8H_{17}	8.62	50[d]	VI	4-$(CH_3)_2N$	7.05
9	I	2,4,4-Me-C_8H_{14}	7.98	51	VI	2,5-Cl_2	7.64
10[d]	I	n-C_9H_{19}	8.43	52	VI	2,6-Cl_2	7.11
11	I	n-$C_7H_{15}CH(CH_3)$	9.77	53	VI	3,4-Cl_2	7.70
12	I	n-$C_{10}H_{21}$	7.70	54	VI	2,4-Me_2	6.04
13	I	n-$C_7H_{15}CH(Et)$	8.24	55[d]	VI	2,5-Me_2	6.77
14	I	n-$C_{11}H_{23}$	7.37	56	VI	2,6-Me_2	6.92
15[d]	I	n-C_7H_{15}CH(n-Pr)	7.89	57	VI	3,4-Me_2	7.96
16	I	n-$C_{12}H_{25}$	7.82	58	VI	3,5-Me_2	5.74
17	I	n-C_7H_{15}CH(n-Bu)	7.16	59	VI	2-Me-4-tBu	8.52
18	I	n-$C_{13}H_{27}$	7.35	60[d]	VI	3-Me-4-Br	7.89
19	I	n-$C_{14}H_{29}$	6.40	61	VI	2,4,5-Cl_3	7.49
20[d]	I	n-$C_{16}H_{33}$	6.20	62	VI	2,3,5,6-F_4	5.92
21	I	n-$C_{18}H_{37}$	6.30	63	VI	2,3,4,5,6-Cl_5	7.80
22	I	Cyclopentyl	5.01	64	VI	2,3,4,5,6-F_5	4.96
23	I	Cyclohexyl	5.28	65[d]	VII-1[e]	H	4.82
24	II	See Figure 10.1	8.35	66	VII-1	2-Cl	5.18
25[d]	III	See Figure 10.1	8.12	67	VII-1	4-Cl	5.34
26	IV	See Figure 10.1	8.51	68	VII-1	2-Me	4.90
27	V	See Figure 10.1	8.49	69	VII-1	3-Me	5.10
28	VI	H	5.06	70[d]	VII-1	4-Me	5.20
29	VI	2-Cl	5.68	71	VII-1	4-$C_{12}H_{25}$	5.80
30[d]	VI	2-Br	6.17	72	VII-2	H	5.49
31	VI	2-Me	5.37	73	VII-3	H	6.77
32	VI	2-Et	5.54	74	VIII	n-C_8H_{17}	7.37
33	VI	2-iPr	6.47	75[d]	VIII	Ph	4.21
34	VI	2-OMe	5.70	76	VIII	Ph(4-tBu)	7.64
35[d]	VI	2-CH_2OH	5.15	77	IX	Ph	3.98
36	VI	2-OPh	6.66	78	IX	n-C_6H_{13}	6.83
37	VI	2-NHPh	7.52	79	IX	n-C_8H_{17}	6.90
38	VI	3-Cl	6.40	80[d]	IX	n-$C_{10}H_{21}$	7.36
39	VI	3-Br	6.06	81	IX	n-$C_{12}H_{25}$	7.21
40[d]	VI	3-CF_3	5.60	82	IX	Ph-C_2H_4	6.31
41	VI	3-Me	6.89	83	X	n-C_3H_6	6.01
42	VI	3-OMe	5.89	84	X	n-C_4H_8	8.44

(continued)

TABLE 10.1 (continued)
Structure and Inhibitory Activity of the 96 TFKs

No.	SF[a]	Substituent[b]	log (1/IC_{50})[c]	No.	SF[a]	Substituent[b]	log (1/IC_{50})[c]
85[d]	X	n-C_5H_{10}	8.15	91	X	n-$C_{11}H_{22}$	7.88
86	X	n-C_6H_{12}	7.82	92	X	n-$C_{12}H_{24}$	8.09
87	X	n-C_7H_{14}	7.77	93	X	$(CH_2)_2O(CH_2)_2$	7.67
88	X	n-C_8H_{16}	9.09	94	X	$(CH_2)_2S(CH_2)_2$	8.31
89	X	n-C_9H_{18}	8.57	95[d]	X	$CH(CH_3)(CH_2)_2$	6.33
90[d]	X	n-$C_{10}H_{20}$	8.34	96	X	$CH(CH_3)(CH_2)_3$	8.09

Chemicals #1 to #82 were obtained from Wheelock et al. [21] and chemicals #83 to #96 were gathered in Székács et al. [19].

[a] Structural feature as described in Figure 10.1.

[b] The formalism used by Wheelock et al. [21] and Székács et al. [19] to characterize the substituents was not changed.

[c] IC_{50} = concentration that results in 50% inhibition, in *M*.

[d] Test set TFKs.

[e] Value of *n* (see Figure 10.1).

An external test set of 19 TFKs was defined by selecting the chemicals numbered with a multiple of five in the tables that were initially organized according to the different functional groups in Székács et al. [19] and Wheelock et al. [21]. By using this simple strategy, the representativeness of the chemical structures was secured as well as the inclusion of the most active chemical in the training set (Table 10.1). It is noteworthy that random selections were also investigated.

10.2.2 Molecular Descriptors

The 96 chemical structures (Table 10.1, Figure 10.1) were geometry optimized with the AM1 semiempirical method in HyperChem (Hypercube, Gainesville) and exported to MOPAC. The structures were then introduced into the Comprehensive Descriptors for Structural and Statistical Analysis (CODESSA) Pro software (http://www.codessa-pro.com) to compute 42 topological, geometrical, and quantum-chemical descriptors considered to be suited to encode the structural and functional characteristics of the studied molecules. In addition, the 1-octanol/water partition coefficients (log *P*) were calculated from the AUTOLOGP software [28,29] for all the TFKs because the log *P* values of Székács et al. [19] and Wheelock et al. [21] were computed with different methods.

The automatic feature selection procedure of the STATISTICA™ software (StatSoft, Fr), which is based on the use of a genetic algorithm [30] and a regression neural network [31], was used for optimizing the number of descriptors. This yielded the selection of the five following descriptors: Balaban index (BI), complementary information content index of order 1 (CIC1), HOMO-LUMO energy gap (GAP), total molecular surface area (quantum) (TMSAQ), and partial positively charged surface area (quantum) (PPSA-1Q).

10.2.3 Statistical Analyses

10.2.3.1 Regression Equation Models

The simple and multiple regression analyses [32] were tested first because they are the most widely used in QSAR [33] due to their straightforward interpretability.

The potential advantages of the partial least squares (PLS) regression were also evaluated. As in multiple regression analysis (MRA), the main goal of PLS is to derive a linear model $\boldsymbol{Y}=\boldsymbol{XB}+\boldsymbol{E}$, where $\boldsymbol{Y}$ is a matrix of n observations by m response variables, $\boldsymbol{X}$ is a matrix of n observations by p predictor variables, $\boldsymbol{B}$ is the matrix of regression coefficients $p \times m$, and $\boldsymbol{E}$ is the error term of the model of same dimension than $\boldsymbol{Y}$. However, PLS regression works by searching for a set of components (called latent vectors) that perform a simultaneous decomposition of $\boldsymbol{X}$ and $\boldsymbol{Y}$ with the constraint that these components explain as much as possible of the covariance between $\boldsymbol{X}$ and $\boldsymbol{Y}$ [34]. To do so, the nonlinear iterative partial least squares (NIPALS) algorithm is commonly used. Calculations are made on standardized variables (i.e., with a mean of zero and a variance unity). PLS regression is more robust than a simple MRA. It is particularly suited when the number of predictors is large compared to the number of observations and in case of multicollinearity [35,36].

The multivariate adaptive regression splines (MARS) [37] was experienced. It is a nonparametric regression method making no assumption about the functional relationship between the dependent and independent variables. Briefly, such a regression technique is based on a "divide and conquer" strategy that partitions the input space into regions characterized by their own regression equation [37]. The technique has become particularly popular in numerous areas including QSAR [38,39] because it does not assume or impose any particular type of relationship between the predictor variables and the dependent variable of interest. Instead, useful models can be derived even in situations where the relationship between the predictors and the dependent variables is non-monotone and difficult to approximate with parametric models. It has a great power and flexibility to model relationships that are nearly additive or involve interactions in at most a few variables. In addition, the model can be represented in a form that separately identifies the additive contributions and those associated with the different multivariable interactions [37].

10.2.3.2 Regression Trees

10.2.3.2.1 Classification and Regression Tree (CART)

Unlike classical regression analysis for which the relationship between the response variable and the predictor variables is prespecified and tests are performed to prove or disprove the relationship, regression tree analysis does not assume such relationship. It is primarily a method constructing a set of *if–then* rules on the predictor variables. The rules are constructed by recursively partitioning the training data set into successively smaller groups with binary splits based on a single predictor variable. Splits for all of the predictors are examined by an exhaustive search procedure and the best split is chosen. For regression trees, the selected split is the one that maximizes the homogeneity of the two resulting groups with respect to the response variable (i.e., the split that maximizes the between-groups sum of squares, as in analysis of variance), although other options exist such as misclassification error or deviance. The output is a tree diagram with the branches determined by the splitting rules and a series of terminal nodes [40,41]. Regression trees are effective in uncovering structure in data with hierarchical or nonadditive variables. Because no *a priori* assumptions are made about the nature of the relationships among the response variable and the predictor, the method is suited for accounting for interactions and nonlinearities among variables [41]. CART analysis has found applications in QSAR (e.g., [42,43]).

10.2.3.2.2 Random Forests for Regression

Proposed by Breiman [44], the random forest algorithm computes a collection (ensemble) of simple tree predictors, each capable of producing a response when presented with a set of predictor values. For regression problems, the tree response is an estimate of the dependent variable given the predictors, while for classification problems, this response takes the form of a class membership, which associates a set of predictor values with one of the categories present in the dependent variable. In practice, bootstrap samples are drawn to construct the trees, each of them being grown with a randomized subset of predictors. This explains the term "random," while "forest" has to be related to the fact that a collection of trees is computed. The number of predictors used to find the best split at each node is a randomly chosen subset of the total number of predictors. To perform a random forest regression analysis, two categories of parameters have to be set, those characterizing the construction of the trees and the stopping parameters, which control the complexity of the individual trees that will be built at each consecutive step. The former category includes the number of predictors for the tree models, the number of simple regression trees to be computed in successive forest building steps, the subsample proportion to be used for drawing the bootstrap learning samples for consecutive steps, and a constant for seeding the random number generator, which is used to select the subsamples for consecutive trees. The latter category includes the minimum n of cases that allows splitting to continue until all terminal nodes contain no more than a specified minimum number of cases or objects; the minimum n in child node, which controls the smallest permissible number in a child node, for a split to be applied; the maximum n of levels, which is a stopping criterion based on the number of levels in a tree; and the maximum n of nodes, which is also a stopping criterion based on the number of nodes in each tree. Random forest trees are increasingly used in structure–activity modeling [45–47].

10.2.3.2.3 Stochastic Gradient Boosting Trees for Regression

Introduced by Friedman [48], the general idea of this method is to compute a sequence of (very) simple trees, where each successive tree is built for the prediction residuals of the preceding tree. Thus, suppose that we limit the complexity of the trees to 3 nodes (1 root node + 2 child nodes). At each step of the boosting algorithm, a simple (best) partitioning of the data is determined, and the deviations of the observed values from the respective means (residuals for each partition) are computed. The next 3-node tree will then be fitted to those residuals, to find another partition that will further reduce the residual (error) variance for the data, given the preceding sequence of trees.

Such "additive weighted expansions" of trees can produce an excellent fit of the predicted values to the observed ones, even if the relationships between the predictors and the dependent variable are complex. However, one of the major problems of the boosting trees is to "know when to stop" to avoid overfitting. To avoid this limitation, two strategies are used. Each consecutive simple tree is built for only a randomly selected subsample of the training set. Thus, each consecutive tree is built for the prediction residuals (from all preceding trees) of an independently drawn random sample. This is called stochastic gradient boosting method. By plotting the prediction error for the training and test sets over the successive boosted trees, the point of overfitting can be identified [48,49]. To run such a method, two types of parameters have to be set. Parameters allowing the construction of the boosted trees include the learning rate, the number of simple regression trees to be computed in successive boosting steps that is called number of additive terms, the subsample proportion to be used for drawing the random learning sample for consecutive boosting steps, and a constant for seeding the random number generator that is used to select the subsamples for consecutive boosting trees. The second group deals with the stopping parameters, which control the complexity of the individual trees that will be built at each consecutive boosting step. They include the minimum *n* of cases, the minimum *n* in child node, the maximum *n* of levels, and the maximum *n* of nodes (see Section 10.2.3.2.2). Boosting tree regressions have found some applications in QSAR [50,51].

10.2.3.3 Artificial Neural Networks

10.2.3.3.1 Three-Layer Perceptron

The three-layer perceptron (TLP) is perhaps the most popular supervised artificial neural network (ANN) in use in QSAR and QSPR (quantitative structure–property relationship) (see, e.g., [22,26,28,52–56]). Because its functioning has been widely described in literature (see, e.g., [31]), only the basic principles are recalled here. A TLP includes one input layer with a number of neurons corresponding to the number of selected molecular descriptors, one output layer of one neuron corresponding to the modeled activity, and one hidden layer, between the two above layers, with an adjustable number of neurons for distributing the information. Too many hidden neurons often lead to overfitting, and hence, to avoid problems, their number has to be limited. This is done from a trial-and-error procedure and also by the use of pruning algorithms. The neurons of each layer are connected in the forward direction (i.e., input to output) and are activated by means of activation functions. Each connection is associated with a weight. The weights are adjusted during the learning process aiming at minimizing an error computed from the target and calculated outputs. Numerous learning algorithms are available, and among them, the backpropagation, the Levenberg–Marquardt, and the Broyden–Fletcher–Goldfarb–Shanno (BFGS) algorithms were tested alone or in combination [31].

10.2.3.3.2 Radial Basis Function Network

A radial basis function network (RBFN) is also a supervised ANN, which has found applications in QSAR (e.g., [24,57]). It is typically configured with a single hidden layer of neurons whose activation function is selected from a class of functions called radial basis functions (RBFs), usually a Gaussian or some other kernel functions. Each hidden unit acts as a locally tuned processor that computes a score for the match between the input vector and its connection weights or centers. The weights connecting the basis units to the outputs are used to take linear combinations of the hidden units to give the final output. More precisely, in an RBFN, the weights into the hidden layer basis units are usually set before the second layer of weights is adjusted. As the input moves away from the connection weights, the activation value falls off. It is the reason why the term of "center" is allocated to the first-layer weights [58]. These center weights can be computed by different statistical techniques. In the present study, K-means and random sampling were tested. They are then used to set the areas of sensitivity for the RBFN hidden units, which then remain fixed. Once the hidden layer weights are set, a second phase of training allows the adjustment of the output weights [58]. An RBFN trains much faster than a classical backpropagation TLP and does not suffer from local minima.

10.2.3.4 Support Vector Regression

Support vector machines (SVMs) [59,60] are rooted on two key ideas. The first one is the notion of maximum margin. The margin is the distance between the boundary of separation and the closest samples. These are called support vectors. The separation boundary is chosen as the one that maximizes the margin. In order to deal with cases where the data are not linearly separable, the second key idea of SVMs is to transform the representation space of the input data into a space of greater dimension (possibly infinite), in which it is likely that there exists a linear separation. This is achieved through a kernel function. The following kernel functions are commonly tested: linear, polynomial, RBF, and sigmoid. To construct an optimal hyperplane, SVM uses an iterative training algorithm that minimizes an error function. According to the form of the error function, the SVM will be a classifier or a regression tool. Two types of support vector regression (SVR) can be computed, the so-called SVR type 1 (ε-SVR) and type 2 (ν-SVR) with their own error functions. Both were tested in the present study. The advantages of SVR are the following: the presence of a global minimum solution resulting from the minimization of a convex programming problem, relatively fast training speed, and sparseness in solution representation. SVR has found numerous applications in QSAR [24,57,61–63].

10.2.3.5 Statistical Analyses and Data Preprocessing

Statistical analyses were performed from the multivariate analysis, neural network, and data mining modules (versions 6, 8, and 10) of the STATISTICA™ software (StatSoft, Fr) and from the Earth (version 3.1–2) and e1071 (version 1.6) programs found in the Comprehensive R Archive Network (http://cran.r-project.org/) to perform regression splines and SVR, respectively. The first step of the data preprocessing consisted of dividing the calculated values of the CIC1, TMSAQ, and PPSA-1Q descriptors by 100 to have all the selected molecular descriptor values with the same order of magnitude. Then, the scaling procedures recommended for the different statistical approaches under study were systematically used when necessary.

10.3 RESULTS AND DISCUSSION

Wheelock et al. [21] in their 2D QSAR analysis noted a high dependence between the inhibition potential of TFKs and their 1-octanol/water partition coefficient (log *P*). Our results contradict their findings. We have found that this hydrophobic parameter only marginally explained the inhibitory activity of TFKs (Figure 10.2). Indeed, the r^2 value calculated from the log $(1/IC_{50})$ values for the 96 studied TFKs and their corresponding log *P* data computed with the AUTOLOGP software equals 0.12. It is noteworthy that a poor correlation is also obtained from the 109 TFKs selected by Wheelock et al. [21] and described by log *P* calculated from the ClogP software (Biobyte Corp., Claremont, CA) with $r^2 = 0.08$ (Figure 10.3). The parabolic model (log *P*, log P^2) only yields an r^2 value of 0.25. Last, a coefficient of determination of only 0.09 was also obtained when the TFK inhibitors are considered in their *gem*-diol form instead of ketone form. This explains why the 1-octanol/water partition coefficient was not retained in our feature selection process.

Use of a forward or backward MRA allowed us to obtain a five-parameter model (Equation 10.1):

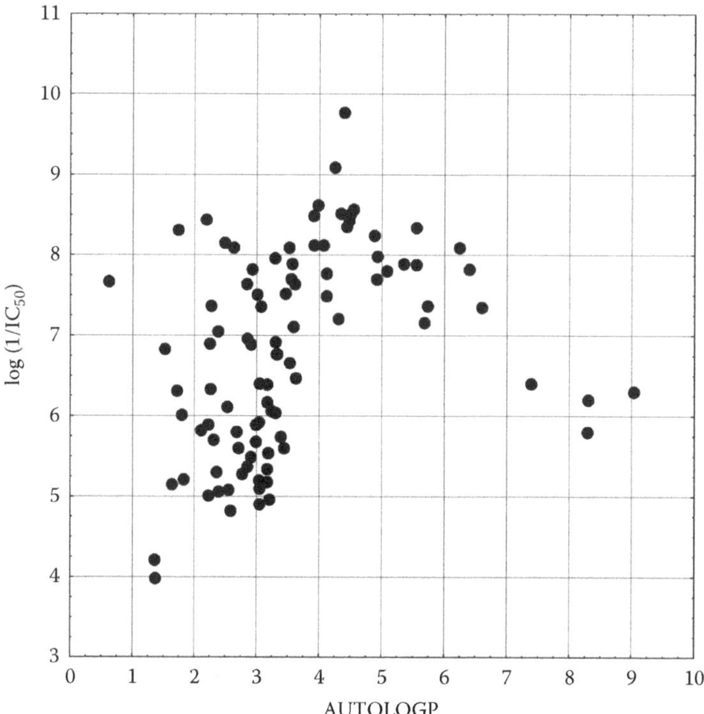

FIGURE 10.2 Plot of JHE inhibition (log $1/IC_{50}$) versus lipophilicity parameter calculated from AUTOLOGP for 96 TFKs (this study).

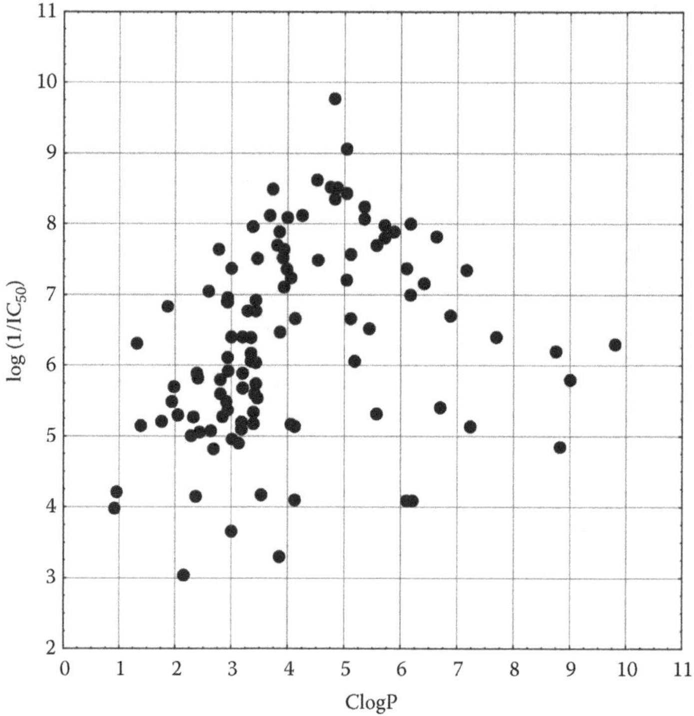

FIGURE 10.3 Plot of JHE inhibition (log $1/IC_{50}$) versus lipophilicity parameter calculated from ClogP for 109 TFKs. (Data from Wheelock, C.E. et al., *Bioorg. Med. Chem.*, 11, 5101, 2003.)

$$\log\left(\frac{1}{IC_{50}}\right) = -0.50\,GAP + 1.35\,BI + 1.07\,TMSAQ - 3.89\,CIC1 + 1.13\,PPSA\text{-}1Q + 0.89 \qquad (10.1)$$

$$n = 77, \quad r^2 = 0.56, \quad s = 0.85, \quad F = 18.25, \quad r_{test}^2 = 0.71, \quad Q^2 = 0.47.$$

where

- n is the number of training set chemicals
- r^2 is the coefficient of determination of the model
- s is the residual standard error
- F is the Fisher's test
- r_{test}^2 is the coefficient of determination for the external test set of 19 TFKs
- Q^2 is the leave-one-out cross-validated correlation coefficient

Inspection of the statistical parameters of Equation 10.1 shows that this five-parameter regression equation presents poor predictive performances. This clearly appears in Figure 10.4, which gives the residuals (observed–calculated log $(1/IC_{50})$ values) for the training set chemicals. With Equation 10.1, the activity of TFK #11 is

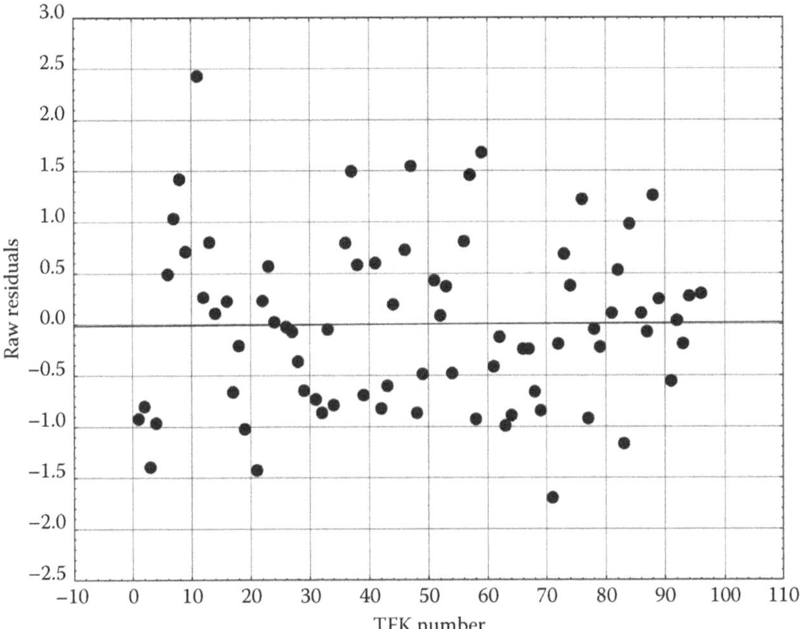

FIGURE 10.4 Residuals obtained with Equation 10.1 for the 77 training set TFKs.

totally underestimated. Use of PLS regression analysis, with the NIPALS algorithm, provides worst results. Only one component is significant yielding an r^2 value equal to 0.30 and a Q^2 value equal to 0.21.

An attempt was made to compute a more powerful model by introducing the five selected descriptors into the Earth program, which is the R implementation of MARS. Because there are only 77 TFKs in the training set and to avoid problems of overfitting, the degree of interactions between the independent (predictor) variables to be included in the model was set to 1 in order to take into account only their main effects in the model. In the same way, the default values proposed in the Earth program for the maximum number of basis functions, the penalty value, and the threshold value were selected. A pruning algorithm was used for deleting the nonsignificant knots, which are the predictors of the local models. The obtained model (Equation 10.2) showed significantly better internal and external predictive performances than Equation 10.1.

$$\log\left(\frac{1}{IC_{50}}\right) = -0.66\max(0;\text{GAP-8.46}) - 1.60\max(0;\text{2.77-BI}) - 2.00\max(0;\text{4.98-TMSAQ}) + 2.97\max(0;\text{0.51-CIC1}) - 0.70\max(0;\text{PPSA-1Q-2.81}) - 1.31\max(0;\text{3.93-PPSA-1Q}) + 9.25 \tag{10.2}$$

$$n = 77,\quad r^2 = 0.79,\quad s = 0.57,\quad r_{test}^2 = 0.89,\quad Q^2 = 0.68$$

However, analysis of the residual values obtained with Equation 10.2 reveals that the most potent TFK is badly predicted. Indeed, the difference between the observed and calculated activity values for chemical #11 equals 1.33. This is not the case for chemical #88 (Table 10.1), which is the second most active chemical (Table 10.1) and for which the residual equals 0.70. In the same way, all the training set TFKs with a log $(1/IC_{50}) > 8$ have their activity correctly estimated by Equation 10.2, except chemical #84, which shows a residual value of 1.01. Four other less active TFKs belonging to the training set present residuals >1 in absolute value. These are chemicals #46 (residual value = 1.03), #57 (1.38), #58 (−1.10), and #69 (−1.29). It is worth noting that all the residual values of the test set compounds were <1 (absolute value).

CART allowed us to obtain r^2 and r^2_{test} values of 0.91 and 0.73, respectively. However, this does not mean that this simple regression tree is able to correctly gauge the influence of the structural characteristics of the TFKs on their JHE inhibitory activity. Indeed, the graphical display of the observed versus calculated log $(1/IC_{50})$ values reveals that the model is able to only predict levels of activities. In addition, TFK #11, located in the right upper part of Figure 10.5, is badly predicted with a residual value equal to 1.15 (absolute value).

The random forest regression model was constructed from a trial-and-error procedure during which the influence of the different parameters was tested on

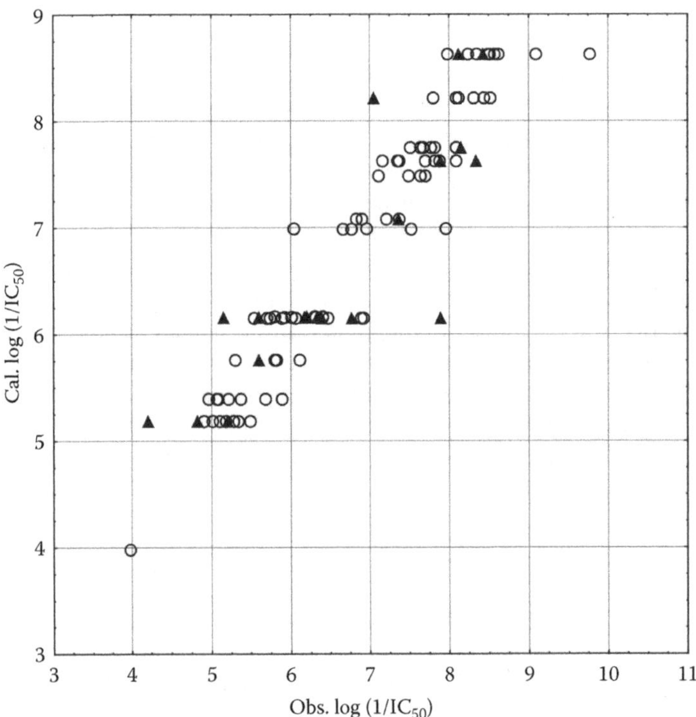

FIGURE 10.5 Observed versus calculated log $(1/IC_{50})$ with the CART model for the training set compounds (circles) and test set compounds (triangles).

its internal and external predictive power. Concomitantly, an attempt was made to reduce as best as possible the number of trees without altering too much the performances of the model. Thus, the number of predictors for the tree models was fixed to 4 and the number of simple regression trees to be computed in successive forest building steps was set to 100 in a first step. The subsample proportion to be used for drawing the bootstrap learning samples for consecutive steps and the constant for seeding the random number generator were fixed to 0.5 and 1, respectively. Regarding the stopping parameters, the minimum *n* of cases was fixed to 5 and the minimum *n* in child node to 1. The maximum *n* of levels and the maximum *n* of nodes were set to 6 and 8, respectively. With such parameter values, the best compromise in the model performances was obtained with 10 trees (Figure 10.6).

A model with 10 trees yielded r^2 and r_{test}^2 values of 0.72 and 0.73, respectively. Inspection of the observed versus calculated log ($1/IC_{50}$) values (Figure 10.7) shows the presence of plateaus, especially for the most potent TFK inhibitors, which are badly predicted by the model.

The stochastic gradient boosting tree regression model was also computed from a trial-and-error procedure. The best results were obtained with a learning rate of 0.095 and by considering 27 trees. The subsample proportion to be used for drawing the random learning sample for consecutive boosting steps was fixed to 0.47, and the constant for seeding the random number generator was set to 1. The minimum *n* of cases, the minimum *n* in child node, the maximum *n* of levels, and the maximum

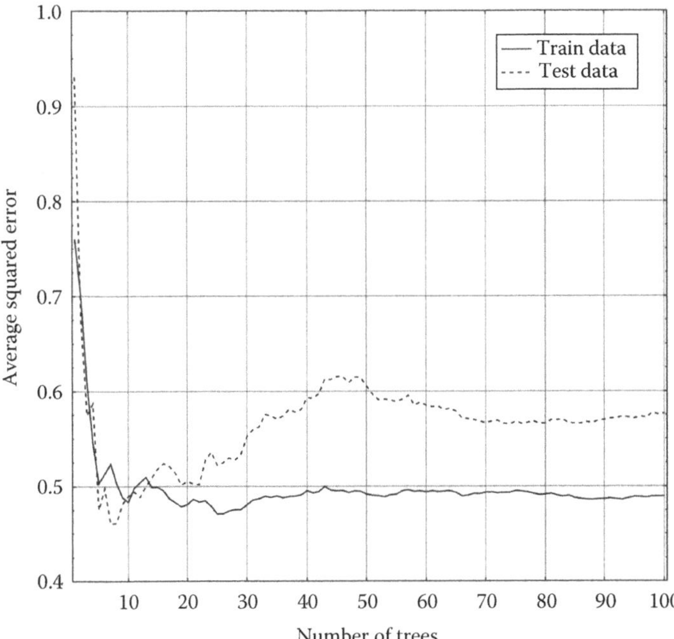

FIGURE 10.6 Average squared errors for the training and test sets over the successive tree addition steps.

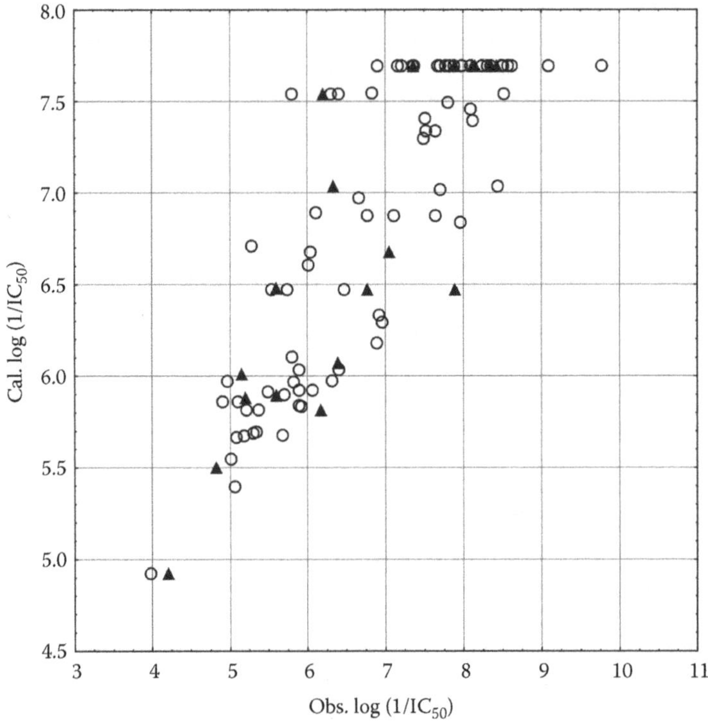

FIGURE 10.7 Observed versus calculated log $(1/IC_{50})$ with the random forest regression model for the training set compounds (circles) and test set compounds (triangles).

n of nodes were also fixed to 5, 1, 6, and 8, respectively. With such a configuration, r^2 and r^2_{test} values of 0.88 and 0.89 were obtained for the training set and test set, respectively. Analysis of the residuals (Figure 10.8) shows fewer large deviations than those observed with the CART (Figure 10.5) and random forest (Figure 10.7) models. However, the TFKs #11 and #88, having an observed log $(1/IC_{50}) > 9$, are badly modeled by the model since their residuals are equal to 1.37 and 0.87, respectively (see the right upper part of Figure 10.8). It is noteworthy that an increase of the number of trees in the random forest regression model to reach the level of complexity of the stochastic gradient boosting tree regression model did not change significantly the results obtained for the training set, but a slight deterioration of the prediction performances on the test set was observed. This can be easily verified on Figure 10.6.

To fully assess the interest of the regression trees for predicting the activity of the TFKs under study, runs were also performed by using a 80%–20% random sampling procedure for the training set and test set, respectively. Only the runs including chemical #11 (Table 10.1) in the training set were considered for deriving the models. No significant changes were observed with CART. Conversely, while it was possible to maintain the r^2 value of the training sets, the performances of the test set were mostly deteriorated with the random forest and stochastic gradient boosting tree

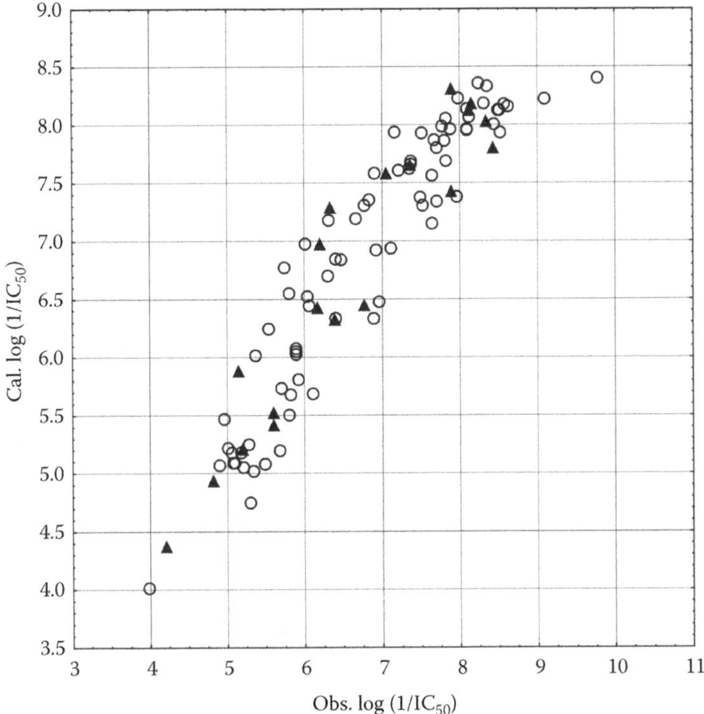

FIGURE 10.8 Observed versus calculated log $(1/IC_{50})$ with the stochastic gradient boosting tree regression model for the training set compounds (circles) and test set compounds (triangles).

regression analyses when the tree construction options and the stopping parameters were not changed. It is noteworthy that the random forests and the stochastic gradient boosting trees are more suited when used on a large data matrix.

The TLP has proved its interest in QSAR modeling and the predictions obtained with this nonlinear device are very often better than those computed with a classical MRA or a PLS regression analysis. Again, this finding was confirmed in the present study. Among all the different architectures, learning algorithms, and parameter values tested, the best results were obtained with a TLP including six neurons on the hidden layer (i.e., 5/6/1), a bias connected to the hidden and output layers, and using the BFGS second-order training algorithm. A hyperbolic tangent function (tanh) and a negative exponential function were used for the hidden and output neurons, respectively. The convergence was obtained after only 108 cycles. With such a configuration, the r^2 and r^2_{test} values were equal to 0.85 and 0.94, respectively. The calculation of the Q^2 parameter is not informative with a TLP [31]. Consequently, to fully assess the interest of a TLP for predicting the activity of the TFKs under study, runs were also performed by using a 80%–20% random sampling for the training set and test set, respectively. As previously indicated, only the runs including chemical #11 (Table 10.1) in the training set

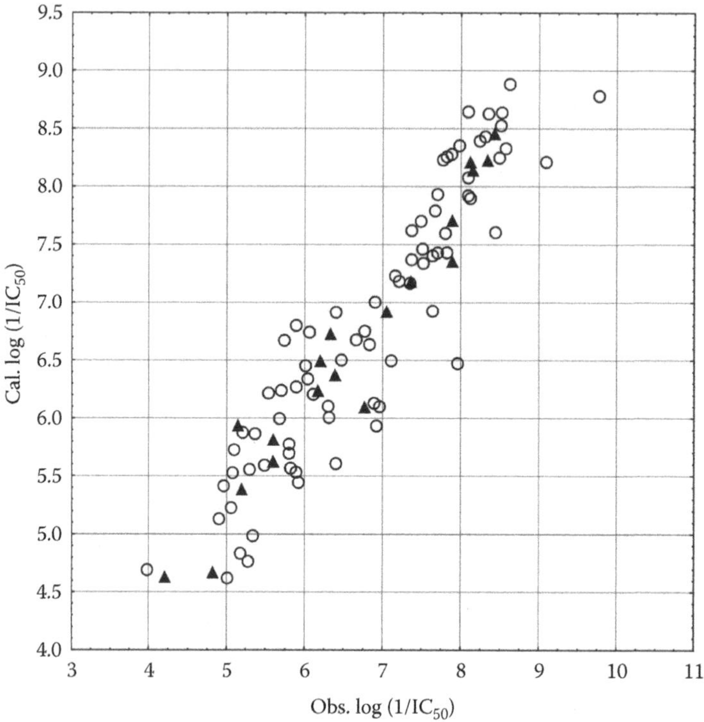

FIGURE 10.9 Observed versus calculated log $(1/IC_{50})$ with the TLP model for the training set compounds (circles) and test set compounds (triangles).

were kept. In these conditions, the r^2 values were broadly the same and r^2_{test} values were generally a little bit lower than those obtained with our selected test set. While the overall prediction performances of the selected TLP are rather satisfying (Figure 10.9), the log $(1/IC_{50})$ value of #11 remains badly predicted by the neural network model with a residual value of 0.99. Chemicals #88 and #84 are also poorly predicted with residual values equal to 0.88 and 0.83, respectively. Finally, all the other training set TFKs with a log $(1/IC_{50}) > 8$ (Table 10.1) show their inhibitory activity correctly estimated, the poorest predicted activity being chemical #7 with a residual value of −0.56. This is also the case for chemicals #10, 25, 85, and 90 belonging to the test set and which present a residual value <0.15 (in absolute value). Notice that the inhibitory activity of chemicals #57 and #58 remains badly predicted by the selected model with residual values equal to 1.49 and −0.93, respectively.

Because the RBFN is also commonly used in QSAR, we also tested the performances of this supervised three-layer neural network on the prediction of the JHE inhibitory activity of TFKs. To have the possibility to directly compare the performances of the two types of supervised ANNs, a 5/6/1 RBFN with a bias on the output layer was used. The best results were obtained with six radial spreads (one per hidden neuron), a Gaussian function, and an identity function for the hidden and output neurons, respectively. With such a configuration, the r^2 equaled 0.73 and a value of 0.85 was computed for the r^2_{test}. The residual value obtained for chemical #11 (Table 10.1) is equal to 1.43. The inhibitory activity of the TFKs #17, 32, 51–53, 57, 58, and 77 is also very badly predicted with residuals >1 (absolute value). Among the 19 test set chemicals (Table 10.1), it is worth noting that the activity of 12 of them is overestimated by the model. In addition, the residual value of TFK #60 equals 1.26. Broadly the same results were obtained with models computed by using an 80/20 random sampling for the training set and test set, respectively.

The SVMs, which can be used for classifications or regressions, have gained attention during the past few years in QSAR because they are at least as efficient as the ANNs. Moreover, a significant advantage of SVMs is that while ANNs can suffer from multiple local minima, the solution to an SVM is global and unique. Two additional advantages of SVMs are that they have a simple geometric interpretation and give a sparse solution. Last SVMs are less prone to overfitting than ANNs.

One of the main ideas of SVMs is to transform the representation space of the input data into a space of greater dimension, in which it is likely that there exists a linear separation. This is achieved through a kernel function and the adjustment of a cost parameter (C), which can introduce some flexibility in the system. In our study, linear, polynomial, RBF, and sigmoid kernels were tested. All needed to be adjusted by a trial-and-error procedure except the linear kernel. To construct an optimal hyperplane, an SVM uses an iterative training algorithm that minimizes an error function. To derive a regression model, two different errors can be computed yielding the so-called ε-SVR (SVR type 1) or ν-SVR (SVR type 2). Each category requires a different parameter to adjust. In our study, SVR type 1 with an RBF as kernel provided the best prediction results with C, ε, and γ set to 27, 0.1, and 0.57, respectively. In that case, the r^2 and r^2_{test} values were both equal to 0.91. The good predictive performances of the SVM model clearly appear in Figure 10.10. The most active TFKs, chemicals #11 and #88 in Table 10.1, are well predicted with residual values equal to 0.47 and 0.27, respectively. All the other training set TFKs with a log ($1/IC_{50}$) > 8 (Table 10.1) show their inhibitory activity correctly estimated with a residual <0.35 (absolute value). Chemicals #57 and #58 remain badly predicted with differences between the observed and calculated log ($1/IC_{50}$) equal to 1.21 and −1.11, respectively. Regarding the test set, all the chemicals are correctly predicted by the model, the highest residual being equal to −0.64 (TFK #40). The observed versus calculated log ($1/IC_{50}$) values of the TFKs belonging to the training and test sets are displayed in Figure 10.10. Use of randomly selected training (80%) and test (20%) sets yielded same results.

The goal of this study was to model the inhibitory activity of TFKs against JHE focusing on the most active chemicals. Previous QSAR studies have shown that an increased hydrophobicity, encoded by the 1-octanol/water partition coefficient (log P), correlated well with the inhibitor potency of TFKs, so it was shown that compounds containing, for example, longer, more hydrophobic alkyl chains were more potent inhibitors [19,21]. These are crude relationships, which work only on structurally diverse data sets including series of molecules varying mainly from each other by their number of carbon atoms on the alkyl chain, the number of chlorine or bromine atoms on the aromatic moiety, and so on. Nevertheless, it has been shown that

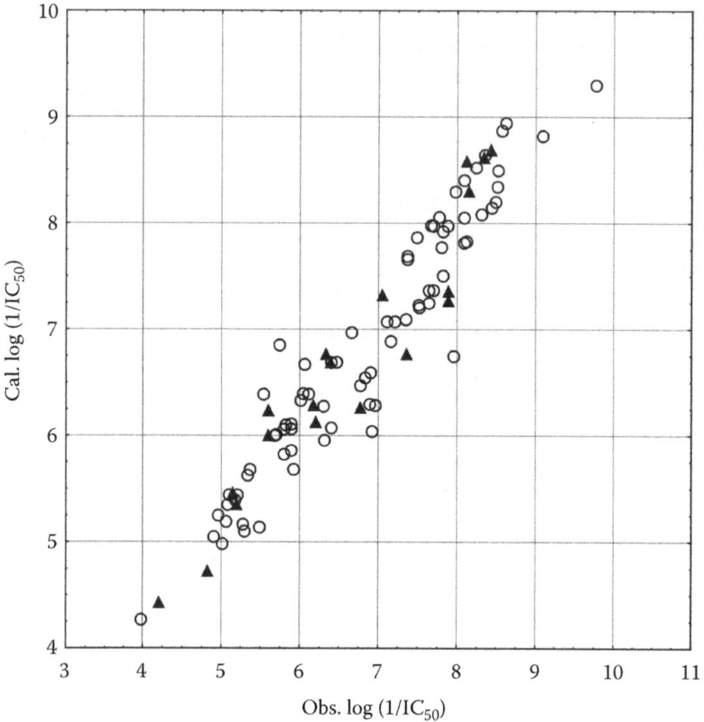

FIGURE 10.10 Observed versus calculated log ($1/IC_{50}$) with the SVR model for the training set compounds (circles) and test set compounds (triangles).

the crystal structure of JHE includes a long, hydrophobic binding pocket with the solvent-inaccessible catalytic triad located at the end [6]. Even if the hydrophobicity of TFKs can influence their inhibitory activity against JHE, our study clearly shows that other factors intervene in the process. A simple inspection of the selected data set (Table 10.1) shows that the inhibitory activity can significantly vary according to the position of a methyl group on the aromatic ring (#54–58) or the length of a carbon backbone (#6–8, 10, 12, 14, 16, 18–21) linked to the thio-trifluoromethylketone group. Our QSAR investigations have revealed that the BI, CIC1, HOMO-LUMO energy gap (GAP), TMSAQ, and PPSA-1Q were particularly suited for modeling the inhibitory activity of TFKs. BI and CIC1 allow to describe all the topological and steric characteristics of the molecules [64], while the other descriptors encode the reactivity potential of the molecules [65,66]. It is noteworthy that in all the QSAR models, the HOMO-LUMO energy gap was the less significant descriptor. Nevertheless, our results corroborate the recent findings of Rayo et al. [67] who indicated that the inhibitory activity of TFKs was a subtle balance between their reactivity and steric effects.

Among all the panel of linear and nonlinear methods tested on the same data set, the best results were obtained with the SVR model. The model outperforms the

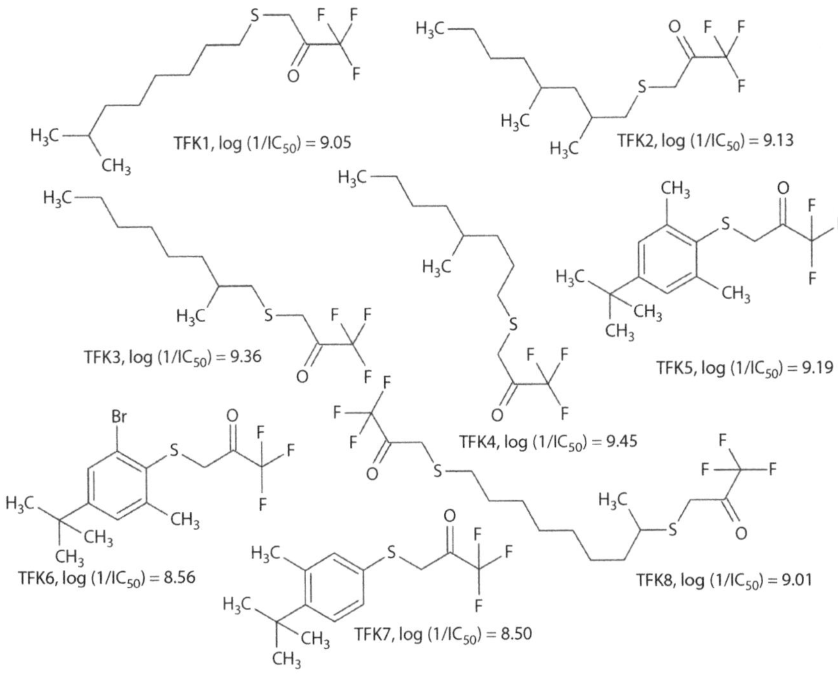

FIGURE 10.11 Predictions from the SVM model of TFK structures having potent JHE inhibitory activity.

others in terms of statistics as well as in its capability to correctly predict the most potent JHE inhibitors.

Consequently, it was used to predict TFK structures potentially having a log $(1/IC_{50}) \geq 9$. Some candidates are displayed in Figure 10.11 with their structural characteristics and predicted inhibitory activity. Obviously, it should be necessary to synthesize them to confirm or infirm these *in silico* predictions.

10.4 CONCLUSION

Our study shows that it is possible to derive a 2D QSAR model allowing to propose TFK structures having potentially a high inhibitory activity on the JHE. Such type of model can be useful to design new pest control chemicals. It can also be helpful to better understand the chemical structures that are able to inhibit the activity of the JHs of the beneficial insects. Thus, for example, Mackert et al. [68] have shown that 3-octylthio-1,1,1-trifluoro-2-propanone (OTFP at 2 mM) inhibited >95% of the JH degradation activity present in honey bee hemolymph as well as in the fat body. They deduced that JH degradation was performed primarily by the JHE, whereas only 5% may be attributed to JH epoxide hydrolase. Consequently, it is anticipated that TFKs could have the potential to disrupt the development of the honey bees by inhibiting the degradation of their JH.

Our study also demonstrates that it is always useful to test different statistical methods on the same data set to select the one that will be able to capture all the key information included in the data matrix. In QSAR, it is now commonplace to start the modeling procedure by using a linear method such as an MRA or a PLS regression analysis and then test whether a nonlinear method (e.g., TLP) could compete favorably. The originality of this study deals with the number and variety of the methods used. The SVR provides better results since it allows to accurately predict the inhibitory activity of the most potent TFKs. However, all generalization from this study would be totally hazardous because the selection of the most suited statistical method depends undoubtedly of the data matrix.

ACKNOWLEDGMENT

The financial support from the French Ministry of Ecology and Sustainable Development is gratefully acknowledged (PNRPE program).

REFERENCES

1. K. Hartfelder, *Insect juvenile hormone: From "status quo" to high society*, Braz. J. Med. Biol. Res. 33 (2000), pp. 157–177.
2. L.I. Gilbert, N.A. Granger, and R.M. Roe, *The juvenile hormones: Historical facts and speculations on future research directions*, Insect Biochem. Mol. Biol. 30 (2000), pp. 617–644.
3. S.R. Palli, *Recent advances in the mode of action of juvenile hormones and their analogs*, in *Biorational Control of Arthropod Pests*, I. Ishaaya and A.R. Horowitz, eds., Springer, New York, 2009, pp. 111–129.
4. S.G. Kamita and B.D. Hammock, *Juvenile hormone esterase: Biochemistry and structure*, J. Pestic. Sci. 35 (2010), 265–274.
5. S.G. Kamita, A.C. Hinton, C.E. Wheelock, M.D. Wogulis, D.K. Wilson, N.M. Wolf, J.E. Stok, B. Hock, and B.D. Hammock, *Juvenile hormone (JH) esterase: Why are you so JH specific?* Insect Biochem. Mol. Biol. 33 (2003), pp. 1261–1273.
6. M. Wogulis, C.E. Wheelock, S.G. Kamita, A.C. Hinton, P.A. Whetstone, B.D. Hammock, and D.K. Wilson, *Structural studies of a potent insect maturation inhibitor bound to the juvenile hormone esterase of* Manduca sexta, Biochemistry 43 (2006), pp. 4045–4057.
7. M. Riba, A. Sans, P. Bau, G. Grolleau, M. Renou, and A. Guerrero, *Pheromone response inhibitors of the corn stalk borer* Sesamia nonagrioides. *Biological evaluation and toxicology*, J. Chem. Ecol. 27 (2001), pp. 1879–1897.
8. G.V.P. Reddy, C. Quero, and A. Guerrero, *Activity of octylthiotrifluoropropan-2-one, a potent esterase inhibitor, on growth, development, and intraspecific communication in* Spodoptera littoralis *and* Sesamia nonagrioides, J. Agric. Food Chem. 50 (2002), pp. 7062–7068.
9. J. Devillers, *Endocrine Disruption Modeling*, CRC Press, Boca Raton, FL, 2009.
10. C. E. Wheelock, G. Shan, and J.A. Ottea, *Overview of carboxylesterases and their role in metabolism of insecticides*, J. Pestic. Sci. 30 (2005), pp. 75–83.
11. M. Hosokawa, *Structure and catalytic properties of carboxylesterase isozymes involved in metabolic activation of prodrugs*, Molecules 13 (2008), pp. 412–431.
12. T.C. Sparks and B.D. Hammock, *Comparative inhibition of the juvenile hormone esterases from* Trichoplusia ni, Tenebrio molitor, *and* Musca domestica, Pestic. Biochem. Physiol. 14 (1980), pp. 290–302.

13. R.J. Linderman, L. Upchurch, M. Lonikar, K. Venkatesh, and R.M. Roe, *Inhibition of insect juvenile hormone esterase by* α,β-unsaturated and α-acetylenic trifluoromethyl ketones, Pestic. Biochem. Physiol. 35 (1989), pp. 291–299.
14. R.J. Linderman, T. Tshering, K. Venkatesh, D.R. Goodlett, W.C. Dauterman, and R.M. Roe, *Organophosphorus inhibitors of insect juvenile hormone esterase*, Pestic. Biochem. Physiol. 39 (1991), pp. 57–73.
15. W. Karcher and J. Devillers, *Practical Applications of Quantitative Structure–Activity Relationships (QSAR) in Environmental Chemistry and Toxicology*, Kluwer Academic Publishers, Dordrecht, the Netherlands, 1990.
16. B.D. Hammock, K.D. Wing, J. McLauglin, V.M. Lovell, and T. Sparks, *Trifluoromethylketones as possible transition state analog inhibitors of juvenile hormone esterase*, Pestic. Biochem. Physiol. 17 (1982), pp. 76–88.
17. A. Székács, B.D. Hammock, Y.A.I. Abdel-Aal, M. Philpott, and G. Matolcsy, *Inhibition of juvenile hormone esterase by transition-state analogs. A tool for enzyme molecular biology*, in *Biotechnology for Crop Protection*, P.A. Hedin, J.J. Menn, and R.M. Hollingworth, eds., ACS Symp. Ser. 379, American Chemical Society, Washington, DC, 1988, pp. 215–227.
18. A. Székács, B. Bordás, G. Matolcsy, and B.D. Hammock, *Quantitative structure–activity relationship study of aromatic trifluoromethyl ketones.* In vitro *inhibitors of insect juvenile hormone esterase*, in *Probing Bioactive Mechanisms*, P.S. Magee, D.R. Henry, and J.H. Block, eds., ACS Symp. Ser. 413, American Chemical Society, Washington, DC, 1989, pp. 169–182.
19. A. Székács, B. Bordás, and B.D. Hammock, *Transition state analog enzyme inhibitors: Structure–activity relationships of trifluoromethyl ketones*, in *Rational Approaches to Structure, Activity, and Ecotoxicology of Agrochemicals*, W. Draber and T. Fujita, eds., ACS Symp. Ser, American Chemical Society, Washington, DC, 1992, pp. 219–249.
20. G. Rosell, S. Herrero, and A. Guerrero, *New trifluoromethyl ketones as potent inhibitors of esterases: 19F NMR spectroscopy of transition state analog complexes and structure–activity relationships*, Biochem. Biophys. Res. Commun. 226 (1996), pp. 287–292.
21. C.E. Wheelock, Y. Nakagawa, M. Akamatsu, and B.D. Hammock, *Use of classical and 3-D QSAR to examine the hydration state of juvenile hormone esterase inhibitors*, Bioorg. Med. Chem. 11 (2003), pp. 5101–5116.
22. D.V. Eldred, C.L. Weikel, P.C. Jurs, and K.L.E. Kaiser, *Prediction of fathead minnow acute toxicity of organic compounds from molecular structure*, Chem. Res. Toxicol. 12 (1999), pp. 670–678.
23. J. Devillers, *Linear versus nonlinear QSAR modeling of the toxicity of phenol derivatives to* Tetrahymena pyriformis, SAR QSAR Environ. Res. 15 (2004), pp. 237–249.
24. X.J. Yao, A. Panaye, J.P. Doucet, R.S. Zhang, H.F. Chen, M.C. Liu, Z.D. Hu, and B.T. Fan, *Comparative study of QSAR/QSPR correlations using support vector machines, radial basis function neural networks, and multiple linear regression*, J. Chem. Inform. Comput. Sci. 44 (2004), pp. 1257–1266.
25. J. Devillers, J.P. Doucet, A. Panaye, N. Marchand-Geneste, and J.M. Porcher, *Structure–activity of a diverse set of androgen receptor ligands*, in *Endocrine Disruption Modeling*, J. Devillers, ed., CRC Press, Boca Raton, FL, 2009, pp. 335–355.
26. J. Devillers, J.P. Doucet, A. Doucet-Panaye, A. Decourtye, and P. Aupinel, *Linear and non-linear QSAR modelling of juvenile hormone esterase inhibitors*, SAR QSAR Environ. Res. 23 (2012), pp. 357–369.
27. B.D. Hammock and T.C. Sparks, *A rapid assay for insect juvenile hormone esterase activity*, Anal. Biochem. 82 (1977), pp. 573–579.
28. J. Devillers, D. Domine, C. Guillon, and W. Karcher, *Simulating lipophilicity of organic molecules with a back-propagation neural network*, J. Pharm. Sci. 87 (1998), pp. 1086–1090.

29. J. Devillers, *Calculation of octanol/water partition coefficients for pesticides. A comparative study*, SAR QSAR Environ. Res. 10 (1999), pp. 249–262.
30. J. Devillers, *Genetic Algorithms in Molecular Modeling*. Academic Press, London, U.K., 1996.
31. J. Devillers, *Strengths and weaknesses of the backpropagation neural network in QSAR and QSPR studies*, in *Neural Networks in QSAR and Drug Design*, J. Devillers, ed., Academic Press, London, U.K., 1996, pp. 1–46.
32. N. Draper and H. Smith, *Applied Regression Analysis*, 2nd ed., Wiley, New York, 1981.
33. J. Devillers, *Application of QSARs in aquatic toxicology*, in *Computational Toxicology. Risk Assessment for Pharmaceutical and Environmental Chemicals*, S. Ekins, ed., Wiley, Hoboken, NJ, 2007, pp. 651–675.
34. H. Abdi, *Partial least squares regression and projection on latent structure regression (PLS Regression)*, WIREs Comput. Stat. 2010, DOI: 10.1002/wics.051.
35. J. Devillers, A. Chezeau, E. Thybaud, and R. Rahmani, *QSAR modeling of the adult and developmental toxicity of glycols, glycol ethers, and xylenes to* Hydra attenuata, SAR QSAR Environ. Res. 13 (2002), pp. 555–566.
36. J. Devillers, A. Chezeau, and E. Thybaud, *PLS-QSAR of the adult and developmental toxicity of chemicals to* Hydra attenuata, SAR QSAR Environ. Res. 13 (2002), pp. 705–712.
37. J.H. Friedman, *Multivariate adaptive regression splines*, Ann. Stat. 19 (1991), pp. 1–67.
38. V. Nguyen-Cong and G. Van Dang, *Using multivariate adaptive regression splines to QSAR studies of dihydroartemisinin derivatives*, Eur. J. Med. Chem. 31 (1996), pp. 797–803.
39. M. Jalali-Heravi and A. Mani-Varnosfaderani, *QSAR modeling of 1-(3,3-diphenylpropyl)-piperidinyl amides as CCR5 modulators using multivariate adaptive regression spline and bayesian regularized genetic neural networks*, QSAR Comb. Sci. 28 (2009), pp. 946–958.
40. L. Breiman, J. Freidman, R. Olshen, and C. Stone, *Classification and Regression Trees*, Wadsworth, Belmont, CA, 1984, p. 358.
41. A.M. Prasad, L.R. Iverson, and A. Liaw, *Newer classification and regression tree techniques: Bagging and random forests for ecological prediction*, Ecosystems 9 (2006), pp. 181–199.
42. S. Izrailev and D. Agrafiotis, *A novel method for building regression tree models for QSAR based on artificial ant colony systems*, J. Chem. Inform. Comput. Sci. 2001, 41, 176–180.
43. M. Atabati, K. Zarei, and E. Abdinasab, *Classification and regression tree analysis for molecular descriptor selection and binding affinities prediction of imidazobenzodiazepines in quantitative structure–activity relationship studies*, Bull. Korean Chem. Soc. 30 (2009), pp. 2717–2722.
44. L. Breiman, *Random forests*, Mach. Learn. 45 (2001), pp. 5–32.
45. V. Svetnik, A. Liaw, C. Tong, J.C. Culberson, R.P. Sheridan, and B.P. Feuston, *Random forests: A classification and regression tool for compound classification and QSAR modeling*, J. Chem. Inform. Comput. Sci. 43 (2003), pp. 1947–1958.
46. V. Svetnik, A. Liaw, C. Tong, and T. Wang, *Application of Breiman's random forest to modeling structure–activity relationships of pharmaceutical molecules*, in *Multiple Classifier Systems, Lecture Notes in Computer Science*, F. Roli, J. Kittler, and T. Windeatt, eds., Springer-Verlag, Berlin, Germany, 3077 (2004), pp. 334–343.
47. H. Hong, W. Tong, Q. Xie, H. Fang, and R. Perkins, *An* in silico *ensemble method for lead discovery: Decision forest*, SAR QSAR Environ. Res. 16 (2005), pp. 339–347.
48. J.H. Friedman, *Stochastic Gradient Boosting*, Stanford University, Stanford, CA, 1999, p. 10. Available at http://www-stat.stanford.edu/~jhf/ftp/stobst.pdf.

49. J. Elith, J. R. Leathwick, and T. Hastie, *A working guide to boosted regression trees*, J. Anim. Ecol. 77 (2008), 802–813.
50. V. Svetnik, T. Wang, C. Tong, A. Liaw, R.P. Sheridan, and Q. Song, *Boosting: An ensemble learning tool for compound classification and QSAR modeling*, J. Chem. Inform. Model. 45 (2005), pp. 786–799.
51. J. Jiao, S.M. Tan, R.M. Luo, and Y.P. Zhou, *A robust boosting regression tree with applications in quantitative structure–activity relationship studies of organic compounds*, J. Chem. Inform. Model. 51 (2011), pp. 816–828.
52. T. Aoyama and H. Ichikawa, *Neural networks as nonlinear structure–activity relationship analyzers. Useful functions of the partial derivative method in multilayer neural networks*, J. Chem. Inf. Comput. Sci. 32 (1992), pp. 492–500.
53. J. Devillers, M.H. Pham-Delègue, A. Decourtye, H. Budzinski, S. Cluzeau, and G. Maurin, *Structure–toxicity modeling of pesticides to honey bees*, SAR QSAR Environ. Res. 13 (2002), 641–648.
54. J. Devillers, *A new strategy for using supervised artificial neural networks in QSAR*, SAR QSAR Environ. Res. 16 (2005), pp. 433–442.
55. J.C. Dearden and M. Hewitt, *QSAR modelling of bioconcentration factor using hydrophobicity, hydrogen bonding and topological descriptors*, SAR QSAR Environ. Res. 21 (2010), pp. 671–680.
56. F. Gharagheizi and M. Sattari, *Estimation of molecular diffusivity of pure chemicals in water: A quantitative structure–property relationship study*, SAR QSAR Environ. Res. 20 (2009), pp. 267–285.
57. A. Panaye, B.T. Fan, J.P. Doucet, X.J. Yao, R.S. Zhang, M.C. Liu, and Z.D. Hu, *Quantitative structure–toxicity relationships (QSTRs): A comparative study of various non linear methods. General regression neural network, radial basis function neural network and support vector machine in predicting toxicity of nitro- and cyano-aromatics to* Tetrahymena pyriformis, SAR QSAR Environ. Res. 17 (2006), pp. 75–91.
58. J.P. Bigus, *Data Mining with Neural Networks*, McGraw-Hill, New York, 1996.
59. V.N. Vapnik, *The Nature of Statistical Learning Theory*, Springer, New York, 1995.
60. N. Cristianini and J. Shawe-Taylor, *An Introduction to Support Vector Machines and other Kernel-based Learning Methods*, Cambridge University Press, Cambridge, U.K., 2000, p. 189.
61. J.P. Doucet, F. Barbault, H.R. Xia, A. Panaye, and B.T. Fan, *Nonlinear SVM approaches to QSPR/QSAR studies and drug design*, Curr. Comput. Aided Drug Des. 3 (2007), pp. 263–289.
62. Y. Wang, M. Zheng, J. Xiao, Y. Lu, F. Wang, J. Lu, X. Luo, W. Zhu, H. Jiang, and K. Chen, *Using support vector regression coupled with the genetic algorithm for predicting acute toxicity to the fathead minnow*, SAR QSAR Environ. Res. 21 (2010), pp. 559–570.
63. R. Darnag, A. Schmitzer, Y. Belmiloud, D. Villemin, A. Jarid, A. Chait, E., Mazouz, and D. Cherqaoui, *Quantitative structure–activity relationship studies of TIBO derivatives using support vector machines*, SAR QSAR Environ. Res. 21 (2010), pp. 231–246.
64. J. Devillers and A.T. Balaban, *Topological Indices and Related Descriptors in QSAR and QSPR*, Gordon and Breach Science Publishers, Amsterdam, the Netherlands, 1999.
65. D.T. Stanton and P.C. Jurs, *Development and use of charged partial surface area structural descriptors in computer-assisted quantitative structure–property relationship studies*, Anal. Chem. 62 (1990), pp. 2323–2329.
66. M. Karelson, *Quantum-chemical descriptors in QSAR*, in *Computational Medicinal Chemistry for Drug Discovery*, P. Bultinck, H. de Winter, W. Langenaeker, and J.P. Tollenaere, eds., Marcel Dekker, New York, 2004.

67. J. Rayo, L. Muñoz, G. Rosell, B.D. Hammock, A. Guerrero, F.J. Luque, and R. Pouplana, *Reactivity versus steric effects in fluorinated ketones as esterase inhibitors: A quantum mechanical and molecular dynamics study*, J. Mol. Model. 16 (2010), pp. 1753–1764.
68. A. Mackert, K. Hartfelder, M.M.G. Bitondi, and Z.L.P. Simões, *The juvenile hormone (JH) epoxide hydrolase gene in the honey bee* (Apis mellifera) *genome encodes a protein which has negligible participation in JH degradation*, J. Insect Physiol. 56 (2010), pp. 1139–1146.

11 Receptor-Guided Structure–Activity Modeling of Inhibitors of Juvenile Hormone Epoxide Hydrolases

Julio Caballero

CONTENTS

ABSTRACT

Juvenile hormone epoxide hydrolase (JHEH) plays a pivotal role in regulating insect juvenile hormone (JH) titer along with JH esterase. Recently, some specific inhibitors have been reported for JHEHs of several pest-causing insects based on their biochemical properties. These JHEH inhibitors are considered as potential insecticides. The structural features and structure–activity relationship of JHEH inhibitors have been investigated using quantitative structure–activity relationship (QSAR) methods. In this chapter, we inform readers about the availability of molecular structures of JHEHs constructed from the crystal structure of the epoxide hydrolase (EH) of the yeast *Aspergillus niger.* This protein structure has been employed as a template to derive structural models for JHEHs using comparative homology modeling. To obtain these models, three-dimensional coordinates of the template structure and the annotated sequences of JHEHs are used. The possibility of obtaining structural models is the basis for accomplishing receptor–guided structure–activity modeling, by docking of inhibitors within JHEH active site, molecular dynamics to get conformational sampling of the complex, and evaluation of binding energies. This consistent methodology could contribute to the future development of new potent JHEH inhibitors. We tested this method in the study of the difference between four urea-like compounds as JHEH inhibitors.

KEYWORDS

Juvenile hormone epoxide hydrolase inhibitors, Receptor–guided structure–activity modeling, Molecular dynamics, MM-GBSA, Comparative or homology modeling

11.1 INTRODUCTION

Juvenile hormones (JHs) are a group of acyclic sesquiterpenoids (Figure 11.1) that regulate many aspects of insect physiology, including development of larval insects,

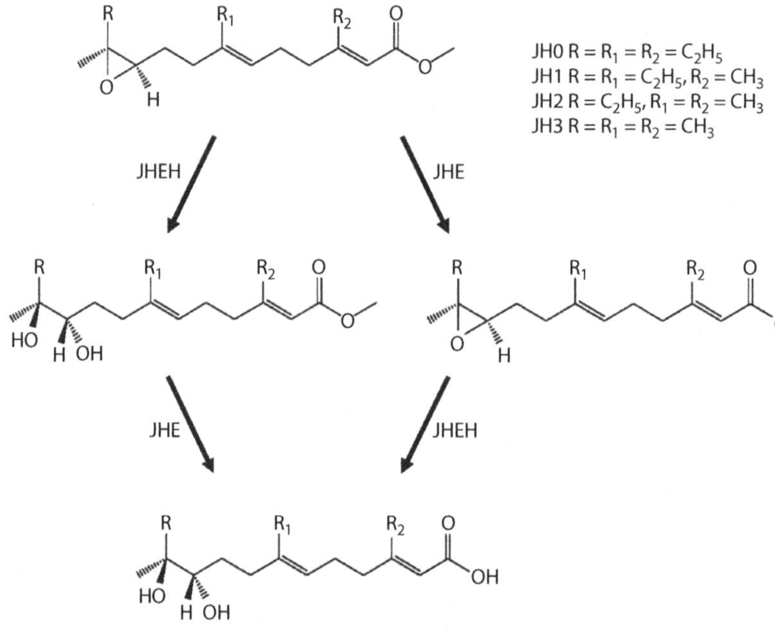

FIGURE 11.1 Structures of JHs and their hydrolysis to JH acid, JH diol, and JH acid–diol.

reproduction of adult insects, diapause, and polyphenisms. They are generated by the endocrine gland corpus allatum and play a crucial role in the metamorphosis [1]. The primary routes of JH catabolism in insects are ester hydrolysis to JH acid in the presence of JH esterase (JHE, EC 3.1.1.1) and epoxide hydration to JH diol in the presence of JH epoxide hydrolase (JHEH, EC 3.3.2.3) [2]. The primary metabolites JH acid and JH diol are also metabolized to JH acid–diol (Figure 11.1); these three compounds are physiologically inactive [3]. The previously mentioned catabolic routes strictly regulate JH concentration. The level of JHs correlates with specific developmental events, such as molting and metamorphosis [4].

Most of the research on JH metabolism in many insects has focused on the mechanism of action of JHE and JHEH. JHE is secreted to the hemolymph, while JHEH is a non-secreted enzyme that works in certain organs and tissues. In this sense, the characteristics of both enzymes in degradation of JHs are quite different. JHEH seems to have a more important role because it acts in certain organs. In addition, catalysis mediated by JHEH is irreversible, while catalysis mediated by JHE is reversible [5].

Since JHs are insect specific and the regulation of their concentration is a key factor for the orderly development of insect species, the enzymes implicated in JH regulation have been identified as potential targets in pest management [6]. Compounds that stimulate or inhibit JH degradation modify JH titer and thereby disturb the physiology of an insect. Chemicals that mimic the action of hormones involved in insect growth, development, and metamorphosis, the ecdysteroids and JHs, can be used as insect control agents, with more selective modes of action and reduced risks for nontarget organisms and the environment [7]. These chemicals have been called insect growth regulators (IGRs) or third-generation insecticides [8]. The design of new IGRs is mandatory, because some insect species representing Diptera, Coleoptera, Homoptera, and Lepidoptera show cross-resistance to IGRs [9].

The design of epoxide hydrolase (EH) inhibitors has been developed mainly to obtain compounds with activity against mammalian soluble and microsomal EHs [10–12]. In general, mammalian EH inhibitors are not as effective against insect EHs. The inhibitors of insect EHs exhibit high variation in activity depending on the insect and substrate used for the assay. Some specific inhibitors of insect JHEHs have been reported. For instance, several analogs of JHs have been identified as JHEH inhibitors [13,14]. In addition, analogs of the compound methyl 10,11-epoxy-11-methyldodecanoate (MEMD) as *Trichoplusia ni* JHEH inhibitors have been identified [15]. In other report, a group of urea- and amide-based inhibitors of the JHEH of *Manduca sexta* were reported [16].

Despite the importance of JHEH inhibitors in the development of possible new insecticides, no crystal structures of JHEHs have been released yet. However, the crystal structure of the EH of the yeast *Aspergillus niger* was released 10 years ago [17], which has ~26% sequence identity with most of JHEHs. In this sense, homology models of JHEHs can be derived and these models can be used to study the complexes between these enzymes and their inhibitors. In this chapter, we orient how structural bioinformatics methods can support the current structural information to derive new useful information for the design of new JHEH inhibitors.

11.2 QSAR METHODS APPLIED TO THE STUDY OF STRUCTURE–ACTIVITY RELATIONSHIP OF JHEH INHIBITORS

Quantitative structure–activity relationship (QSAR) is the process by which chemical structures are quantitatively correlated with biological activities. QSAR methods have contributed in different areas, including medicinal chemistry and prediction of toxicity [18,19]. They represent predictive models derived from application of statistical tools correlating activity or action of chemicals with descriptors representative of molecular structures or physicochemical properties. QSAR offers the advantages of higher speed and lower costs, especially when compared to experimental testing.

Very few computational studies have been reported for describing the structure–activity relationship of EH inhibitors. The vast majority of these reports were developed to investigate inhibitors of mammalian EHs [20–22]. To our knowledge, the only study till date, which applied QSAR methodology to JHEH inhibitors, was reported by our group last year [23]. In that work, the structure–activity relationship of 47 urea-like compounds as *M. sexta* JHEH inhibitors was reported. First, we made docking experiments to obtain 3D orientations of compounds within the active site of *M. sexta* JHEH, and then we applied comparative molecular field analysis (CoMFA) [24] to derive quantitative relationships between the structures of the studied compounds and their JHEH inhibitory activities. The theoretical model of *M. sexta* JHEH was obtained from ModBase (database of comparative protein structure models) [25]. CoMFA models with adequate statistics and predictive abilities were obtained. Other interesting results were accounted: Inhibitor-binding orientations were similar to that observed for the binding of dialkylurea inhibitors in complexes with murine and human soluble EHs solved by x-ray crystallography [26,27], with carbonyl urea close to hydroxyl of Tyr298 and Tyr372 and urea's nitrogen atoms close to side chain of Asp227 within the active site, while the hydrophobic chains were placed inside pockets of the "L"-shaped hydrophobic tunnel. We also observed that small inactive compounds did not establish these interactions.

11.3 STRUCTURES OF JHEHS: THE CURRENT STATE OF THE ART

11.3.1 Availability of Structures of JHEHS: Comparative Models

The knowledge of protein 3D structures provides valuable insights into the protein function at a molecular level, allowing a proper design of experiments such as the structure-based design of specific inhibitors. The known protein 3D structures are stored in the Protein Database (PDB) [28], which is a freely accessible repository for the 3D structural data of large biological molecules, including proteins and nucleic acids, typically obtained by x-ray crystallography or NMR spectroscopy, and submitted by biologists and biochemists from around the world. The methods to determine experimental structures are time-consuming without guaranteed success; therefore, the number of 3D structurally characterized proteins is low compared to the number of known protein sequences. The known protein sequences are stored in UniProtKB/Swiss-Prot [29], which is a manually annotated protein sequence database that contains information extracted mainly from scientific literature. Currently, about 70,000 protein structures are stored in PDB (most of these correspond to the same protein crystallized in different studies, e.g., cocrystallized with ligands, forming dimers), and UniProtKB/Swiss-Prot contains about 500,000 sequence entries. Clearly, no structural information is available for most of protein sequences.

The computational methods for protein structure prediction have gained much interest in the last decades in order to obtain the structure of proteins that have not yet been crystallized. Among all current theoretical approaches, comparative homology modeling (CHM) is the sole method that can reliably generate a 3D model of a target protein from its amino acid (AA) sequence [30]. CHM complements experimental structure elucidation in the exploration of the protein structure space. The main requirement for model building using CHM is at least one experimentally solved 3D structure (template) that has a significant AA sequence similarity to the target protein sequence; for this reason, CHM is only applicable when a template is detected. CHM is carried out in four sequential steps: identifying the template (or templates) related to the target protein sequence, aligning the target sequence with the templates, building models, and assessing the models.

As more experimental structures become available, more reliable models become accessible to the researchers. Web resources that assist in analyzing protein structures and structural models and evaluating their reliability are also accessible. There are also databases of comparative protein structure models such as ModBase [25] and SWISS-MODEL portal [31].

In the last decade, genome sequencing projects revealed different numbers of JHEH genes in diverse species. Three genes have been reported in *Drosophila melanogaster* and *Aedes aegypti*, two genes in *Ctenocephalides felis* [32], and five genes in *Tribolium castaneum* [33], and only one gene has been reported in lepidopteran insects *T. ni* [34] and *M. sexta* [16]. Very recently, several JHEHs were reported for the Lepidoptera *Bombyx mori* by using the silkworm genome database KAIKOBLAST [35].

The crystal structure of the EH of *A. niger* can be used as a template for CHM of JHEHs. In fact, several JHEH structures of the previously mentioned organisms are accessible for researchers in ModBase and SWISS-MODEL portal. The website and UniProt IDs of the sequences of these JHEHs are listed in Table 11.1. As can be seen, the models in both databases were constructed using the crystal structures of EH of *A. niger* with the PDB codes 3g02 or 1qo7. JHEH sequences have between 23% and 29% sequence identity with *A. niger* EH. With such reduced percentages, the interpretation of these models has to be made with care. Other sequences available in UniProt database listed in Table 11.2 can be used to derive homology models of these proteins.

The alignments of these sequences using the structure of the EH of *A. niger* assume that the two proteins are basically similar over the entire length of one another; however, researchers should consider that the sequences of JHEHs and the EH of *A. niger* differ in several parts. The degree of success to be expected in predicting the structure of a target protein from its sequence using the known 3D structure of a

TABLE 11.1
Available Structural Models of JHEHs Constructed Using CHM

Organism	Website	ID UniProt	Template	JHEH Classification (If Exists)
Ae. aegypti	ModBase	Q8MMJ6	1qo7	JHEH
Ae. aegypti	ModBase	Q86HE6	1qo7	JHEH 2
B. mori	ModBase, SWISS-MODEL	Q6U6J0	3g02/1qo7	JHEH
C. felis	ModBase	Q8MZR6	1qo7	JHEH 1
C. felis	ModBase/SWISS-MODEL	Q8MZR5	1qo7/1qo7	JHEH 2
Culex quinquefasciatus	SWISS-MODEL	B0W1S8	1qo7	JHEH 1
D. melanogaster	ModBase, SWISS-MODEL	Q7JRC3	1qo7/3g02	JHEH 1
D. melanogaster	ModBase, SWISS-MODEL	Q7KB18	1qo7/3g02	JHEH 2, isoform A
D. melanogaster	ModBase	A1ZBF3	1qo7	JHEH 2, isoform C
D. melanogaster	ModBase/SWISS-MODEL	Q7K1W4	1qo7/3g02	JHEH 3
D. melanogaster	ModBase, SWISS-MODEL	Q8MMJ5	1qo7/1qo7	JHEH 3
M. sexta	ModBase	Q25489	1qo7	JHEH

TABLE 11.2
Sequences Available in UniProt Database That Can Be Used to Derive Homology Models

Organism	ID UniProt	JHEH Classification (If Exists)
Ae. aegypti	Q16QD6	JHEH 2
Pediculus humanus subsp. *corporis*	E0VZD2	JHEH, putative
Athalia rosae	Q2Z1T2	JHEH
Helicoverpa armigera	C0KH33	JHEH
Cu. quinquefasciatus	B0W1T0	JHEH 1
Cu. quinquefasciatus	B0W1S9	JHEH 1
Glossina morsitans morsitans	D3TQ58	JHEH
Spodoptera exigua	Q1W696	JHEH
Harpegnathos saltator	E2BM47	JHEH 2
Camponotus floridanus	E2A3G1	JHEH 1
H. saltator	E2BM46	JHEH 1
Acromyrmex echinatior	F4WRP1	JHEH 2
Ac. echinatior	F4WV40	JHEH 1
Ca. floridanus	E2A149	JHEH 2
Melipona scutellaris	B2ZGJ2	JHEH
Ca. floridanus	E2A150	JHEH 2

template depends of the sequence identity [36]. A template provides a close general model for other proteins with which its sequence homology is >50%, but sequences falling below a 20% sequence identity can have very different structures and are impossible to predict. However, the active sites of distantly related proteins can have very similar geometries because of the coupling of the structural changes that has occurred during evolution [37]. Thus, naturally occurring homologous proteins have similar protein structure, and the structure of the active site in a template protein may provide a good model for those in related target proteins even if the overall sequence homologies are low. Model quality declines with decreasing sequence identity; the errors are significantly higher in the loop regions, where the AA sequences of the target and template proteins may be completely different. Errors in side-chain packing and position also increase with decreasing identity [38]. Taken together, these various atomic-position errors are significant and make the use of homology models difficult for purposes that require atomic-resolution data, such as drug design and protein–protein interaction predictions.

Considering these aspects, researchers should be cautious since the JHEH models can contain inaccuracies in the alignment of these different parts. The global alignment of the structures attempts to match them to each other from end to end, even though parts of the alignment are not very convincing. There are segments of the two sequences that have good similarity and match well in the alignment, but the reliability of the parts with marked differences is low. The only way to get more reliable 3D structures of the JHEHs is to obtain a closer template (for instance, the crystal structure of one JHEH of any insect species). Researchers interested in the design of JHEH inhibitors should focus on solving 3D structures of JHEHs; however, they can work today with the available models, considering their limitations.

11.3.2 Specificities of JHEH Binding Sites

The alignment of the 3D structural models of JHEHs and their comparison with the template (EH of *A. niger*) allows the analysis of the specificities of the binding sites of JHEHs. In general, EHs contain a hydrophobic tunnel with several hydrophobic pockets (denoted as pockets P-I, P-II, P-III, and P-IV in this manuscript) surrounding a hydrophilic central pocket [23]. Figure 11.2 shows the alignment and a representation of the binding site for seven sequences of JHEHs. The conserved AAs in all JHEH structures are represented with red letters; the AAs that are also conserved in EH of *A. niger* are marked with an asterisk. Non-conserved AAs are represented with black letters and AAs that occupy the same position in other JHEHs are indicated. Finally, AAs that are located in sequence motifs where the alignment is unreliable are indicated with a question mark.

The central pocket contains the conserved aspartate and tyrosine residues (Asp227, Tyr298, and Tyr372 in *M. sexta* JHEH) that can establish HB interactions with ligands. Asp227 interacts with the partial positive charge on the electrophilic carbon of an epoxide substrate, acting as catalytic nucleophile, and can interact with the partial positive charge on donor groups of inhibitors. The phenolic hydroxyl groups of Tyr298 and Tyr372 donate HBs to acceptor groups of substrates and

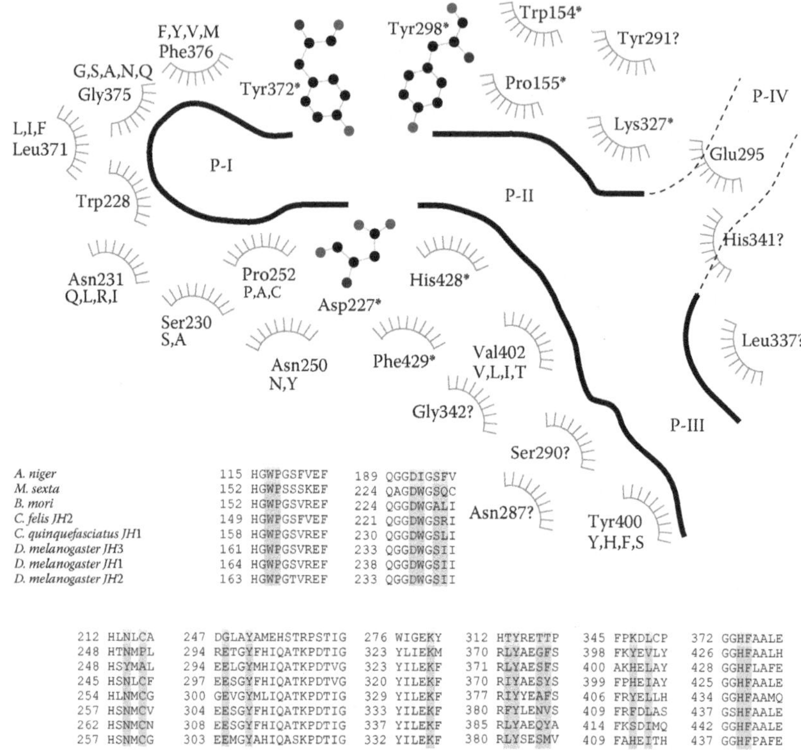

FIGURE 11.2 **(See color insert.)** (Top) Binding site of JHEHs; a number of residues correspond to *M. sexta* JHEH, pockets P-I-P-IV are labeled, and the meaning of colored letters, asterisks, and question marks is explained in the text. (Bottom) Alignment of relevant motifs between several JHEHs and the *A. niger* EH; residues in the top part of the figure are highlighted in cyan in the sequences.

inhibitors; in this context, they are general acid catalysts that facilitate epoxide ring opening in the first step of the hydrolysis reaction.

The pocket P-I is delimited by the previously mentioned residues Asp227 and Tyr372 and the residues Trp228, Ser230, Asn231, Asn250, Pro252, Leu371, Gly375, and Phe376. The residue Trp228 is conserved in all JHEHs (motif GD<u>W</u>G) and soluble EHs [26]; however, there is an isoleucine at this position in *A. niger* EH (Figure 11.2). Seino et al. [35] found that the conserved tryptophan (Trp228) is important for substrate selectivity; these authors observed that Trp228 is not found in *B. mori* JHEHr3 and r4, which do not have JHEH activity. The size of the pocket P-I in different species is larger or smaller depending on the non-conserved AAs. In addition, the more hydrophobicity/hydrophilicity of the pocket P-I also depends on the non-conserved AAs.

The pocket P-II is delimited by the previously mentioned residue Tyr298; the conserved residues Trp154, Pro155, His428, Phe429, and Lys327; and the non-conserved residues Val402, Tyr291, and Gly342. The residues Trp154 and Pro155 (motif HG<u>WP</u> in JHEHs) are involved in the formation of what is known as the oxyanion hole.

The oxyanion hole stabilizes the negative charge that evolves on the nucleophilic aspartate during hydrolysis of the covalent enzyme–substrate intermediate [39]. His428 is one of the residues of the catalytic triad (Asp227-His428-Glu401, numbers from *M. sexta* JHEH) [40]. The basicity of His428 is enhanced by the partial negative charge of the phenyl group of Phe429 (a tryptophan in soluble EHs) establishing a cation–π interaction and the negatively charged carboxylate side chain of Glu401. The conserved residue Lys327, part of the motif YXXEKX in JHEHs (Figure 11.2), is located at the entrance of the pocket P-II. The side chain of this residue faces the phenyl group of Tyr298. The role of this conserved residue in JHEHs should be investigated in the future. The non-conserved residue Val402 is changed by other hydrophobic residues in other JHEHs. Finally, the residues Tyr291 and Gly342 from *M. sexta* JHEH are located at segments that do not match well in the alignment with the template and alignments between JHEHs; therefore, their position in the CHM model is not trustworthy (the same for residues His341, Leu337, Asn287, and Ser290 that delimit the pocket P-III according to our model).

The pockets P-III and P-IV are at the entrance of the tunnel. The pocket P-IV is delimited by the residues Lys327, Glu295, and His341. The residue Glu295, part of the motif ETGYXXIQA that contains the residue Tyr298, is conserved in JHEHs (Figure 11.2). The role of this AA should be investigated in the future. According to our model, the pocket P-III is delimited by the non-conserved residues Tyr400, Asn287, Ser290, and Leu337. These AAs are different for each JHEH.

11.4 RECEPTOR-GUIDED MODELING OF JHEH INHIBITORS

11.4.1 Molecular Dynamics for Receptor-Guided Modeling

The availability of JHEH structural models allows the use of structural bioinformatics methods to investigate the binding of inhibitors to JHEHs. The orientation of the ligands within the binding site can be predicted by docking experiments; however, the result of docking is a frozen structure and does not include the solvent media. A more realistic picture of a protein–inhibitor complex should include an analysis of the movement of the system inside a water box. Molecular dynamics (MD) simulations allow to study the dynamics of ligand–enzyme complexes with detail on scales where motion of individual atoms can be tracked (traditional MDs can only simulate events efficiently up to pico- or nanoseconds) [41]. When applying an MD simulation, the pose obtained by docking is perturbed and the system can access to new ligand conformations that can be originated when new favorable interactions with the receptor are established.

MD is a method that works with a classic description of the electrostatic interactions (Coulomb interactions between point-like atomic charges) rather than a quantum description of the electrons and uses empirical parameterizations of the molecular structures. MD is adequate to simulate biomolecular systems because these systems involve many particles. To derive macroscopic properties from MD simulations, the individual structures have to represent an ensemble of structures at the given experimental conditions, e.g., constant temperature and pressure, or constant temperature and volume.

The first step of an MD simulation is the energy minimization of the system. When the starting configuration (e.g., the docking solution) is very far from equilibrium, large forces can cause the simulation to crash the system. The steepest descent algorithm [42] is usually employed for energy minimization; this method moves each atom a short distance in the direction of decreasing energy. MD simulations evaluate the potential energy of the studied system as a function of point-like atomic coordinates and compute the motion of every single atom within the simulation system, which is determined by the interaction forces between all the atoms. Typically, these forces are described by a force field. Several force fields for proteins, DNA, and RNA have been developed, such as CHARMM [43], AMBER [44], OPLS [45], and GROMOS [46] that are widely used. A force field consists of both the set of equations used to calculate the potential energy and forces from particle coordinates and a collection of parameters used in the equations. Figure 11.3 shows interaction terms from which force fields are typically composed. They describe bonded interactions (covalent bond-stretching, angle-bending, torsion potentials when rotating around bonds and out-of-plane improper torsion potentials) and nonbonded interactions (Pauli repulsion and van der Waals [VDW] interactions described by a Lennard–Jones potential and electrostatic interactions described by a Coulomb potential).

The force fields serve to calculate the trajectory of the system via the classical Newtonian equations of motion. Given the potential and force (negative gradient of potential) for all atoms, the coordinates are updated for the next step; then, the updated coordinates are used to evaluate the potential energy in the new step.

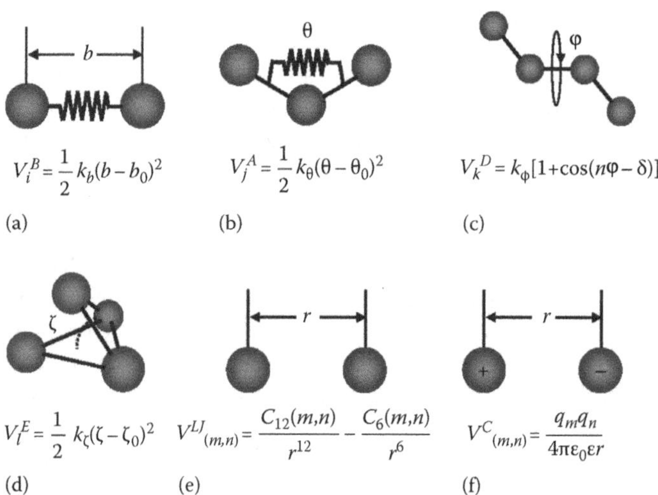

FIGURE 11.3 Interaction functions in force fields. Bonded interactions include (a) covalent bond-stretching, (b) bond angle-bending, (c) torsion rotation around bonds (dihedral angles), and (d) out-of-plane or "improper" torsions (extraplanar angles). Nonbonded interactions consist of (e) Lennard-Jones attraction and repulsion (VDW interactions) as well as (f) Coulomb electrostatic interactions.

There are some important details of MD simulations that we briefly mention here. The treatment of the system boundaries, or their periodic continuation, is an important issue. The use of periodic boundary conditions avoids surface artifacts, so that the water molecules or ions that exit to one side reappear on the opposite side. The system box must be sufficiently large to avoid the interactions between the molecules and their periodic copies. Typically, periodic boundary conditions are used in conjunction with particle-mesh Ewald method [47] to account for nonbonded interactions in the system. Other important issues are the freezing of fast bond vibrations [48], the treatment of protonable residues, the proper placement of salt ions in the vicinity of the biomacromolecule, the definition of the canonical thermodynamic ensembles via appropriate coupling to heat and pressure baths [49–51], the description of electronic polarizability, etc. [52].

In the last decades, MD simulations have emerged as an indispensable method in computational chemistry of macromolecules due to the exponential increase of available computing resources. For MD simulations of ligand–enzyme complexes, it is necessary to obtain suitable starting molecular structures (docking results are adequate starting points), immerse them in a solvent box, choose good simulation parameters, and accomplish the energy minimization of the system. Then, the production MD can be achieved. The analysis of the results must be carried out using only the trajectory obtained in the last step. The trajectory obtained from an MD simulation contains the positions and velocities of the simulated particles at discrete time steps.

All the commonly used force fields include the topologies and parameters of the most important biomacromolecules: proteins, DNA, RNA, and lipids. The topologies and parameters of small organic molecules (information concerning bonds, bond angles, and charges of the ligands) are not available, and their construction used to be rather laborious. Some programs such as PRODRG [53], Antechamber [54], SwissParam [55], ParamChem [56], and PrimeX [57] have been released to generate the topologies and parameters of many small organic molecules. These programs automatically construct the topologies and parameters of the ligands from their 3D coordinates; therefore, they facilitate the study of the binding interactions between drugs and biomacromolecules.

The ligand binds to the receptor active site by establishing mainly VDW interactions and hydrogen bonds (HBs). The analysis of the MD trajectory allows in determining the stability and propensity of these interactions and the conformational landscape of the complex. The analysis of the simulations can also give potential energies, structural fluctuations, geometrical features, coordinate stability, and the interactions with the solvent. MD simulations provide a way to explain experimental observations and permit to make predictions for the development of new experiments.

It is important to take into account the limitations of the MD method for an adequate application. MD simulations give valuable information about the movement of the molecular system under study, but it cannot reproduce quantum effects such as bond formation or breaking and electron-derived properties. The theoretical framework of MD lacks of quantum effects because it is based on statistical mechanics. In addition, MD simulations have a limited timescales accessible.

11.4.2 Protein–Ligand-Binding Free Energy Estimation

The binding of drug molecules to their protein targets is determined by the affinity and the stability of associating two molecules together. In rational bioactive molecular design, it is desired an accurate estimate for the binding energy to justify the difference in affinity between compounds. Scoring functions implemented within docking programs are not enough accurate to evaluate binding energies. The most accurate current techniques used for the calculation of binding free energies are MD-derived methods [58–60]. MD samples the conformational space of the complete system; such sampling is necessary to obtain a correct distributions of the conformers and, consequently, to obtain a more realistic protein–ligand-binding free energy value.

To determine the free energy of binding of a ligand to the receptor, the MD-derived methods free energy perturbation (FEP) and thermodynamic integration (TI) are among the most rigorous methods currently available. FEP and TI methods can be used to predict the relative binding strength of different complexes. The difference in binding free energy between two given ligands L_1 and L_2 and the receptor R can be computed using the thermodynamic cycle in Figure 11.4. Instead of calculating the individual binding energies $\Delta G_{bind}(L_1)$ and $\Delta G_{bind}(L_2)$ to determine the relative binding free energy ($\Delta\Delta G_{bind}$), the energies of the alchemical transformations $L_1 \rightarrow L_2$ in solution (ΔG^{w}_{mut}) and when the ligands bound to the protein (ΔG^{p}_{mut}) are estimated using the following equation:

$$\Delta\Delta G_{bind} = \Delta G_{bind}\left(L_2\right) - \Delta G_{bind}\left(L_1\right) = \Delta G^{p}_{mut} - \Delta G^{w}_{mut} \tag{11.1}$$

To accomplish the alchemical transformation, the ligands L_1 and L_2 are linearly combined using the coupling parameter λ, and MD simulations are used to slowly transform one ligand (L_1, $\lambda = 0$) into the other (L_2, $\lambda = 1$) in both the free- and receptor-bound forms. Relative free energies are obtained, considering that the free energy is a state function that can be calculated by any reversible path between the initial and final states.

The main limitations of these methods are that they require an exhaustive conformational sampling to obtain a proper averaged ensemble, they have slow convergences, and they are computationally expensive [60]. Despite providing very accurate free energies, TI and FEP are not widely applied. They are inefficient in configurational sampling because the appearance and disappearance of atoms restrict their use only to small transformations, which limits the analysis to a few closely related compounds.

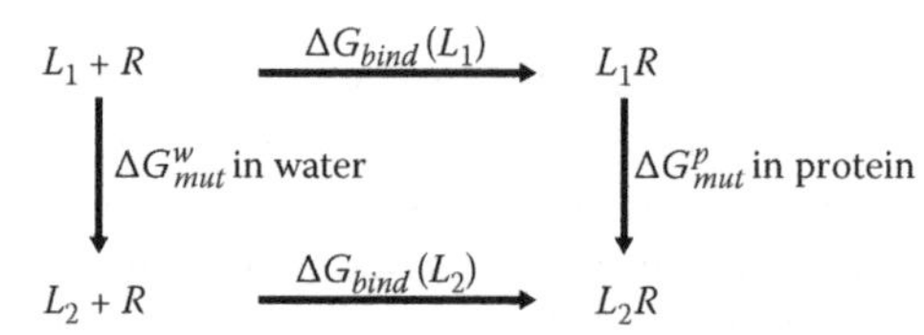

$$\begin{array}{ccc} L_1 + R & \xrightarrow{\Delta G_{bind}(L_1)} & L_1R \\ \downarrow \Delta G^{w}_{mut} \text{ in water} & & \downarrow \Delta G^{p}_{mut} \text{ in protein} \\ L_2 + R & \xrightarrow{\Delta G_{bind}(L_2)} & L_2R \end{array}$$

FIGURE 11.4 Thermodynamic cycle used to compute differences in free energy of binding between related inhibitors.

The approaches linear interaction energy (LIE) [61], molecular mechanic–Poisson–Boltzmann surface area (MM-PBSA) [62], and molecular mechanic–generalized Born surface area (MM-GBSA) [59] provide relatively good energy values at a moderate cost. The semiempirical MD approach LIE [61] estimates the binding energy ΔG_{bind} considering electrostatic and VDW components, assuming that ΔG_{bind} can be extracted from simulations of the free and bound states of the ligand. The ΔG_{bind} value is calculated using the following equation:

$$\Delta G_{bind} = \alpha \left\langle V_{bound}^{elect} - V_{free}^{elect} \right\rangle + \beta \left\langle V_{bound}^{VDW} - V_{free}^{VDW} \right\rangle + \gamma \tag{11.2}$$

where the first term represents the averaged change in electrostatic energy and the second term represents the averaged change in VDW energy in going from the aqueous solution to the receptor binding site, while α, β, and γ are empirically determined constants. The energies are calculated achieving an MD simulation of the ligand bound to the protein and another of the free ligand in water. The main shortcoming of the LIE method is the use of empirically derived constants that may need to be modified for each particular system. This requirement restricts the broad application of the LIE method. On the other hand, MM-PBSA and MM-GBSA methods combine molecular mechanics (MM) and continuum solvent approaches to estimate binding energies [63]. A thermally average ensemble of structures is required that can be obtained using MD simulations in explicit solvent. MD snapshots, after removing all water and counterion molecules, can be used to derive the total binding free energy of the system. The ΔG_{bind} value is calculated using the following equation:

$$\Delta G_{bind} = \Delta E_{MM} + \Delta G_{solv} - T\Delta S \tag{11.3}$$

where ΔE_{MM} is the change of the gas-phase *MM* energy upon binding and includes $\Delta E_{internal}$ (bond, angle, and dihedral energies), ΔE_{elect} (electrostatic), and ΔE_{VDW} (VDW) energies. ΔG_{solv} is the change of the solvation free energy upon binding and includes the electrostatic solvation free energy ΔG_{solvPB} or ΔG_{solvGB} (polar contribution calculated using Poisson–Boltzmann or generalized Born model) and the nonelectrostatic solvation component ΔG_{solvSA} (nonpolar contribution estimated by solvent accessible surface area). Finally, $T\,\Delta S$ is the change of the conformational entropy upon binding. The MM-PBSA and MM-GBSA methods produce accurate free energies at a moderate computational cost without using adjustable parameters. A single MD simulation for the complete system can be used to determine all energy values. The main drawback of the methods is that the changes in internal energy of the ligand and receptor upon complex formation are neglected, which would produce significant errors in flexible systems where the induced-fit effect is important.

The free energy of binding can also be described for a reaction coordinate including the ligand inside and outside the protein active site. Thereby, the progress of the process, when the ligand is physically separated from the protein receptor, can be investigated using the potential of mean force (PMF) concept [64]. PMF is essentially the free energy profile along the defined reaction coordinate and is determined through the Boltzmann-weighted average over all degrees of freedom other than the reaction coordinate. PMF captures the energetics of the studied process. The motion along the reaction coordinate is well approximated as a diffusive motion with all the other degrees of freedom averaged out. Various methods have been proposed for calculating PMFs such as umbrella sampling [65], steered molecular dynamics (SMD) [66], metadynamics [67,68], and adaptive biasing force (ABF) [69]. The computation of the PMF for a binding event between proteins and ligands is an expensive approach, but it is useful to construct enzyme-inhibitor binding processes [70–73].

11.4.3 Differences in the JHEH Inhibitory Activities between Urea-Like Compounds Using MD and MM-GBSA Calculations

MM-GBSA and MM-PBSA methods are the most widely used free energy calculation methods to estimate inhibitor affinities due to the encouraging results obtained with these methodologies [74–76]. When compared to docking scoring functions, the MM-GBSA and MM-PBSA procedures provide improved enrichment in the description of the binding of small molecules and better correlation between calculated binding affinities and experimental data.

Urea-like JHEH inhibitors represent potential new insecticides that can be used in agricultural applications [16]. Recently, we reported 3D models of 47 urea-like derivatives in complex with JHEH of *M. sexta* by using docking method [23]. To understand the inhibitor–JHEH interactions in detail, we studied the movement and binding energy of four complexes including the compounds in Table 11.3 by using MD and MM-GBSA calculations. These compounds show a clear dependency between the size of the R substituent and the IC_{50} value. IC_{50} values represent the compound μM concentrations that inhibit the JHEH activity by 50% in *trans*-diphenylpropene oxide (*t*-DPPO) assays [16].

TABLE 11.3
Structure of Urea-Like Inhibitors of *M. sexta* JHEH

O
N H N H R

Compound	R	IC_{50} (μM)
1	n-$C_{16}H_{33}$	0.09
2	n-$C_{12}H_{25}$	0.18
3	n-$C_{9}H_{19}$	0.78
4	n-$C_{8}H_{17}$	1.56

The MD of the complexes was studied using the CHARMM force field [77] in an explicit solvent with the TIP3 water model [78] within the NAMD software [79]. The initial coordinates for the MD calculations were taken from the docking experiments achieved in Garriga and Caballero [23]. The water molecules were then added (the dimensions of each orthorhombic water box were approximately 96 Å×87 Å×81 Å, which ensured that the entire surface of each complex was covered by the solvent model), and the systems were neutralized by adding Cl^- counterions to balance the net charges of the systems. Before equilibration and long production MD simulations, the systems were minimized. Then, a first 1 ns equilibration MD simulation was performed on each complex system, and each system was followed for a 5 ns long production MD simulation. The CGenFF force field [56] was used along with the ParamChem website (https://www.paramchem.org/preview.php) to provide and check the necessary force field parameters for the ligands. During the MD simulations, the equations of motion were integrated with a 1 fs time step in the NPT ensemble. The SHAKE algorithm was applied to all hydrogen atoms; the VDW cutoff was set to 9 Å. The temperature was maintained at 300 K, employing the Nose–Hoover thermostat method with a relaxation time of 1 ps. Long-range electrostatic forces were taken into account by means of the particle-mesh Ewald (PME) approach. Data were collected every 1 ps during the MD runs. Visualization of protein–ligand complexes and MD trajectory analysis was carried out with the VMD software package [80].

The convergences of the calculations were tested by analyzing the root-mean-square deviation (RMSD) values of the positions for all backbone atoms of the protein from their initial configuration as a function of simulation time. The RMSD values remained within 0.56 Å for all systems; this demonstrates the conformational stabilities of the protein structures. Along the simulations, the studied compounds were in the expected orientations. In all MD simulations, the studied compounds adopted the same binding mode (Figure 11.5). They placed the hydrophobic chains inside pockets P-I and P-III, establishing VDW interactions with the residues in these pockets. In addition, the carbonyl oxygen of urea established HB interactions with the hydroxyls from side-chain groups of Tyr298 and Tyr372, and urea's nitrogen atoms established HB interactions with the side-chain carboxylate of Asp227. To trace the motion and the dynamics for HB interactions, the interatomic distances describing them were monitored during the four developed MD simulations. Figure 11.5 shows the distance between the carbon of urea and the carboxylic carbon from side-chain group of Asp227 (D_{CD227}) and the distances between the carbonyl oxygen of urea and the oxygen atoms from the hydroxyl groups of Tyr298 and Tyr372 (D_{OY298} and D_{OY372}) for all of the complexes. D_{CD227} takes values around 5 Å when HB is formed between one of the oxygen atoms from side-chain carboxylate group of Asp227 and one of the NH groups of the urea group from the inhibitor. We observed that this HB interaction is stable for all the complexes. D_{OY298} and D_{OY372} take values around 3 Å when HBs are formed between the inhibitor and residues Tyr298 and Tyr372. From the analysis of these distances, we observed that the studied JHEH inhibitors can establish HB interactions with Tyr298 or Tyr372 or can establish HB interactions with both residues at the same time. We also

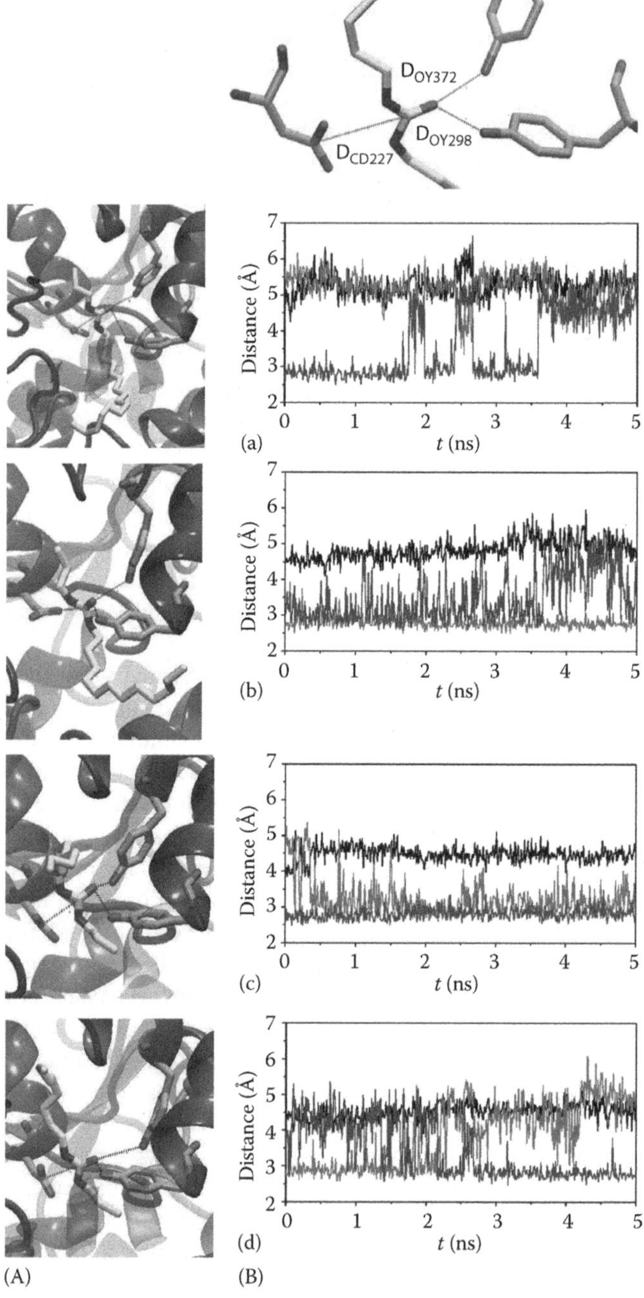

FIGURE 11.5 **(See color insert.)** MD of complexes containing compounds **1–4** (a–d). (A) Orientation of the inhibitor (yellow) within JHEH's active site; Asp227, Tyr298, and Tyr372 are in cyan and stick representation. (B) Distances between groups of inhibitor and residues of JHEH for the complexes extracted from MD simulations. D_{CD227}, D_{OY298}, and D_{CD372} are represented in black, red, and blue, respectively.

observed that the most active compound **1** had conformations where the compound established an HB interaction with Asp227 and did not establish HB interactions with Tyr298 or Tyr372. This compound, however, has a large hydrophobic chain that establishes VDW interactions with the major part of the hydrophobic tunnel of the protein.

The binding free energy (ΔG_{bind}) of each ligand was estimated using Prime MM-GBSA method (Prime, version 2.1, Schrodinger, LLC, New York, 2009). The calculations were performed on each complex system using twenty snapshots from MD simulations. The equation 3 was used to obtain ΔG_{bind} values; the term $T\,\Delta S$ was calculated using normal-mode analysis RRHO contained in MacroModel module (MacroModel, version 9.5, Schrodinger, LLC, New York, 2007). The averaged energy values obtained from MM-GBSA calculations are reported in Table 11.4. The calculated ΔG_{bind} values were in correlation with experimental binding free energies (Figure 11.6) and showed that interaction is more favorable for complexes between JHEH and compounds that have large substituents (R = n-$C_{16}H_{33}$ and n-$C_{12}H_{25}$). According to the energy components of the binding free energies (Table 11.4), the major favorable contributors to ligand binding are VDW, electrostatic, and polar solvation (ΔG_{solvGB}) terms, whereas $\Delta E_{internal}$ and entropy terms oppose binding. It is noticeable that larger R substituent of the ligands results in more favorable contribution of VDW and nonpolar solvation (ΔG_{solvSA}) terms, while the electrostatic term is similar for all the complexes. After the whole analysis, we can conclude that favorable values of electrostatic and total solvation terms are essential for binding of the ligands, while VDW energy term plays a main role in the differential potency of the studied JHEH inhibitors. The electrostatic term reflects the HB interactions between the urea group of the inhibitors and the residues Asp227, Tyr298, and Tyr372, and the solvation terms reflect the preference of inhibitors to stay within JHEH binding site. Meanwhile, the VDW energy term reflects the hydrophobic interactions between the inhibitors and the residues of the

TABLE 11.4
Calculated Binding Free Energies for JHEH–Ligand Complexes Using MM-GBSA for the Snapshots of MD Simulations

Complex	$\Delta E_{internal}$ (kcal/mol)	ΔE_{elect} (kcal/mol)	ΔE_{VDW} (kcal/mol)	ΔG_{solvGB} (kcal/mol)	ΔG_{solvSA} (kcal/mol)	$T\,\Delta S$ (kcal/mol)	ΔG_{bind} (kcal/mol)
Compound **1**	3.819	−17.025	−61.638	−20.553	−5.306	−22.409	−78.292
JHEH	±1.602	±3.382	±2.077	±4.729	±1.013	±0.848	±6.999
Compound **2**	2.024	−26.345	−51.454	−21.133	−2.145	−19.991	−79.061
JHEH	±0.985	±3.393	±1.796	±4.722	±1.289	±0.943	±5.927
Compound **3**	1.810	−24.266	−41.877	−25.309	0.397	−18.175	−71.071
JHEH	±1.125	±3.099	±1.786	±5.362	±0.939	±1.095	±6.516
Compound **4**	2.373	−22.050	−40.560	−22.027	1.655	−17.598	−63.010
JHEH	±1.417	±3.253	±1.682	±5.724	±0.832	±1.069	±6.892

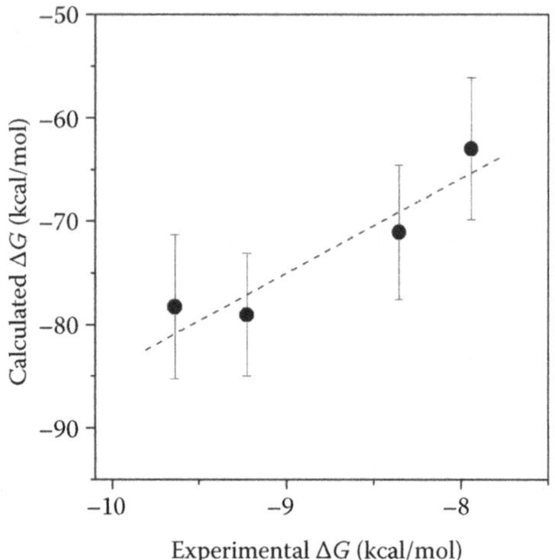

FIGURE 11.6 Free energy of binding ΔG_{bind} for each JHEH inhibitor as calculated using MM-GBSA versus the experimentally measured value.

"L"-shaped hydrophobic tunnel of JHEH; these interactions are better for larger hydrophobic molecules.

11.5 CONCLUSIONS

This chapter attempts to encourage researchers to use available or derivable structural information to study the interactions between JHEHs and their inhibitors. It has been demonstrated that docking methods are a good tool for describing the binding of JHEH inhibitors. In addition, MD simulations can be employed to complement and improve the docking protocol and provide an ensemble of structures that give a more realistic representation of the behavior of the complexes in solution. MD simulations allow the identification of conformational changes, which is of great importance because they occur during the interaction between the protein and its ligand *in vivo*. They can also help in the study of the complexes in order to observe the movement of the ligands within the protein active site and can derive new valuable information about affinity, selectivity, interactions with the water media, etc. MD can be combined with more advanced methods (e.g., free energy calculations) with the aim to estimate and predict the ligand-binding free energy.

The available information can be summarized in (1) sequences of JHEHs and (2) the crystal structure of *A. niger* EH that can be used as a template for CHM. This information is the basis of the application of structural bioinformatics methods to derive new valuable information. We believe that structural bioinformatics methods, and in particular MD simulations, should reasonably be expected to significantly improve the design of potent JHEH inhibitors in the future. Better success in this issue will come, in part, with an increased understanding of the structure of the complexes, and simulations will increasingly make useful contributions here.

REFERENCES

1. L.M. Riddiford, *Molecular aspects of juvenile hormone action in insect metamorphosis*, in *Metamorphosis*, L.I. Gilbert, J.R. Tata, and B. G. Atkinson, eds., Academic Press, Inc., New York, 1996, pp. 223–251.
2. R.M. Roe and K. Venkatesh, *Metabolism of juvenile hormones: Degradation and titer regulation*, in *Morphogenetic Hormones of Arthropods*, G.A. Rutgers, ed., Vol. 1, University Press, New Brunswick, NJ, 1990, pp. 126–179.
3. B.D. Hammock, *Regulation of juvenile hormone titer: Degradation*, in *Comprehensive Insect Physiology, Biochemistry and Pharmacology*, G.A. Kerkut and L.I. Gilbert, eds., Pergamon Press, Oxford, U.K., 1985, pp. 431–472.
4. L.I. Gilbert, R. Rybczynski, and S.S. Tobe, *Endocrine cascade in insect metamorphosis*, in *Metamorphosis: Post-Embryonic Reprogramming of Gene Expression in Amphibian and Insect Cells*, L.I. Gilbert, J.R. Tata, and B. G. Atkinson, eds., Academic Press, Inc., San Diego, CA, 1996, pp. 59–107.
5. A. Tan, H. Tanaka, T. Tamura, and T. Shiotsuki, *Precocious metamorphosis in transgenic silkworms overexpressing juvenile hormone esterase*, Proc. Natl. Acad. Sci. USA. 102 (2005), pp. 11751–11756.
6. C. Minakuchi and L.M. Riddiford, *Insect juvenile hormone action as a potential target of pest management*, J. Pestic. Sci. 31 (2006), pp. 77–84.
7. K.H. Hoffmann and M.W. Lorenz, *Recent advances in hormones in insect pest control*, Phytoparasitica 26 (1998), pp. 323–330.
8. G.B. Staal, *Insect control with insect growth regulators based on insect hormones*, in *Natural Products and the Protection of Plants*, G.B. Marini-Bettolo, ed., Pontificia Academia Scientiarum, Roma, Italy, 1977, pp. 253–283.
9. L.P. Pedigo, *Entomology and Pest Management*, Prentice Hall, Upper Saddle River, NJ, 1996.
10. C. Morisseau, M.H. Goodrow, D. Dowdy, J. Zheng, J.F. Greene, J.R. Sanborn, and B.D. Hammock, *Potent urea and carbamate inhibitors of soluble epoxide hydrolases*, Proc. Natl. Acad. Sci. U. S. A. 96 (1999), pp. 8849–8854.
11. B. Borhan, A.D. Jones, F. Pinot, D.F. Grant, M.J. Kurth, and B.D. Hammock, *Mechanism of soluble epoxide hydrolase. Formation of an alpha-hydroxy ester-enzyme intermediate through Asp-333*, J. Biol. Chem. 270 (1995), pp. 26923–26930.
12. H.-F. Tzeng, L.T. Laughlin, S. Lin, and R.N. Armstrong, *The catalytic mechanism of microsomal epoxide hydrolase involves reversible formation and rate-limiting hydrolysis of the alkyl-enzyme intermediate*, J. Am. Chem. Soc. 118 (1996), pp. 9436–9437.
13. J. Casas, L.G. Harshman, and B.D. Hammock, *Epoxide hydrolase activity on juvenile hormone in* Manduca sexta, Insect Biochem. 21 (1991), pp. 17–26.
14. L.G. Harshman, J. Casas, E.C. Dietze, and B.D. Hammock, *Epoxide hydrolase activities in* Drosophila melanogaster, Insect Biochem. 21 (1991), pp. 887–894.
15. R.J. Linderman, R.M. Roe, S.V. Harris, and D.M. Thompson, *Inhibition of insect juvenile hormone epoxide hydrolase: asymmetric synthesis and assay of glycidol-ester and epoxy-ester inhibitors of* Trichoplusia ni *epoxide hydrolase*, Insect Biochem. Mol. Biol. 30 (2000), pp. 767–774.
16. T.F. Severson, M.H. Goodrow, C. Morisseau, D.L. Dowdy, and B.D. Hammock, *Urea and amide-based inhibitors of the juvenile hormone epoxide hydrolase of the tobacco hornworm* (Manduca sexta: *Sphingidae*), Insect Biochem. Mol. Biol. 32 (2002), pp. 1741–1756.
17. J. Zou, B.M. Hallberg, T. Bergfors, F. Oesch, M. Arand, S.L. Mowbray, and T.A. Jones, *Structure of* Aspergillus niger *epoxide hydrolase at 1.8 A resolution: implications for the structure and function of the mammalian microsomal class of epoxide hydrolases*, Structure 8 (2000), pp. 111–22.

18. J. Caballero and M. Fernández, *Artificial neural networks from MATLAB in medicinal chemistry. Bayesian-regularized genetic neural networks (BRGNN): application to the prediction of the antagonistic activity against human platelet thrombin receptor (PAR-1)*, Curr. Top. Med. Chem. 8 (2008), pp. 1580–1605.
19. M.T. Cronin, J.S. Jaworska, J.D. Walker, M.H. Comber, C.D. Watts, and A.P. Worth, *Use of QSARs in international decision-making frameworks to predict health effects of chemical substances*, Environ. Health Perspect. 111 (2003), pp. 1391–1401.
20. P.D. Mosier and P.C. Jurs, *QSAR/QSPR studies using probabilistic neural networks and generalized regression neural networks*, J. Chem. Inform. Comput. Sci. 42 (2002), pp. 1460–1470.
21. Y. Nakagawa, C.E. Wheelock, C. Morisseau, M.H. Goodrow, B.G. Hammock, and B.D. Hammock, *3-D QSAR analysis of inhibition of murine soluble epoxide hydrolase (MsEH) by benzoylureas, arylureas, and their analogues*, Bioorg. Med. Chem. 8 (2000), pp. 2663–2673.
22. K.H. Kim, *Outliers in SAR and QSAR: Is unusual binding mode a possible source of outliers?*, J. Comput. Aided Mol. Des. 21 (2007), pp. 63–86.
23. M. Garriga and J. Caballero, *Insights into the structure of urea-like compounds as inhibitors of the juvenile hormone epoxide hydrolase (JHEH) of the tobacco hornworm* Manduca sexta: *Analysis of the binding modes and structure-activity relationships of the inhibitors by docking and CoMFA calculations*, Chemosphere 82 (2011), pp. 1604–1613.
24. R.D. Cramer, D.E. Patterson, and J.D. Bunce, *Comparative molecular field analysis* (CoMFA). 1. *Effect of shape on binding of steroids to carrier proteins*, J. Am. Chem. Soc. 110 (1988), pp. 5959–5967.
25. U. Pieper, N. Eswar, B.M. Webb, D. Eramian, L. Kelly, D.T. Barkan, H. Carter, P. Mankoo, R. Karchin, M.A. Marti-Renom, F.P. Davis, and A. Sali, MODBASE, *a database of annotated comparative protein structure models and associated resources*, Nucleic Acids Res. (2008), pp. gkn791.
26. M.A. Argiriadi, C. Morisseau, M.H. Goodrow, D.L. Dowdy, B.D. Hammock, and D.W. Christianson, *Binding of alkylurea inhibitors to epoxide hydrolase implicates active site tyrosines in substrate activation*, J. Biol. Chem. 275 (2000), pp. 15265–15270.
27. G.A. Gomez, *Human soluble epoxide hydrolase: Structural basis of inhibition by 4-(3-cyclohexylureido)-carboxylic acids*, Protein Sci. 15 (2006), pp. 58–64.
28. H.M. Berman, T.N. Bhat, P.E. Bourne, Z. Feng, G. Gilliland, H. Weissig, and J. Westbrook, *The Protein Data Bank and the challenge of structural genomics*, Nat. Struct. Biol. 7 Suppl (2000), pp. 957–959.
29. B. Boeckmann, A. Bairoch, R. Apweiler, M.C. Blatter, A. Estreicher, E. Gasteiger, M.J. Martin, K. Michoud, C. O'Donovan, I. Phan, S. Pilbout, and M. Schneider, *The SWISS-PROT protein knowledgebase and its supplement TrEMBL in 2003*, Nucleic Acids Res. 31 (2003), pp. 365–370.
30. M.A. Marti-Renom, A.C. Stuart, A. Fiser, R. Sanchez, F. Melo, and A. Sali, *Comparative protein structure modeling of genes and genomes*, Annu. Rev. Biophys. Biomol. Struct. 29 (2000), pp. 291–325.
31. F. Kiefer, K. Arnold, M. Kunzli, L. Bordoli, and T. Schwede, *The SWISS-MODEL Repository and associated resources*, Nucleic Acids Res. 37 (2009), pp. D387–D392.
32. K.C. Keiser, K.S. Brandt, G.M. Silver, and N. Wisnewski, *Cloning, partial purification and* in vivo *developmental profile of expression of the juvenile hormone epoxide hydrolase of* Ctenocephalides felis, Arch. Insect Biochem. Physiol. 50 (2002), pp. 191–206.
33. T. Tsubota, T. Nakakura, and T. Shiotsuki, *Molecular characterization and enzymatic analysis of juvenile hormone epoxide hydrolase genes in the red flour beetle* Tribolium castaneum, Insect Mol. Biol. 19 (2010), pp. 399–408.

34. S. Harris, D.M. Thompson, R.J. Linderman, M.D. Tomalski, and R.M. Roe, *Cloning and expression of a novel juvenile hormone-metabolizing epoxide hydrolase during larval-pupal metamorphosis of the cabbage looper*, Trichoplusia ni, Insect Mol. Biol. 8 (1999), pp. 85–96.
35. A. Seino, T. Ogura, T. Tsubota, M. Shimomura, T. Nakakura, A. Tan, K. Mita, T. Shinoda, Y. Nakagawa, and T. Shiotsuki, *Characterization of juvenile hormone epoxide hydrolase and related genes in the larval development of the silkworm* Bombyx mori, Biosci. Biotechnol. Biochem. 74 (2010), pp. 1421–1429.
36. A.M. Lesk and C. Chothia, *The response of protein structures to amino-acid sequence changes*, Philos. Trans. R. Soc. Lond. A 317 (1986), pp. 345–356.
37. A.M. Lesk and C. Chothia, *How different amino acid sequences determine similar protein structures: the structure and evolutionary dynamics of the globins*, J. Mol. Biol. 136 (1980), pp. 225–270.
38. S.Y. Chung and S. Subbiah, *A structural explanation for the twilight zone of protein sequence homology*, Structure 4 (1996), pp. 1123–1127.
39. J.K. Beetham, D. Grant, M. Arand, J. Garbarino, T. Kiyosue, F. Pinot, F. Oesch, W.R. Belknap, K. Shinozaki, and B.D. Hammock, *Gene evolution of epoxide hydrolases and recommended nomenclature*, DNA Cell Biol. 14 (1995), pp. 61–71.
40. K.H. Hopmann and F. Himo, *Theoretical study of the full reaction mechanism of human soluble epoxide hydrolase*, Chem. Eur. J. 12 (2006), pp. 6898–6909.
41. D.W. Borhani and D.E. Shaw, *The future of molecular dynamics simulations in drug discovery*, J. Comput. Aided Mol. Des. 26 (2012), pp. 15–26.
42. M.C. Payne, M.P. Teter, D.C. Allan, T.A. Arias, and J.D. Joannopoulos, *Iterative minimization techniques for ab initio total-energy calculations: Molecular dynamics and conjugate gradients*, Rev. Mod. Phys. 64 (1992), pp. 1045–1097.
43. B.R. Brooks, R.E. Bruccoleri, B.D. Olafson, D.J. States, S. Swaminathan, and M. Karplus, *CHARMM: A program for macromolecular energy, minimization, and dynamics calculations*, J. Comput. Chem. 4 (1983), pp. 187–217.
44. W.D. Cornell, P. Cieplak, C.I. Bayly, I.R. Gould, K.M. Merz, Jr., D.M. Ferguson, D.C. Spellmeyer, T. Fox, J.W. Caldwell, and P.A. Kollman, *A second generation force field for the simulation of proteins, nucleic acids, and organic molecules*, J. Am. Chem. Soc. 117 (1995), pp. 5179–5197.
45. W.L. Jorgensen, D.S. Maxwell, and J. Tirado-Rives, *Development and testing of the OPLS all-atom force field on conformational energetics and properties of organic liquids*, J. Am. Chem. Soc. 118 (1996), pp. 11225–11236.
46. L.D. Schuler, X. Daura, and W.F. van Gunsteren, *An improved GROMOS96 force field for aliphatic hydrocarbons in the condensed phase*, J. Comput. Chem. 22 (2001), pp. 1205–1218.
47. T.E. Cheatham, III, J.L. Miller, T. Fox, T.A. Darden, and P.A. Kollman, *Molecular dynamics simulations on solvated biomolecular systems: The particle mesh Ewald method leads to stable trajectories of DNA, RNA, and proteins*, J. Am. Chem. Soc. 117 (1995), pp. 4193–4194.
48. J.P. Ryckaert, G. Ciccotti, and H.J.C. Berendsen, *Numerical integration of the Cartesian equations of motion of a system with constraints: molecular dynamics of n-alkanes*, J. Comput. Phys. 23 (1977), pp. 327–341.
49. S. Nose, *A unified formulation of the constant temperature molecular-dynamics methods*, J. Chem. Phys. 81 (1984), pp. 511–519.
50. W.G. Hoover, *Canonical dynamics: Equilibrium phase-space distributions*, Phys. Rev. A 31 (1985), pp. 1695–1697.
51. H.J.C. Berendsen, J.P.M. Postma, W.F. van Gunsteren, A. DiNola, and J.R. Haak, *Molecular dynamics with coupling to an external bath*, J. Chem. Phys. 81 (1984), pp. 3684–3690.

52. W.F. van Gunsteren and H.J.C. Berendsen, *Computer simulation of molecular dynamics: Methodology, applications, and perspectives in chemistry*, Angew. Chem. Int. Ed. 29 (1990), pp. 992–1023.
53. D.M. van Aalten, R. Bywater, J.B. Findlay, M. Hendlich, R.W. Hooft, and G. Vriend, *PRODRG, a program for generating molecular topologies and unique molecular descriptors from coordinates of small molecules*, J. Comput. Aided Mol. Des. 10 (1996), pp. 255–262.
54. J. Wang, W. Wang, P.A. Kollman, and D.A. Case, *ANTECHAMBER: An accessory software package for molecular mechanical calculations*, Abstr. Pap. Am. Chem. Soc. 222 (2001), p. U403.
55. V. Zoete, M.A. Cuendet, A. Grosdidier, and O. Michielin, *SwissParam: A fast force field generation tool for small organic molecules,* J. Comput. Chem. 32 (2011), pp. 2359–2368.
56. K. Vanommeslaeghe, E. Hatcher, C. Acharya, S. Kundu, S. Zhong, J. Shim, E. Darian, O. Guvench, P. Lopes, I. Vorobyov, and A.D. Mackerell, Jr., *CHARMM general force field: A force field for drug-like molecules compatible with the CHARMM all-atom additive biological force fields*, J. Comput. Chem. 31 (2010), pp. 671–690.
57. C.A. Lipinski, F. Lombardo, B.W. Dominy, and P.J. Feeney, *Experimental and computational approaches to estimate solubility and permeability in drug discovery and development settings*, Adv. Drug Deliv. Rev. 23 (1997), pp. 3–25.
58. S. Huo, J. Wang, P. Cieplak, P.A. Kollman, and I.D. Kuntz, *Molecular dynamics and free energy analyses of cathepsin D-inhibitor interactions: insight into structure-based ligand design*, J. Med. Chem. 45 (2002), pp. 1412–1419.
59. P.A. Kollman, I. Massova, C. Reyes, B. Kuhn, S. Huo, L. Chong, M. Lee, T. Lee, Y. Duan, W. Wang, O. Donini, P. Cieplak, J. Srinivasan, D.A. Case, and T.E. Cheatham, 3rd, *Calculating structures and free energies of complex molecules: combining molecular mechanics and continuum models*, Acc. Chem. Res. 33 (2000), pp. 889–897.
60. T.P. Straatsma and J.A. McCammon, *Computational alchemy*, Annu. Rev. Phys. Chem. 43 (1992), pp. 407–435.
61. J. Aqvist, C. Medina, and J.E. Samuelsson, *A new method for predicting binding affinity in computer-aided drug design*, Protein Eng. 7 (1994), pp. 385–391.
62. J.M. Swanson, R.H. Henchman, and J.A. McCammon, *Revisiting free energy calculations: A theoretical connection to MM/PBSA and direct calculation of the association free energy*, Biophys. J. 86 (2004), pp. 67–74.
63. J. Srinivasan, T.E. Cheatham, P. Cieplak, P.A. Kollman, and D.A. Case, *Continuum solvent studies of the stability of DNA, RNA, and phosphoramidate-DNA helices*, J. Am. Chem. Soc. 120 (1998), pp. 9401–9409.
64. M.S. Lee and M.A. Olson, *Calculation of absolute protein-ligand binding affinity using path and endpoint approaches*, Biophys. J. 90 (2006), pp. 864–877.
65. G.M. Torrie and J.P. Valleau, *Nonphysical sampling distributions in Monte Carlo free-energy estimation: umbrella sampling*, J. Comput. Phys. 23 (1977), pp. 187–199.
66. S. Park, F. Khalili-Araghi, E. Tajkhorshid, and K. Schulten, *Free energy calculation from steered molecular dynamics simulations using Jarzynski's equality*, J. Chem. Phys. 119 (2003), p. 3559.
67. A. Laio and M. Parrinello, *Escaping free-energy minima*, Proc. Natl. Acad. Sci. USA. 99 (2002), pp. 12562–12566.
68. M. Iannuzzi, A. Laio, and M. Parrinello, *Efficient exploration of reactive potential energy surfaces using Car-Parrinello molecular dynamics*, Phys. Rev. Lett. 90 (2003), p. 238302.
69. E. Darve, D. Rodriguez-Gomez, and A. Pohorille, *Adaptive biasing force method for scalar and vector free energy calculations*, J. Chem. Phys. 128 (2008), p. 144120.

70. D. Zhang, J. Gullingsrud, and J.A. McCammon, *Potentials of mean force for acetylcholine unbinding from the alpha7 nicotinic acetylcholine receptor ligand-binding domain*, J. Am. Chem. Soc. 128 (2006), pp. 3019–3026.
71. S. Izrailev, S. Stepaniants, M. Balsera, Y. Oono, and K. Schulten, *Molecular dynamics study of unbinding of the avidin-biotin complex*, Biophys. J. 72 (1997), pp. 1568–1581.
72. B. Isralewitz, S. Izrailev, and K. Schulten, *Binding pathway of retinal to bacterio-opsin: A prediction by molecular dynamics simulations*, Biophys. J. 73 (1997), pp. 2972–2979.
73. X. Huang, F. Zheng, and C.G. Zhan, *Human butyrylcholinesterase-cocaine binding pathway and free energy profiles by molecular dynamics and potential of mean force simulations*, J. Phys. Chem. B 115 (2011), pp. 11254–11260.
74. J. Du, H. Sun, L. Xi, J. Li, Y. Yang, H. Liu, and X. Yao, *Molecular modeling study of checkpoint kinase 1 inhibitors by multiple docking strategies and prime/MM-GBSA calculation*, J. Comput. Chem. 32 (2011), pp. 2800–2809.
75. T. Yang, J.C. Wu, C. Yan, Y. Wang, R. Luo, M.B. Gonzales, K.N. Dalby, and P. Ren, *Virtual screening using molecular simulations*, Proteins 79 (2011), pp. 1940–1951.
76. C. Muñoz, F. Adasme, J.H. Alzate-Morales, A. Vergara-Jaque, T. Kniess, and J. Caballero, *Study of differences on the VEGFR2 inhibitory activities between semaxanib and SU5205 by using 3D-QSAR, docking, and molecular dynamics simulations*, J. Mol. Graph. Model. 32 (2012), pp. 39–48.
77. A.D. MacKerell, D. Bashford, M. Bellott, R.L. Dunbrack, J.D. Evanseck, M.J. Field, S. Fischer, J. Gao, H. Guo, S. Ha, D. Joseph-McCarthy, L. Kuchnir, K. Kuczera, F.T.K. Lau, C. Mattos, S. Michnick, T. Ngo, D.T. Nguyen, B. Prodhom, W.E. Reiher, B. Roux, M. Schlenkrich, J.C. Smith, R. Stote, J. Straub, M. Watanabe, J. Wiórkiewicz-Kuczera, D. Yin, and M. Karplus, *All-Atom empirical potential for molecular modeling and dynamics studies of proteins*, J. Phys. Chem. B 102 (1998), pp. 3586–3616.
78. E. Neria, S. Fisher, and M. Karplus, *Simulation of activation free energies in molecular systems*, J. Chem. Phys. 105 (1996), pp. 1902–1921.
79. L. Kale, R. Skeel, M. Bhandarkar, R. Brunner, A. Gursoy, N. Krawetz, J. Phillips, A. Shinozaki, K. Varadarajan, and K. Schulten, *NAMD2: greater scalability for parallel molecular dynamics*, J. Comput. Phys. 151 (1999), pp. 283–312.
80. W. Humphrey, A. Dalke, and K. Schulten, *VMD: Visual molecular dynamics*, J. Mol. Graph. 14 (1996), pp. 33–38.

12 Structural Studies of Juvenile Hormone Binding Proteins

Agnieszka J. Pietrzyk, Mariusz Jaskolski, and Grzegorz Bujacz

CONTENTS

ABSTRACT

Juvenile hormones are involved in the control of insect metamorphosis. These highly hydrophobic molecules are transported from the site of their synthesis to target tissues by juvenile hormone binding proteins (JHBPs), which also protect the labile hormone molecules from degradation by esterases in the insect hemolymph. Structural studies of JHBP proteins, started only a few years ago, have already furnished data, from which a fascinating picture of protein–hormone interactions is emerging, answering some questions but opening new ones as well. This chapter summarizes all the available structural information about JHBP proteins, including the description of their fold, ligand binding sites, and interactions with the hormone molecules.

KEYWORDS

Juvenile hormones, Juvenile hormone binding protein, Juvenile hormone binding site, Metamorphosis, Insects

12.1 INTRODUCTION

Insects are tiny six-legged creatures with an incredible spread and diversity around the world. The main common feature of their morphology is the division of their body into three segments, the head, thorax, and abdomen. However, due to adaptation to various environments, the overall body shape and appearance could be completely different from one species to another. The number of insect species is enormous and only a fraction of them have been thoroughly studied and described. Despite this scarcity of information, those several well-characterized insect species are ideal models for evolutional, genomic, and physiological studies, and therefore, they are commonly used in molecular biology laboratories [1].

A highly successful mechanism of biological adaptation of some insects is their metamorphosis. This strategy allows an adaptation of a particular insect development stage to its development role. However, not all insects undergo metamorphosis and sometimes a metamorphosis cycle is incomplete. It is usually related to different food habitats and environment [2]. Considering the complete metamorphosis, development can be divided into three stages: the larva, the pupa, and the adult form. The larva maintains its form through several instars (growth steps with molting), after which it reaches the pupal stage. Finally, the pupa is transformed into the adult form [3].

The process of metamorphosis is controlled by two families of hormones, the ecdysteroids and the juvenile hormones (JHs) (Figure 12.1); the latter term refers collectively to juvenile hormone and its isoforms [4]. The presence of JH in the hemolymph maintains the larval stage of the insect [5]. There are several molts during the larval stage, corresponding to the larva instars. After each molt, the larva retains its morphology [6]. During the last larval instar, a shift in hormone secretion occurs, manifested by a decrease of the JH titer in the hemolymph and an increase of ecdysteroids (especially 20-hydroxyecdysone). This triggers the metamorphosis. Finally, the ecdysteroids drive the processes of adult differentiation [7].

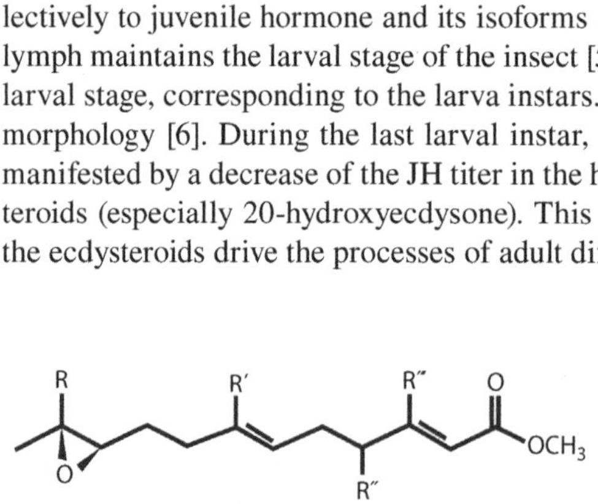

JH 0: R = R′ = R‴ = CH_2CH_3, R″ = 0
JH I: R = R′ = CH_2CH_3, R″ = 0, R‴ = CH_3
JH II: R = CH_2CH_3, R′ = R‴ = CH_3 R″ = 0
JH III: R = R′ = R‴ = CH_3, R″ = 0
iso-JH 0: R = R′ = CH_2CH_3, R″ = R‴ = CH_3

FIGURE 12.1 Chemical structures of JH isoforms. "0" indicates no substituent (only hydrogen atom).

12.2 JHBPs: A GENERAL OVERVIEW

JHs are highly hydrophobic molecules synthesized in the corpora allata and secreted to the hemolymph, for transport of the hormonal signal to distant target tissues [8]. In the aqueous milieu of the hemolymph, the JHs are easily hydrolyzed by juvenile hormone esterases (JHEs) [9]. Therefore, they require a protective agent to prevent degradation and aid delivery from the site of synthesis to target tissues. This protective/carrier role is played by juvenile hormone binding proteins (JHBPs) [8].

There are four classes of JHBPs in insects. The first group comprises lipophorins of high molecular weight, which are the most widely distributed JH carriers. Another class, found in *Orthoptera*, includes high molecular weight hexameric lipoproteins [10]. These proteins were first recognized in the hemolymph of *Locusta migratoria* [11] and *Gomphocerus rufus* [12]. The third group, comprising low molecular weight JHBPs, was described a few years earlier by Trautmann [13] and Whitmore and Gilbert [14]. Over the next few decades, low molecular weight JHBPs from *Manduca sexta* [15,16], *Galleria mellonella* [17,18], *Heliothis virescens* [19], and *Bombyx mori* [20,21] were purified and characterized. The last class consists of high molecular weight JHBPs, which were found in the diapausing *Busseola fusca* [22].

Lepidopteran low molecular weight JHBPs have been thoroughly characterized biochemically, but structural studies are scarce. Several biochemical tests aiming at characterization of the JH binding site in JHBPs have been performed, and it was determined that low molecular weight JHBPs are single-polypeptide proteins and bind only one JH molecule [23,24]. Moreover, several experiments suggested that a conformational change occurs upon JH binding. For instance, JH binding made the protein more resistant to enzymatic degradation, lowered its mobility in native electrophoresis [25], and caused changes in the circular dichroism (CD) and ultraviolet (UV) difference spectra [26]. Other interesting analyses addressed the number and position of disulfide bridges. The cysteine content of Lepidopteran JHBPs with known sequences is similar and varies between four Cys residues in *G. mellonella* [27] and six Cys residues in *M. sexta* [16]. Despite this variation, it is likely that the pattern of disulfide bonds is conserved in the Lepidopteran proteins. For instance, in JHBP from *M. sexta*, only four of the six cysteine residues are involved in disulfide bridges [28]. The first (N-terminal) disulfide bridge could play a critical role in JH binding, as indicated by mutational studies of JHBP from *H. virescens*. When the Cys9 and Cys16 residues were replaced by Ala, the protein showed a reduced hormone-binding affinity [19].

To gain insight into the interactions between the amino acid side chains and the JH molecule in its binding site, structural studies at the atomic level have been undertaken. The first crystal structure of a JHBP molecule was solved for a protein isolated from its natural source, the hemolymph of the larvae of wax moth, *G. mellonella* (GmJHBP; PDB code: 2rck) [29], followed by a preliminary crystallographic study of a natural-source JHBP from mulberry silkworm, *B. mori* (BmJHBP) [30]. Recently, several crystal structures of recombinant BmJHBP have been solved, either in the *apo* form without a JH molecule bound (PDB code: 3aot) or in complex with JH II (PDB code: 3aos) or 2-methyl-2,4-pentanediol (MPD) (PDB code: 3a1z) [31]. In addition, a solution structure determined by nuclear magnetic resonance (NMR) spectroscopy for a JH III complex (PDB code: 2rqf) has also been published [32]. The structural characterization of BmJHBP–JH complexes has allowed a detailed analysis of the protein–hormone interactions and of the molecular recognition mechanism.

In this chapter, we focus mainly on the structural data and on comparisons of all available JHBP structures, to highlight their similarities and differences. The JH binding sites of JHBPs and JHE are also compared. The structural studies on JHBPs were performed assuming that a set of known structures will help in leading to new insect control methods (Table 12.1).

12.3 OVERALL STRUCTURE OF JUVENILE HORMONE BINDING PROTEINS

GmJHBP was the first JHBP to have its crystal structure solved. The protein was isolated from its natural source, the hemolymph of the wax moth larvae. The structure was determined using the multiwavelength anomalous diffraction (MAD) method for phase calculation [29].

Low molecular weight JHBPs are formed by a single-polypeptide chain with a molecular mass of about 30 kDa, folded as a single domain. The JHBP fold will be described in this chapter according to the original GmJHBP structure (Figure 12.2). It consists of a very long C-terminal helix α4 wrapped in a highly curved antiparallel β-sheet with a right-handed helical twist. Two β-strands, β-4 and β-5, located in the middle of the sheet are extremely long. Additionally, three shorter helices are present. Two of them (α2, α3) form a right-angle motif connected with the long helix α4. The N-terminal helix α1 in GmJHBP is structurally close to the C-terminus of helix α4. The N-terminus of the protein does not form any regular secondary structure. However, in one of the molecules (A) in the asymmetric unit, all N-terminal residues have been included in the model although they do not interact with other structural elements of the protein but instead are exposed to the solvent [29].

The GmJHBP contains two disulfide bridges and no free cysteine residues [34]. The disulfide bridges link Cys10 with Cys17 and Cys151 with Cys195 [35]. The unstructured N-terminal peptide is connected with helix α1 by the Cys10–Cys17 disulfide bridge. The Cys151–Cys195 bond is located between the central part of helix α4 and the middle part of strand β6 from the β-wrap [29].

Previous biochemical and biophysical indications that GmJHBP is glycosylated [35,36] have been confirmed by the structural study. Since the crystals of GmJHBP used in the x-ray diffraction experiment were obtained from natural-source protein, isolated directly from *G. mellonella* hemolymph, it was possible to identify the post-translational glycosylation in the crystal structure. One *N*-acetylglucosamine residue was found to be attached to Asn94 on the "outer" surface of the β-wrap [29].

When the JHBP molecule is oriented with the long helix α4 aligned horizontally and with the N- and C-termini located on the left, the axial poles of the molecule are defined as its west (W) and east (E) ends (Figure 12.2), according to Kołodziejczyk et al. [29]. At each pole there is a clearly discernible cavity formed between the long helix α4 and a set of β-strands from the β-sheet. The main entrance to the W cavity is 6 by 10 Å wide and is formed by nonpolar residues. On the other hand, the entrance

TABLE 12.1
Summary of Structural Information Available for JHBP and Related Proteins

PDB Code	2rck	2rqf	3a1z	3aos	3aot	3e8t	3e8w	2fj0
Protein	GmJHBP	BmJHBP	BmJHBP	BmJHBP	BmJHBP	EpTo1	EpTo1	MsJHE
Ligand	—	JH III	MPD	JH II	—	Ubiquinone	Ubiquinone	OTFP[a]
Protein source	Natural	Recombinant	Recombinant	Recombinant	Recombinant	Recombinant	Recombinant	Recombinant
Method	XRD[b]	NMR	XRD	XRD	XRD	XRD	XRD	XRD
Resolution (Å)	2.44	—	2.59	2.20	2.20	1.30	2.50	2.70
R_{work}/R_{free} (%)	19.4/26.4	—	22.5/29.0	24.2/29.1	23.5/25.5	14.4/18.2	18.7/26.0	20.4/25.1
Space group	$P3_121$	—	$P2_12_12_1$	C2	$P6_322$	$P2_1$	$P2_12_12_1$	$P4_12_12$
References	[29]	[32]	[31]	[31]	[31]	[44]	[44]	[41]

[a] 3-Octylthio-1,1,1-trifluoropropan-2-one.
[b] X-ray diffraction.

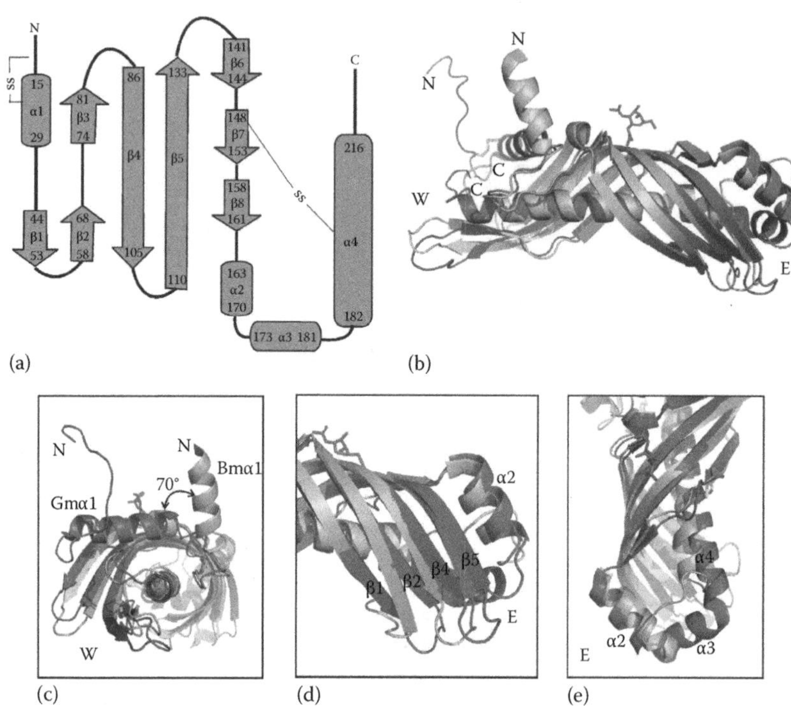

FIGURE 12.2 (See color insert.) Overall structure of JHBP and structural comparison of GmJHBP (PDB code: 2rck) and BmJHBP (3aot). (a) Topology diagram of GmJHBP. Cylinders and arrows represent α-helices and β-strands, respectively. The secondary structure elements are numbered from α1 to α4 and from β1 to β8. All numbers correspond to amino acid residues in the GmJHBP sequence. (b) A cartoon representation of JHBP fold. The GmJHBP structure shown in red is superposed on the apo-BmJHBP structure (3aot) shown in green. The Cys10–Cys17 disulfide bridge (yellow) is visible near the N-terminus. At the top of the GmJHBP molecule, one *N*-acetylglucosamine residue is attached to Asn94. The axial poles of the molecule are defined as its W and E ends. (c) The W pole of the JHBP structures is shown illustrating different conformations of helix α1 in GmJHBP (Gmα1) and apo-BmJHBP (Bmα1). (d) The strands β1–β2 and β4–β5 are curved differently in the two structures. (e) The BmJHBP structure has an extended loop instead of the helix α3 present in GmJHBP. This and all other structural illustrations were prepared in PyMol. (From W. L. DeLano, The PyMOL Molecular Graphics System, DeLano Scientific, San Carlos, CA (2002), http://www.pymol.org.)

to the E cavity is 6 by 13 Å wide and contains four charged residues. The main division and indeed the wall separating the two cavities are formed by the internal disulfide bridge between the helix α4 and the β-wrap [29].

The interior of both cavities is hydrophobic [29], and therefore, this property alone could not be used to identify which of the two is the JH binding site. However, by correlating the crystal structure of GmJHBP with the results of photoaffinity labeling performed for JHBP from *M. sexta* using a radioactive hormone analog [37], it was possibly concluded that the JH binding site is located in the W cavity

[29]. Specifically, the photoaffinity labeling experiment implicated peptides 1–34, 117–127, and 181–226 in JH binding [37], and the same peptides are involved in the formation of the W cavity. Moreover, at the entrance to the W cavity, only nonpolar residues are present, whereas the entrance to the E cavity is decorated with positively charged residues. The calculated volumes of the W and E cavities are 632 and 668 $Å^3$, respectively. The molecular volumes of the three JH homologs JH I, JH II, and JH III were also calculated as 516, 488, and 460 $Å^3$, respectively. It can be, therefore, concluded that the volume of cavity W is sufficient for effective JH binding [29].

Attempts to prepare GmJHBP–JH complexes have been unsuccessful as crystals of GmJHBP rapidly disintegrated on contact with JH. Kołodziejczyk et al. [29] explained this difficulty by crystal packing, as access to the W cavities in the crystal was largely blocked, therefore necessitating a packing rearrangement to allow hormone binding. Another reason could be a conformational change of the protein, postulated to be induced by hormone binding.

12.4 COMPARISON OF THE GmJHBP AND BmJHBP STRUCTURES

The level of amino acid sequence identity between BmJHBP and GmJHBP is only 41.0% (Figure 12.3). To gain a better insight into the evolutionary relationships among JHBP and takeout (To) proteins (also described in this chapter), a phylogenetic tree based upon sequence similarity (Table 12.2) has been constructed (Figure 12.4).

Although the overall JHBP folding motif defined by the GmJHBP structure, namely, a long helix wrapped in a curved β-sheet, is conserved in the BmJHBP models, there are also several notable structural differences. The first BmJHBP structure (PDB code: 3a1z) was solved by experimental phasing using selenomethionine single wavelength anomalous diffraction (Se-Met SAD) data. Two further BmJHBP structures (PDB codes: 3aos, 3aot) were determined by molecular replacement, using the coordinates 3a1z as the starting model for the 3aos structure, which then served as a reference model for the 3aot structure [31]. It is interesting to note that all the discussed structures were solved in different space groups (Table 12.1). This is very likely related to the fact that in all cases different crystallization conditions were used.

In all the crystal structures of BmJHBP, the N-terminal residues are not visible (Figure 12.2c). The complete N-terminal peptide is present in the NMR structure 2rqf, where residues 1–8 interact with residues in helix α1 and with the residues at the C-terminus. Some of these interactions are also observed in the crystal structure 3aos, where only the first three N-terminal residues are missing [31]. As noted earlier, the N-terminal residues in the GmJHBP structure are exposed to solvent and do not interact with the rest of the protein.

The most interesting and complicated case for comparing structures is presented by helix α1. However, this aspect of structural comparisons is strictly connected with the conformational change of the protein on hormone binding and with the binding mechanism itself, and therefore, it will be discussed in detail in Section 12.5.

The ends of the β-sheet formed by strands β1-β2 and β4-β5 deviate in the BmJHBP fold from the conformation found in the GmJHBP structures. As illustrated in Figure 12.2d, the BmJHBP conformation is characterized by a much higher β-sheet curvature.

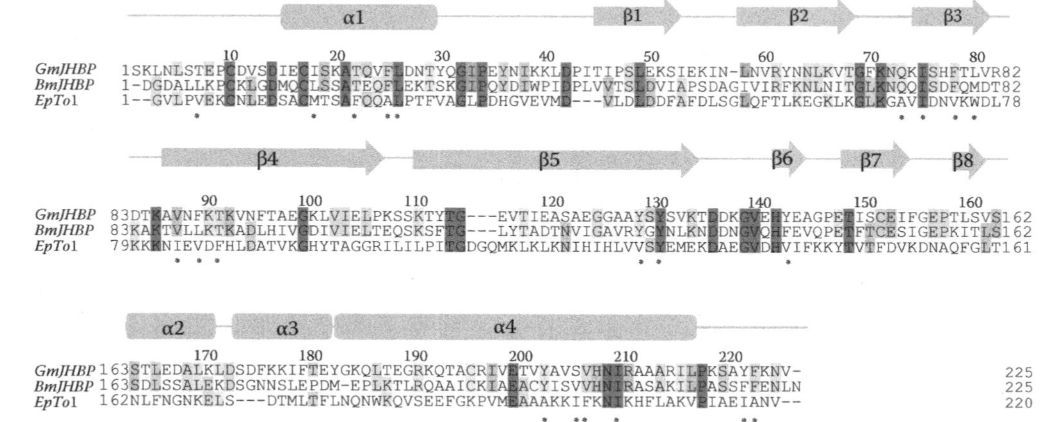

FIGURE 12.3 Alignment of JHBPs and To sequences calculated in ClustalW. The sequences of BPI and CETP were not included in the alignment due to the low level of its sequence similarity to JHBPs. The corresponding secondary structure elements are annotated as in the GmJHBP structure. Cylinders and arrows represent α-helices and β-strands, respectively. Level of residue conservation is indicated by the darkness of the lettering background. The residues forming the W cavity are indicated by asterisks. (From Thompson, J.D. et al., *Nucleic Acids Res.*, 22, 4673, 1994.)

TABLE 12.2
Sequence Homology among the Proteins Discussed in This Chapter

	BmJHBP	EpTo1	BPI N-Terminal Domain	BPI C-Terminal Domain	CEPT N-Terminal Domain	CEPT C-Terminal Domain
Sequence alignment with GmJHBP[a]	63 (41)[b]	47 (23)	nssf[c]	nssf[c]	nssf[c]	41 (22)
Score[d]	39.0	15.0	5.0	5.0	4.0	9.0

There is no significant sequence similarity between GmJHBP and N-terminal domains of BPI and CEPT or C-terminal domain of BPI. The low score for alignment of these sequences also indicates a very low level of sequence homology. However, these proteins are structurally related.

[a] The similarities and identities between GmJHBP sequence and sequences of the other proteins discussed in this chapter were calculated using BLAST [38].

[b] The similarities (identities) are in %.

[c] No significant similarity found.

[d] The score for alignment of sequences calculated using ClustalW [39].

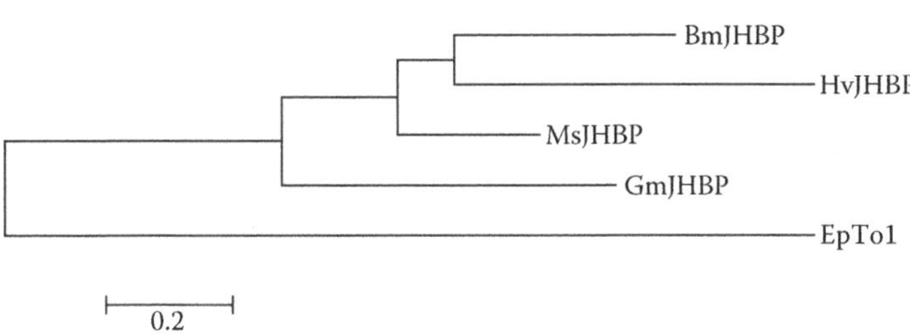

FIGURE 12.4 A phylogenetic tree constructed for the amino acid sequences of four JHBPs (from *B. mori*, *G. mellonella*, *M. sexta*, and *H. virescens*) and To protein from *E. postvittana*. The phylogenetic tree was prepared with Jalview. (From Waterhouse, A.M. et al., *Bioinformatics*, 25, 1189, 2009.)

In contrast to the situation found in the GmJHBP structure, residues 174–184 of BmJHBP do not adopt a helical fold (Figure 12.2e). Instead, they form an extended loop between helices α2 and α4 (PDB codes: 3aos, 3aot). This difference results in significant changes in the architecture of the E cavity. Its main entrance, already described for GmJHBP, becomes smaller in BmJHBP, only about 6 Å across, and is formed mainly by three nonpolar leucine residues, Leu161, Leu178, and Leu185. On the other hand, the accessory second entrance created by Gln108, Asp172, and Met182 is much wider, about 10–12 Å in diameter.

Interestingly, there are five cysteine residues in the BmJHBP sequence. Four of them are paired in the same fashion as in GmJHBP (Cys9–Cys16 and Cys151–Cys194), whereas Cys200 has a free sulfhydryl group [31].

No posttranslational modifications are observed in the BmJHBP structures as the protein used in these studies was obtained using recombinant expression in *Escherichia coli* cells [31,32].

12.5 CASE OF HELIX α1

A superposition of the Cα atoms of all the JHBP structures reveals that the orientation of helix α1 is significantly different in the apo-BmJHBP structure 3aot. This is a surprising observation because in the other two JHBP structures, BmJHBP (PDB code: 3a1z) and GmJHBP without JH bound, helix α1 has exactly the same conformation, also reproduced in the BmJHBP–JH complexes. A possible explanation is provided by the NMR structure of BmJHBP in solution, which reveals that helix α1 is highly mobile and can have multiple conformations, from an open to an entirely closed form. During this movement helix α1 is capable of turning around the W cavity by about 70° (Figure 12.2c). It appears that JH binding fosters a tightly closed conformation of helix α1 [31,32]. In the 3a1z structure, one MPD molecule from the crystallization solution is bound in the W cavity, and its binding site overlaps the epoxy group of JH. Most likely, MPD binding stabilizes the closed conformation of the α1 helix as well. Intriguingly, in the GmJHBP structure with no ligand in the W cavity, the helix α1 conformation is exactly the same as in the BmJHBP–JH complexes. It can be speculated that the closed conformation of helix α1 is caused in this case by crystal packing. This would be consistent with the argument about the role of crystal packing in obstructing the GmJHBP–JH complex formation by crystal soaking [29]. Furthermore, it could be concluded that helix α1 functions as a gating element, regulating JH binding in the W cavity. In the packing mode of the GmJHBP crystal, there is insufficient space for a movement of helix α1 into the open state. The necessary space is occupied by a loop between helix α1 and strand β1 from a symmetry-related molecule. An opening of the W cavity forced by JH soaking would certainly damage the ordered packing of the GmJHBP crystal.

12.6 HORMONE BINDING MECHANISM

The gating movement of helix α1 provides a confirmation of the earlier speculations about a conformational change of JHBP upon ligand binding in the W cavity. In addition, the crystal structure of BmJHBP in complex with JH II (PDB code: 3aos) [31] provides an excellent confirmation of the prediction based on the apo-GmJHBP structure that the hormone molecule is bound only in the W cavity [29].

The process of JH binding has been compared by Suzuki et al. [31] to a gate-and-latch mechanism. As explained earlier, the gate function is played by helix α1, while the latch element is provided by the C-terminal tail, which in the closed conformation interacts with the N-terminal peptide of the protein. When the gate is wide open, the empty cavity W is exposed to the solvent. In this state, helix α1 and the N-terminus are highly mobile and can have many different conformations. As soon as JH is captured by the protein, the gate is closed and the W cavity becomes completely sealed. This protein conformation is then stabilized by a number of interactions. In particular, polar interactions mediated mainly by water molecules cause an association

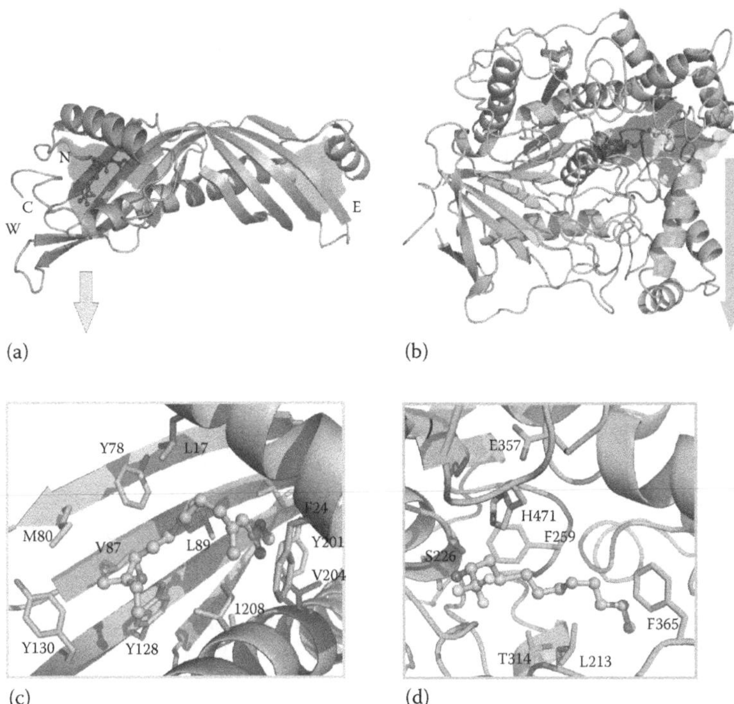

FIGURE 12.5 **(See color insert.)** The JH binding sites of BmJHBP and MsJHE. (a) The BmJHBP–JH II complex (PDB code: 3aos), with the JH II molecule bound in the W cavity. (b) The overall fold of the MsJHE protein (PDB code: 2fj0) and the location of JH binding site, with a JH analog (OTFP) bound. The ligands are shown as orange sticks and balls. The binding pockets W and E in BmJHBP and the JH binding site in MsJHE are in gray. (c) The W cavity of BmJHBP. The majority of residues forming the pocket are shown in green in stick representation. (d) The JH binding site in the MsJHE structure, the catalytic triad (Ser226, E357, and His471), and the residues interacting with an JH analog (orange) are shown in green in stick representation.

of helix α1 and strand β2. The closed conformation is also maintained by hydrophobic interactions between residues Leu17, Phe24, and Leu25 of helix α1 and the α,β-unsaturated ester moiety of the JH molecule. Finally, two hydrogen bonds formed between the N-terminal and the C-terminal residues arrest the latch in the closed state. The JH molecule is then entirely buried in the W cavity (Figure 12.5a) [31].

Despite the detailed structural characterization of BmJHBP–JH complexes, the mechanism of JH dissociation from the complex remains still unclear. It was observed that GmJHBP binds JH in a pH-dependent manner. At pH 7, the JH binding is the highest, whereas no binding could be observed at pH 4 [17]. The same observation was reported for BmJHBP. Moreover, in a further experiment, the BmJHBP–JH II complex was prepared at pH 6, then the pH was lowered to 4, and NMR spectra were measured for the sample. The results did not indicate any significant differences in backbone amide chemical shifts between pH 6 and 4.

It was concluded that the JH binding is very tight and the pH decrease is not sufficient to enable the dissociation of the JH molecule. Probably some other factors are essential to release the hormone [31].

12.7 JUVENILE HORMONE BINDING CAVITY

The crystal structure of the BmJHBP–JH II complex shows that the JH molecule fits the W cavity perfectly. A number of residues are involved in JH II binding. The residues participating in nonpolar interactions with JH II are Phe24, Leu25, Val204, Val205, Ile208, and Phe220. There is also a direct hydrogen bond between the oxygen atom of the epoxy group of the hormone and the hydroxyl group of Tyr128. Furthermore, the W cavity matches almost ideally the van der Waals surface of the bound JH II molecule. The entire hormone molecule is fully buried in the cavity and is therefore fully protected from degradation by esterases. In conclusion, the residues forming the walls of this pocket, and its architecture, seem to have evolved to play a key role in the recognition of the characteristic molecular and functional groups of the hormone [31]. The sequence alignment (Figure 12.3) of BmJHBP and GmJHBP indicates that the residues interacting with hormone are conserved and the majority of them are identical in both sequences.

The E cavity of the BmJHBP molecule differs significantly from the corresponding pocket of GmJHBP. Its shape is incompatible with binding a JH hormone molecule.

12.8 JH BINDING SITES OF JHBPs AND JUVENILE HORMONE ESTERASE

To this date, only one crystal structure of JHE was determined (PDB code 2fj0) [41]. The structure of JHE from the tobacco hornworm *M. sexta* (MsJHE) was solved in complex with a transition-state analog inhibitor 3-octylthio-1,1,1-trifluoropropan-2-one (OTFP) covalently bound in the active site. The enzyme hydrolyzes the ester bond present in the JH molecule and the reaction is carried out in the hemolymph, where the enzyme is located [41]. Normally, JHBPs protect the JH from degradation by these esterases [9]. MsJHE is a very efficient enzyme, with a k_{cat}/K_m of at least 3×10^7 M^{-1} s^{-1} [42]. The hydrolysis is a multiple-step reaction with a covalent acyl-enzyme intermediate. MsJHE is a high molecular weight protein composed of 551 amino acid residues, of which 530 could be mapped in the crystal structure [41]. MsJHE belongs to α/β-hydrolases and has the canonical fold of this family [43]. The structure consists of 16 α-helices and 14 β-strands. Strands 8–11 form a β-sheet core characteristic of the α/β-hydrolase family (Figure 12.5b). Interestingly, the substrate-binding cavity of MsJHE is unusual for α/β-hydrolase proteins (Figure 12.5d). The long, hydrophobic substrate-binding pocket matches the shape of the substrate, making the enzyme very selective for JH molecules. The bound JH molecule is entirely buried in the cavity [41], exactly as in the case of the hormone-binding cavity of JHBP. The main difference between these two pockets is the presence of catalytic residues at one end of the JHE-binding cavity. The interior of the binding pocket is mainly formed by hydrophobic residues, except for the catalytic triad and the residues forming the oxyanion hole. The efficiency of this enzyme can be linked to the tight fit of the substrate in the cavity. The catalytic triad is created by Ser226, His471, and Glu357. The catalytic mechanism involves the stabilization of the tetrahedral transition state by the oxyanion hole formed by the main chain nitrogen atoms of Gly146, Gly147, and Ala227. These residues as well as the residues of the catalytic triad are conserved in the sequences of Lepidopteran JHEs. Among the noncatalytic residues, only Thr314 is conserved [41]. In conclusion, the proteins that selectively and efficiently bind JH molecules have a binding cavity that matches the hormone in shape and hydrophobic character and in addition create a hydrogen-bond link targeting the epoxy group of the ligand.

12.9 ROLE OF THE E CAVITY

The role of the E cavity, which replicates the JH-binding cavity at the opposing pole of the JHBP molecule, remains a mystery. It is tempting to speculate that it could serve as a binding site for some unknown molecule or class of molecules. The authors of the GmJHBP structure reported that some fragmented residual electron density was detected inside the E cavity and postulated that it could correspond to an unidentified ligand of low molecular weight [29]. According to the volume of the E cavity, the unknown ligand is probably of similar size as the JH molecule. It is likely that this ligand is more polar than JH, as suggested by the polar residues located at the entrance to the E cavity [29]. It can be further speculated that the E cavity binds a ligand that could trigger the dissociation mechanism of the JHBP–JH complex. In this scenario, an interaction with an additional factor or receptor would facilitate the release of the JH molecule. Identification of the elusive second ligand is a challenge for JHBP research and could change our understanding of JH signaling mechanisms in insects.

12.10 STRUCTURAL COMPARISONS

The canonical JHBP fold was established by the crystal structure of GmJHBP [29]. The JHBP fold is very unusual and there are only a few proteins with similar structure. In particular JHBP-type topology is found in human proteins binding hydrophobic ligands and in To proteins [44] with transport functions. The structural similarity reflects the phylogenetic relations of these proteins (Figure 12.3) although the level of sequence identity is low (Table 12.2). The root mean square deviation (RMSD) values for superposition of GmJHBP with all discussed crystal structures were calculated and summarized in Table 12.3. Detailed structural comparisons of JHBP with human lipid-binding proteins [29] and with To proteins [44] have been presented before. Therefore, in this section only the most essential highlights will be summarized.

12.10.1 Structural Comparison with Takeout Proteins

The structural topology of To family members is similar to that of JHBPs. To date, only two structures of a To protein from *Epiphyas postvittana* (EpTo1; PDB codes:

TABLE 12.3
RMSD Calculated for GmJHBP (Chain A) Superposed with BmJHBP, EpTo1, BPI, and CEPT (Chain A) Crystal Structures

PDB Code	3a1z	3aos	3aot	3e8t	3e8w	1bp1	2obd
Protein	BmJHBP	BmJHBP	BmJHBP	EpTo1	EpTo1	BPI	CEPT
RMSD[a] (Å)	1.66	2.00	2.13	2.09	2.14	4.33	3.24
Cα number[b]	194	196	172	175	178	162	131

The RMSD value calculated for GmJHBP (chain A) superposed with all BmJHBP (chain A) crystal structures is the highest, 2.13 Å, for the structural comparison of GmJHBP with apo-BmJHBP (3aot) that is connected to the different conformation of helix α1 in this structure. According to the RMSD values, it is clear that the proteins belonging to the To family are close structural relatives of JHBPs. The RMSD values for GmJHBP in comparison with BPI and CEPT proteins are much higher than 2.0 Å, but it is related to the fact that the BPI and CEPT are proteins containing two domains and only one of these domains could be superposed with GmJHBP, when we want to compare the whole structure.

[a] The RMSD value for the superposition of the Cα atoms of the GmJHBP (2rck) with crystal structures listed in the table.

[b] The number of the Cα atoms of GmJHBP superposed with particular structures.

3e8t, 3e8w) are available in the PDB. The disulfide bond located at the N-terminal fragment of JHBP is conserved in the EpTo1 structure. The second disulfide bridge, riveting the long helix α4 to the β-wrap, is absent in the EpTo1 structure. The number and volume of the internal ligand-binding cavities present in JHBP-type structures are strictly dependent on the presence or absence of the latter disulfide bridge. Therefore, instead of the two smaller cavities located at the poles of JHBP molecules, the EpTo1 structure contains only one cavity, which is actually a large central tunnel in the interior of the protein. According to electron density maps, the tunnel of the EpTo1 molecules in the crystal structures is occupied by a ligand molecule, identified by mass spectrometry as ubiquinone-8. These results indicate that To proteins can bind ligands with long aliphatic chains [44].

Two conserved C-terminal motifs present in the EpTo1 sequence are unique to the To family [45]. These motifs are clustered together in the EpTo1 structure. The first motif is formed by residues 97–113 from the C-terminal end of strand β3 (corresponding to β4 in the GmJHBP structure) and the N-terminal end of strand β4 (β5 in the GmJHBP structure) with the connecting loop. These two β-strands are deformed relative to their corresponding JHBP conformation structures in order to fully enclose the ligand in the binding pocket of EpTo1. The second characteristic EpTo1 motif is created by residues 160–180 [44]. This motif consists in EpTo1 of an extended loop and an α-helix. The helix corresponds to helix α3 in GmJHBP, but its conformation is different. Also this motif serves to close the long internal tunnel of EpTo1 [44]. Interestingly, this site is completely different in EpTo1, GmJHBP, and BmJHBP (Figure 12.6). In the GmJHBP structure, there is an additional helix α2 instead of the long loop in front of helix α3 [29]. What is more interesting, the BmJHBP structure lacks helix α3 entirely, and residues 173–182 are creating a long loop.

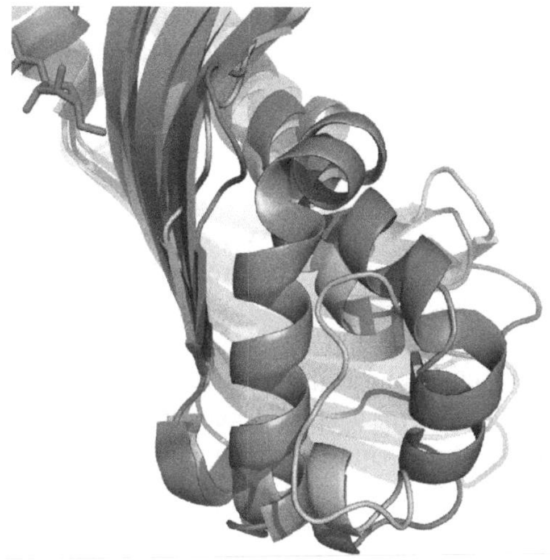

FIGURE 12.6 (See color insert.) Structural comparison of GmJHBP (red), BmJHBP (green), and EpTo1 (blue) in the area of the E pole, where the molecules show the highest degree of structural divergence.

12.10.2 Structural and Evolutionary Relation to Human Lipid-Binding Proteins

According to the Structural Classification of Proteins database [46], JHBP can be classified as an α- and β-protein with "super roll" topology [29]. This folding class includes two human lipid-binding proteins, namely, bactericidal/permeability-increasing protein (BPI; PDB code: 1bp1) [47] and cholesteryl ester transfer protein (CETP; PDB code: 2obd) [48]. These proteins are also members of the lipid transfer protein (LTP)/lipopolysaccharide (LPS) family [49]. Both BPI and CETP consist of two similar domains, each domain having the JHBP fold. Structural superpositions indicate a closer similarity to the JHBP structure of the N-terminal domains of BPI and CETP. This is also related to the presence of a disulfide bridge between the longest α-helix and one of the β-strands, which is present in the N-terminal domains but not in the C-terminal domains of BPI and CETP. The most important observation is that in both BPI and CETP, there is a hydrophobic pocket located between a short α-helix, the C-terminal end of the long helix, and part of the curved β-sheet in both domains. It corresponds exactly to the W cavity of JHBP and is evidently linked with the hydrophobic ligand-binding properties of these proteins [29]. In both BPI and CEPT, the two cavities are not separated and form a long tunnel, which starts in one domain and ends in the other one. Each of the BPI domains binds a molecule of phosphatidylcholine [48]. A similar situation can be observed in the CETP structure, where each domain contains a molecule of cholesteryl oleate and a molecule of phosphatidylcholine [49]. In conclusion, exactly the same ligands could be ascribed to one of each domain of BPI and CEPT, which indicates that the domains are highly structurally similar and have the same binding properties. This similarity might suggest that the BPI and CEPT genes were duplicated in the past and supports the hypothesis that it is connected to evolutionary divergence. Probably, an ancestral protein of JHBP fold existed and was a relative of JHBP, BPI, and CEPT [29].

12.11 SUMMARY

This chapter provides a comprehensive overview of the available structures of insect JHBPs. All structural studies on JHBPs and related proteins are not older than 10 years and much progress has been made. Information important for understanding of the structure–function relationship was provided and supplements the results of previous biochemical and biophysical experiments. To conclude, the thoroughly described [29,31] JHBP fold seems to be well conserved in Lepidoptera. The observed conformational change [31] is obviously an interesting characteristic of JHBPs and is connected with its JH binding properties. The BmJHBP–JH complexes enabled the detailed analysis of a JH binding mechanism. The shape and hydrophobic interior of the JH-binding cavity are similar in JHBPs [31] and in JHE [41], which are strictly connected to its ligand specificity.

However, the role of the E cavity, its potential ligand, and the mechanism of JH release still remain unclear. More biochemical and structural studies are required to solve these questions and further studies should focus on these issues. Finally, detailed analysis of JH binding site and mechanism is a good basis for designing new, specific insecticides interfering in insect development.

REFERENCES

1. M.J. Klowden, *Physiological Systems in Insects*, Academic Press, New York, 2007.
2. J.W. Truman and L.M. Riddiford, *The origins of insect metamorphosis*, Nature 401 (1999), pp. 447–452.
3. N.P. Kristensen, *The phylogeny of hexapod "orders". A critical review of recent accounts*, Z. Zool. Syst. Evol. 13 (1975), pp. 1–44.
4. H.F. Nijhout, *Insect Hormones*, Princeton University Press, Princeton, NJ, 1994.
5. L.M. Riddiford and J.W. Truman, *Biochemistry of insect hormones and insect growth regulators*, in *Biochemistry of Insects*, M. Rockstein, ed., Academic Press, New York, 1978, pp. 308–355.
6. L.M. Riddiford, *Cellular and molecular actions of juvenile hormone. I. General considerations and premetamorphic action*, Adv. Insect Physiol. 24 (1994), pp. 213–274.
7. L.M. Riddiford, *Hormone receptors and the regulation of insect metamorphosis*, Receptor 3 (1993), pp. 203–209.
8. W. Goodman and L.I. Gilbert, *Haemolymph protein binding of juvenile hormone in* Manduca sexta, Am. Zool. 14 (1978), p. 1289.
9. K. Touhara, K.A. Lerro, B.C. Bonning, B.D. Hammock, and G.D. Prestwich. *Ligand binding by a recombinant insect juvenile hormone binding protein*, Biochemistry 32 (1992), pp. 2068–2075.
10. S.C. Trowell, *High affinity juvenile hormone carrier protein in the hemolymph of insects*, Comp. Biochem. Physiol. 103B (1992), pp. 795–807.
11. M.G. Peter, S. Gunawan, G. Gellisen, and H. Emmerich, *Differences in hydrolysis and binding of homologous juvenile hormones in* Locusta migratoria *hemolymph*, Z. Naturforsch. 34 (1979), pp. 588–598.

12. R. Hartmann, *The juvenile hormone-carrier in the haemolymph of the acridine grasshopper* Gomphocerus rufus *L.: blocking of the juvenile hormone's action by means repeated injections of an antibody to the carrier*, Wilhelm Roux' Arch. Dev. Biol. 184 (1978), 301–324.
13. K.H. Trautmann, In vitro *study of the carrier proteins of 3 H-labelled juvenile hormone active compounds in the haemolymph of* Tenebrio molitor *L. larvae*, Z. Naturforsch B 27 (1972), pp. 263–273.
14. E. Whitmore and L.I. Gilbert, *Haemolymph lipoprotein transport of juvenile hormone*, J. Insect Physiol. 18 (1972), pp. 1153–1167.
15. K.J. Kramer, P.E. Dunn, R.C. Peterson, H.L. Seballos, L.L. Sanburg, and J.H. Law, *Purification and characterization of the carrier protein for juvenile hormone from the hemolymph of the tobacco hornworm* Manduca sexta *Johannson (Lepidoptera: Sphingidae)*, J. Biol. Chem. 251 (1976), pp. 4979–4985.
16. K.A. Lerro and G.D. Prestwich, *Cloning and sequencing of a cDNA for the hemolymph juvenile hormone binding protein of larval* Manduca sexta, J. Biol. Chem. 265 (1990), pp. 19800–19806.
17. A. Ożyhar and M. Kochman, *Juvenile-hormone binding protein from the hemolymph of* Galleria mellonella *(L). Isolation and characterization*, Eur. J. Biochem. 162 (1987), pp. 675–682.
18. E. Wieczorek, J.M. Parkitna, J. Szkudlarek, A. Ożyhar, and M. Kochman, *Immunoaffinity purification of juvenile hormone-binding protein from* Galleria mellonella *hemolymph*, Acta Biochim. Pol. 43 (1996), pp. 603–610.
19. H. Wojtasek and G.D. Prestwich, *Key disulfide bonds in an insect hormone binding protein: cDNA cloning of a juvenile hormone binding protein of Heliothis virescens and ligand binding by native and mutant forms*, Biochemistry 34 (1995), pp. 5234–5241.
20. K. Kurata, M. Nakamura, T. Okuda, H. Hirano, and H. Shinbo, *Purification and characterization of a juvenile hormone binding protein from hemolymph of the silkworm,* Bombyx mori, Comp. Biochem. Physiol. B Biochem. Mol. Biol. 109 (1994), pp. 105–114.
21. A.M. Vermunt, M. Kamimura, M. Hirai, M. Kiuchi, and T. Shiotsuki, *The juvenile hormone binding protein of silkworm haemolymph: gene and functional analysis*, Insect Mol. Biol. 10 (2001), pp. 147–154.
22. E.O. Osir, G.C. Unnithan, and G.D. Prestwich, *Juvenile hormone binding to a 500-K diapause-associated protein of the stem-borer,* Busseola fusca, Comp. Biochem. Physiol. 99B (1991), pp. 165–169.
23. K.J. Kramer, P.E. Dunn, R.C. Peterson, and J.H. Law, *Interaction of juvenile hormone with binding proteins in hemolymph*, in *The Juvenile Hormones*, L.I. Gilbert, ed., Plenum Press, New York, 1976, pp. 327–341.
24. W. Goodman, P.A. O'Hern, R.H. Zaugg, and L.I. Gilbert, *Purification and characterization of a juvenile hormone binding protein from the hemolymph of the fourth instar tobacco hornworm,* Manduca sexta, Mol. Cell. Endocrinol. 11 (1978), pp. 225–242.
25. E. Wieczorek and M. Kochman, *Conformational change of the haemolymph juvenile-hormone-binding protein from* Galleria mellonella *(L)*, Eur. J. Biochem. 201 (1991), pp. 347–353.
26. D. Krzyżanowska, M. Lisowski, and M. Kochman, *UV-difference and CD spectroscopy studies on juvenile hormone binding to its carrier protein*, J. Pept. Res. 51 (1998), pp. 96–102.
27. J.M. Rodriguez Parkitna, A. Ozyhar, J.R. Wisniewski, and M. Kochman, *Cloning and sequence analysis of* Galleria mellonella *juvenile hormone binding protein a search for ancestors and relatives*, Biol. Chem. 383 (2002), pp. 1343–1355.

28. Y.C. Park and W.G. Goodman, *Analysis and modification of thiols in the hemolymph juvenile hormone binding protein of* Manduca sexta, Arch. Biochem. Biophys. 302 (1993), pp. 12–18.
29. R. Kołodziejczyk, G. Bujacz, M. Jakób, A. Ozyhar, M. Jaskolski, and M. Kochman, *Insect juvenile hormone binding protein shows ancestral fold present in human lipid-binding proteins*, J. Mol. Biol. 377 (2008), pp. 870–881.
30. A.J. Pietrzyk, A. Bujacz, M. Łochyñska, M. Jaskólski, and G. Bujacz, *Isolation, purification, crystallization and preliminary X-ray studies of two 30 kDa proteins from silkworm hemolymph*, Acta Cryst. F67 (2011), pp. 372–376.
31. R. Suzuki, Z. Fujimoto, T. Shiotsuki, W. Tsuchiya, M. Momma, A. Tase, M. Miyazawa, and T. Yamazaki, *Structural mechanism of JH delivery in hemolymph by JHBP of silkworm,* Bombyx mori, Sci. Rep. 1 (2011), p. 133; doi:10.1038/srep00133.
32. R. Suzuki, A. Tase, Z. Fujimoto, T. Shiotsuki, and T. Yamazaki, *NMR assignments of juvenile hormone binding protein in complex with JH III*, Biomol. NMR Assign. 3 (2009), pp. 73–76.
33. W. L. Delano, *The PyMol Molecular Graphics System,* Delano Scientific, San Carlos, CA, USA (2002), http://www.pymol.org
34. R. Kołodziejczyk, P. Dobryszycki, A. Ozyhar, and M. Kochman, *Two disulphide bridges are present in juvenile hormone binding protein from* Galleria mellonella, Acta Biochim. Pol. 48 (2001), pp. 917–920.
35. J. Debski, A. Wysłouch-Cieszyñska, M. Dadlez, K. Grzelak, B. Kłudkiewicz, R. Kołodziejczyk, A. Lalik, A. Ozyhar, and M. Kochman, *Positions of disulfide bonds and N-glycosylation site in juvenile hormone binding protein*, Arch. Biochem. Biophys. 421 (2004), pp. 260–266.
36. M. Duk, H. Krotkiewski, E. Forest, J.M. Rodriguez Parkitna, M. Kochman, and E. Lisowska, *Evidence for glycosylation of the juvenile-hormone-binding protein from* Galleria mellonella *hemolymph*, Eur. J. Biochem. 242 (1996), pp. 741–746.
37. K. Touhara and G.D. Prestwich, *Binding site mapping of a photoaffinity-labeled juvenile hormone binding protein*, Biochem. Biophys. Res. Commun. 182 (1992), pp. 466–473.
38. S.F. Altschul, W. Gish, W. Miller, E.W. Myers, and D.J. Lipman, *Basic local alignment tool*, J. Mol. Biol. 215 (1990), pp. 403–410.
39. J.D. Thompson, D.G. Higgins, and T.J. Gibson, *CLUSTAL W: improving the sensitivity of progressive multiple sequence alignment through sequence weighting, position-specific gap penalties and weight matrix choice*, Nucleic Acids Res. 22 (1994), pp. 4673–4680.
40. A.M. Waterhouse, J.B. Procter, D.M.A. Martin, M. Clamp, and G.J. Barton, *Jalview Version 2—A multiple sequence alignment editor and analysis workbench*, Bioinformatics 25 (2009), pp. 1189–1191.
41. M. Wogulis, C.E. Wheelock, S.G. Kamita, A.C. Hinton, P.A. Whetstone, B.D. Hammock, and D.K. Wilson, *Structural studies of a potent insect maturation inhibitor bound to the juvenile hormone esterase of* Manduca sexta, Biochemistry 45 (2006), pp. 4045–4057.
42. A.C. Hinton and B.D. Hammock, In vitro *expression and biochemical characterization of juvenile hormone esterase from* Manduca sexta, Insect Biochem. Mol. Biol. 33, (2003), pp. 317–329.
43. D.L. Ollis, E. Cheah, M. Cygler, B. Dijkstra, F. Frolow, S.M. Franken, M. Harel, S.J. Remington, I. Silman, J. Schrag, J.L. Sussman, K.H.G. Verschueren, and A. Goldman, *The alpha/beta-hydrolase fold*, Protein Eng. 5 (1992), pp. 197–211.
44. C. Hamiaux, D. Stanley, D.R. Greenwood, E.N. Baker, and R.D. Newcomb, *Crystal structure of* Epiphyas postvittana *takeout 1 with bound ubiquinone supports a role as ligand carriers for takeout proteins in insects*, J. Biol. Chem. 284 (2009), pp. 3496–3503.
45. W.V. So, L. Sarov-Blat, C.K. Kotarski, M.J. McDonald, R. Allada, and M. Rosbash, *Takeout, a novel* Drosophila *gene under circadian clock transcriptional regulation*, Mol. Cell. Biol. 20 (2000), pp. 6935–6944.

46. A.G. Murzin, S.E. Brenner, T. Hubbard, and C. Chothia, *SCOP: A structural classification of proteins database for the investigation of sequences and structures*, J. Mol. Biol. 247 (1995), pp. 536–540. http://scop.mrc-lmb.cam.ac.uk/scop/index.html
47. L.J. Beamer, S.F. Carroll, and D. Eisenberg, *Crystal structure of human BPI and two bound phospholipids at 2.4 angstrom resolution*, Science 276 (1997), pp. 1861–1864.
48. X. Qiu, A. Mistry, M.J. Ammirati, B.A. Chrunyk, R.W. Clark, Y. Cong, J.S. Culp, D.E. Danley, T.B. Freeman, K.F. Geoghegan, M.C. Griffor, S.J. Hawrylik, C.M. Hayward, P. Hensley, L.R. Hoth, G.A. Karam, M.E. Lira, D.B. Lloyd, K.M. McGrath, K.J. Stutzman-Engwall, A.K. Subashi, T.A. Subashi, J.F. Thompson, I.K. Wang, H. Zhao, and A.P. Seddon, *Crystal structure of cholesteryl ester transfer protein reveals a long tunnel and four bound lipid molecules*, Nat. Struct. Mol. Biol. 14 (2007), pp. 106–113.
49. L.J. Beamer, *Structure of human BPI (bactericidal/permeability-increasing protein) and implications for related proteins*, Biochem. Soc. Trans. 31 (2003), pp. 791–794.

13 *In Silico* Stereoelectronic Profile and Pharmacophore Similarity Analysis of Juvenile Hormone, Juvenile Hormone Mimics (IGRs), and Insect Repellents May Aid Discovery and Design of Novel Arthropod Repellents

Apurba K. Bhattacharjee

CONTENTS

ABSTRACT

The chapter presents a comparative analysis of quantum chemically calculated stereo-electronic properties of DEET, DEET-like compounds, juvenile hormone mimics (IGRs), and juvenile hormone (JH). The results led to unravel several interesting molecular level information, most significant of which was electrostatic bio-isosterism between the compounds. Observation of similarity of stereo-electronic properties of the amide/ester moiety, negative electrostatic potential regions beyond the van der Waals surface, and large distribution of weak electrostatic potential (hydrophobic) regions in the compounds provided a model for molecular recognition with the receptor at a distance. In addition, the chapter discusses how these stereo-electronic attributes could be organized to develop a feature based on three-dimensional pharmacophore model for arthropod repellent activity and demonstrates how the model could be utilized for search of compound databases to identify new repellent compounds and aid in the design of novel repellents for custom synthesis. Since identity of the target for arthropod repellent activity still remains uncertain, the *in silico* approaches presented in the chapter should be useful for discovery and design of potentially well-tolerated, target-specific arthropod repellents for practical use.

KEYWORDS

In silico stereo-electronic properties, Quantum chemical calculations, 3D pharmacophore, DEET, Juvenile hormone mimics (IGRs), Juvenile hormone (JH)

13.1 INTRODUCTION

The goal of this chapter is to illustrate how *in silico* analysis of stereoelectronic properties of juvenile hormone (JH), juvenile hormone mimics (JH mimics) (insect growth regulators, IGRs), DEET (*N,N*-diethyl-1,3-toluamide or *N,N*-diethyl-*m*-toluamide or *m*-toluamide), and DEET-like insect repellents can be organized to develop a pharmacophore model for repellent activity that can enable database searches for identification, design, and custom synthesis of new arthropod repellents.

Arthropods are well-known vectors of lethal human diseases that include malaria, leishmaniasis, African trypanosomiasis, dengue fever, filariasis, viral encephalitis, and a number of other well-known diseases [1,2]. The direct effects of arthropods include tissue damage due to stings, bites, and exposure to vesicating fluids; infestation of tissue by the arthropods themselves; annoyance; and entomophobia (an uncontrolled fear of arthropods). The indirect effects on human health include disease transmission and allergic reactions due to bites and stings and to arthropod skins or emanations. Arthropods also destroy property and material used by the military, and there is even some concern that arthropods could potentially be used as biological weapons. In terms of disease transmission, among the arthropods, mosquitoes are the world's most notorious insect vectors [1]. Since mosquitoes feed on blood, they cause more human suffering than any other organism and responsible for over one million human deaths from mosquito-borne diseases every year. Malaria results from infection of a protozoan (*Plasmodium falciparum*) carried by mosquitoes, and according to the World Health Organization, it continues to cause as many as one million deaths annually [3].

Mosquitoes can also transmit the arboviruses responsible for yellow fever, dengue hemorrhagic fever, epidemic polyarthritis, and several forms of encephalitis. *Bancroftian filariasis* is caused by a nematode transmitted by mosquito bite. Increased international travels have enhanced this problem all over the world [3]. Thus, the quest to repel arthropods continued for decades of scientific research particularly focusing on mosquito behavior and control.

Although numerous methods for controlling arthropod-vectored diseases have been investigated ranging from development of insecticides to therapeutics for prevention and treatment of diseases, use of repellent remained by far the most effective personal protection measure in reducing bites of blood-sucking arthropods and preventing vector-borne disease transmissions. Traditionally, discovery of new repellents involves initial screening of thousands of compounds using various methods with very few successes. However, despite the obvious desirability of finding an effective nontoxic repellent, no ideal repellent has been identified yet [1]. The process is more complicated because of the required properties for an ideal arthropod repellent, which include effectiveness against a broad spectrum of species, long duration of protection, no toxicity or side effects, resistant to abrasion, greaseless, and odorless. In addition, lack of proper understanding of the mode of repellent action has further complicated the process.

The most effective personal protection system [2] currently being used is a controlled-release formulation of DEET (*N,N*-diethyl-1,3-toluamide) as a topical repellent, which is found to defend against several types of biting arthropods [2]. However, DEET has many shortcomings; the general consensus is that an ideal repellent agent should repel multiple species of biting arthropods, remain effective for 8–10 h, cause no irritation to the skin or mucous membranes, have no systemic toxicity, be resistant to abrasion and rub-off, and be greaseless and odorless. As a repellent for human use, DEET is not equally effective against all insects and arthropod vectors of diseases. Recent studies have indicated that it can cause toxic encephalopathy (a diffuse disease that alters brain structure or function), seizure, acute manic psychosis, cardiovascular toxicity, and dermatitis after heavy or excessive exposure [4]. However, no available insect repellent meets all of the aforementioned required criteria for a desirable repellent. Efforts to find such a compound face numerous challenges and variables that affect the inherent repellency property of any chemical. Repellents do not all share a single mode of action, and surprisingly little is known about how repellents act on the target insects. Therefore, more studies on the mechanism of action of repellents are necessary to design better repellents.

Since it is desirable that an insect repellent or a pesticide be of natural origin that ostensibly has limited or no adverse effects on the environment, many research efforts were directed toward biorational pesticides or biopesticides and repellents of biological origin that are derived from biological sources, including viruses, bacteria, and fungi, or from biochemicals such as IGRs or JHs. IGRs or JHs and its analogs (JH mimics) are a class of insecticides that have acquired lot of attention by the environmental agencies, which were designed specifically to disrupt endocrine-regulated processes unique to insects [5].

13.2 JUVENILE HORMONES, JH MIMICS (INSECT GROWTH REGULATORS), AND ARTHROPOD REPELLENTS

13.2.1 Juvenile Hormones and Insect Growth Regulators

JH is a key hormone for regulating metamorphosis and previtellogenic ovarian development in mosquitoes [5,6]. It is naturally synthesized and released from the corpora allata (CA), a pair of endocrine glands with nervous connections to the brain [7], and these endogenous hormones influence both metamorphosis and development of insects. These hormones are transported by binding proteins and they enter cells by diffusion across the cell membrane. The products of the CA in the cell membrane interact in some way with genome, probably via nuclear receptors of the steroid family [8]. During normal insect development, the concentration of JH decreases in the final larval instar stage, allowing development of pupal and adult stages.

JHs belong to a group of sesquiterpenoid compounds constituting a family of derivatives of farnesoic acid that are structurally unique among animals and have been identified only in arthropods. JHs (Figure 13.1a) are ubiquitous growth regulators in the arthropod world and serve as a rational source for design of synthetic

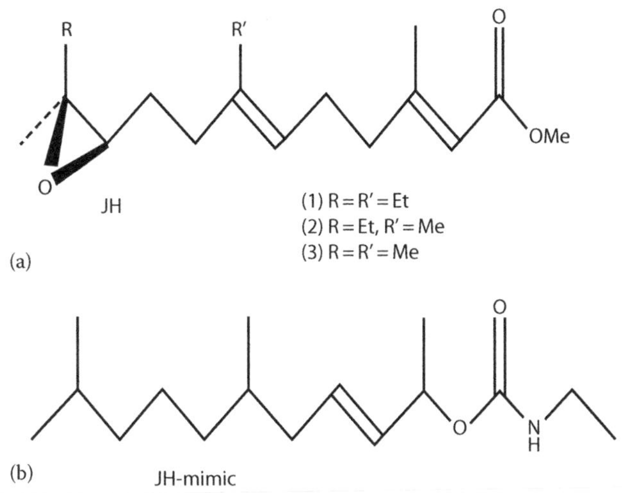

(1) R = R′ = Et
(2) R = Et, R′ = Me
(3) R = R′ = Me

FIGURE 13.1 (a) Chemical structure of the natural JH. (b) Chemical structure of JH mimic (IGR) (From Bhattacharjee, A.K. et al., *J. Mol. Recognit.*, 13, 213, 2000.)

arthropod growth regulators [9–11]. A number of studies have shown that chemical compounds containing specific functional groups or features are more effective arthropod repellents as measured by duration of protection [9,12]. Meola et al. [13,14] have shown that JH induction occurs at the first biting cycle of the species and continues with alternating periods of JH production, cyclic pattern of biting, and host-seeking behavior throughout the life of the mosquito. These researchers also demonstrated that JH deprivation caused by surgical removal of CA shortly after adult emergence blocked the initiation of biting behavior in *Culex pipiens* and *Culex quinquefasciatus*. Reimplantation of CA or injection of synthetic JH corrected the JH deficiency and restored the biting behavior of mosquitoes. Thus, JH is a target to stop biting behavior of the mosquitoes. Recently Olmstead et al. [5] demonstrated that the crustacean JH, methyl farnesoate, programs oocytes of the crustacean *Daphnia magna* to develop into males probably mimicking the action of methyl farnesoate producing altered sex ratios of offspring. Clearly, JH and its analogs were designed as a class of insecticides specifically to disrupt endocrine-regulated processes.

Like the JHs, chitin synthesis is another vital component of the insect endocrine system, regulating the biting behavior, reproductive maturation, sexual behavior, and reproductive diapauses of insects. Thus, another group of insecticides emerged as IGRs that comprised diverse groups of chemical compounds and found to be highly effective against larvae of mosquitoes and other arthropods (Figure 13.1b). A large number of IGRs, juvenoids, juvenile mimics, and inhibitors of chitin synthesis have been evaluated for vector control programs, but only a few, such as diflubenzuron, methoprene, and fenoxycarb, are found effective to be commercially feasible [15–18]. Diflubenzuron is a benzoylurea IGR, used in agriculture, horticulture, forestry, and public health applications. Although it has low solubility in water, it is stable in aqueous solution and to photolysis and has a half-life significantly shorter at less acidic conditions. It was not observed to cause any carcinogenic, mutagenic, teratogenic, or neurotoxic effects and showed low toxicity to other wildlife, other than insects, with *D. magna* being the most sensitive species reported. Methoprene had been used to control some species of mosquitoes [19–22]. Fenoxycarb is a carbamate IGR and is used as fire ant bait as well as for flea, mosquito, and cockroach controls. Fenoxycarb is also used to control butterflies and moths (*Lepidoptera*), scale insects, and sucking insects on olives, vines, cotton, and fruit as well as for control of these pests on stored products. It blocks the ability of an insect to change into an adult from the juvenile stage (metamorphosis) as it interferes with the molting of larvae. Although fenoxycarb is nearly nontoxic to mammals, acute inhalation of it in rats (greater than 480 mg/m^3) showed toxicity and, thus, carries the signal word CAUTION on its label [18]. In general, chitin inhibitors cause sterilization of insect adults and enable mortality at every molt of larvae or nymphs, whereas the JH analogs cause sterilization of adults, but mortality occurs only at pupation or adult emergence [23].

However, the IGRs in general can be used at a relatively very low dose compared to conventional insecticides. For example, ecdysone agonists are hormonally active IGRs that disrupt development of larvae but found to be active against *Aedes aegypti, Anopheles gambiae,* and *C. quinquefasciatus* at low doses [23]. Moreover, the IGRs are also known to have a good margin of safety to most nontarget biota including invertebrates, fish, birds, and other wildlife. These are relatively safe to human and domestic animals too. The IGR compounds do not induce quick mortality in the preimaginal stages, but mortality occurs after several days of treatment. This is indeed a desirable feature of the control agents because larvae of mosquitoes and other vectors are important sources of food for fish and wildlife. On account of these advantages and high level of activity against target species, the IGRs play a very important role in the vector control programs. Furthermore, since the IGRs interfere with the hormonal mechanisms of target organisms that result in various kinds of morphological, anatomical, and physiological changes in order that the target species does not reach the final stage of development, the IGRs are less likely to develop resistance [24].

However, despite JH being an attractive target for control of insect pests and vectors of diseases, the minute size of the CA, the glands that synthesize JH, has made it difficult for large-scale utility. Therefore, novel compounds that mimic the activity of JH as IGRs had become an efficient alternative approach for discovery of arthropod control agents. For example, the JH mimic pyriproxyfen with relatively low mammalian toxicity has been successfully used as an IGR, which was first registered in Japan in 1991 for controlling public health pests (Sumilarv-Public Health). With the success of pyriproxyfen, several structure–activity studies on JH analogs were performed that enabled the discovery of many JH-like compounds [9,12] through mimicking the morphogenetic activity of hormones with the aim to control insect population. However, designing of insect repellents rationalizing the JH activity is unknown.

13.2.2 Arthropod Repellents

Although considerable efforts had been made to understand [25] why humans are attractive to insects, especially mosquitoes, and many chemicals have been discovered to have repellent activity, the mode of action of repellents still remained poorly understood. The most effective repellent currently being used is a controlled-release formulation of DEET (*N,N*-diethyl-1,3-toluamide) as a topical repellent [2]. The physical requirements for repellent activity of DEET and another well-known insect repellent, *N,N*-diethylphenylacetamide (DEPA), against anopheles *A. aegypti* mosquitoes had been proposed [26] on the basis of their lipophilicity, vapor pressure, and molecular length. The olfactory sensation of insects had been invoked for repellent action and it has been recognized to be controlled by JH responses or JH activity [27–29]. These observations suggest that an ideal repellent should be volatile, must come in contact with the mosquito's olfactory organ, and should have some degree of lipid solubility to trigger the olfactory sensation. Therefore, determining the gas phase or molecular-level properties of repellents is an important first step for understanding the mechanism of interaction with the olfactory organ and hence JH activity.

The following sections will illustrate how *in silico* methods can be useful for determining the intrinsic molecular-level information of JH, JH mimics (IGRs), and DEET analogs to facilitate our understanding of the mechanism of repellent action and how the information can guide in the development of pharmacophore models to enable identification and design of novel repellents for custom synthesis. Emphasis of this chapter had been on quantum chemically derived stereoelectronic properties, quantitative structure–activity relationships (QSAR), and pharmacophore modeling to enable an early stage of discovery, design, and synthesis of novel repellents for further development.

13.3 COMPUTER-AIDED MOLECULAR MODELING

Computer-aided molecular modeling (CAMM) or *in silico* methods have made remarkable progress in mechanistic drug design and discovery of novel bioactive chemical entities in recent years [30–32]. Commonly used computational approaches include ligand-based drug design (3D pharmacophore modeling based on the arrangement of chemical features that are essential for biological activity), structure-based drug design for known x-ray crystallographic structure of the target protein (drug-target docking), and the QSAR/QSPR (quantitative structure–activity relationship and quantitative structure–property relationship) studies. CAMM techniques provide five major types of molecular informations that are useful for mechanistic design of new bioactive chemical entities. The informations are as follows: (1) 3D structure of a molecule, (2) physicochemical properties and feature characteristics of a molecule, (3) comparison of molecular structures and properties, (4) graphical visualization of stereoelectronic properties and complexes formed between the bioactive molecule and the target protein, and (5) predictions about how related molecules match with the new ones along with an estimate of activity.

With the advent of modern computers and graphic techniques, computations and visualization of structures ranging from small to large biomolecules, such as proteins, can be accomplished with greater speed and precision. The current advances in these methodologies allow direct applications ranging from accurate ab initio quantum chemical calculations of stereoelectronic properties, generation of 3D pharmacophores, virtual screening of databases to identify bioactive agents, simulation of proteins, and docking drug molecules at the active sites of proteins that are crucial and directly related to biomedical research.

Molecular modeling has now become an indispensable part of research activities that not only aid discovery of novel therapeutics or novel chemical entities (NCEs) but also provide molecular bases for understanding the environmental, biochemical, and biological processes. *In silico* methodologies are routinely used as decision tools for direct experimental investigations. Although *in silico* (computational) technologies are relatively new, their remarkable success in discovery of novel compounds and prediction of ADME/toxicity properties [33] has revolutionized pharmaceutical industries because of their cost- and time effectiveness. It takes almost 10 years and several hundreds of millions of dollars to bring a new drug to the market. Thus, historically any technology that can improve the efficiency of the process is highly valuable to the industry, and that is why the *in silico* technologies made such remarkable accomplishments. A few pharmaceutical success stories involving these technologies include Aricept for Alzheimer's disease (Eisai), Trusopt for glaucoma treatment (Merck), Crixivan for AIDS (Merck), Viracept for AIDS (Lilly and Agouron), and Zomig for migraine treatment (Wellcome, Zeneca). The list is ever growing. Although *in silico* technologies started developing from the mid-1980s, rapid growths started to emerge only with the advent of powerful modern computers. Since a drug takes approximately 10–12 years from the discovery stage to the market, we are just beginning to see the success stories [33].

Discovery of a successful new insect repellent is also as complex as drug discovery process and ever changing with the emergence of new technologies. Although no model is perfect, regardless of whatever it represents, virtual screening of hundreds of compounds in a short period of time and simulations of 3D protein structures in a computer using high levels of theory have pushed modern technologies to the cutting edge of discovery [30].

The ability of a bioactive molecule to interact with the complementary sites of a receptor protein results from a combination of steric and electronic properties. Therefore, the study of stereoelectronic properties of known bioactive molecules can provide valuable information about the "interaction pharmacophore" not only for better understanding the mechanism of action but also aid in developing pharmacophore models. Since quantum chemical methods can provide accurate estimation of stereoelectronic properties of molecules, these are commonly used for of determining the "interaction pharmacophore" profiles to assess interactions with the target receptors [32].

Thus, determining stereoelectronic properties of known repellents can be a very useful first step toward development of *in silico* 3D pharmacophore models. Once a reliable pharmacophore model can be generated, it can be used for rapid identification of potential new candidates. The pharmacophore modeling methods have made great advances in recent years [30,34–36]. *In silico* "3D pharmacophore" is perceived as an ensemble of steric and electronic properties that are necessary for optimal interaction with a specific receptor to trigger or inhibit its biological response [36]. That is to say, when a target receptor (a protein active site) recognizes the critical stereoelectronic features of the bioactive molecule, the protein immediately triggers interactions with its complimentary sites, and as a consequence, the specific biological response occurs. The pharmacophores are represented by a geometric distribution of chemical features such as hydrogen-bond acceptors and donors, aliphatic and aromatic hydrophobic sites, ring aromaticity, and ionizable sites in the 3D space of a molecule. The advantage of the pharmacophore is that it transcends the structural class and captures only features that are necessary for intrinsic activity of potentially active compounds and hence provides a powerful template for search of compounds of new chemical classes or chemo-types. Calculated stereoelectronic profiles, particularly the molecular electrostatic potentials at the van der Waals surface and at different distances from this surface, can provide accurate estimation of electron densities in the surrounding space of a molecule, which in turn can guide in the selection of the features for performing the pharmacophore generation. For example, a large electron density region in the surrounding space of a molecule would indicate the presence of H-bond acceptor characteristic of a particular atom in the region; similarly, the most positive potential region in a molecule would indicate H-bond donor characteristic of the atom, and very weak electrostatic potential regions would indicate hydrophobic regions. Pharmacophore modeling methods for generation of pharmacophores are based on multiple conformations from a set of structurally diverse molecules, and the validation of the model is carried out using statistical methods.

In recent years, several additional studies based on quantitative derivations [37–41] have shown that the pharmacophore recognition process can be analyzed from three types of 3D molecular field- or property-based similarity evaluations: (1) steric, (2) electrostatic, and (3) hydrophobic. It has been well documented that bioactive agents (ligands) will bind to a receptor in a similar manner by aligning their common molecular field or property characteristics [38,40]. This concept is known as bioisosterism where atoms or functional groups with similar properties are used for ligand designing [41] and has been one of the most common practices in the discovery of new leads in pharmaceutical research. This method of selection of pharmacophores is mainly based on simple superposition principles using the analogy of complementarity. The degree of complementarity between the molecular fields of the bioactive agent and its receptor should be directly related to the binding strength and relative activity of the agent [39,40]. Similarity may also be determined by comparing the molecular graphs [38]. The word "similarity" in the present discussion may be viewed as having a common bioisosteric group.

13.4 STEREOELECTRONIC FEATURES OF JH, JH MIMIC (IGRs), AND ARTHROPOD REPELLENTS

In pursuit of the goals for determining the "interaction pharmacophore" profiles, we focused [42–45] primarily on the calculated stereoelectronic properties of JH and a JH mimic (IGR) and compared the results with DEET and DEET-like repellents. These calculated properties eventually enabled us to develop a pharmacophore model for database search and design of new repellents for custom synthesis. These stereoelectronic property studies discussed here are presented in two sections: (1) steric property analysis and (2) electronic property analysis, respectively. Comparative analysis of stereoelectronic properties between natural insect JH, a synthetic insect JH mimic (IGR) undecen-2-yl carbamate, and *N*,*N*-diethyl-*m*-toluamide (DEET) together with several of its analogs revealed several significant molecular similarities. The observed similarities were important guidelines for developing the pharmacophore model for repellent activity.

Semiempirical quantum chemical (AM1) calculations [43,44] were performed to determine the stereoelectronic properties for analysis between the JH, the JH mimic (IGR), and the DEET analogs.

13.4.1 Steric Property Analysis

Structure of the DEET analogs is shown in Table 13.1. The stereoelectronic properties of JH were discussed separately in a later section. The calculated properties of

TABLE 13.1
Structure and Mosquito Repellent Protection Time of DEET and Its Analogs

a = aromatic ring or s = saturated ring

Compound	Structure	Protection Time (h)	Ring	R	$R_1 = R_2$	
1a	*m*-Toluamide (DEET)	5	a	3-CH_3	C_2H_5	
1b	Cyclohexamide	4	s	H	C_2H_5	
1c	*p*-Anisamide	1	a	4-OCH_3	C_2H_5	
2a	*o*-Chlorobenzamide	5	a	2-Cl	CH_3	
2b	*m*-Toluamide	3	a	3-CH_3	CH_3	
2c	*p*-Anisamide	1	a	4-OCH_3	CH_3	
3a	Benzamide	3	a	H	iC_3H_7	
3b	*p*-Anisamide	1.17	a	4-OCH_3	iC_3H_7	
3c	*p*-Toluamide	0.5	a	4-CH_3	iC_3H_7	
					R_1	R_2
4a	*m*-Toluamide	0.67	a	3-CH_3	H	C_2H_5
4b	Cyclohexamide	0.5	s	H	H	C_2H_5
4c	*o*-Ethoxybenzamide	0.08	a	2-OC_2H_5	H	C_2H_5
					N, R_1, R_2	
5a	Benzamide	3	a	H	Piperidine	
5b	*m*-Toluamide	1.42	a	3-CH_3	Piperidine	
5c	*p*-Anisamide	0.75	a	4-OCH_3	Piperidine	

Sources: Ma, D. et al., *Am. J. Trop. Med. Hyg.*, 60, 1, 1999; Bhattacharjee, A.K. et al., *J. Mol. Recognit.*, 13, 213, 2000.

TABLE 13.2
Similarities of a Few Selected Steric Properties between JH Mimic (IGR) and the Analogs: Nonbonded Distances and Total Surface Area Containing the C_7, O, N, and C_{R1} Atoms

	N···O=C (Å)	C=O···C_{R1}, Å	Surface area (Å^2) C_7, O, N, C_{R1} Atoms
JH mimic	2.345	2.871	11.5
1a	2.296	2.751	13.6
1b	2.282	2.707	13.6
1c	2.296	2.746	13.6
2a	2.293	2.761	13.6
2b	2.286	2.749	13.6
2c	2.285	2.744	13.6
3a	2.297	2.810	11.7
3b	2.296	2.806	11.7
3c	2.296	2.809	11.7
4a	2.296	2.811	11.7
4b	2.292	2.815	11.7
4c	2.298	2.803	11.7
5a	2.293	2.761	13.6
5b	2.293	2.761	13.6
5c	2.289	2.757	13.6

Source: Bhattacharjee, A.K. et al., *J. Mol. Recognit.*, 13, 213, 2000.

the JH mimic (IGR) and the DEET analogs are shown in Tables 13.2 and 13.3. The chemical structures of JH, JH mimic (IGR), and the DEET analogs are shown in Figures 13.1 and 13.2, respectively. Structurally, DEET, its analogs (Table 13.2), and the JH mimic have the N-C=O fragment (Table 13.2, Figure 13.2), and therefore, the compounds are likely to have a similar steric recognition effect at the receptor. Although JH is somewhat structurally different with a –O-C=O fragment, the steric recognition effect at the receptor will be similar because both N-C=O and –O-C=O fragments have bent geometry with the carbonyl oxygen atom at the apex. However, similarity in molecular electronic shape, not solely the similarity in chemical structure, had long been recognized as the dominant factor for olfactory sensations [46]. Thus, utilizing data from calculated stereoelectronic properties of one of our earlier published study [43,44] on predicting mosquito repellent activity, we performed similarity analysis between JH, JH mimic, and the repellents. The following sections describe details of the analysis.

13.4.1.1 Conformational Energies

The lowest energy conformers of the compounds were identified by systematic rotation of the single bonds. This procedure generated several hundred conformers for each of the compounds (JH mimic and the DEET compounds) identifying the low energy

TABLE 13.3
Similarities of a Few Selected Steric Properties between JH Mimic (IGR) and the Analogs: A Few Selected Structural Parameters of JH-Mimic and DEET Compounds

	Bond Distance (Å)		Bond Angle (°)			Dihedral Angle (°)	
Compound	$C_7{=}O$	N-C_7	N-C_{R1}	N-$C_7{=}O$	C_{R1}-N-C_7	C_{R1}-N-$C_7{=}O$	C_{R1}-N-C_7-C_1 or C_{R1}-N-C_7-O
JH mimic	1.241	1.388	1.442	127.8	119.1	23.0	−161.1
1a	1.247	1.392	1.447	120.8	119.5	2.6	179.9
1b	1.248	1.393	1.448	119.4	118.5	4.6	−178.6
1c	1.247	1.393	1.447	120.7	120.7	2.2	179.5
2a	1.246	1.387	1.437	120.9	120.2	7.8	−174.9
2b	1.248	1.387	1.436	120.2	120.4	3.2	−176.9
2c	1.248	1.388	1.437	120.0	120.3	−3.2	−176.6
3a	1.248	1.384	1.436	121.4	122.2	8.1	−173.1
3b	1.248	1.385	1.436	121.2	122.1	8.0	−173.2
3c	1.248	1.385	1.436	121.3	122.2	8.1	−173.1
4a	1.248	1.384	1.437	121.3	122.3	7.5	−173.5
4b	1.247	1.380	1.437	121.3	122.8	7.0	−174.9
4c	1.249	1.384	1.439	121.4	121.5	11.6	−173.5
5a	1.247	1.396	1.450	120.5	119.7	−10.7	173.0
5b	1.247	1.391	1.494	120.6	119.7	−10.4	173.3
5c	1.248	1.391	1.492	120.2	119.7	−12.5	171.2

Source: Bhattacharjee, A.K. et al., *J. Mol. Recognit.*, 13, 213, 2000.

conformers along with their corresponding Boltzmann population densities. The lowest energy conformer ranges from 75.6% to 80% in population densities, whereas the higher energy conformers were present in varying population densities ranging from 6% to 0.05%, with energies greater than 5.0 kcal/mol from the lowest energy conformer. In the lowest energy conformers, the amide moiety in both the JH mimic and the DEET compounds [43,44] was planar and superimposable with each other with the nonbonded distances, N···O and C=O···C_{R1}, within 0.1 Å of each other (Table 13.2).

The surface area and volume of the amide-containing portion of the JH mimic and the DEET compounds showed significant similarity. The surface area containing the amide C_7, O, N, and C_{R1} atoms of the JH mimic and the DEET compounds ranged from 11.5 to 13.6 Å^2 (Table 13.3), respectively, whereas the calculated steric bulk of this amide portion in JH mimic was 8.7 Å^3, and for the DEET compounds, it was 8.3 Å^3. Because steric complementarity is a prerequisite for ligand–receptor recognition, active biological agents with a common receptor binding site should possess sterically similar binding surfaces [47]. This contribution reflects molecular size and overall shape, features important to the steric complementarity of ligand at the binding site [47]. Bond distances and angles of the amide moiety in DEET and its

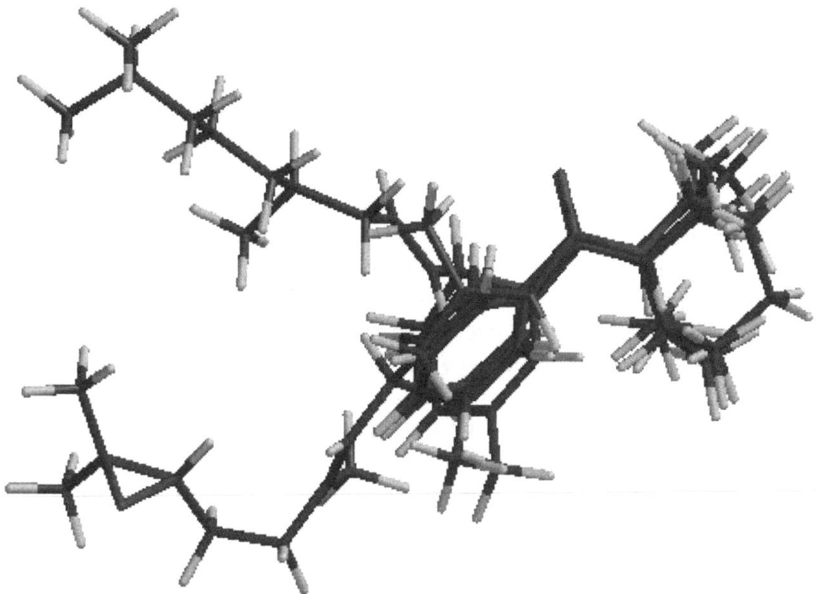

FIGURE 13.2 Superimposition of JH, JH mimic, and DEET to illustrate the steric similarity of the pharmacophore. (From Bhattacharjee, A.K. et al., *J. Mol. Recognit.*, 13, 213, 2000.)

analogs were also similar to those in the JH mimic (Table 13.3). The bond distances of JH mimic, DEET, and DEET's most repellent analogs were within 0.007 Å of each other. The bond angles and the dihedral angles differ from each other by up to 8° and 7° to about 19°, respectively, a reasonable variation keeping in mind the large intrinsic differences in the geometry of the molecules.

13.4.1.2 Steric Properties of Natural JH

Conformational search calculations on the structure of natural JH molecule (where R = methyl group) identified three conformers of significant abundance with a relatively small energy difference of 3.9 kcal/mol between the maximum and the minimum energy conformer. An energy barrier of 3.9 kcal/mol can easily be surmounted in biological systems and statistically the distribution of all three conformers cannot be ruled out. The ester moiety was observed to be flat in all the three conformers, and the nonbonded distances O···O and $C{=}O \cdots C_{R1}$ were equal to 2.23 and 2.58 Å, respectively. The three conformers differed in the conformation of the alkyl chain due to rotations about the single bonds [43,44]. Although JH has an epoxide moiety at one end of the molecule, this functionality does not appear to be important for growth regulator activity since mimics lacking this functionality are potent growth inhibitors [12].

Thus, despite JH, JH mimic (IGR), and the DEET molecules having many degrees of conformational freedom of their structure, the main bioactive conformer, the ester, the carbamate moiety, and the amide group, were found to be superimposable, clearly suggesting steric similarity between the molecules (Figure 13.2).

13.4.2 Electronic Property Analysis

13.4.2.1 Molecular Electrostatic Potential and Hydrophobic Properties

The electronic similarity analyses between JH, JH mimic (IGR), and DEET compounds were primarily based on the analysis of electrostatic potential profiles, molecular orbital, and hydrophobic properties of the common amide fragments. Electrostatic potential is a very important experimentally determinable property of a molecule, which can provide a wealth of information about its intrinsic reactivity. Electrostatic potential surfaces that a molecule's nuclei and electrons create in the surrounding space have been well documented as key recognizable features that can promote this type of recognition interactions between molecules at longer distances [40]. Quantum chemical methods provide an accurate estimate of these properties. The electrostatic potential profiles on the van der Waals surface and beyond the van der Waals surface are believed to be the most important features responsible for recognition interaction between an approaching molecule and its receptor at longer distances of separation [48–52]. It is through this potential that a molecule recognizes its receptor and accordingly promotes interaction between the complimentary sites. Although JH mimic is more structurally similar to the DEET compounds than the natural JH as JH has an ester rather than an amide moiety, the electrostatic characteristics of the DEET compounds, JH, and the JH mimic were found to be similar (Figures 13.3 through 13.5) and, therefore, likely to result in a similar recognition interaction with the receptor to promote binding with the receptor. The electrostatic characteristics include Mulliken charges, electrostatic potentials at the van der Waals surface, dipole moment, and the profiles of electrostatic potential beyond the van der Waals surface. This complementarity essentially means that the charge distribution of a substrate has to find its counterpart at the binding sites in order to allow

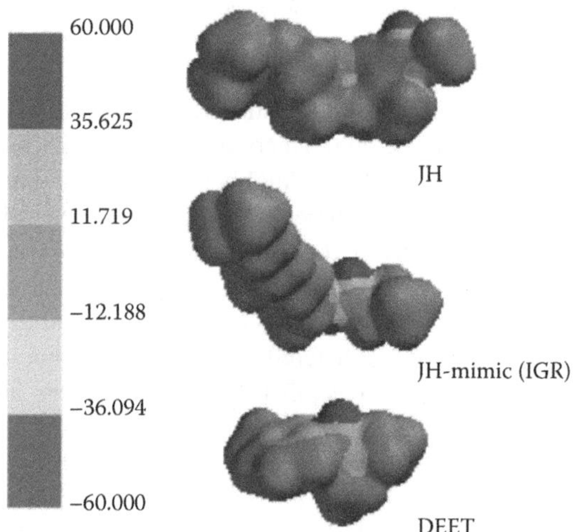

FIGURE 13.3 **(See color insert.)** MEP maps of JH, JH mimic, and DEET on the van der Waals surface showing similarity of the single nucleophilic site and large area of hydrophobic similarity.

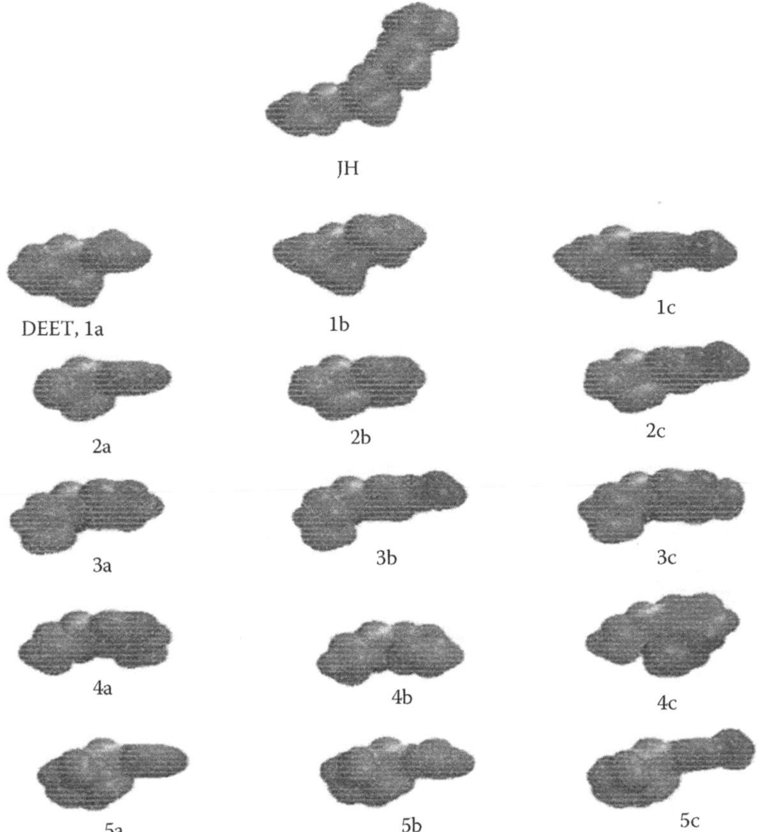

FIGURE 13.4 **(See color insert.)** MEP maps of JH and DEET-like compounds showing electronic similarity of the carbonyl group. (From Bhattacharjee, A.K. et al., *J. Mol. Recognit.*, 13, 213, 2000.)

maximum interaction with the receptor [51]. It works like a magnet between them and, thereby, contributes to the binding affinity. The calculated Mulliken charges and the electrostatic potential at the amide atoms in JH mimic and DEET and its analogs are presented in Table 13.4. The calculated charge densities on the amide nitrogen atom and the C7 atom were found to be close between JH mimic and the DEET compounds, keeping in mind the diverse nature of the substituents in other parts of the molecules. The charge of the carbonyl oxygen atom of JH mimic was −0.02 electrons more negative than for DEET and its analogs, whereas the negative potential by the carbonyl oxygen atom, −73.2 kcal/mol, was within the same range as calculated for DEET and its analogs, −73.1 to −77.3 kcal/mol (Table 13.4).

The carbonyl oxygen atom was also the site for the most negative potential, represented as the deepest red color (Figures 13.3 and 13.4) in DEET, its analogs, JH, and JH mimic, suggesting the carbonyl oxygen atom to be the most nucleophilic site in all the molecules as this site showed the maximum localized electron density. Although the calculated dipole moment was somewhat lower in JH mimic than in

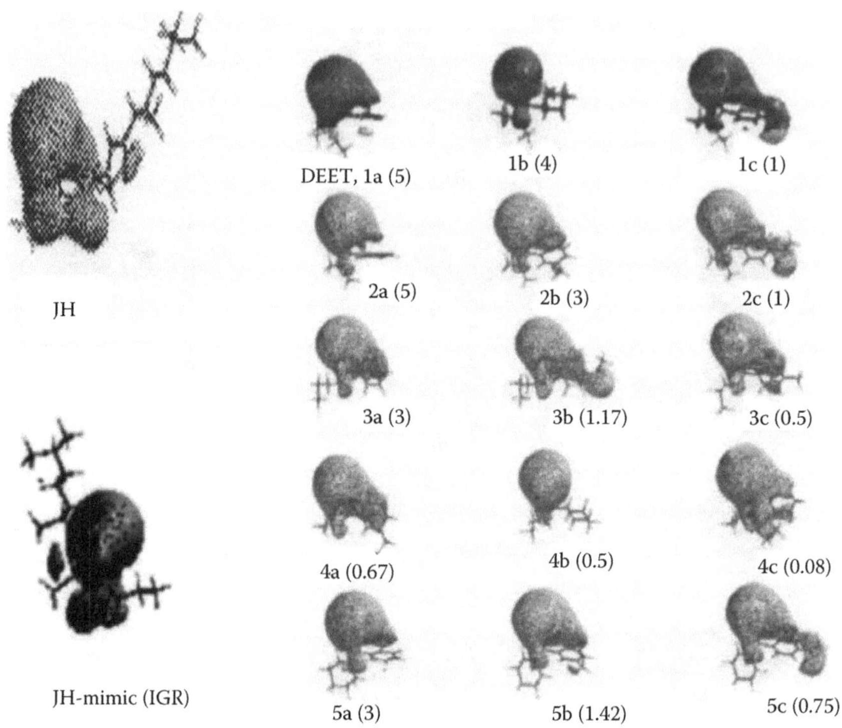

FIGURE 13.5 MEP maps of JH, JH mimic, DEET, and DEET-like compounds beyond the van der Waals surface (approximately 1.4 A away) showing similarity of the large extended area by the carbonyl group as the common recognition feature. (From Bhattacharjee, A.K. et al., *J. Mol. Recognit.*, 13, 213, 2000.)

the DEET compounds, the dipole moment was pointing toward the carbonyl oxygen atom in all the compounds. Thus, the carbonyl oxygen atom can be considered to be the most reactive site in JH, JH mimic and the DEET compounds.

Conversely, the site for the most positive potential is the most electrophilic or acidic because this site has the minimum electron density. In JH mimic, the most positive potential was found to be located by the amide hydrogen atom with a value of 37.3 kcal/mol, whereas in DEET and its analogs, it was scattered around different hydrogen atoms in the molecules, in the range of 15.4–35.2 kcal/mol. However, a large area on the molecular surface of JH, JH mimic, and DEET showed weak electrostatic potentials (green-colored regions) ranging from −12.0 to 11.0 kcal/mol (Figures 13.3 and 13.4) indicating strong hydrophobic similarity. This characteristic pattern was also observed on the molecular surface of other DEET analogs (Figure 13.4). Large hydrophobic regions in the molecule appeared to be necessary for both recognition as well as potent repellent activity. Hydrophobic effects are the result of averaged electrostatic interaction of the molecule with its surroundings, solvent, and protein environment. Sites of nonpolar or weakly polar region in different molecules tend to come together in order to escape contact with water and to minimize the dehydration free energies [39].

TABLE 13.4
Similarities of a Few Selected Electronic Properties between JH Mimic (IGR) and the Analogs: Comparison of Dipole Moments, Mulliken Charges (Electrons), and MEPs

	Carbonyl O Atom		Amide N Atom		C_7 Atom	
Compound	**Charge**	**MEP[a]**	**Charge**	**MEP[a]**	**Charge**	**Dipole Moment Debye**
JH mimic	−0.39	−73.2	−0.32	−37.3	0.38	2.02
1a	−0.35	−75.0	−0.33	−23.2	0.34	3.68
1b	−0.37	−74.5	−0.33	−25.1	0.30	3.25
1c	−0.36	−75.5	−0.33	−30.0	0.35	3.55
2a	−0.35	−73.1	−0.33	−22.8	0.35	3.82
2b	−0.36	−76.2	−0.34	−17.0	0.35	3.45
2c	−0.37	−75.7	−0.34	−18.7	0.35	4.37
3a	−0.36	−75.9	−0.37	−20.3	0.34	3.63
3b	−0.37	−76.7	−0.37	−27.8	0.34	3.25
3c	−0.36	−77.3	−0.37	−22.1	0.34	3.74
4a	−0.36	−74.5	−0.37	−25.8	0.34	3.22
4b	−0.37	−75.4	−0.38	−26.7	0.30	3.46
4c	−0.37	−75.7	−0.35	−38.6	0.35	3.55
5a	−0.35	−73.8	−0.31	−33.0	0.34	3.52
5b	−0.35	−75.6	−0.31	−33.1	0.34	3.56
5c	−0.36	−75.7	−0.32	−30.2	0.35	3.27

Source: Bhattacharjee, A.K. et al., *J. Mol. Recognit.*, 13, 213, 2000.

[a] Most negative electrostatic potential located by indicated atom expressed in kcal/mol superimposed on the isodensity surface (0.002 e/au^3) of the molecule

Thus, matching of the nonpolar regions of ligands and receptor sites gives a reasonable measure of hydrophobic complementarity and represents the stabilization of the enzyme-substrate or ligand–receptor complex. Polarity of a certain region in the molecule can be regarded as proportional to the electrostatic field. A strong electrostatic field of a molecule attracts molecules having large dipoles, such as water, while the weak electrostatic field regions of the molecule do not attract water molecules and are, therefore, hydrophobic [39,40]. Different approaches have recently appeared to theoretically represent hydrophobic interactions in terms of local solute–solvent electrostatics [52]. However, a simple assessment of hydrophobic similarity may be carried out by determining the distribution of charges or electrostatic potentials at different regions on the van der Waals surface of the molecule. The observed low-dipole moments (<4 Debye) of JH mimic and the DEET analogs also support the lipophilic nature of the compounds. Because olfactory sensations of the insects require some degree of lipid solubility [26], hydrophobicity or lipophilicity of the repellents is likely to be an important factor for potent repellent activity.

The profiles of electrostatic potential beyond the van der Waals surface at a constant potential of –10.0 kcal/mol were also observed to be comparable with a large negative potential region localized by the carbonyl moiety in JH, JH mimic, DEET, and DEET analogs (Figure 13.5). The carbonyl oxygen atom of all the JH conformers was the most nucleophilic site, being the most negative potential site on the van der Waals surface in the molecule. It varied between −66.1 and –71.3 kcal/mol for the maximum energy conformers. The electrostatic potential feature beyond the van der Waals surface by the carbonyl oxygen atom of both JH and JH mimic was observed to be similar to the DEET analogs as it had the largest –10 kcal/mol potential surface (Figure 13.5).

The electrostatic potentials of functional groups that are commonly found in diphenylether and terpenoid JH mimics were found to be similar to each other. This electrostatic bioisosterism had enabled an understanding of universality of active structures and aided the design of new active analogs [48]. Therefore, the observed electrostatic similarity between JH, JH mimic (IGR), and DEET analogs was very significant as it clearly demonstrated electrostatic bioisosterism between these compounds and, thus, provided a powerful template for design of novel insect repellents.

In addition, another feature was noteworthy and that was the localized negative potential region by the amide moiety in DEET and its analogs. The compounds appeared to be qualitatively linked to their potent repellent activity, with the less potent repellent compounds having a more extended and, therefore, more diffuse, negative potential zone (Figure 13.5). It appeared that a more localized negative potential region toward the amide group, as seen with JH mimic, was consistent with higher protection times. Because of the similarity of negative potential profiles at −10.0 kcal/mol, a long-range electrostatic bioisosterism would exist in these molecules that would likely be responsible for repellent activity and should be useful for design of this class of insect repellents.

13.4.2.2 Molecular Orbital Properties

The DEET compounds showed similar energy gaps between the highest occupied molecular orbital (HOMO) and the lowest unoccupied molecular orbital (LUMO) as seen by the η values in Table 13.5, an index of intrinsic reactivity [53]. These η values indicated that the DEET compounds were nearly similar in intrinsic reactivity. The energy of HOMO and LUMO orbitals plays a major role in governing chemical reactions. The energy difference between the orbitals is known as the electronic bandgap and is often responsible for the formation of many charge–transfer complexes [53]. Table 13.5 shows a relatively constant E_{HOMO} with a large negative value and a more variable and much smaller in magnitude E_{LUMO}, implying a greater role of LUMO or electron acceptor ability of the compounds than their electron-donating power. Therefore, the electron transfer from a suitable receptor molecular orbital to the LUMO of the DEET compounds, rather than a donation of electrons from the DEET compounds, seemed a more plausible mechanism for the compounds.

The HOMO, LUMO, and reactivity index of JH range from −9.574 to −9.602 eV, −0.038 to 0.025 eV, and 4.76 to 4.81 kcal/mol, respectively. The dipole moment of the JH conformers varied from 1.83 to 4.45 Debye. Thus, clearly there exists an

TABLE 13.5
Similarities of a Few Selected Electronic Properties between JH Mimic (IGR) and the Analogs: Comparison of HOMO and LUMO Eigenvalues

	Eigenvalues (eV)		
Compound	**HOMO**	**LUMO**	$\eta = (E_{LUMO} - E_{HOMO})/2$
JH mimic	−9.847	0.98	5.41
1a	−9.542	0.146	4.84
1b	−9.555	1.514	5.53
1c	−9.207	0.137	4.67
2a	−9.589	−0.187	4.70
2b	−9.518	0.027	4.77
2c	−9.256	0.011	4.63
3a	−9.854	−0.090	4.88
3b	−9.274	−0.063	4.60
3c	−9.599	−0.105	4.74
4a	−9.593	−0.051	4.77
4b	−9.889	1.536	5.71
4c	−9.319	−0.047	4.63
5a	−9.514	0.111	4.81
5b	−9.495	0.151	4.82
5c	−9.236	0.111	4.67

Source: Bhattacharjee, A.K. et al., *J. Mol. Recognit.*, 13, 213, 2000.

electronic similarity between the natural JH molecule, the JH mimic, and the DEET analogs that further strengthened the idea of electronic bioisosterism between the molecules, suggesting a useful clue for designing new repellents.

13.4.3 Calculation of Stereoelectronic Properties

Computational calculations were performed using SPARTAN version 5.0 [54] running on a Silicon Graphics Octane workstation. Detailed conformational search calculations of JH, JH mimic, and the DEET compounds were performed by multiple rotations of single bonds in the compounds, thereby generating several low-energy conformers with varying population densities. The most abundant and the lowest energy conformers were identified. The geometry of these conformers was optimized, and the electronic properties were calculated on the optimized geometry. Geometry optimization and energy calculations were performed in the gas phase state of the compounds using the semiempirical AM1 quantum chemical method as implemented in SPARTAN. 3D MEP maps for all compounds were generated using SPARTAN graphics [54] at the AM1-optimized geometry of the molecules.

The MEPs were sampled over the entire accessible surface of a molecule (corresponding roughly to the van der Waals contact surface) and into space extending beyond the molecular surface, providing a measure of charge distribution from the point of view of an approaching reagent. The regions of negative potential indicate areas of excess negative charges and, therefore, suitable attraction sites for the positively charged test probe or the nucleophilic site in the molecule. Conversely, the regions of positive potential indicate areas of excess positive charges and, therefore, suitable repulsion sites in the molecule for the positively charged test probe or the electrophilic site in the molecule.

13.5 PHARMACOPHORE MODEL FOR ARTHROPOD REPELLENT PROPERTY

Since JH, JH mimic, and DEET compounds showed the previously discussed stereoelectronic similarities, particularly the electrostatic potential bioisosterism, we continued our efforts to utilize these observations for discovery and design of new insect repellents by developing a pharmacophore model. We utilized the 3D QSAR-CATALYST methodology [55] on known structurally diverse insect repellent compounds, including DEET to develop the model [45].

13.5.1 Pharmacophore Analysis

The generated chemical feature-based 3D pharmacophore for arthropod repellent activity was found to contain two aliphatic hydrophobic functions, one aromatic hydrophobic (aromatic ring) function and one hydrogen-bond acceptor function in specific geometric locations surrounding the molecular space (Figure 13.6). This implies that an arthropod repellent compound needs to have the aforementioned physiochemical characteristics for potent activity.

Although the model was generated from only DEET and its analogs, it was found to be consistent with the observed quantum chemically calculated stereoelectronic properties of JH, JH mimic (IGR), and the DEET compounds. The large hydrophobic regions (weak electrostatic potentials) observed in JH, JH mimic, and DEET compounds (Figure 13.3 through 13.5) clearly corroborate with the three hydrophobic functions of the model, and the hydrogen-bond acceptor feature corresponds to the localized negative potential region (nucleophilic) by the carbonyl oxygen atom in these molecules.

The pharmacophore model was generated by creating a training set of 11 structurally diverse known arthropod repellent compounds (Chart 13.1) having a broad range of repellent activities as shown in Table 13.6. The repellent activity of the 11 repellent compounds in the training set that includes DEET covered a broad range of activity, from an ED_{50} of about 1 μg/cm^2 to about 50 mg/cm^2.

Earlier reported [42–44] results of stereoelectronic properties of these compounds provided guidance for selection of these physiochemical features. A plot of the protection (repelling) time conferred by the compounds in the training set and their predicted protection (repelling) time demonstrated a good correlation (r=0.91) within the range of uncertainty 3, indicating the predictive power of the

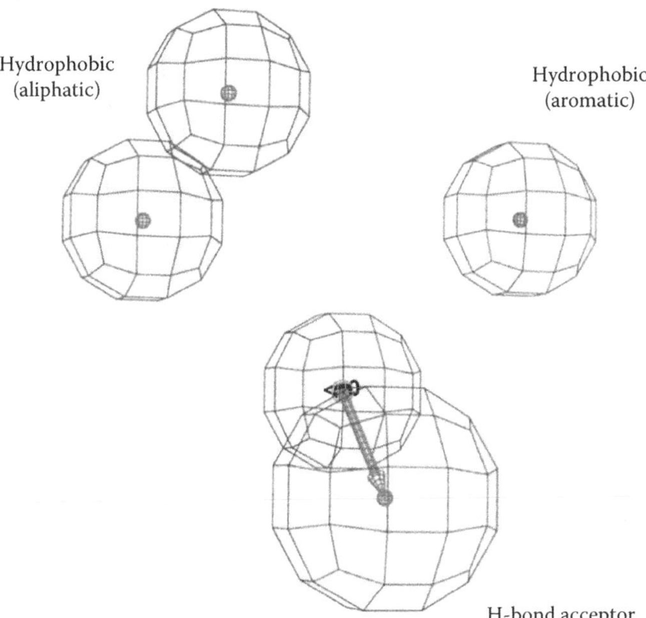

FIGURE 13.6 Pharmacophore model for insect repellent activity.

pharmacophore [45]. Using the model, we discovered several new compounds through our in-house database searches to validate the predictive power of the pharmacophore model [45,56–58]. All these compounds mapped the pharmacophore in varying degrees [45,56]. To further examine the validity of the pharmacophore, we mapped it onto an interesting literature reported compound, a novel 18-carbon acid, isolated from samples of greasy gaur hair, which was found to have insect repellent activity that can function both as a landing and feeding deterrent to the mosquitoes [45,57]. Surprisingly, the pharmacophore mapped extremely well on this molecule proving the consistency in the predictive power insect repellent activity of the model (Figure 13.7).

For the database search, each compound in the database was converted into 3D multiconformations and stored in the database. After screening, the down selection of the identified compounds was carried out by evaluating the *in silico* ADME/toxicity properties and choosing only those compounds that have favorable properties. ADME/toxicity evaluations were carried out by using Cerius2 [59] and TOPKAT [60] methodologies. The procedures for compound identification and down selection were carried out by generating several shape-based pharmacophore templates on the potent analogs of the training set (Chart 13.1). This approach allowed addition of steric attributes along with the pharmacophore features necessary for binding of the repellents at the yet unknown target. Database searches based on this procedure resulted in the identification of both similar and new compounds. Ultimately, we were able to short-list four compounds that were found to exhibit remarkable repellent activity fulfilling the important goals of an ideal repellent [45,56].

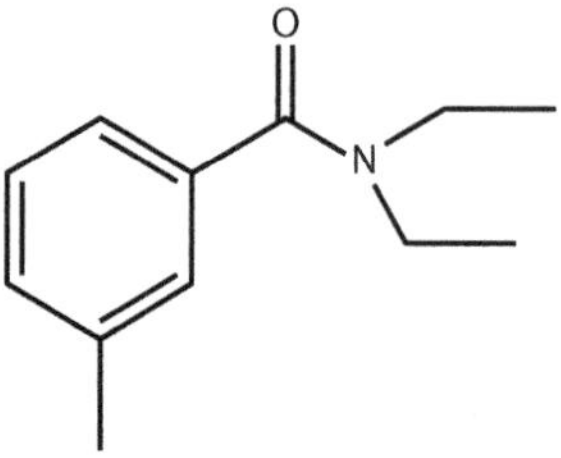

(a)

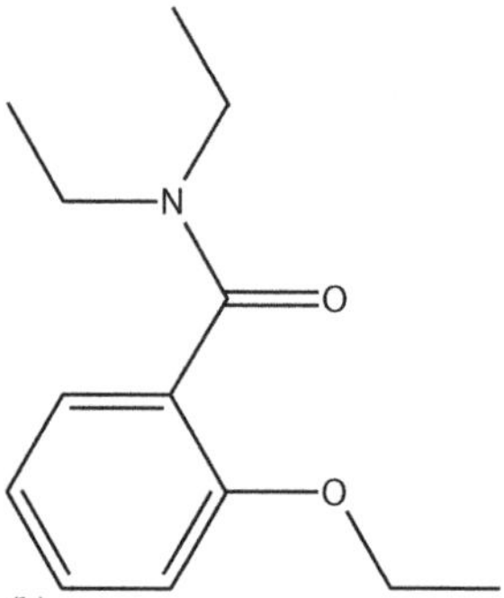

(b)

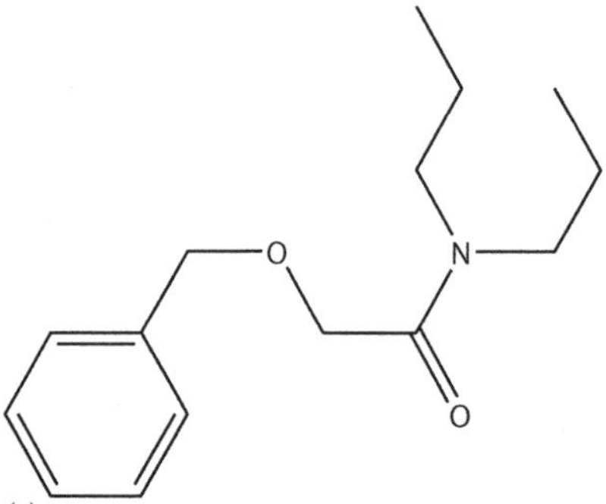

(c)

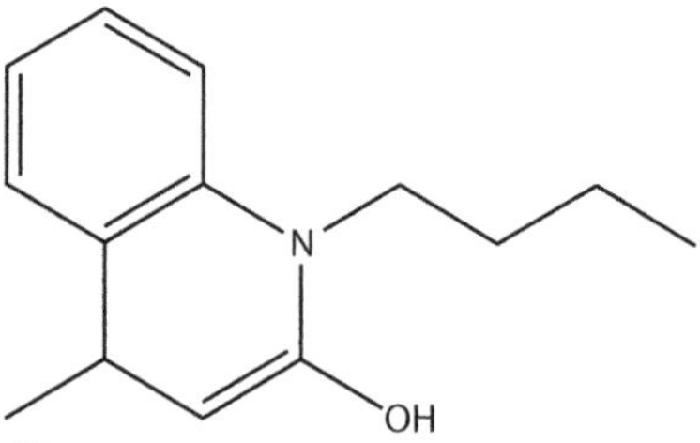

(d)

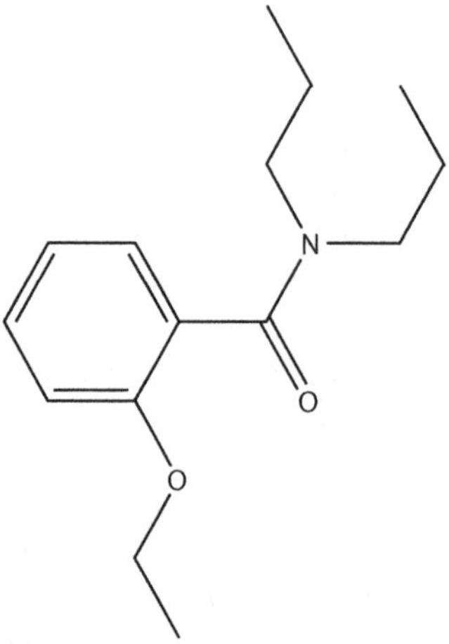

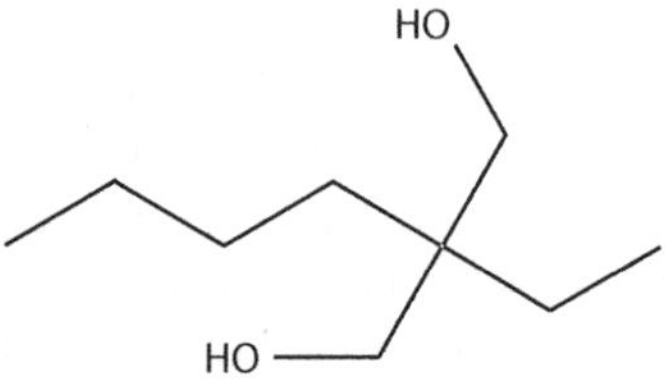

(f)

(e)

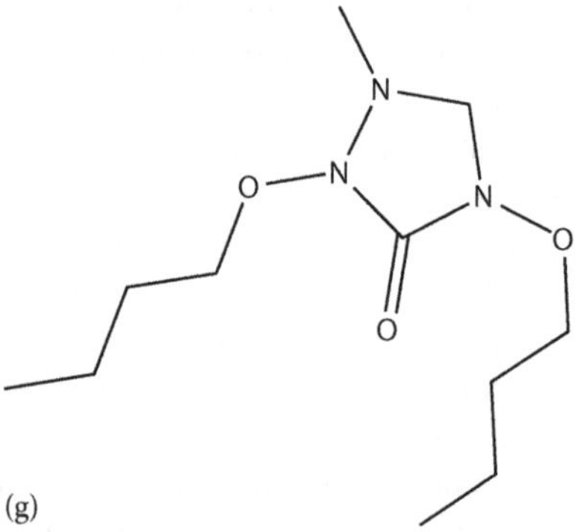

(g)

CHART 13.1 Structure of 11 diverse arthropod repellent compounds for creating the training set to develop the pharmacophore model. Names and repellent activities of these compounds are shown in Table 13.6.

(continued)

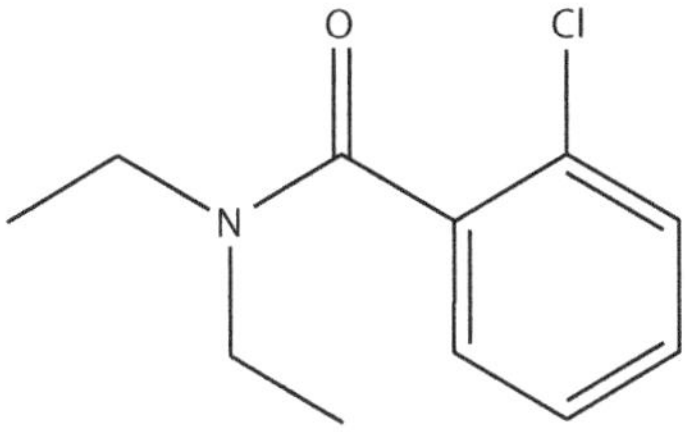

(h) 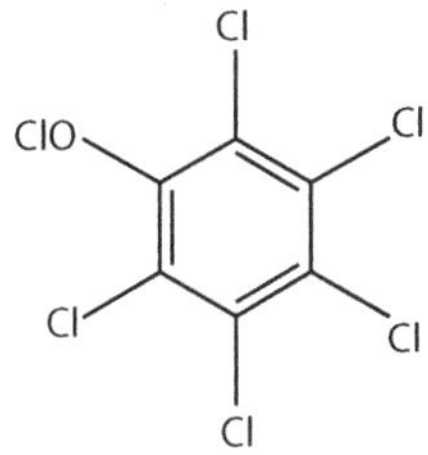 (i)

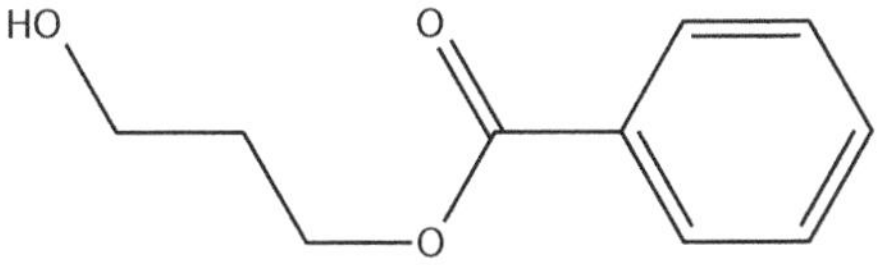

(j)

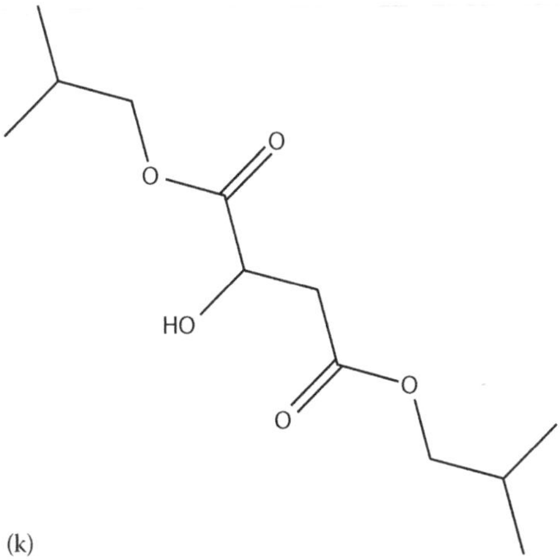

(k)

CHART 13.1 (continued) Structure of 11 diverse arthropod repellent compounds for creating the training set to develop the pharmacophore model. Names and repellent activities of these compounds are shown in Table 13.6.

13.5.2 Methodology for Pharmacophore Development

Pharmacophore development was performed using CATALYST version 4.8 software [55]. The algorithm treats molecular structures as templates composed of chemical functions localized in space that will bind effectively with complementary functions on the respective binding proteins. The most relevant biological features are extracted from a small set of compounds that cover a broad range of activity [61]. This enables the use of structure and activity data for a set of lead compounds to generate a pharmacophore characterizing the activity of the lead set. Structures of the arthropod repellent compounds (Table 13.6) were edited within CATALYST and energy minimized to the closest local minimum using the generalized CHARMM-like force field as implemented in the program. Molecular flexibility was taken into

TABLE 13.6
Names and Activity of the Repellents Used to Create the Training Set for Pharmacophore Development

Compound	Name	Repellent Activity (in h of PT[a])
1	DEET (*N*,*N*-diethyl-*m*-toluamide)	1.0
2	*N*,*N*-Diethyl-2-ethoxybenzamide	0.5
3	*N*,*N*-Dipropyl-2-benzyloxyacetate	0.5
4	1-Butyl-4-methylcarbostyril	2.0
5	*N*,*N*-Dipropyl-2-ethoxybenzamide	0.3
6	2-Butyl-2-ethyl-1,3-propanediol	1.7
7	1,3-Bisbutoxymethyl-2-imidazol	0.6
8	*N*,*N*-Diethyl-2-chlorobenzamide	1.2
9	Hexachlorophenol	0.2
10	1,3-Propanediolmonobenzoate	7.5
11	Diisobutylmalate	2.5

Source: Bhattacharjee, A.K. and Gupta, R.K., *J. Am. Mosq. Control Assoc.*, 21, 23, 2005.

[a] PT is protection time in hours afforded by the repellent compounds. Nulliparous female *A. aegypti* mosquitoes (red-eye Liverpool strain) were used [64]. These were laboratory-reared and maintained at 28°C and 80% RH under a photoperiod of 12:12 (L:D) h using standard mosquito-rearing procedures [67,70].

account by considering each compound as an ensemble of conformers representing different accessible areas in a 3D space. The "best searching procedure" was applied to select representative conformers within 10 kcal/mol of the global minimum [62].

Conformational models of the training set of 11 repellents (Table 13.6) were generated that emphasize representative coverage within a range of permissible Boltzmann population with significant abundance (within 10.0 kcal/mol) of the calculated global minimum. This conformational model was used for pharmacophore generation to identify the best 3D arrangement of chemical functions, such as hydrophobic regions, hydrogen-bond donor, hydrogen-bond acceptor, and positively and/or negatively ionizable sites, distributed over a 3D space explaining the activity variations among the training set. The hydrogen-bonding features were vectors, whereas all other functions were points.

Details of the pharmacophore model generation were reported earlier [45]. Briefly, the model was developed by setting the default parameters in the automatic generation procedure in CATALYST [55] such as function weight 0.302, mapping coefficient 0, resolution 260 pm, and activity uncertainty 3. An uncertainty "Δ" indicates an activity value lying somewhere in the interval from "activity divided by Δ" to "activity multiplied by Δ." The statistical relevance of the obtained pharmacophore was assessed on the basis of the cost relative to the null hypothesis and the correlation coefficient [35,55]. The pharmacophores were then used to estimate the activities of the training set and were derived from the best conformation generation

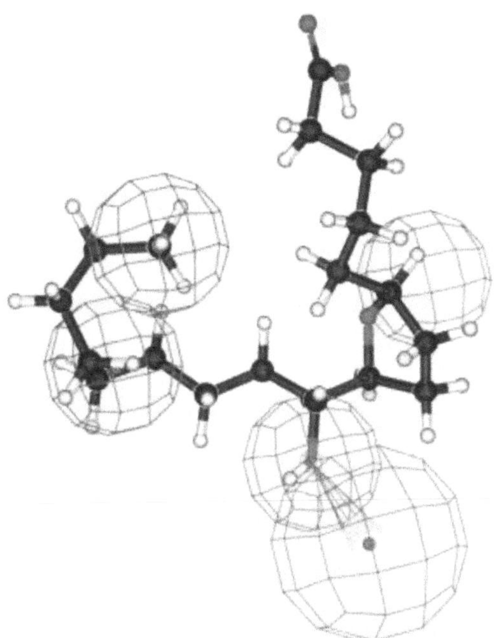

FIGURE 13.7 Mapping of the pharmacophore onto 5-[5-(1-hydroxy-nonyl)-tetrahydro-furan-2-yl]-pentanoic acid (a compound recently reported in the C & E News [57]) having repellent property in order to further cross validate the pharmacophore model.

model of the conformers displaying the smallest root-mean-square (RMS) deviations when projected onto the pharmacophore [45].

13.6 DESIGN, SYNTHESIS, AND EVALUATION OF NEW REPELLENTS

The aforementioned pharmacophore model provided a powerful template to design new repellents. From the feature characteristics of the model, the hydrogen-bond acceptor feature adjacent to an aromatic ring appeared significant. In addition, an aliphatic group (of different chain lengths) extending from the hydrogen-bond acceptor appeared to be necessary for repellent activity. This chain length could be modified to increase or decrease volatility and to affect the duration of repellent efficacy. Together, we organized the structural requirements of an insect repellent and hypothesized that three-substituted furans to have insect repellent properties as furans have a hydrogen-bond acceptor adjacent to the aromatic furan ring at the three-position with an aliphatic chain extending from it. The hydrogen-bond acceptor feature requirement was satisfied by an ether, ketone, or ester moiety in the furans. Indeed, the furans are known to have insect repellent properties [25–27]. Although the aliphatic chain length could be varied to obtain an overall molecular weight ranging from 182 to 251 g/mol, this would also vary the boiling point from as low as 180°C and to as high as 275°C.

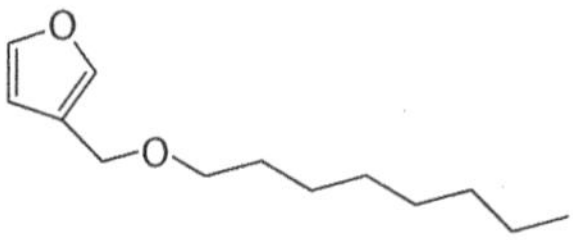

1 (3-furylmethyl octyl ether)

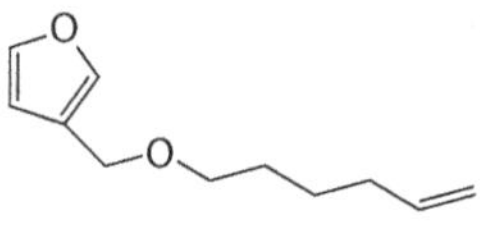

2 (3-furylmethyl-5-hexenyl ether)

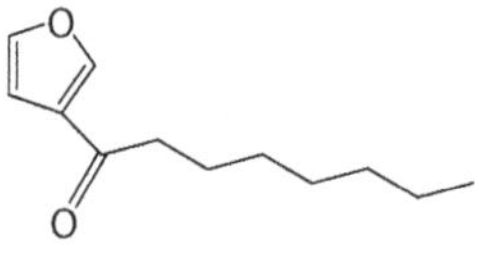

3 (3-furyl-heptyl ketone)

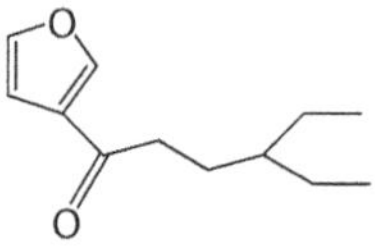

4 (3-furyl-(3-ethyl)pentyl ketone)

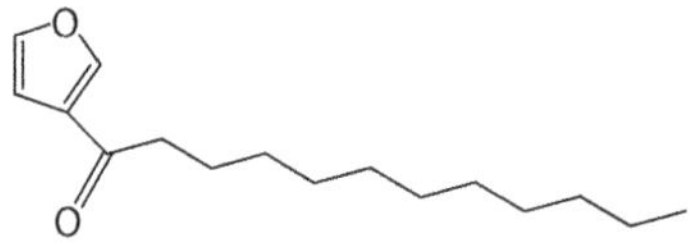

5 (3-furyl-undecyl ketone)

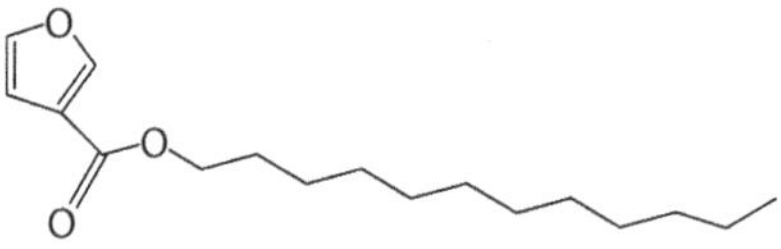

6 (Furan-3-carboxylic acid dodecyl ester)

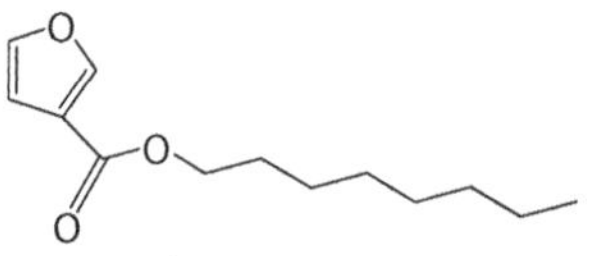

7 (Furan-3-carboxylic acid octyl ester)

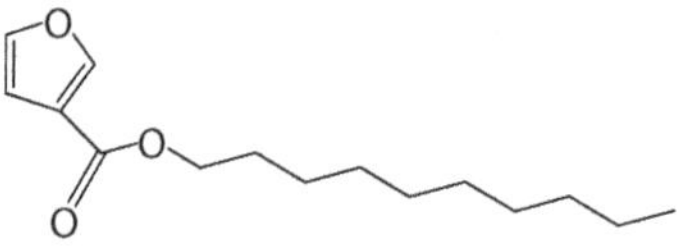

8 (Furan-3-carboxylic acid decyl ester)

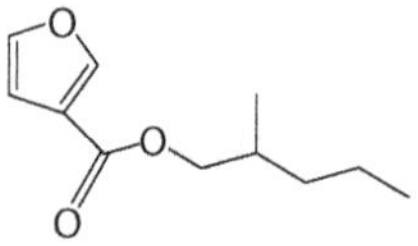

9 (Furan-3-carboxylic acid 2-methyl-pentyl ester)

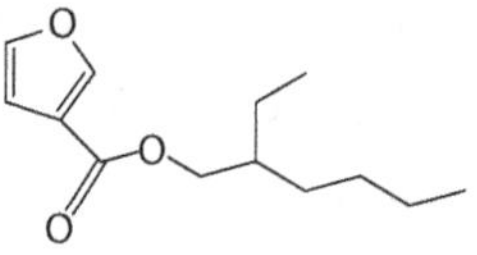

10 (Furan-3-carboxylic acid 2-ethyl-hexyl ester)

CHART 13.2 Structure of 10 new candidate repellents designed from the pharmacophore model and custom synthesis. (Repellent activities of these compounds are shown in Table 13.7).

Based on the aforementioned design strategy, 10 new candidate repellents (Chart 13.2) were custom synthesized [63]. Brief methods of synthesis of the compounds have been described in Appendix 13.A.1. These repellents include ethers, esters, and ketones. The compounds were as follows: **1** (3-furylmethyl octyl ether), **2** (3-furylmethyl-5-hexenyl ether), **3** (3-furyl-heptyl ketone), **4** (3-furyl-(3-ethyl) pentyl ketone), **5** (3-furyl-undecyl ketone), **6** (furan-3-carboxylic acid dodecyl ester), **7** (furan-3-carboxylic acid octyl ester), **8** (furan-3-carboxylic acid decyl ester),

TABLE 13.7
***In Vitro* Repellency of DEET and 10 Newly Synthesized Repellents against *A. aegypti* (mg/cm²)**

Repellents	ED_{50}[a]	ED_{95}[b]	R^2	G[c]
DEET	0.017±0.027–0.036	0.129±0.184–0.337	0.99	0.09
1	0.035±0.047–0.060	0.203±0.326–0.775	0.99	0.1
2	0.041±0.063–0.10	0.197±0.415–3.42	0.97	0.25
3	0.000034±0.001	0.083±0.131–0.400	0.97	0.28
4	0.021±0.032–0.042	0.082±0.109–0.170	0.98	0.11
5	0.046±0.055–0.067	0.390±0.639–1.39	0.99	0.05
6	0.019±0.032–0.043	0.082±0.113–0.195	0.98	0.14
7	0.012±0.023–0.033	0.080±0.110–0.187	0.98	0.14
8	0.029±0.034–0.038	0.240±0.303–0.410	0.99	0.02
9	0.047±0.052–0.057	0.329±0.411–0.541	0.99	0.01
10	0.023±0.0502–0.092	0.090±0.164–1.49	0.95	0.43

Sources: Dheranetra, W. et al., Comparative study of four membranes for evaluation of new insect/arthropod repellents using *Aedes aegypti*, in *Arthropod Borne Viral Infections—Current Status and Research*, D. Raghunath and C. Durga Rao, eds., *The Eighth Sir Dorabji Tata Symposium*, Tata McGRAW Hill Publishing Company Ltd, New Delhi, India, pp. 418–424, 2008.

[a] ED_{50} is the concentration required to repel 50% of the mosquito population.

[b] ED_{95} is the concentration required to repel 95% of the mosquito population.

[c] G is probability indicating the statistical confidence for the model. The median effective dose to repel 50% of the mosquito test population was calculated by the method of Goldstein (chemical sensitivity levels calculated from the dose–response regression equation) for single curves with graded responses [70]. Significant differences were then determined by comparing the 95% confidence level among effective doses. Statistical differences between the four membranes compared using a one-way analysis of variance (ANOVA) [64].

9 (furan-3-carboxylic acid 2-methyl-pentyl ester), and **10** (furan-3-carboxylic acid 2-ethyl-hexyl ester) (Chart 13.2). The repellent activity (ED_{50} and ED_{95} values) of these 10 compounds is shown in Table 13.7 and the *in vitro* method for evaluation of repellency [64] is presented in Appendix 13.A.2.

Inspection of Table 13.7 indicates that five of the ten repellent compounds—**3** (3-furyl-heptyl ketone), **4** (3-furyl-(3-ethyl)pentyl ketone), **6** (furan-3-carboxylic acid dodecyl ester), **7** (furan-3-carboxylic acid octyl ester), and **10** (furan-3-carboxylic acid 2-ethyl-hexyl ester)—exhibited superior repellency compared to DEET. Candidate repellent compound **3** (3-furyl-heptyl ketone) provided the best repellent activity using the *in vitro* test system (Table 13.7). Although compound 3 showed the best efficacy, there was no significant difference of repellent activity observed in compound 4, 6, 7, and 10. The repellent efficacy of these compounds was found to be either better or equal to DEET, suggesting the predictive power of the pharmacophore model. However, repellency potential of three other compounds—**2** (3-furylmethyl-5-hexenyl ether), **5** (3-furyl-undecyl ketone), and **9** (furan-3-carboxylic acid 2-methyl-pentyl ester)—was less than DEET.

The pharmacophore mappings onto the 10 new candidate repellents are shown in Figure 13.8a and b. Mapping of all the features (one hydrogen-bond acceptor site, two aliphatic sites, and one aromatic ring site) of the pharmacophore in the 10 aforementioned custom-designed compounds and their observed repellent activity, thus, validated the potential of the model. In addition, the experimental ED_{50} values indicated the predictive power of the pharmacophore for design and selection of new repellent compounds. Thus, in summary, the study validated our *in silico* approaches starting from stereoelectronic analysis of JH, JH mimic (IGR), and DEET analogs to the development of a pharmacophore model, discovery of new repellents through database searches, and design for custom synthesis of new compounds to have repellent activity an *in vitro* test system.

13.7 IMPORTANCE OF THE APPROACH

Thus far, no known literature has been reported where the design and custom synthesis of insect repellents have been presented rationalizing the stereoelectronic properties and pharmacophore development from the electrostatic bioisosterism similarity analysis of JH, JH mimics (IGRs), and DEET analogs. Prerequisite for developing a reliable model for insect repellent property is the correlation of reproducible biological activity to structural information of known repellent compounds. The conformational model of known repellents generated by using the quantum chemical methods enabled us to develop the best 3D arrangement of chemical functions that in turn allowed the pharmacophore development to predict repellent activities. The generated model not only allowed search and identification of novel repellents from compound databases but also assisted in the design for custom synthesis of new repellent compounds.

13.8 CONCLUSIONS

The chapter provides details of how analysis of *in silico* stereoelectronic properties of JH, JH mimic (IGR), and DEET repellents established an electrostatic bioisosterism between the compounds and how those observations could be utilized for generation of a 3D pharmacophore model for arthropod repellent activity. Furthermore, this chapter describes how the pharmacophore model was used for search of the in-house Walter Reed Army Institute of Research Chemical Information System (WRAIR-CIS) database [58] to identify new repellent candidates [45] and for design and custom synthesis of novel repellents [63,64]. Majority of the newly synthesized compounds showed efficacy to be either better or equal to DEET in the *in vitro* test system, demonstrating the predictive power of the pharmacophore model.

The comparative analysis of stereoelectronic properties of DEET compounds with JH mimic (IGR) and JH unraveled three important molecular-level information, specifically, (1) an amide moiety on one end of the molecule, which contains a charge

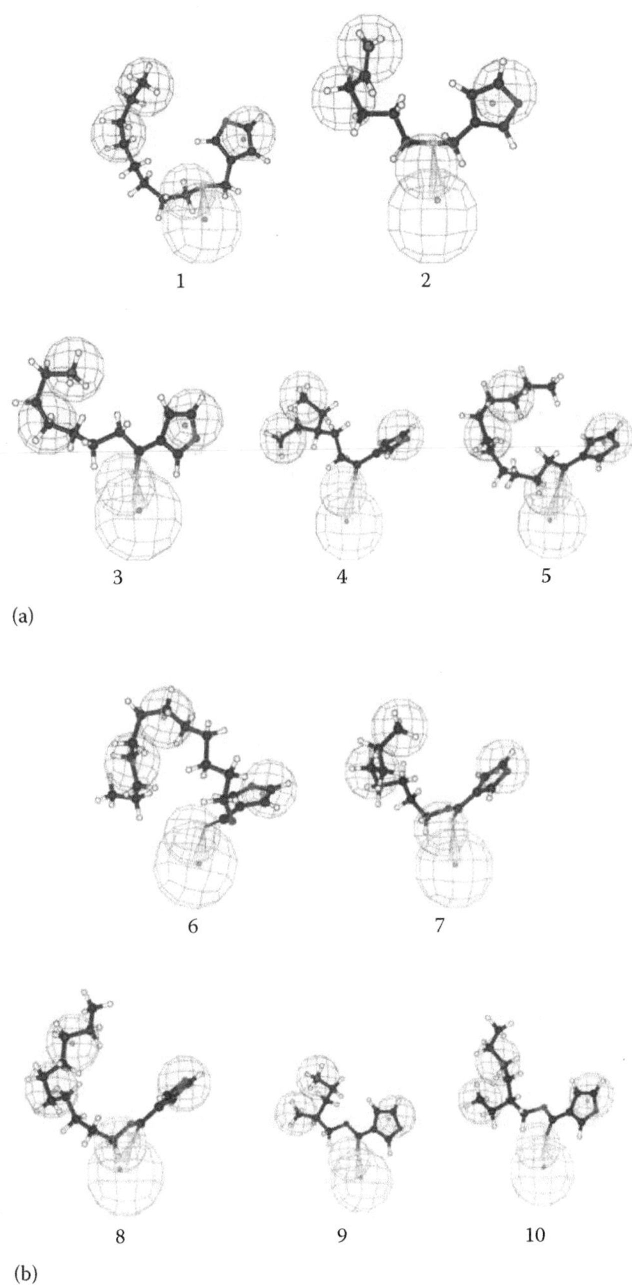

FIGURE 13.8 (a) Mapping of the pharmacophore onto the newly synthesized repellents (1–5). (b) Mapping of the pharmacophore onto the newly synthesized repellents (6–10).

separation between the oxygen and nitrogen atoms to facilitate a strong electronic interaction with the receptor; (2) electrostatic similarity to JH or its mimic molecules (IGRs); and (3) a large weakly charged region to facilitate optimum hydrophobic interaction with the receptor, which may be a long-chain hydrocarbon and need not be an aromatic ring. Thus, the study not only established the remarkable electrostatic bioisosterism between the JH, its mimic (IGR), and the DEET repellents but also provided an important insight about the probable mode of action of insect repellent activity.

Moreover, the stereoelectronic guidance led to the development of a model whose validity extended to structurally different classes of compounds and thereby provided a powerful template for future identification of novel arthropod repellent candidates. Because the identity of target for arthropod repellent activity still remains uncertain, the *in silico* approaches presented in this chapter should be useful for design of well-tolerated, target-specific arthropod repellents and could be a great benefit for synthetic efforts of better repellents for practical use. However, possible pitfalls for designing potent repellents based on these criteria may not be ruled out if the designed molecules are either too volatile or insufficiently volatile with substituted derivatives too bulky for fitting into the receptor site.

Although discovery and development of arthropod repellent is a long and continuous endeavor, *in silico* technologies can undoubtedly help in reducing the rapidly increasing costs for discovery of new chemical entities (NCEs). Furthermore, since understanding the molecular basis of repellent action is expanding in different frontiers and is likely to continue unabated, the *in silico* modeling techniques are uniquely positioned to maximize the usefulness of molecular-level information for discovery and development of new arthropod repellents.

13.A APPENDIX

13.A.1 Brief Procedure for Synthesis of the New Repellents

Details of synthesis of the newly designed repellents were presented earlier [63]. However, the design procedure of repellents from the pharmacophore model and a few outlines of the synthesis are briefly described here. From our generated pharmacophore model [44], it was obvious that a hydrogen-bond acceptor feature adjacent to an aromatic ring would be necessary along with an aliphatic group (of different chain lengths) extending from the hydrogen-bond acceptor for repellent activity. Thus, we organized the structural requirements accordingly and hypothesized three-substituted furans to have insect repellent properties as furans have a hydrogen-bond acceptor adjacent to the furan ring (aromatic) at the three-position with an aliphatic chain extending from it. Furthermore, the hydrogen-bond acceptor group could also be found by an ether, ester, or a ketone moiety and the chain length could be modified to increase or decrease volatility and to affect the duration of repellent efficacy. From these considerations, three-substituted furans with oxygen functionality were predicted to have intrinsic repellency properties and literature survey revealed that three-substituted furans have arthropod repellent properties [25–27]. Accordingly, the three-substituted furans with derivatives of ethers, esters, and ketones were synthesized and different synthetic pathways were followed. The development of the synthetic pathway, explanation of the synthetic mechanism, selection of starting materials, and synthesis of each class is briefly described in the appropriate sections as follows.

13.A.1.1 Ethers

The ethers were synthesized using the classic organic reaction, the Williamson ether synthesis. The reaction involved the deprotonation of an alcohol with sodium hydride, followed by nucleophilic attack on an alkyl halide or alkyl sulfonate by the oxide anion produced. This reaction proceeded via an S_N2 mechanism. The Williamson synthesis was chosen because of the available starting material, 3-furan methanol [63]. A general overall reaction may be visualized as shown in the following:

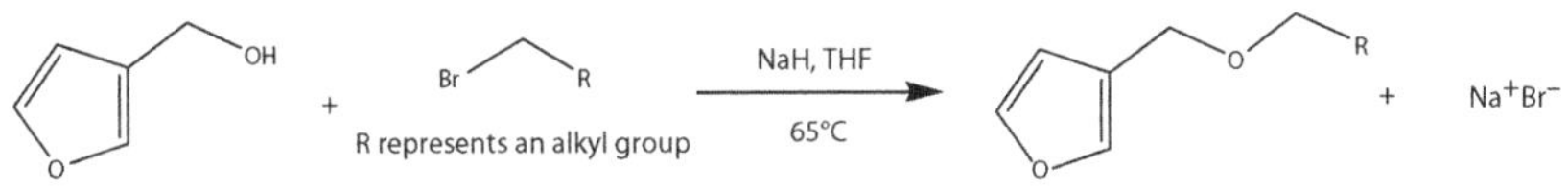

OH + Br R — R represents an alkyl group — NaH, THF, 65°C → O R + Na^+Br^-

Or more elaborately,

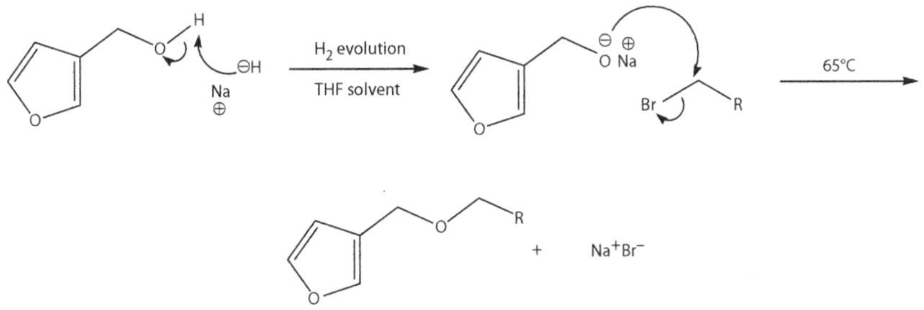

H, O, Na ⊕, ⊖H — H_2 evolution, THF solvent → ⊖ ⊕ O Na, Br R — 65°C →

O R + Na^+Br^-

13.A.1.2 Esters

The esters were synthesized following the synthetic method developed by Mitsunobu [63]. The reaction is a condensation of an alcohol and an acidic component (carboxylic acid), requiring the use of triphenyl phosphine (PPh_3) and an alkyl diazocarboxylate. The reaction proceeds with inversion of configuration at the alcohol. According to the pharmacophore model, structure of the new repellents would require the ester group to be attached to the furan ring at the C3 carbon. To accomplish this, the mechanism (described as follows) required that the starting material of the furan component be acidic. Accordingly, the starting material selected was 3-furoic acid. Thus, the general scheme for the reaction to form esters [60] may be depicted as follows:

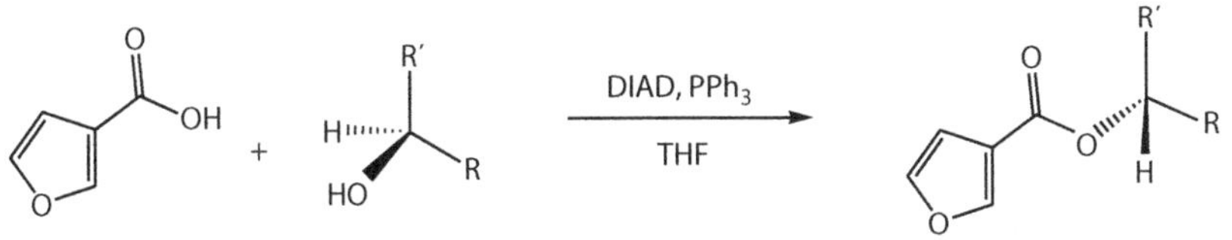

O, OH + R′, H, HO, R — DIAD, PPh_3, THF → O, O, R′, R, H

13.A.1.3 Ketones

The ketone derivatives of the repellents were synthesized from a β-ketoester following a two-step synthetic procedure involving deprotonation of the α-hydrogen followed by the conjugation of an alkyl chain (acetoacetic ester synthesis) and then a β-decarboxylation. The reaction was known earlier, but a mechanistic study was published by Kluger and others in 1986 [65]. The overall reaction scheme is shown as follows:

O O O Br R + 1. NaOEt, iPrOH 2. NaOH 3. H_3O^+, heat O O R

The first set of steps involved alkylation of the α-carbon of the β-ketoester in an acetoacetic acid ester synthesis. The second set of steps involved the basic hydrolysis of the ester followed by decarboxylation [65].

13.A.2 *In Vitro* Method for Evaluation of Repellent Activity

An *in vitro* bioassay system for evaluation of repellent efficacy was developed and reported earlier [64]. The salient features of the procedure are briefly described here. The newly synthesized compounds were tested for repellent efficacy against anopheles *A. aegypti* using an *in vitro* blood feeding system following the literature procedure [66]. The *in vitro* test system provided an estimate of the amount of repellent that must be applied to produce a given level of effectiveness against an arthropod test population [67]. Mosquitoes were reared under standardized conditions and held in a cage at 27°C and 75% relative humidity until testing [68]. This test system consisted of a mosquito blood feeder, a constant-temperature water circulator, and a specially designed cage. The mosquito blood feeder contained five circular blood reservoirs, each of which was filled with outdated human blood and covered with the candidate repellent-treated Baudruche membrane. To start with, the repellents were diluted in ethanol to provide concentrations of 0.02, 0.04, 0.08, and 0.16 mg/cm^2. The test materials, including the control, were applied on the membrane at random. Then 250 female mosquitoes (aged 5–15 days) were admitted to the blood reservoirs on a "free choice" basis by way of a sliding door in the floor of the test cage [69]. The number of mosquitoes probing and feeding on each well was noted at 2 min intervals. The test was terminated at the end of 20 min. The test results were the totals of 10 feeding counts. The effective dose was then calculated from a probit analysis of the feeding count obtained in the respective tests. The statistical distribution of tested chemical sensitivity levels within the strain of *A. aegypti* was calculated from the dose–response regression equation [70].

ACKNOWLEDGMENTS

Material has been reviewed by the WRAIR. There is no objection to its presentation and/or publication. The opinions or assertions contained herein are the private views of the author and are not to be construed as official or reflecting true views of the Department of the Army or the Department of Defense.

REFERENCES

1. D. Brewste, *The story of mankind's deadliest foe*, Bio. Med. J. 323 (2001), pp. 289–290.
2. B.F. Eldridge and J.D. Edman (eds), *Medical Entomology: A Textbook on Public Health and Veterinary Problems Caused by Arthropods*, Kluwer Academic Press, Dordrecht, the Netherlands, 2000.
3. World Malaria Report (2009), World Health Organization, Geneva, Switzerland. Available at http://malaria.who.int/whosis/whostat/EN_WHS09_Full.pdf (accessed 13 February 2009)
4. G. Koren, D. Matsui, and B. Bailey, *DEET-based insect repellents: Safety implications for children and pregnant and lactating women*, Can. Med. Assoc. J. 169 (2003), pp. 209–212.
5. A.W. Olmstead and G.A. LeBlanc, *Insecticidal juvenile hormone analogs stimulate the production of male offspring in the crustacean* Daphnia magna, Environ. Health Perspect. 111 (2003), pp. 919–924.
6. M.J. Klowden, *Endocrine aspects of mosquito reproduction*, Arch. Insect Biochem. Physiol. 35 (1997), pp. 491–512.
7. Y.P. Li, S. Hernandez-Martinez, G.C. Unnithan, R. Feyereisen, and F.G. Noriega, *Activity of the corpora allata of adult female* Aedes aegypti: *Effects of mating and feeding*, Insect Biochem. Mol. Biol. 33 (2003), pp. 1307–1315.
8. K.G. Davey, *Do thyroid hormones function in insects?* Insect Biochem. Mol. Biol. 30 (2000), pp. 877–884.
9. A.H. Nakayama, A. Iwamura, Y. Niwa, Y. Nakagawa and T. Fujita, *Development of insect juvenile hormone active oxime o-ethers and carbamates*, J. Agric. Food Chem. 33 (1985), pp. 1034–1041.
10. B.J. Bergot, G.C. Jamieson, M.A. Ratcliff, and D.A. Schooley, *JH zero: New naturally occurring insect juvenile hormone from developing embryos of the tobacco hornworm*, Science 210 (1980), pp. 336–338.
11. K.J. Judy, D.A. Schooley, L.L. Dunham, M.S. Hall, B.J. Bergot, and J.B. Siddall, *Isolation, structure, and absolute configuration of new natural insect juvenile hormone from* Manduca sexta, Proc. Natl. Acad. Sci. USA 70 (1973), pp. 1509–1513.
12. A. Nakayama and W.G. Richards, *A quantum chemical study of insect juvenile hormone mimics: The active conformation and the electrostatic similarities*. Quant. Struct. Act. Relat. 6 (1987), pp. 153–157.
13. R.W. Meola and J. Readio, *Juvenile hormone regulation of the second biting cycle in* Culex pipiens, J. Insect Physiol. 33 (1987), pp. 751–754.
14. R.W. Meola and R.S. Petralia, *Juvenile induction of biting behavior in Culex mosquitoes*, Science 209 (1980), pp. 1548–1550.
15. B.K. Tyagi, M. Kalyanasundaram, P.K. Das, and N. Somachary, *Evaluation of a new compound (VCRC/INS/A-23) with juvenile hormone activity against mosquito vectors*, Indian J. Med. Res. 82 (1985), pp. 9–13.
16. B.K. Tyagi, N. Somachari, V. Vasuki, and P.K. Das, *Evaluation of three formulations of a chitin synthesis inhibitor (fenoxycarb) against mosquito vectors,* Indian J. Med. Res. 85 (1987), pp. 161–167.

17. D. Amalraj and P.K. Das, *Estimation of predation by the larvae of Toxorhynchites splendens on the aquatic stages of* Aedes aegypti, Southeast Asian J. Trop. Med. Public Health 29 (1998), pp. 177–183.
18. V. Vasuki, *Role of Insect Growth Regulators in Vector Control,* Proceedings of 2nd Symposium on Vector –Borne Diseases, Trivandrum, Kerala, 1988, pp. 204–218.
19. D.H. Ross, P. Cohle, P.R. Blase, J.B. Bussard, and K. Neufeld, *Effects of the insect growth regulator (S)-methoprene on the early life stages of the fathead minnow* Pimephales promelas *in a flow through laboratory system,* J. Am. Mosquito Control Assoc. 10 (1994), pp. 211–221.
20. D.H. Ross, D. Judy, B. Jacobson, and R. Howell, *Methoprene concentrations in freshwater microcosms treated with sustained-release Altosid formulations*, J. Am. Mosquito Control Assoc. 10 (1994), pp. 202–210.
21. S.A. Ritchie, M. Asnicar, and B.H. Kay, *Acute and sublethal effects of (S)-methoprene on some Australian mosquitoes,* J. Am. Mosquito Control Assoc. 13 (1997), pp. 153–155.
22. A.E. Pinkney, P.C. McGowan, D.R. Murph, T.P. Lowe, D.W. Sparling, and L.C. Ferrington, *Effects of mosquito larvicides temephos and methoprene on insect populations in experimental ponds*, Environ. Toxicol. Chem. 19 (2000), pp. 678–684.
23. N.E. Beckage, K.M. Marion, W.E. Walton, M.C. Wirth, and F.F. Tan, *Comparative larvicidal toxicities of three ecdysone against on the mosquitoes* Aedes aegypti, Culex quinquefasciatus, and Anopheles gambiae, Arch. Insect Biochem. Physiol. 57 (2004), pp. 111–122.
24. D. Amalraj, V. Vasuki, M. Kalyanasundaram, B.K. Tyagi, and P.K. Das, *Laboratory and field evaluation of three insect growth regulators against mosquito vectors,* Indian J. Med. Res. 87 (1988), pp. 24–31.
25. W.A. Skinner and H.L. Johnson, *The design of insect repellents*, Drug Design 10 (1980), pp. 277–302.
26. E.T. McCabe, W.F. Barthel, S.I. Gertler, and S.A. Hall, *Insect repellents. III. N,N-diethylamides, J.* Org. Chem. 19 (1954), pp. 493–498.
27. H.L. Johnson, W.A. Skinner, H.I. Maibach, and T.R. Pearson, *Repellent activity and physical properties of ring-substituted N,N-diethylbenzamides*, J. Econ. Entomol. 60 (1967), pp. 173–176.
28. S.B. McIver, *A model for the mechanism of action of the repellent DEET on* Aedes Aegypti *(Diptera: Culicidae)*, J. Med. Entomol. 18 (1981), pp. 357–361.
29. S.S. Rao and K.M. Rao, *Insect repellent N,N-diethylphenylacetamide: An update*, J. Med. Entomol. 28 (1991), pp. 303–306.
30. I.M. Kapetanovic, *Computer-aided drug discovery and development (CADDD):* In silico-*chemico-biological approach*, Chem. Biol. Interact. 171 (2008), pp. 165–176.
31. D. Janseen, *The power of prediction*, Drug Discov. (2002), p. 38.
32. B.L. Podlogar, I. Muegge, and L.J. Brice, *Computational methods to estimate drug development parameters,* Curr. Opin. Drug Discov. 12 (2001), pp. 102–109.
33. H. Kubinyi, *Success stories of computer-aided design*, in *Computer Applications in Pharmaceutical Research and Development*, S. Ekins and B. Wang, eds., Wiley-Interscience, Hoboken, NJ, 2006, pp. 377–424.
34. O. Dror, A. Shulman-Peleg, R. Nussinov, and H.J. Wolfso, *Predicting molecular interactions* in silico: *I. A guide to pharmacophore identification and its applications to drug design*, Curr. Med. Chem. 11 (2004), pp. 71–90.
35. O.F. Güner, *Manual pharmacophore generation: Visual pattern recognition*, in *Pharmacophore, Perception, Development, and Use in Drug Design*, O.F. Güner, ed., University International Line (IUL Biotechnology Series), San Diego, CA, 2000, pp.17–20.
36. A.R. Leach, V.J. Gillet, R.A. Lewis, and R. Taylor, *Three-dimensional pharmacophore methods in drug discovery,* J. Med. Chem. 53 (2010), pp. 539–558.

37. A.K. Bhattacharjee and J.M. Karle, *Molecular electronic properties of a series of 4-quinolinecarbinolamines define antimalarial activity profile*, J. Med. Chem. 39 (1996), pp. 4622–4629.
38. J. Mestres, D.C. Rohrer, and G.M. Maggiora, *MIMIC: A molecular-field matching program: Exploiting applicability of molecular similarity approaches*, J. Comput. Chem. 18 (1997), pp. 934–954.
39. G. Naray-Szabo and T. Balogh, *Viewpoint 7—The average molecular electrostatic field as a QSAR descriptor. Part 4. Hydrophobicity scales for amino acid residues,* J. Mol. Struct. (Theochem.) 284 (1993), pp. 243–248.
40. T. Balogh and G. Naray-Szabo, *Application of the average molecular electrostatic field in quantitative structure-activity relationships*, Croat. Chem. Acta. 66 (1993), pp. 129–140.
41. A.R. Leach, *Molecular Modelling, Principles and Applications*, A.W. Longman Ltd., Essex CM20 2JE, England, U.K., 1998.
42. D. Ma, A.K. Bhattacharjee, R.K. Gupta, and J.M. Karle, *Predicting mosquito repellent potency of N,N-diethyl-m-toluamide (DEET) analogs from molecular electronic properties*, Am. J. Trop. Med. Hyg. 60 (1999), pp. 1–6.
43. A.K. Bhattacharjee, R.K. Gupta, D. Ma, and J.M. Karle, *Molecular similarity analysis between insect juvenile hormone and N, N-diethyl-m-toluamide (DEET) analogs may aid design of novel insect repellents*, J. Mol. Recognit. 13 (2000), pp. 213–220.
44. A.K. Bhattacharjee and R.K. Gupta, *Analysis of molecular stereo-electronic similarity between N,N-diethyl-m-toluamide (DEET) analogs and insect juvenile hormone to develop a model pharmacophore for insect repellent activity*, J. Am. Mosq. Control Assoc. 21 (2005), pp. 23–29.
45. A.K. Bhattacharjee, W. Dheranetra, D.A. Nichols, and R.K. Gupta, *3D pharmacophore model for insect repellent activity and discovery of new repellent candidates*, QSAR Comb. Sci. 24 (2005), pp. 593–602.
46. M. Randic and G. Krilov, *On characterization of molecular surfaces*, Int. J. Quant. Chem. 65 (1997), pp. 1065–1076.
47. P.G. Mezey, *Shape in Chemistry: An Introduction to Molecular Shape and Topology*, VCH, New York, 1993.
48. A.M. Richard, *Quantitative comparison of molecular electrostatic potentials for structure-activity studies,* J. Compt. Chem. 12 (1991), pp. 959–969.
49. A.C. Good, S.S. So, and W.G. Richards, *Structure-activity relationships from molecular similarity matrices*, J. Med. Chem. 36 (1993), pp. 433–436.
50. J. Avery, *A model for biological specificity*, Int. J. Quant. Chem. 26 (1984), pp. 843–855.
51. J.S. Murray, B.A. Zilles, K. Jayasuriya, and P. Politzer, *Comparative analysis of the electrostatic potentials of dibenzofuran and some dibenzo-p-dioxins*, J. Am. Chem. Soc. 108 (1986), pp. 915–918.
52. R.G.A. Bone and H.O. Villar, *Discriminating D1 and D2 agonists with a hydrophobic similarity index*, J. Mol. Graph. 13 (1995), pp. 201–208.
53. R.G. Pearson, *The principle of maximum hardness*, Acc. Chem. Res. 26 (1993), pp. 250–255.
54. SPARTAN SGI Version 5.1.2., SPARTAN user's manual, version 5.0., Wavefunction, Inc., Irvine, CA, 1998; software available at http://www.wavefunction.com
55. CATALYST Version 4.8 software, Accelrys Inc., San Diego, CA, 2003; software available at http://www.accelrys.com
56. R.K. Gupta, D. Ma, and A.K. Bhattacharjee, U.S. Patent: US7897162, 2011.
57. J.E. Oliver and K.S. Patterson, *Wild ox Bugs Mosquitoes*, C. & E. News, 2003, p. 49.
58. The Chemical Information System, Division of experimental therapeutics, Walter Reed Army Institute of Research, 503 Robert Grant Avenue, Silver Spring, MD, pp. 20910–7500.

59. Cerius^2 4.9, Accelrys Inc., San Diego, CA, 2003; software available at http://www.accelrys.com
60. TOPKAT 6.1, Accelrys Inc., San Diego, CA, 2003; software available at http://www.accelrys.com
61. M. Grigorov, *A QSAR study of the antimalarial activity of some synthetic 1,2,4-trioxanes*, J. Chem. Inf. Comput. Sci. 35 (1995), pp. 285–304.
62. P.A. Greenridge and J. Weiser, *A comparison of methods for pharmacophore generation with the catalyst software and their use for 3D-QSAR: Application to a set of 4-aminopyridine thrombin inhibitors*, Mini Rev. Med. Chem. 1 (2001), pp. 79–87.
63. B.T. Mott, *Synthesis of 3-subsituted furans as novel insect repellents*, M.S. Thesis, George Mason University, Fairfax, VA, 2003.
64. W. Dheranetra, K.L. Lawrence, J.P. Benante, M.A. Potter, K.R. Chauhan, N. Bathini, C.E. White, B. Mott, D.A. Nichols, A.K. Bhattacharjee, and R.K. Gupta, *Comparative study of four membranes for evaluation of new insect/arthropod repellents using* Aedes aegypti, in *Arthropod Borne Viral Infections—Current Status and Research*, D. Raghunath and C. Durga Rao, eds., *The Eighth Sir Dorabji Tata Symposium*, Tata McGRAW Hill Publishing Company Ltd, New Delhi, India, 2008, pp. 418–424.
65. R. Kluger and M.J. Brandl, *Acetoacetic ester condensation*, J. Org. Chem. 51 (1986), pp. 3964.
66. M. Debboun, D. Strickman, T.A. Klein, J.A. Glass, E. Wylie, E. Laughinghouse, R.A. Writz and R.K. Gupta, *Laboratory evaluation of AI3–37220, AI3–35765, CIC-4, and DEET repellents against three species of mosquitoes*, J. Am. Mosq. Control Assoc. 15 (1999), pp. 342–347.
67. L.M. Rueda, L.C. Rutledge, and R.K. Gupta, *Effect of skin abrasions on the efficacy of the repellent deet against* Aedes aegypti, J. Am. Mosq. Control Assoc. 14 (1998), pp. 178–182.
68. E.J. Gerberg, D.R. Barnard, and R.A. Ward, *Manual for mosquito rearing and experimental techniques*, Am. Mosq. Control Assoc. (1994), Bull. No. 5.
69. L.C. Rutledge, M.A. Moussa, and C.J. Belletti, *An* in vitro *blood feeding system for quantitative testing of mosquito repellents*, Mosq. News 24 (1976), pp. 407–419.
70. A. Goldstein, *Biostatistics: An Introductory Text*, Macmillan, New York, 1964.

14 Use of Multicriteria Analysis for Selecting Candidate Insecticides for Vector Control

James Devillers, Laurent Lagadic, Ohri Yamada, Frédéric Darriet, Robert Delorme, Xavier Deparis, Jean-Philippe Jaeg, Christophe Lagneau, Bruno Lapied, Françoise Quiniou, and André Yébakima

CONTENTS

ABSTRACT

The SIRIS multicriteria analysis was applied to 129 insecticides to identify those which could be potentially used as larvicides and/or adulticides in vector control against mosquitoes. Their selection was made without *a priori* restrictions, even on the basis of criteria of toxicity or regulatory interdictions. The unique condition was that they had to be already marketed. Each insecticide was described by six toxicity/ecotoxicity criteria and five exposure/environmental fate criteria identified as important by a panel of experts with complementary skills. A scenario for the larvicides and another one for the aldulticides were considered. The toxicity/ecotoxicity criteria were ranked differently in the two scenarios while the same hierarchy was used in both scenarios for the exposure/environmental fate criteria. In addition, only 114 substances were considered for the adulticide scenario because some insecticides can only be used as larvicides on mosquitoes. Each insecticide being characterized by two SIRIS scores, one score value per hierarchy, a map was designed for each scenario. The substances located in the left bottom part of each map are the most interesting because they present low SIRIS score values on both the Toxicity/Ecotoxicity and Exposure/ Environmental fate scales. Their interest as potential candidates in vector control was discussed. Nevertheless, the need to complete this multicriteria identification by a full risk assessment study is a necessary condition before their use.

KEYWORDS

Multicriteria analysis, SIRIS method, Vector control, Mosquitoes, Insecticide selection, Biocides, French regulation

14.1 INTRODUCTION

Since the past decades, there has been a significant recrudescence of several vector-borne diseases (VBDs) worldwide. VBDs are emerging or resurging as a result of insecticide resistance, genetic modifications in pathogens, climate changes, societal behaviors, and/or changes in public health policy [1].

Dengue fever, which is caused by Flaviviridae viruses transmitted by *Aedes aegypti* and *Aedes albopictus*, is endemic in Africa, Asia, Caribbean, and Latin America [2]. In Europe, except a massive outbreak in Greece at the end of the 1920s [3,4], the continent was spared till the 1970s. Then, the ever growing development of international trade exchanges has favored both the expansion of the distribution area of the *Aedes* [5] and the importation of contaminated travelers inside Europe [2,6]. The situation has been exactly the same with the chikungunya fever, which is also transmitted to humans through the bite of virus-infected mosquitoes [7,8]. To date, cases of autochthonous dengue and chikungunya fevers have been diagnosed in south European countries, including metropolitan France [9,10].

There is no vaccine available for protecting the human and animal populations against these viruses, and the prevention relies entirely on personal protection and mosquito control.

To date, vector control is mainly made through the use of insecticides despite their potential adverse effects on the biota.

In the early 2000s, implementation of the European Directive 98/8/EC [11] has yielded the obligation for manufacturers to perform and submit to the European Commission a complete toxicological, ecotoxicological, and biological evaluation of all substances and biocidal products marketed in Europe, especially those intended for public health. These substances include insecticides, acaricides, and other compounds used to fight against arthropods as well as the insecticides used for mosquito control.

The cost of such an approach and its requirements in terms of safety to the humans and the environment, which reflect societal demands in terms of hazard and risk, are so important that biocides, although effective, but which do not fit these demands, are not supported and, hence, are withdrawn from the market. However, if the withdrawal of existing products is not followed by the introduction of biocides showing more favorable (eco)toxicological profiles, this will undoubtedly lead to the reduction of the range of molecules available and suitable for vector control. This situation is particularly annoying because the alternate use of insecticides is necessary to have a more effective vector control action. In addition, this is the best strategy to reduce the phenomena of resistance to insecticides.

Currently, in France, vector control is made almost exclusively with *Bacillus thuringiensis* subspecies *israelensis* (*Bti*) and deltamethrin to fight larvae and adults of mosquitoes, respectively. Unfortunately, cases of resistance to organophosphorus and pyrethroids have been observed [12–14]. In the same way, resistance to *Bti* toxins has been evidenced in strains of mosquitoes reared in laboratory [15,16]. These resistances decrease the toxicity of the substances against the targeted organisms and, as a result, reduce the efficacy of the treatment campaigns.

The search for alternatives to insecticides used today can be focused either on the development of new molecules more selective against the mosquitoes or on the use of molecules already marketed for other usages, such as in agriculture or veterinary medicine, standing the fact to be able to develop formulations suited for vector control.

The first strategy mainly relies on the use of quantitative structure–activity relationship (QSAR) models aiming to relate the physicochemical properties and/or structural characteristics of the molecules to their activity by means of a linear or nonlinear statistical method in order to design a computational device able to predict the activity of unsynthesized molecules. Then, on the basis of the prediction, a chemist can decide which molecules to go ahead and make [17]. Thus, for example, genetic function approximation with linear and spline options was used by Begum et al. [18] for deriving a QSAR model on chalcone derivatives tested for mosquito larvicidal activity against the third instar larvae of *Culex quinquefasciatus* and described by various spatial, electronic, and physicochemical parameters. QSAR models were developed [19] using different series of organotins tested for their larvicidal activities against *A. aegypti* and *Anopheles stephensi* mosquito larvae. The most important descriptors for encoding toxicity were the hydrophobic (π) and the Hammett electronic (σ^+) parameters of the substituents. A data set of 304 chemicals with their juvenile hormone activity tested on *Culex pipiens* larvae was used for deriving QSAR models from different statistical approaches. Chemicals were described by various descriptors encoding topological and physicochemical information [20].

Besides these *in silico* approaches, collective expertise analyses are also commonly undertaken to evaluate whether molecules already marketed for specific usages could be used in vector control. Thus, in 2006, Afsset (French Agency for Environmental and Occupational Health Safety) conducted a collective expertise to identify potential substitutes to *Bti* and deltamethrin. On the basis of an extensive bibliographical investigation performed by Institute of Research and Development (IRD) [21] and discussions between experts, pyriproxyfen and spinosad were proposed as potential larvicides, while pyrethrin and naled were identified as potential adulticides [22,23]. Naled was later excluded due to its human toxicity [24].

Similar works have been also made by groups of experts and international organizations. Thus, the Global Collaboration for Development of Pesticides for Public Health (GCDPP) established by World Health Organization Pesticide Evaluation Scheme (WHOPES) [25] provides an international forum for the exchange of information and ideas between application equipment manufacturers, governmental agencies, organizations, universities, and research institutions on issues related to the development and use of pesticides within the context of disease control strategies [26]. The International Public Health Pesticides Workshop (IPHPW) identifies new approaches, processes, and implementation strategies that could lead to development and approval of new public health pest control tools [27]. The Innovative Vector Control Consortium (IVCC) has developed a project of screening of existing chemical libraries for identifying potential vector control insecticides. The most interesting compounds will be used to direct analogue synthesis to optimize their activity against adult mosquitoes while minimizing their potential toxicity [28].

To our knowledge, all these interesting initiatives are not based on the use of statistical tools to rationalize the selection of potential insecticides to vector control. In this context, an attempt was made to use a multicriteria analysis (MCA) [29] to help in the ranking of insecticides potentially interesting in vector control. Indeed, the identification of such insecticides requires to account for various information, expressed in a qualitative or quantitative way by experts with complementary skills. This must lead to a decision resulting from the optimal use of all available information and knowledge at a given time. Multicriteria methods are particularly suited to provide a rational and structured help to the stakeholders in their decision making. These techniques have been developed specifically to solve problems involving quantitative, semiquantitative, and/or qualitative criteria within a decision-making process. Reaching an agreement between experts on the relative importance of different criteria can obviously be a complex and difficult task. MCA can help in assessing the relative importance of all the criteria selected by experts and reflects this importance in the final decision process. Multicriteria methods allow us to incorporate at best the views of the experts involved in the analysis. Each expert gives his/her own opinion and contributes in a distinct and identifiable way in the search of an optimal solution, which can be subject to changes. Indeed, multicriteria methods are tools that facilitate dialogue between experts and between experts and decision makers, allowing to test new hypotheses in an easy and transparent way, add constraints, and so on [29–31]. Consequently, MCA has found applications in numerous domains including disease control program evaluations [32,33] and the hazard and risk assessment of chemicals [34–36]. In this chapter, an MCA was used for selecting candidate insecticides for vector control on the basis of criteria selected by a group of experts and dealing with their toxicity, ecotoxicity, exposure, and environmental fate.

14.2 MATERIALS AND METHODS

14.2.1 List of Insecticides

The first job of the expert group consisted in establishing a broad list of potential insecticides. No *a priori* restrictions were made even on the basis of criteria of toxicity or regulatory interdictions. The unique condition was to be already marketed. This means that all the insecticides in development were excluded.

Since 1995, WHOPES [25] collects data on the insecticides used in vector control within the UN member states. The fourth and latest edition of these statistics was published in 2009 and covers the period 2006–2007. WHOPES has also published lists of recommended insecticides for indoor residual spraying, spatial nebulization, treatment of breeding sites, and treatment of mosquito nets. The insecticides belonging to these lists were selected in priority.

The authorization of usage of the vector control insecticides in France is subject to the European regulation on biocides. Consequently, the active substances that were authorized or under evaluation as PT18 (insecticides, acaricides, and products to control other arthropods) in October 2010 were also considered. In the same way, all the active substances identified as insecticides in the European phytosanitary legislation were selected whatever their status in September 2010. One of the populations that consumes large amounts of vector control insecticides is the U.S. military troop during its operations in tropical areas. Information on these insecticides being available on the Armed Forces Pest Management Board (AFPMB) website [37], this source of information was also considered.

The last way, which was interesting to consider for gathering potential insecticides for vector control, was the Internet resources dealing with the human and veterinary antiparasitic drugs. Investigations were conducted as part of the work of the Observatory on Pesticide Residues (ORP) [38], especially regarding the effort to develop a protocol to better characterize population exposure to pesticides. Information was mainly retrieved from the WHO Collaborating Centre for Drug Statistics Methodology database [39], the EudraPharm European database [40], and the online Vidal dictionary [41] including the external antiparasitic drugs used in dermatology and the systemic antiparasitic drugs used in infectiology/parasitology.

Regarding the veterinary antiparasitic drugs (livestock and pets) certified for sale in France, the ANMV [42] and the EudraPharm European database [40] were investigated, targeting only on the chemicals for external use.

The previously mentioned sources of information allowed us to establish a first list of 338 active substances.

To remove the active substances *a priori* ineffective against mosquitoes, a first screening based on their effectiveness was performed. An active substance was considered to be potentially effective against mosquitoes when, at least, it was described as effective on dipterans, if not on mosquitoes, or when it was or had been already used in vector control.

As previously indicated, the insecticides recommended by WHOPES [25] and those reported by this organization as being used worldwide have been immediately considered as effective.

The other active substances were reviewed one by one in order to exclude those that did not satisfy either of the following criteria:

- Usage against dipterans claimed in the biocide European regulatory dossier available on the Communication & Information Resource Administrator (CIRCA) website [43].
- Usage against dipterans authorized in the e-phy database [44] of the French Ministry of Agriculture.
- Usage against dipterans authorized in the ACTA index [45].
- Effects on dipterans indicated in the Pesticide Manual [46].
- Existence of publications reporting studies of effectiveness against mosquitoes. It is important to note that due to time constraints, no systematic literature investigations were performed. Only the main papers identified by the experts were considered.

These different constraints allowed us to reduce the list of active substances from 338 to 129 (Table 14.1).

14.2.2 Multicriteria Analysis

Because the system of integration of risk with interaction of scores (SIRIS) method has clearly shown its interest and potentialities in environmental and occupational toxicology [34–36], it was used in this study. Briefly, the method needs first to select the number of criteria (variables) necessary to correctly describe the studied phenomenon. Obviously, this number depends on the complexity of the problem. The more complex the problem, the larger the number of variables it is necessary to consider, at least in a first step. Indeed, in all the modeling processes, it is preferable to optimize the number of variables because their number should not be multiplied needlessly (parsimony principle).

The selected variables, which can be qualitative or quantitative, are then transformed into modalities coded as favorable (f) or unfavorable (d) or as favorable (f), moderately favorable (m), or unfavorable (d). If need be, it is possible to define more modalities. The threshold limits are generally determined from expert judgments.

TABLE 14.1
List of Insecticides with Their CAS Registry Number, Chemical Family, Indication on Their Use in Vector Control, Recommendation in WHOPES, and Efficacy against Mosquitoes, Studied or Claimed

No.	Name	CAS RN	Family	Vector *Control*	WHOPES	Efficacy against *Mosquitoes*
1	Diflubenzuron	35367-38-5	Benzoylurea	Yes	Yes	Yes
2	Triflumuron	64628-44-0	Benzoylurea	Yes	Yes	Yes
3	Bendiocarb	22781-23-3	Carbamate	Yes	Yes	Yes
4	*Bs* (*Bacillus sphaericus* str. 2362)	na[a]	Microorganism	Yes	Yes	Yes
5	*Bti* (*Bacillus thuringiensis* sp. *israelensis*, str. SA3A, AM65-52)	na	Microorganism	Yes	Yes	Yes
6	Alpha-cypermethrin	52315-07-8	Pyrethroid	Yes	Yes	Yes
7	Bifenthrin	82657-04-3	Pyrethroid	Yes	Yes	Yes
8	Cyfluthrin	68359-37-5	Pyrethroid	Yes	Yes	Yes
9	Cypermethrin	52315-07-8	Pyrethroid	Yes	Yes	Yes
10	Cyphenothrin	39515-40-7	Pyrethroid	Yes	Yes	Yes
11	Deltamethrin	52918-63-5	Pyrethroid	Yes	Yes	Yes
12	D-Phenothrin (sumithrin)	26002-80-2	Pyrethroid	apNo[b]	No	Yes
13	Etofenprox	80844-07-1	Pyrethroid	Yes	Yes	Yes
14	Lambda-cyhalothrin	91465-08-6	Pyrethroid	Yes	Yes	Yes
15	Permethrin	52645-53-1	Pyrethroid	Yes	Yes	Yes
16	Pyriproxyfen	95737-68-1	Juvenoid	Yes	Yes	Yes
17	Spinosad	131929-60-7	Spinosyn	Yes	Yes	Yes
18	Novaluron	116714-46-6	Benzoylurea	Yes	Yes	Yes
19	Propoxur	114-26-1	Carbamate	Yes	Yes	Yes
20	Methoprene	40596-69-8	Juvenoid	Yes	Yes	Yes
21	DDT	50-29-3	Organochlorine	Yes	Yes	Yes
22	Chlorpyrifos	2921-88-2	Organophosphorus	Yes	Yes	Yes
23	Chlorpyrifos-methyl	5598-13-0	Organophosphorus	Yes	No	Yes
24	Diazinon	333-41-5	Organophosphorus	Yes	No	Yes
25	Fenitrothion	122-14-5	Organophosphorus	Yes	Yes	Yes
26	Fenthion	55-38-9	Organophosphorus	Yes	Yes	Yes
27	Malathion	121-75-5	Organophosphorus	Yes	Yes	Yes
28	Pirimiphos-methyl	29232-93-7	Organophosphorus	Yes	Yes	Yes
29	Temephos	3383-96-8	Organophosphorus	Yes	Yes	Yes
30	Bioresmethrin	28434-01-7	Pyrethroid	apNo	Yes	Yes
31	d,d-trans-Cyphenothrin	na	Pyrethroid	apNo	Yes	Yes
32	Resmethrin	10453-86-8	Pyrethroid	apNo	Yes	Yes

(*continued*)

TABLE 14.1 (continued)
List of Insecticides with Their CAS Registry Number, Chemical Family, Indication on Their Use in Vector Control, Recommendation in WHOPES, and Efficacy against Mosquitoes, Studied or Claimed

No.	Name	CAS RN	Family	Vector *Control*	WHOPES	Efficacy against *Mosquitoes*
33	Chlorfenapyr	122453-73-0	Arylpyrrole	apNo	No	Yes
34	Abamectin	71751-41-2	Avermectin	apNo	No	apYes[c]
35	Hexaflumuron	86479-06-3	Benzoylurea	apNo	No	Yes
36	*S*-Methoprene	65733-16-6	Juvenoid	Yes	Yes	Yes
37	Acetamiprid	135410-20-7	Neonicotinoid	apNo	No	Yes
38	Clothianidin	210880-92-5	Neonicotinoid	apNo	No	Yes
39	Imidacloprid	138261-41-3	Neonicotinoid	apNo	No	Yes
40	Thiamethoxam	153719-23-4	Neonicotinoid	apNo	No	Yes
41	Azamethiphos	35575-96-3	Organophosphorus	apNo	No	Yes
42	Dichlorvos	62-73-7	Organophosphorus	apNo	No	Yes
43	Naled	300-76-5	Organophosphorus	Yes	No	Yes
44	Piperonyl butoxide	51-03-6	Synergist	Yes	No	Yes
45	Indoxacarb	173584-44-6	Oxadiazine	apNo	No	Yes
46	Fipronil	120068-37-3	Phenyl pyrazole	apNo	No	Yes
47	Pyrethrins	8003-34-7	Natural pyrethrin	Yes	No	Yes
48	d-Allethrin	na	Pyrethroid	apNo	No	Yes
49	d-Tetramethrin	1166-46-7	Pyrethroid	apNo	No	Yes
50	Empenthrin	54406-48-3	Pyrethroid	apNo	No	Yes
51	Esbiothrin	84030-86-4	Pyrethroid	apNo	No	Yes
52	Esfenvalerate	66230-04-4	Pyrethroid	apNo	No	Yes
53	Imiprothrin	72963-72-5	Pyrethroid	apN	No	apY
54	Metofluthrin	240494-70-6	Pyrethroid	apN	No	Yes
55	Prallethrin	23031-36-9	Pyrethroid	apN	No	Yes
56	Tetramethrin	7696-12-0	Pyrethroid	apN	No	Yes
57	Transfluthrin	118712-89-3	Pyrethroid	apN	No	Yes
58	Cyromazine	66215-27-8	Growth regulator	apN	No	Yes
59	Chlorantraniliprole	500008-45-7	Anthranilic	apN	No	apY
60	Teflubenzuron	83121-18-0	Benzoylurea	apN	No	Yes
61	Formetanate	22259-30-9	Carbamate	apN	No	apY
62	Methiocarb	2032-65-7	Carbamate	apN	No	apY
63	Methomyl	16752-77-5	Carbamate	apN	No	apY
64	Thiacloprid	111988-49-9	Neonicotinoid	apNo	No	apY
65	Dimethoate	60-51-5	Organophosphorus	apNo	No	apY
66	Ethoprophos	13194-48-4	Organophosphorus	apNo	No	apY
67	Phosmet	732-11-6	Organophosphorus	apNo	No	apY
68	Metaflumizone	139968-49-3	Semicarbazone	apNo	No	apY
69	Beta-cyfluthrin	68359-37-5	Pyrethroid	apN	No	apY
70	Gamma-cyhalothrin	76703-62-3	Pyrethroid	apN	No	apY
71	Zeta-cypermethrin	52315-07-8	Pyrethroid	apN	No	apY

TABLE 14.1 (continued)
List of Insecticides with Their CAS Registry Number, Chemical Family, Indication on Their Use in Vector Control, Recommendation in WHOPES, and Efficacy against Mosquitoes, Studied or Claimed

No.	Name	CAS RN	Family	Vector *Control*	WHOPES	Efficacy against *Mosquitoes*
72	Spinetoram	187166-40-1 187166-15-0	Spinosyn	apN	No	apY
73	Fenoxycarb	79127-80-3	Juvenoid	apN	No	Yes
74	Hydroprene	41205-09-8 41096-46-2	Juvenoid	apN	No	apY
75	Lindane	58-89-9	Organochlorine	apN	No	Yes
76	Coumaphos	56-72-4	Organophosphorus	apNo	No	apY
77	Phoxim	14816-18-3	Organophosphorus	apNo	No	apY
78	Propetamphos	31218-83-4	Organophosphorus	apNo	No	Yes
79	Allethrin	584-79-2	Pyrethroid	apNo	No	Yes
80	Bioallethrin (d-trans allethrin)	260359-57-7	Pyrethroid	apNo	No	Yes
81	Cyhalothrin	68085-85-8	Pyrethroid	apNo	No	apY
82	Esdepallethrin (S-bioallethrin)	28434-00-6	Pyrethroid	apNo	No	Yes
83	Fenvalerate	51630-58-1	Pyrethroid	apNo	No	apY
84	Phenothrin	26002-80-2	Pyrethroid	apNo	No	Yes
85	Tau-fluvalinate	102851-06-9	Pyrethroid	apNo	No	apY
86	Dicyclanil	112636-83-6	Growth regulator	apN	No	apY
87	Pyraclofos	77458-01-6	Organophosphorus	apNo	No	apY
88	Azadirachtin	11141-17-6	Growth regulator	apN	No	Yes
89	Aldicarb	116-06-3	Carbamate	apN	No	apY
90	Benfuracarb	82560-54-1	Carbamate	apN	No	apY
91	Carbofuran	1563-66-2	Carbamate	apN	No	apY
92	Carbosulfan	55285-14-8	Carbamate	apN	No	Yes
93	Chlordane	57-74-9	Organochlorine	apN	No	apY
94	Dinotefuran	165252-70-0	Neonicotinoid	apNo	No	Yes
95	Nitenpyram	150824-47-8	Neonicotinoid	apNo	No	apY
96	Endosulfan	115-29-7	Organochlorine	apN	No	apY
97	Methoxychlor	72-43-5	Organochlorine	apN	No	apY
98	Azinphos-methyl	86-50-0	Organophosphorus	apNo	No	apY
99	Chlorfenvinphos	470-90-6	Organophosphorus	apNo	No	apY
100	Cyanophos	2636-26-2	Organophosphorus	apNo	No	Yes
101	Ethion (diethion)	563-12-2	Organophosphorus	apNo	No	apY
102	Fonofos	944-22-9	Organophosphorus	apNo	No	apY
103	Heptenophos	23560-59-0	Organophosphorus	apNo	No	apY
104	Mecarbam	2595-54-2	Organophosphorus	apNo	No	apY
105	Omethoate	1113-02-6	Organophosphorus	apNo	No	apY

(continued)

TABLE 14.1 (continued)
List of Insecticides with Their CAS Registry Number, Chemical Family, Indication on Their Use in Vector Control, Recommendation in WHOPES, and Efficacy against Mosquitoes, Studied or Claimed

No.	Name	CAS RN	Family	Vector *Control*	WHOPES	Efficacy against *Mosquitoes*
106	Phenthoate	2597-03-7	Organophosphorus	apNo	No	Yes
107	Phorate	298-02-2	Organophosphorus	apNo	No	apY
108	Phosphamidon	13171-21-6	Organophosphorus	apNo	No	apY
109	Pirimiphos-ethyl	23505-41-1	Organophosphorus	Yes	No	Yes
110	Pyridaphenthion	119-12-0	Organophosphorus	apNo	No	apY
111	Quinalphos	13593-03-8	Organophosphorus	apNo	No	apY
112	Sulfotep	3689-24-5	Organophosphorus	apNo	No	apY
113	Tebupirimfos	96182-53-5	Organophosphorus	apNo	No	apY
114	Terbufos	13071-79-9	Organophosphorus	apNo	No	apY
115	Trichlorfon	52-68-6	Organophosphorus	apNo	No	apY
116	Bromophos-ethyl	4824-78-6	Organophosphorus	apNo	No	apY
117	Dichlofenthion	97-17-6	Organophosphorus	apNo	No	apY
118	Fenchlorphos	299-84-3	Organophosphorus	apNo	No	apY
119	Formothion	2540-82-1	Organophosphorus	apNo	No	apY
120	Iodofenphos	18181-70-9	Organophosphorus	apNo	No	Yes
121	Ethiprole	181587-01-9	Phenyl pyrazole	apNo	No	ApY
122	Beta-cypermethrin	65731-84-2	Pyrethroid	apNo	No	Yes
123	Cycloprothrin	63935-38-6	Pyrethroid	apNo	No	apY
124	Fenpropathrin	39515-41-8 64257-84-7	Pyrethroid	apNo	No	apY
125	Flucythrinate	70124-77-5	Pyrethroid	apNo	No	apY
126	Kadethrin	58769-20-3	Pyrethroid	apNo	No	Yes
127	Silafluofen	105024-66-6	Pyrethroid	apNo	No	apY
128	Tefluthrin	79538-32-2	Pyrethroid	apNo	No	apY
129	Tralomethrin	66841-25-6	Pyrethroid	apNo	No	Yes

[a] Not applicable.
[b] *a priori* no.
[c] *a priori* yes.

The variables are ranked according to their relative importance. Indeed, it is obvious that in a multicriteria decision system, all the selected variables do not have the same importance with respect to the final decision. In other words, they do not have the same weight. When facing this kind of situation, it is of common use to introduce coefficients in the calculation procedure to modify the final weight of some variables. This strategy suffers from a lack of transparency and is too rigid. In SIRIS, the variables are ranked by the decreasing order of importance and this order will have an impact on the results. It is noteworthy that two variables can have the same importance. Last, a min/max scale of scores is calculated according to specific incremental rules, which have been extensively described elsewhere [34]. An illustrative example of calculation is given in Figure 14.1.

In this study, two different categories of criteria were considered, namely, those characterizing the toxicity and ecotoxicity of the insecticides and criteria describing their exposure and environmental fate. Both types of criteria, and their corresponding modalities, are described in the next section.

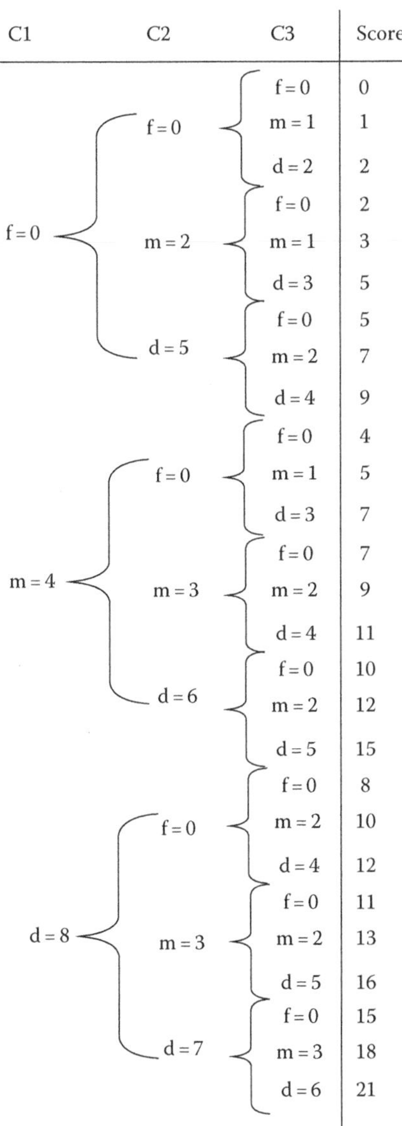

FIGURE 14.1 Scale of SIRIS scores calculated for a system at three criteria (C1, C2, C3) presenting three modalities and with C1> C2> C3.

14.2.3 Selection of the Criteria and Their Modalities

14.2.3.1 Toxicity Criteria

The operators are primarily exposed to vector control treatments. The exposure, which is predominantly dermal and respiratory, is common, especially during epidemics. The general population is also exposed to vector control insecticides during the treatments (dermal and respiratory routes) and to the residues after them, by dermal contact and possibly oral ingestion. As a result, there is a need to characterize the toxicity of the vector control insecticides by all these routes of exposure. This is a complex task requiring to take into account various *in vitro* and *in vivo* assays encoding different phenomena. In addition, it is noteworthy that extrinsic parameters, such as the formulation or the route of exposure, also intervene in the toxicity response. In this study, three representative and important toxicity endpoints were selected. They are described in the following.

14.2.3.1.1 Oral Lethal Dose 50% in Rats

The lethal dose 50% (LD_{50}) is a standard measure of acute toxicity in regulatory dossiers for chemicals. It is the amount of a substance that will kill half of the sample population of a specific test animal, usually rats or mice, in a specified period of time through single exposure via ingestion, skin contact, or injection. An LD_{50} value is measured in micrograms (or milligrams) of the material per kilogram of the test animal's body weight; the lower the amount, the more toxic the substance [47]. This endpoint is suited to compare and rank chemicals. The threshold values for delineating the three modalities were selected in such a way to have balanced classes. The LD_{50} values were classified as unfavorable, moderately favorable, or favorable if they were <100, ≥ 100 but <2000, ≥ 2000 mg/kg, respectively. It is worth noting that the upper limit (2000 mg/kg) represents the limit for classification of substances according to the European regulation 1272/2008 and European Directive 67/548/EEC [48]. The LD_{50} values were mainly extracted from the Footprint Pesticide Properties Database (FOOTPRINT PPDB) [49].

14.2.3.1.2 Carcinogenicity

The phenomenon of carcinogenesis involves the formation of neoplastic lesions induced by chemical substances. Most cancers (>90%) arise from epithelial tissues, such as the inside lining of the lung or prostate. These are called carcinomas and usually affect older people. Sarcomas are tumors arising from mesenchymal tissues such as bone or connective tissue. They occur in young people as well as in adults and comprise less than 1% of all cancers. If the human data, coming from epidemiological studies, are of invaluable value for determining the carcinogenicity of a substance, their number is unfortunately too limited. As a result, *in vivo* tests on rats and mice are used in the frame of the regulation of chemicals.

The Carcinogenic Potency Database (CPDB) [50] was investigated to collect insecticides identified as carcinogens or noncarcinogens from rat and mouse laboratory test results. Data on both genders were collected when available. Data gathering was also made from Devillers et al. [51], who compiled a large amount of results from original publications.

A chemical was considered as carcinogenic (unfavorable) if at least a positive rodent test result was obtained regardless of species and of gender. Conversely, to be classified as noncarcinogenic (favorable), the laboratory test results of a chemical had to be negative on rats and mice of both sexes. These rules were always used except for *Bs* and *Bti* that were directly classified as noncarcinogenic because their carcinogenic potential was considered unlikely in view of their mechanism of action.

14.2.3.1.3 Endocrine Disruption Potential

Endocrine disruptors refer to chemicals, either natural or man-made, which can alter the structure or function(s) of the endocrine system of humans and/or wildlife, and can cause adverse effects to an organism, its progeny, and/or a (sub)population [52,53]. The endocrine system, which operates through different glands and their messengers (hormones), controls pivotal physiological functions in metazoans (e.g., homeostasis, reproduction, development). The hormones need to bind to compatible receptors for initiating specific responses. These receptors, belonging to the nuclear receptor superfamily, include receptors for hydrophobic molecules such as steroid hormones (e.g., estrogens, androgens, progesterone, glucocorticoids, mineralocorticoids, vitamin D3), retinoic acids (all-*trans*, 9-*cis* isoforms), thyroid hormones, fatty acids, leukotrienes, and prostaglandins [54]. The cognate ligand for each receptor is generally dependent on the supply of a dietary for its synthesis, for example, iodine and tyrosine for thyroid hormone [54]. This means that the holistic study of endocrine disruption effects needs to consider a huge number of phenomena, which are, in addition, very often interrelated [55]. Among all these possible effects, the action on sex hormone receptors is the best (or least bad) known. These effects are mainly measured through the use of *in vitro* gene reporter screening tests [56–60]. Such results, extracted directly from the original papers (e.g., [56–60]) or retrieved from databases [49,61,62], were used to characterize the endocrine disruption potential of the 129 insecticides (Table 14.1). Insecticides providing negative results in the *in vitro* tests and indicated as nonendocrine disruptors in the databases were classified favorably. Otherwise, they were classified unfavorable for this criterion. Note that in case of missing data, the insecticide was considered as endocrine disruptor (unfavorable).

14.2.3.2 Ecotoxicity Criteria

Exposure of the environmental compartments depends on the type of treatment (larvicide or adulticide) and its location (indoor, urban, or rural). As a result, exposure of nontarget organisms depends on the dose and modes of application of the insecticides as well as their environmental fate. Indoor treatments do not expose the environmental compartments, except in case of incidental contamination. Only the operators and the residents of the treated homes are exposed. Conversely, the adulticide and larvicide treatments, in urban or rural areas, induce exposure of the different environmental compartments. Hazard assessment of vector control insecticides has to be made from the use of normalized tests. In this study, one terrestrial and two aquatic tests were selected. They are described in the following subsections.

14.2.3.2.1 *Median Oral and Contact Lethal Dose (LD_{50}) in Honey Bees*

Honey bees (*Apis mellifera*) are beneficial arthropods playing a key role in pollinating wild and crop plants. They actively participate in the maintenance of the biodiversity in the terrestrial ecosystems. As a result, whenever honey bees are likely to be exposed to chemical plant protection products, during or after treatments, laboratory toxicity tests are requested by the regulatory authorities. In these laboratory tests, pesticides are administered by oral and contact routes to represent the different types of exposure under field conditions, and the mortality is recorded after 24 or 48 h of exposure in order to determine an LD_{50} value [63]. Indeed, absorption through the bee's integument is the basis for contact toxicity. The physicochemical properties of a pesticide and especially its formulation are largely responsible for the relative hazard related to this mode of entry into the bee. Ingestion of contaminated pollen and nectar offers yet another route of entry. The alimentary tract may become altered or paralyzed, making feeding impossible, or the bee's gut may cease to function [64]. Oral and contact LD_{50} values were mainly collected in FOOTPRINT PPDB [49]. When both LD_{50} values were available, the most toxic was selected to determine the modality.

The threshold limits selected for the SIRIS analysis were determined to be, at best, in agreement with the regulations of chemicals but also to ensure the balance of the three classes considered. They were the following:

Favorable: $LD_{50} \geq 1$ µg/bee

Moderately favorable: <1 µg/bee $LD_{50} \geq 0.01$ µg/bee

Unfavorable: LD_{50} <0.01 µg/bee

It is important to note that while the determination of the oral and contact LD_{50} values of insecticides is a regulatory request, in practice, this is not enough to fully assess the adverse effects of pesticides to bees. Indeed, sublethal effects have to be considered [65,66]. Thus, for example, while the juvenile hormone mimics pyriproxyfen (#16 in Table 14.1), methoprene (#20), S-methoprene (#36), fenoxycarb (#73), and hydroprene (#74) were considered as nontoxic against bees (favorable), adverse effects on their development have been observed [67–69].

14.2.3.2.2 *Median Effective Concentration (EC_{50}) in the Water Flea*

The water flea, *Daphnia magna*, plays a key ecological role in surface waters being an efficient herbivore and aiding in the transfer of energy from autotrophs to the top of the food web [70,71]. Consequently, this small freshwater organism, easy to rear, is used worldwide for estimating the aquatic toxicity of chemicals [72,73]. Briefly, young daphnids (<24 h) are exposed to a range of concentrations of the test substance (five concentrations) for a period of 48 h. The immobilization is recorded at 24 and 48 h and compared to control values obtained under the same experimental conditions. The results are used to calculate a 48 h median effective concentration (EC_{50}) value. The EC_{50} values for the 129 insecticides listed in Table 14.1 were mainly collected in FOOTPRINT PPDB [49]. The threshold limits selected for the SIRIS analysis were determined to be, at best, in agreement with the regulations of chemicals but also to ensure the balance of the three classes considered. They were the following:

Favorable: $EC_{50} \geq 1$ mg/L

Moderately favorable: <1 mg/L $EC_{50} \geq 0.01$ mg/L

Unfavorable: EC_{50} <0.01 mg/L

14.2.3.2.3 Median Lethal Concentration (LC_{50}) in Fish

Fish species also play a crucial role in aquatic ecosystems. Normalized tests are available for estimating the concentration of substance that kills half the members of a fish population, mainly *Danio rerio* or *Oncorhynchus mykiss*, after 96 h of exposure [74–78]. The threshold limits selected for the SIRIS analysis were determined to be, at best, in agreement with the regulations of chemicals but also to ensure the balance of the three classes considered. They were the following:

Favorable: $LC_{50} \geq 1$ mg/L

Moderately favorable: <1 mg/L $LC_{50} \geq 0.01$ mg/L

Unfavorable: LC_{50} <0.01 mg/L

14.2.3.3 Exposure and Environmental Fate Criteria

Knowing the physicochemical properties of xenobiotics is a prerequisite to estimate their bioactivity, bioavailability, transport, and distribution between the different compartments of the environment. Four of them were used to describe the environmental behavior of the 129 insecticides under study (Table 14.1). In addition, the applied dose was also considered. These five criteria are described in the following.

14.2.3.3.1 1-Octanol/Water Partition Coefficient

The 1-octanol/water partition coefficient is defined as the ratio of a chemical's concentration in the octanol phase to its concentration in the aqueous phase of a two-phase octanol/water system. Values of Kow are thus unitless and are conveniently expressed in a logarithmic form (i.e., log Kow or log P). This hydrophobic parameter is a good indicator of the tendency of an organic compound to accumulate in biota and adsorb to soil. Experimental log Kow values, generally measured at 20°C, were retrieved from literature [79,80]. For many pesticides, different log Kow values were available for the same molecule. In these conditions, the value recommended by the author (e.g., "star value" [80]) or the one most consistent with the structure of the molecule was chosen. Because experimental log Kow values for d-phenothrin (#12) and kadethrin (#126) are not available in the literature, they were estimated from the KoWin model version 1.67 [81]. It is impossible to measure or compute the 1-octanol/water partition coefficient for *Bs* and *Bti*. Consequently, a log Kow value of 4 was arbitrarily selected. It corresponds to an average persistence.

Because, in the regulatory dossiers, it is admitted that a chemical with a log Kow <3 does not bioconcentrate, this value was selected as the threshold limit to define the favorable category. A log Kow value ≥ 5 was selected as upper threshold limit to define the unfavorable category. Insecticides with a log Kow value between these two limits were considered as being moderately favorable regarding this criterion.

14.2.3.3.2 Vapor Pressure

The vapor pressure (VP) determines the potential of a chemical to volatilize from its condensed or dissolved phases and to therefore exist as a gas [82]. This physico-chemical parameter is particularly suited for estimating the affinity of a molecule for the air compartment. VP strongly depends on the temperature as expressed in the classical Clausius–Clapeyron equation [83]. A molecule with a VP $<10^{-5}$ Pa being considered as poorly volatile [84], this value was selected as threshold limit of the favorable modality. The threshold value of 10^{-2} Pa being used by the European regulation on biocides to require an inhalation toxicity test, this value was chosen as limit of the unfavorable class. A VP value between these two limits was considered as moderately favorable. Experimental VP values, at 20°C–25°C, for the 129 insecticides listed in Table 14.1, were collected from literature [46,85,86]. Because the experimental VP value of ethiprole (#121 in Table 14.1) was missing, it was estimated at 3.61×10^{-4} Pa from Grain [87]. Lastly, due to their characteristics, the VP values of *Bs* and *Bti* were estimated as belonging to the favorable category.

14.2.3.3.3 Dosage

The dosage to use in vector control can be defined as the amount of active substance necessary to be effective. The higher this amount, the lower the active substance can be considered effective. This measurement is available only for a rather limited number of insecticides [25]. Setting the threshold values involved three steps. First, active substances were grouped according to their activity on mosquitoes. The most active ones were the pyrethroids deltamethrin, lambda-cyhalothrin, and cyfluthrin. The moderately active substances included the other pyrethroids and the less active ones included all other insecticides. The threshold values for dosage were set so as to respect those three groups. In that case, 2.5 and 25 g/ha were selected as threshold values to define the favorable, moderately favorable, and unfavorable classes, respectively. Finally, in case of missing information, the dose of application in agriculture was used. To set the threshold values for dose of application in agriculture, active substances characterized for both vector control dose and agricultural dose were used. The active substance grouping according to vector control doses had to match the one according to agricultural doses. As a result, threshold values of 25 and 250 g/ha were selected to define the favorable, moderately favorable, and unfavorable classes, respectively. In case of conflict between the two scales and for the few missing data, the selection of the category was made by expert judgment. The data of agricultural doses were collected from the e-phy database [44], the ACTA index [45] (as well as previous editions), and the Pesticide Manual [46].

14.2.3.3.4 Biodegradation

Biodegradation is an important mechanism for eliminating xenobiotics by biotransforming them into simple organic and inorganic products. Two types of biodegradation can be considered. The primary biodegradation denotes a simple transformation not leading to a complete mineralization. The biodegradation products are specifically measured from chromatographic methods and the results are expressed by means of kinetic parameters such as biodegradation rate constant (k) and half-life ($T_{1/2}$). The ultimate (or total) biodegradation totally converts chemicals into simple molecules such as CO_2 and H_2O. Because experimental biodegradation data were not available for all the studied insecticides (Table 14.1), BioWin 3 and BioWin 4 [88], which are quantitative structure–biodegradation relationship (QSBR) models commonly used in regulation [89,90], were used for estimating their ultimate and primary biodegradation, respectively. Briefly, both models are based upon a survey of 17 biodegradation experts conducted by the Environmental Protection Agency (EPA), in which the experts were asked to evaluate 200 structurally diverse compounds in terms of the time required to achieve ultimate and primary biodegradation in a typical or "evaluative" aquatic environment. Each expert rated the ultimate and primary biodegradation of each compound on a scale from 1 to 5. The ratings correspond to the following time units: 5-hours, 4-days, 3-weeks, 2-months, and 1-longer. The ratings for each compound were averaged to obtain a single value for modeling. Chemicals were described by 36 structural fragments and molecular weight [91]. The models were in the form of the following regression equation: $\text{BioD rating} = \Sigma(a_i \times \text{nb fragment}_i) + b\,\text{MW} + c$. Based on the guidelines provided in the user manual of the models [88], ≥ 2.75 and <1.75 were considered as the threshold values for defining the favorable and unfavorable categories with BioWin 3. A calculated value between these two limits was considered as moderately favorable. Regarding BioWin 4, the values delineating the favorable and unfavorable categories were >3.75 and ≤ 3.25, respectively. Due to their environmental behavior, *Bs* and *Bti* were positioned in the unfavorable category for primary biodegradability and moderately favorable regarding their ultimate biodegradability.

14.2.4 Selection of the Hierarchies

The SIRIS method allows us to simultaneously consider variables encoding very different information [34]. However, our practical experience shows that it is generally preferable to construct several scales of SIRIS scores (one per category of criteria) instead of a unique scale [92,93]. This was also verified in the present study. Thus, the toxicity/ecotoxicity and exposure/environmental fate criteria were considered separately yielding the construction of two scales of SIRIS scores. The hierarchy of the variables in the SIRIS method being problem dependent, they were ranked by their order of importance accounting for two different scenarios that are the adulticides and larvicides.

Scenario for adulticides is shown in the following:

Toxicity/ecotoxicity scale
Rat > bee > daphnid > fish > endocrine disruption > carcinogenicity
Exposure/environmental fate scale
Dosage > log P > vapor pressure > ultimate biodegradation > primary biodegradation

Scenario for larvicides is shown in the following:

Toxicity/ecotoxicity scale
Rat > daphnid > fish > bee > endocrine disruption > carcinogenicity
Exposure/environmental fate scale
Dosage > log P > vapor pressure > ultimate biodegradation > primary biodegradation

As previously indicated, with the SIRIS method, the different criteria have to be ranked in decreasing order of importance. However, when a criterion cannot be correctly described due to a lack of experimental data, it is preferable to place it at the end of the hierarchy rather than at the beginning in order to limit its weight in the calculation of the SIRIS scores. Thus, although carcinogenicity is a very important toxicological endpoint, the criterion was located at the end of the selected hierarchy for the two toxicity/ecotoxicity scales because the experimental carcinogenicity data were missing for a significant number of insecticides, and as a result, the corresponding chemicals were classified unfavorably for that criterion.

14.3 RESULTS

14.3.1 Larvicides

The typology of the 129 insecticides characterized by their two SIRIS scores calculated from the two hierarchies selected for the larvicide scenario is displayed in Figure 14.2. The SIRIS scores are ranged from 0 to 81 on the toxicity/ecotoxicity scale and from 10 to 52 on exposure/environmental fate scale. The insecticides located in the left bottom part of Figure 14.2 are the most interesting because they present low score values on the toxicity/ecotoxicity and exposure/environmental fate scales. This is the case, for example, for the insecticides #58 (cyromazine) and #86 (dicyclanil). At the opposite, the insecticides located in the top right part of Figure 14.2 are not suited for the selected scenario due to the high values of their SIRIS scores on both scales. Thus, for example, insecticide #117 (dichlofenthion) shows bad scores on both the toxicity/ecotoxicity and exposure/environmental fate scales. A substance can be correctly classified on one scale and with a high score on the other scale. Thus, formetanate (#61) is well located on the exposure/environmental fate scale but shows a rather high score on the toxicity/ecotoxicity scale. This is also the case for deltamethrin (#11) located in the right bottom part of Figure 14.2. *Bs* (#4) and *Bti* (#5) are perfectly ranked on the toxicity/ecotoxicity scale, having SIRIS scores equal to zero, but they are a little bit less well ranked on the other scale. This is even more the case for the insecticides #88 (azadirachtin) and #127 (silafluofen). The insecticides located in the middle part of Figure 14.2, such as metaflumizone (#68) or tau-fluvalinate (#85), are less interesting because they show rather high SIRIS scores on both scales. However, this does not mean that they have to be excluded from a future selection. Indeed, a map of SIRIS scores has to be considered as decision support tool. There are obvious selections for insecticides with low SIRIS score values on both scales, revealing favorable modalities for most of the criteria. At the opposite, a rejection is also obvious for

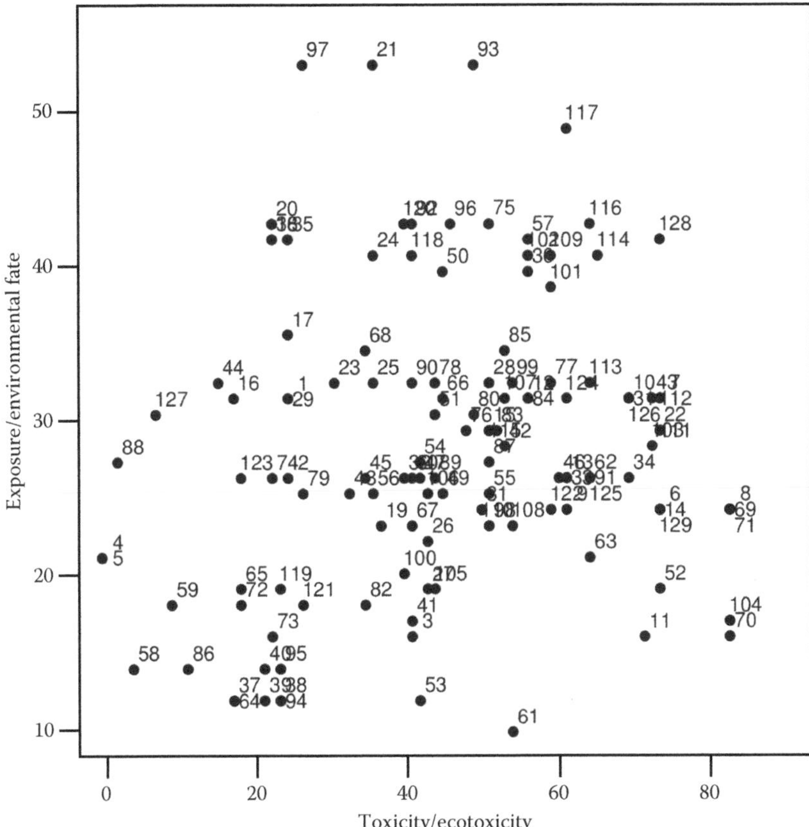

FIGURE 14.2 SIRIS map of the 129 insecticides used or potentially usable as larvicides. See Table 14.1 for the correspondence between the numbers and the chemical names.

insecticides with high SIRIS score values on the toxicity/ecotoxicity and exposure/environmental fate scales. Between these two extremities, all the intermediate situations exist. An insecticide with a low SIRIS score only on one scale can be selected after a reasoned decision. In this case, the SIRIS map allows to pinpoint the criteria showing the worst modalities and, hence, for which it will be necessary to pay a special attention. It is noteworthy that when data were missing, a worst-case approach was applied yielding an attribution to unfavorable category for the corresponding criterion (e.g., carcinogenicity). Therefore, the position of some molecules on the SIRIS map would probably change if new data could replace the missing data. Conversely, it is obvious that regarding the well-documented molecules, the situation will not change. This is the case for deltamethrin (#11).

In order to interpret more thoroughly the SIRIS map, specific information can be projected on it. Thus, for example, a collection of maps has been drawn, each representing a chemical family (Figure 14.3). For clarity, only the selected family has been represented on the initial typology of the larvicides (i.e., Figure 14.2). Thus, Figure 14.3a shows that the organophosphorus and carbamate insecticides are spread out on the SIRIS map. This is also the case for the pyrethroids (Figure 14.3b), the diverse neurotoxics (Figure 14.3c), and the benzoylureas (Figure 14.3d). Conversely, some chemical families form strong clusters. Thus, the neonicotinoid insecticides are all located in the left bottom part of the SIRIS map (Figure 14.3e) as well as the microorganisms, which are better located on the toxicity/ecotoxicity scale but a little bit less on the exposure/environmental fate scale (Figure 14.3f). The growth regulators also show a specific typology (Figure 14.3g). They are all well ranked on the toxicity/ecotoxicity scale, but conversely, they are spread out along the exposure/environmental fate scale. This dispersion is due to the heterogeneity of this family including chemicals with very different structures and, hence, very different environmental fate behaviors. Among this group, cyromazine (#58) and dicyclanil (#86) present the best SIRIS scores on both scales.

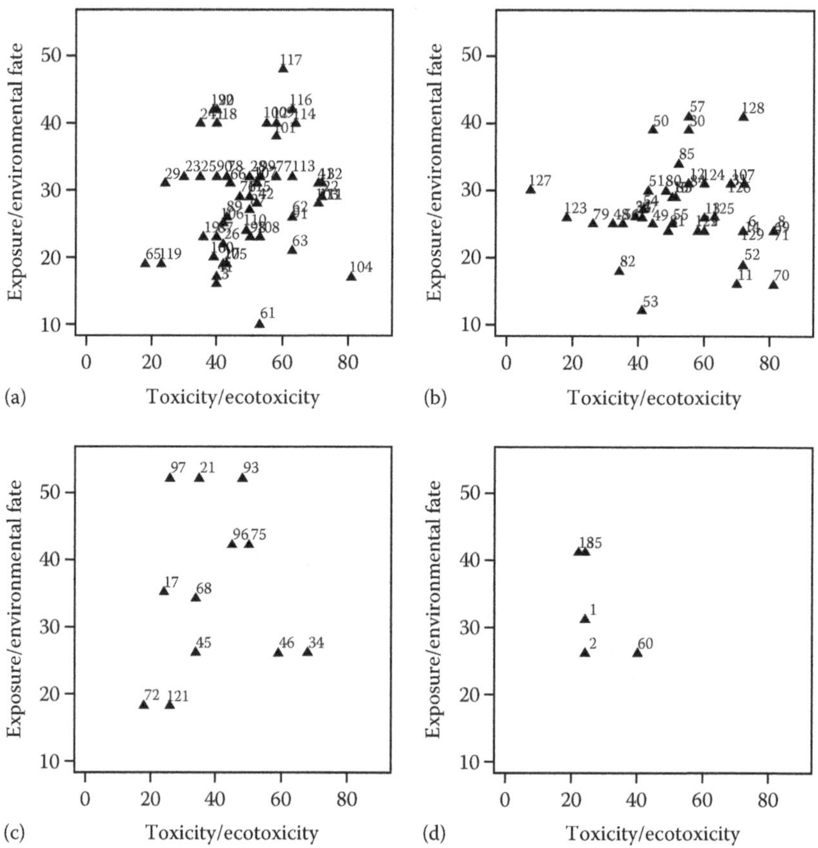

FIGURE 14.3 SIRIS maps characterizing the main chemical families of the 129 insecticides used or potentially usable as larvicides. (a) Organophosphorus and carbamate insecticides; (b) pyrethroids; (c) diverse neurotoxics; (d) benzoylureas. Each time, only the selected chemical family is represented.

(continued)

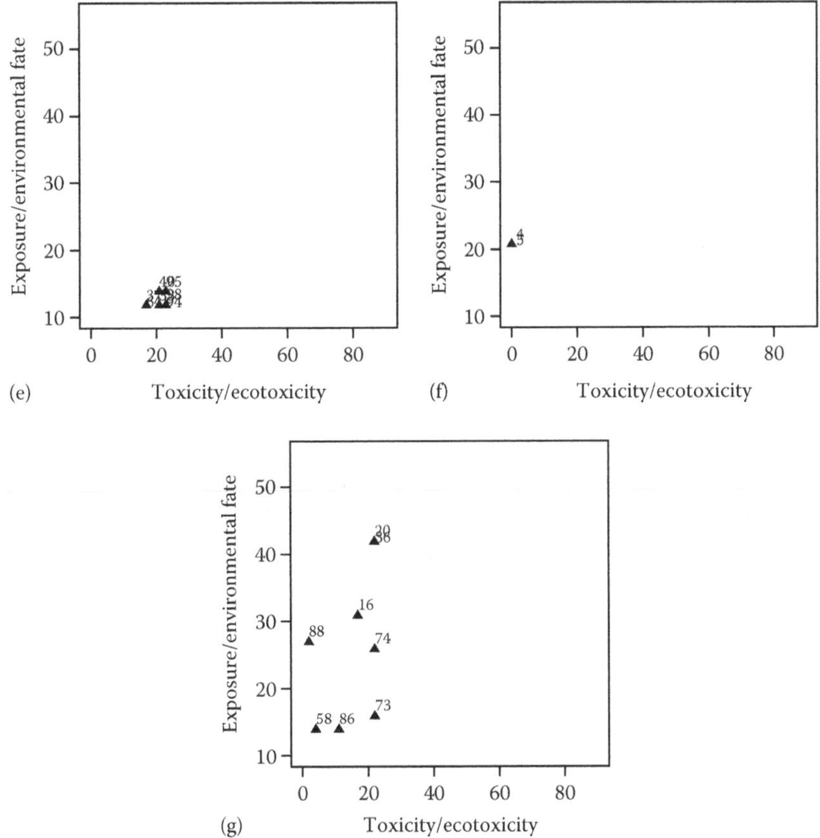

FIGURE 14.3 (continued) SIRIS maps characterizing the main chemical families of the 129 insecticides used or potentially usable as larvicides. (e) Neonicotinoid insecticides; (f) microorganisms; (g) growth regulators including juvenoids. Each time, only the selected chemical family is represented.

In Figure 14.4, insecticides having a well-known activity on mosquitoes (Table 14.1) are represented by black triangles. Inspection of Figure 14.4 shows that eight molecules recognized as being active against mosquitoes are located in the left bottom part of the SIRIS map. These insecticides are *Bs* (#4), *Bti* (#5), acetamiprid (#37), clothianidin (#38), imidacloprid (#39), thiamethoxam (#40), cyromazine (#58), and fenoxycarb (#73). In the same way, in Figure 14.5, insecticides recommended by WHOPES [25] as larvicides are represented by black triangles. Inspection of Figure 14.5 shows that among the 129 insecticides selected as potential candidates for vector control, only a few are recommended by WHOPES [25] as larvicides. Even more, among the insecticides presenting the best SIRIS scores on both scales, only *Bti* (#5) is recommended by WHOPES [25].

Figure 14.6 shows the insecticides that are noncarcinogenic in rats and mice of both genders (black triangles), noncarcinogenic for at least a male or female rat or mouse (empty triangles), carcinogenic for at least one species irrespective of its gender

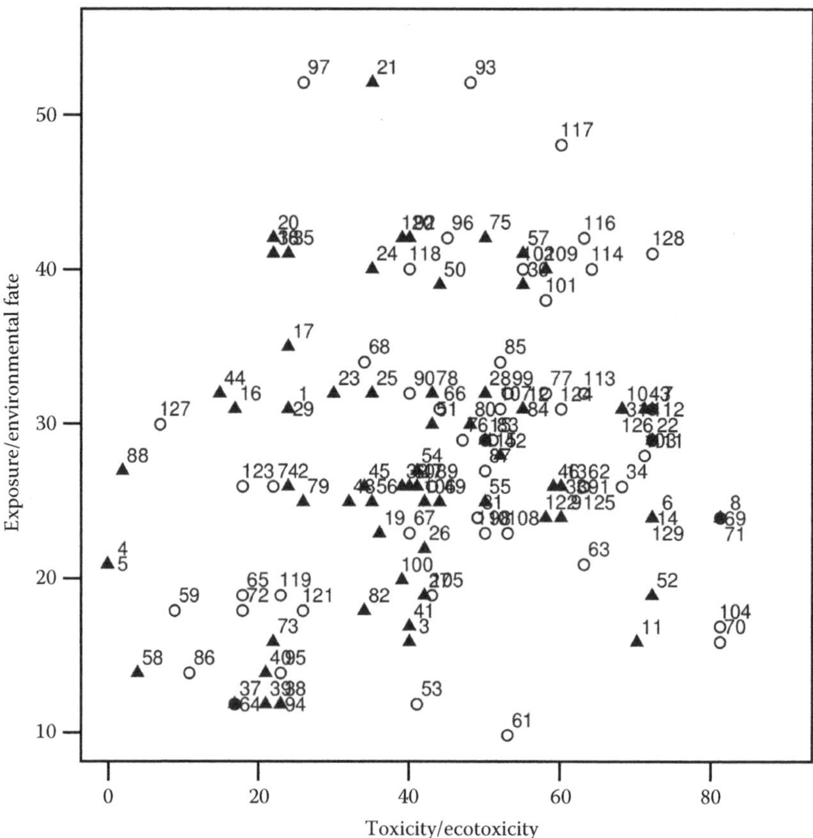

FIGURE 14.4 SIRIS map of the 129 insecticides used or potentially usable as larvicides. Those having a well-known activity on mosquitoes (Table 14.1) are represented by black triangles.

(empty circles), and for which there exists no information on their carcinogenicity (black circles). As previously stressed, for most of the insecticides, there is no information on their potential carcinogenicity. This means that there is an urgent need to obtain this crucial information. Because when the information was not available for carcinogenicity, the insecticide was ranked unfavorably for that criterion, and this means that the general typology was not affected by this worst-case choice. This is also true because this criterion was located at the end of the hierarchy. It is interesting to note that clothianidin (#38), cyromazine (#58), and dimethoate (#65), which are located in the left bottom part of the SIRIS map, are noncarcinogenic in rats and mice of both genders. This is also the case for pyriproxyfen (#16) and spinosad (#17), which are of interest, especially due to their good scores on the toxicity/ecotoxicity scale. Note that this is particularly true for the former chemical. Conversely, while piperonyl butoxide (#44) was also a candidate of interest, its carcinogenicity excludes any future selection (Figure 14.6).

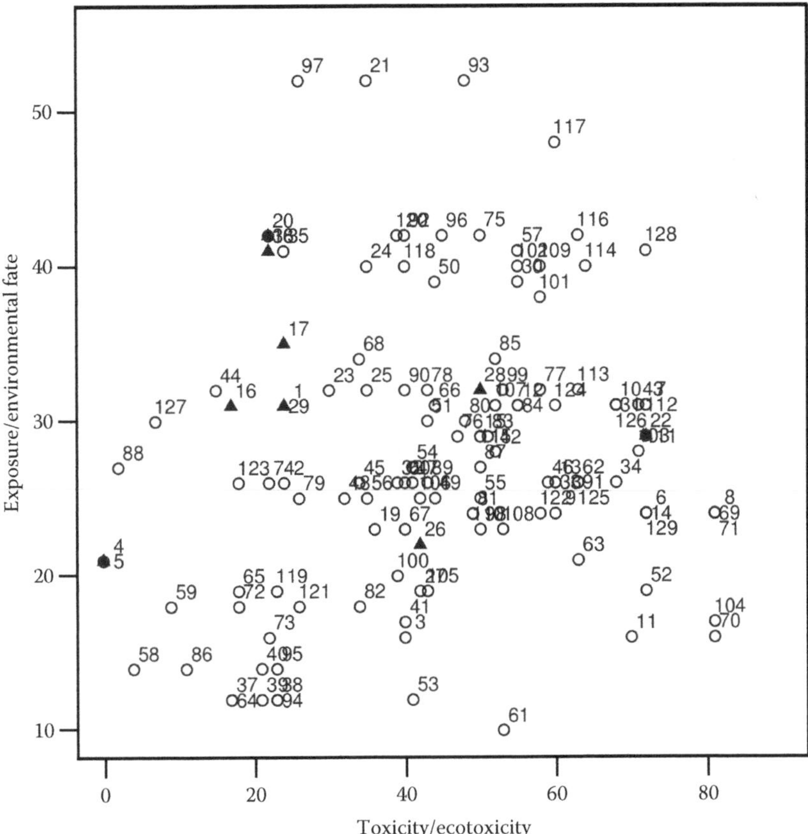

FIGURE 14.5 SIRIS map of the 129 insecticides used or potentially usable as larvicides. Those recommended by WHOPES as larvicides are represented by black triangles.

The same type of comments can be made when the information available on the endocrine disruption potential of the insecticides is reported on the SIRIS map obtained with the larvicide scenario. In Figure 14.7, the insecticides classified as nonendocrine disruptors on the basis of the set of *in vitro* tests investigated (see Section 14.2.3.1.3) are represented by black triangles. Those, which were found positive, under the same conditions, are represented by empty triangles. Lastly, pesticides for which the information was not available are represented by black circles. Inspection of Figure 14.7 shows that for numerous insecticides, the information on their endocrine disruption potential is missing. Conversely, a rather high number are positive. Thus, while dimethoate (#65) was noncarcinogenic (Figure 14.6), unfortunately, this organophosphorus insecticide affects the endocrine system (Figure 14.7). Fenoxycarb (#73), which is also located in the most interesting part of the SIRIS map, has been also identified as an endocrine disruptor. Notice that, on the basis of our criteria of selection, pyriproxyfen (#16), methoprene (#20), S-methoprene (#36), and hydroprene (#74) were not classified

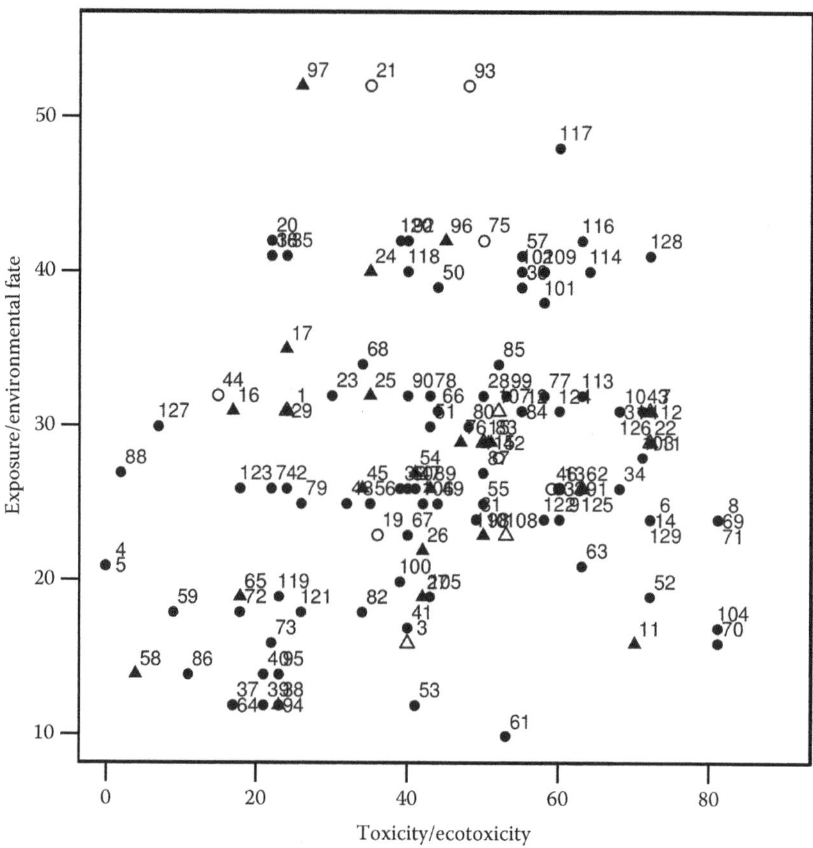

FIGURE 14.6 SIRIS map of the 129 insecticides used or potentially usable as larvicides and characterized by their potential carcinogenicity. Noncarcinogenic on rats and mice of both genders (black triangles); noncarcinogenic for at least a male or female rat or mouse (empty triangles); carcinogenic on at least one species irrespective of its gender (empty circles); information missing (black circles).

as endocrine disruptors, while being juvenile hormone mimics, by nature, they disrupt the endocrine system of the insects and other arthropods.

14.3.2 Adulticides

Because some insecticides can only be used as larvicides on mosquitoes, they were excluded from the analysis. These were diflubenzuron (#1), triflumuron (#2), *Bs* (#4), *Bti* (#5), pyriproxyfen (#16), novaluron (#18), methoprene (#20), hexaflumuron (#35), S-methoprene (#36), cyromazine (#58), teflubenzuron (#60), fenoxycarb (#73), hydroprene (#74), dicyclanil (#86), and azadirachtin (#88). The typology of the 114 insecticides characterized by their two SIRIS scores calculated from the two hierarchies selected for the adulticide scenario is displayed in Figure 14.8. The group of insecticides showing the best scores on the toxicity/ecotoxicity and exposure/

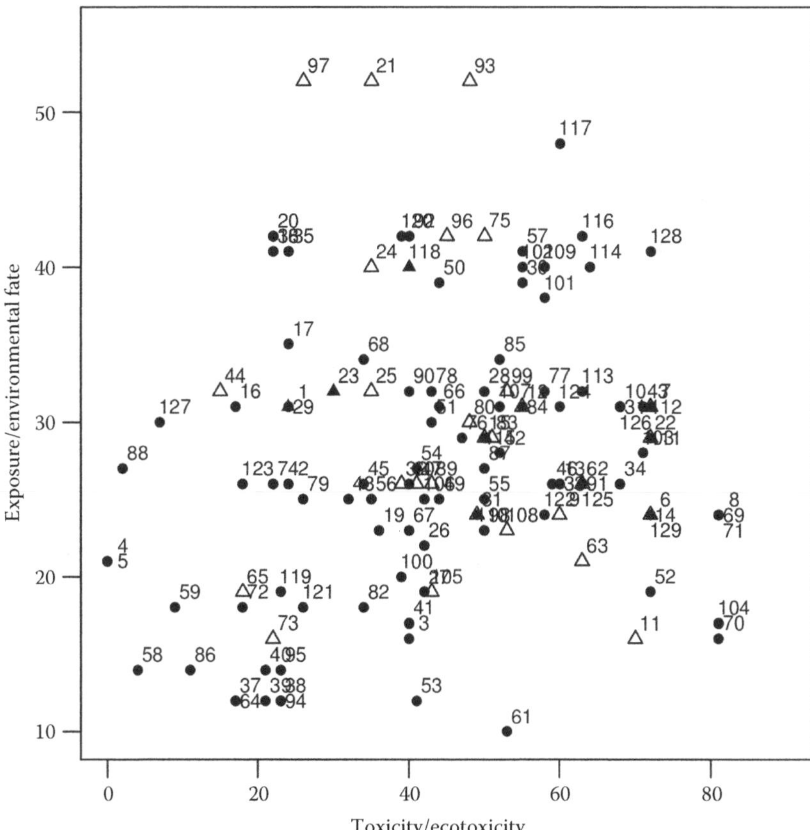

FIGURE 14.7 SIRIS map of the 129 insecticides used or potentially usable as larvicides and characterized by their endocrine disruption potential. Nonendocrine disruptor on the basis of the set of *in vitro* tests investigated (black triangles); positive under the same conditions (empty triangles); information missing (black circles).

environmental fate scales are chlorantraniliprole (#59); neonicotinoids, especially acetamiprid (#37) and thiacloprid (#64); two organophosphorus insecticides (dimethoate (#65) and formothion (#119)); a pyrethroid (S-bioallethrin, #82); spinetoram (#72); and ethiprole (#121).

In Figure 14.9, insecticides having a well-known activity on mosquitoes (Table 14.1) are represented by black triangles. Inspection of Figure 14.9 shows that five molecules recognized as being active on mosquitoes are located in the left bottom part of the SIRIS map. These insecticides are acetamiprid (#37), clothianidin (#38), imidacloprid (#39), thiamethoxam (#40), and S-bioallethrin (#82). In the same way, in Figure 14.10, insecticides recommended by WHOPES [25] as adulticides are represented by black triangles. Inspection of Figure 14.10 shows that among the 114 insecticides selected as candidates for vector control, only a few are recommended by WHOPES [25]. Even more, among the insecticides presenting the best SIRIS scores on both scales, none is recommended by WHOPES [25].

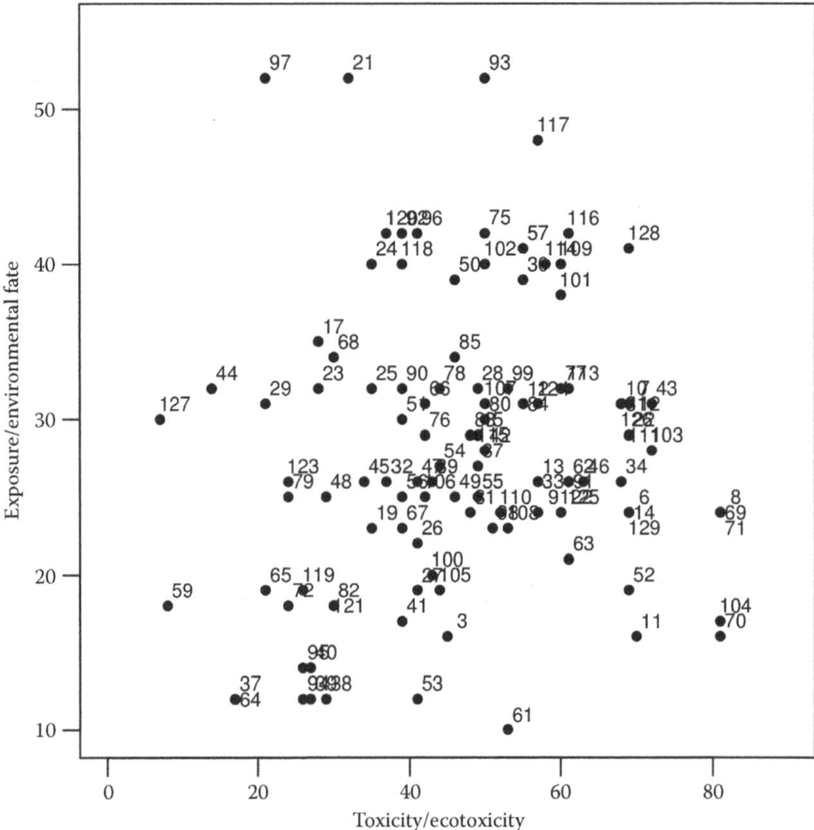

FIGURE 14.8 SIRIS map of the 114 insecticides used or potentially usable as adulticides. See text for the excluded chemicals and Table 14.1 for the correspondence between the numbers and the chemical names.

Projections of the information related to the carcinogenicity and endocrine disruption potentials were also performed (maps not shown). Broadly the same comments can be made as for the larvicides (Figures 14.6 and 14.7). However, it is interesting to stress that deltamethrin (#11), which is recommended by WHOPES [25] and used in France as adulticide, has been detected as endocrine disruptor.

14.4 DISCUSSION

The SIRIS MCA applied to 129 insecticides described by six toxicity/ecotoxicity criteria and five exposure/environmental fate criteria allowed us to identify substances that could potentially be used as larvicides or adulticides in vector control:

- *Bti* (#5), which is currently the most widely used substance as larvicide in France, and *Bs* (#4) show low SIRIS scores, especially on the toxicity/ecotoxicity scale.

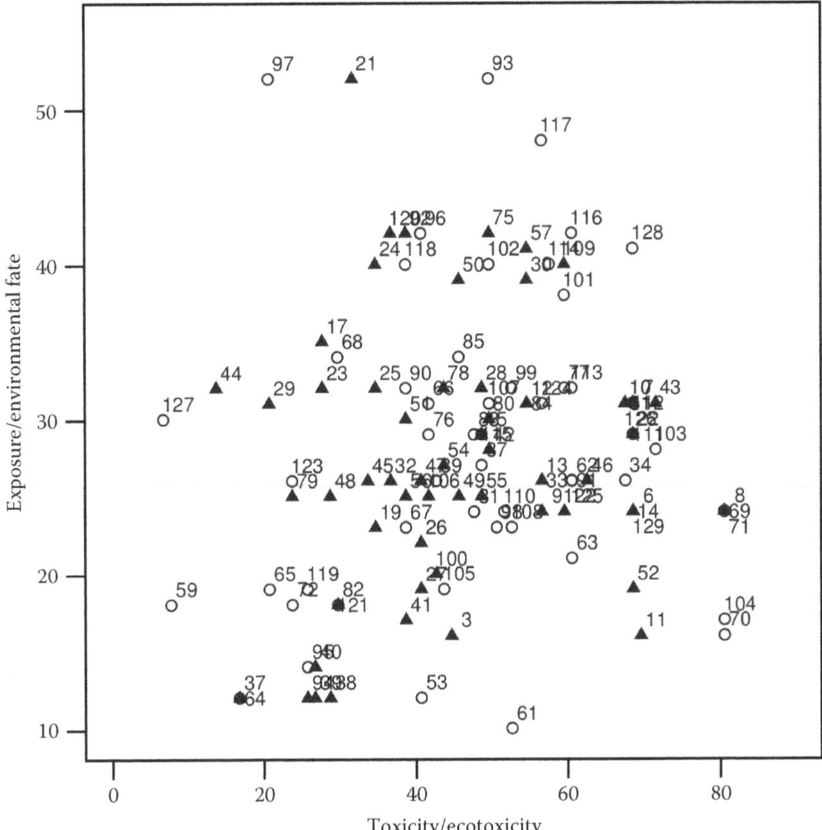

FIGURE 14.9 SIRIS map of the 114 insecticides used or potentially usable as adulticides. Those having a well-known activity on mosquitoes (Table 14.1) are represented by black triangles.

- Among the other larvicides *sensu stricto*, some growth regulators, especially diflubenzuron (#1) and pyriproxyfen (#16) (both recommended by WHOPES), present interesting possibilities. Others, such as cyromazine (#58), triflumuron (#2), and hydroprene (#74), are also of interest.
- Spinosad (#17) and indoxacarb (#45) are less well positioned on the SIRIS map than the previously mentioned substances but could also be candidates.
- Pyrethroids are also an interesting group for vector control, although the use of deltamethrin (#11) could be challenged in the future due to resistance issues as well as its toxicity and ecotoxicity resulting in a high SIRIS score on the toxicity/ecotoxicity scale. The substitution of this substance by type I pyrethroids (imiprothrin (#53), allethrin (#79)) or type II (e.g., cycloprothrine (#123)) or silafluofen (#127), whose SIRIS scores are less penalizing, could be an interesting option to keep this family in vector control strategies.

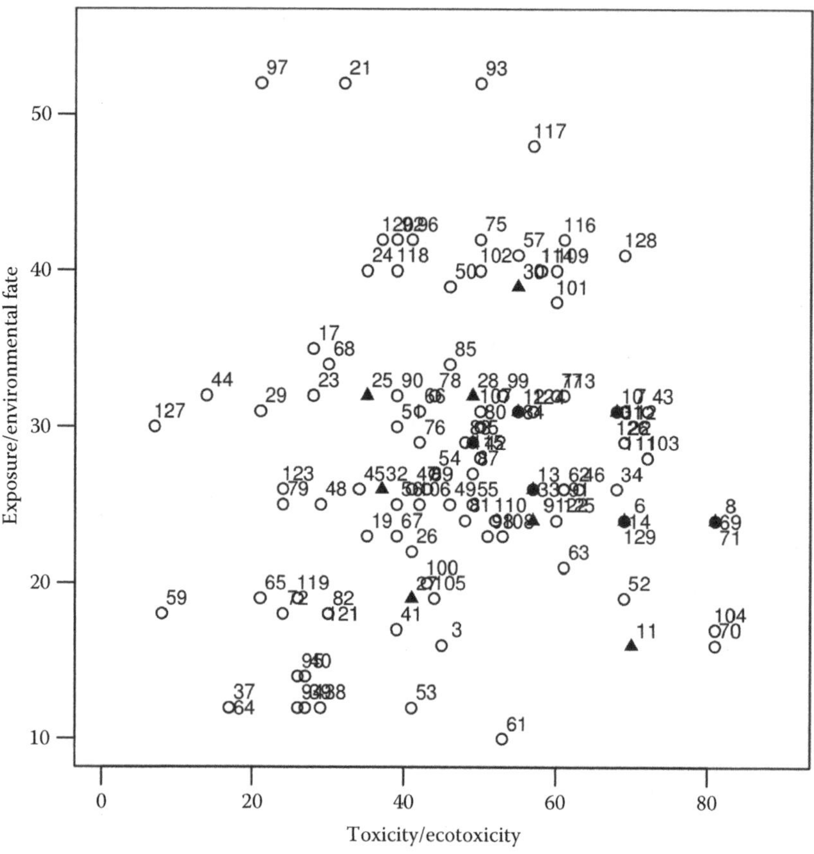

FIGURE 14.10 SIRIS map of the 114 insecticides used or potentially usable as adulticides. Those recommended by WHOPES as adulticides are represented by black triangles.

- Among the organophosphorus insecticides, formothion (#119) and dimethoate (#65) show the best SIRIS scores on both scales. Temephos (#29) presents a high score on the exposure/environmental fate scale, while for fenthion (#26), malathion (#27), and the carbamate bendiocarb (#3), the SIRIS scores of the toxicity/ecotoxicity scale are higher.
- The neonicotinoids (acetamiprid (#37), clothianidin (#38), imidacloprid (#39), thiamethoxam (#40), thiacloprid (#64), dinotefuran (#94), nitenpyram (#95)) cannot be neglected in vector control because they show good SIRIS scores on both the exposure/environmental fate and toxicity/ecotoxicity scales. In addition, their mechanism of action is different from those of the insecticides currently used in vector control. This represents a definitive advantage in the management of insecticide resistance. Nevertheless, it is likely that used alone, these substances would not be sufficiently effective. In fact, the interest would be to combine them with another insecticide of a different chemical family to obtain a synergistic effect that remains to be evaluated.

- Other molecules such as spinetoram (#72), chlorantraniliprole (#59), dicyclanil (#86), and ethiprole (#121) are located in the left bottom part of the SIRIS maps showing low SIRIS scores on both scales. However, they suffer from a lack of data. In addition, while the activity on dipterans is known, their efficacy against mosquitoes remains to be evaluated.

Although the SIRIS method has clearly shown its interest for selecting candidate insecticides for vector control, it is necessary to recall some basic principles sustaining all the multicriteria methods. The results obtained with such methods depend on the quantity and quality of the information used as input and are directly linked to the choices and hypotheses made by the experts prior to run them. Thus, in the present study, missing data were very often coded unfavorably, except in few cases dealing with isomers. This means that if the data gaps are filled, the typologies could change. Obviously, this depends on the importance given to the corresponding criterion in the analysis. Selection of other threshold values in the definition of the favorable, moderately favorable, and unfavorable categories for a criterion affects the results. Obviously, the more the criteria will have their threshold values changed, the more the typologies will be changed. It is also necessary to be aware of the relative significance of a threshold value with such a crisp MCA. Thus, for example, a log Kow value ≥ 5 was selected as the upper threshold limit to define the unfavorable category of this criterion. It is noteworthy that an experimental log Kow ranging from 4.70 to 5.30 for the same insecticide is common place, while yielding a change in the category allocation from moderately favorable to unfavorable or the converse. This is the reason why a specific attention was paid to selecting the log Kow values. When a QSAR or quantitative structure–property relationship (QSPR) model is used to generate missing data, it is also necessary to account for its particularities and limitations because they can impact the outputs of the MCA.

From a practical point of view, this means that the results obtained with the SIRIS method or another MCA must not be considered as immutable. This is particularly true for the objects having mean score values. In this study, these correspond to the insecticides located in the central part of the SIRIS maps. A refinement on the numerical value of a criterion can change the allocation of the modality and move the location of the corresponding insecticide to the favorable part of the map or to the unfavorable part.

Irrespective of the quality of a typology, it is important to stress that the SIRIS method, as any other MCA or multivariate analysis, is only able to pinpoint potential candidates, but their risk assessment will be always necessary before to envision their use.

It should be recalled that the current period of regulation of biocidal products, and therefore for the insecticides used in vector control, is transient since all products on the market have not been evaluated yet. Ultimately, any use of an insecticide in vector control will be subject to authorization on the market. This authorization will be issued on the basis of a full assessment of its effectiveness and risk according to the regulation of Directive 98/8/EC on biocides.

14.5 CONCLUSIONS

The use of the SIRIS method allowed proposing a list of candidate insecticides, which could be potentially used as adulticides and/or larvicides in vector control. Obviously, the different insecticides, which have emerged from the SIRIS analysis, are only candidates, and risk assessment procedures are required before to envision their use.

Besides the identification of insecticides with attractive profiles in the SIRIS analysis, the present study also provides indications on how to use them in vector control. In particular, the use of combination of insecticides from different families may help at controlling resistance development. Synergistic combinations may also result in reducing the applied doses, thus limiting unwanted effects on nontarget species.

Our work was focused on the treatments made in vector control by public operators against *Anopheles*, *Aedes*, and *Culex*. The same methodological approach could be applied to other usages in vector control such as the treatment of mosquito nets or indoor residual spraying as well as other vector targets. Obviously, in these cases, it should be necessary to define new scenarios for the SIRIS analysis.

ACKNOWLEDGMENTS

This study was conducted by a working group directed by ANSES in the frame of an expert appraisal demanded by the French ministries for ecology, health, and labor.

REFERENCES

1. D.J. Gubler, *Resurgent vector-borne diseases as a global health problem*, Emerg. Infect. Dis. 4 (1998), pp. 442–450.
2. U. Bulugahapitiya, S. Siyambalapitiya, S.L. Seneviratne, and D.J.S. Fernando, *Dengue fever in travellers: A challenge for European physicians*, Eur. J. Int. Med. 18 (2007), pp. 185–192.
3. S.B. Halstead and G. Papaevangelou, *Transmission of dengue 1 and 2 viruses in Greece in 1928*, Am. J. Trop. Med. Hyg. 28 (1980), pp. 635–637.
4. L. Rosen, *Dengue in Greece in 1927 and 1928 and the pathogenesis of dengue hemorrhagic fever: New data and a different conclusion*, Am. J. Trop. Med. Hyg. 35 (1986), pp 642–653.
5. A.P. Almeida, Y.M. Goncalves, M.T. Novo, C.A. Sousa, M. Melim, and A.J. Gracio, *Vector monitoring of* Aedes aegypti *in the autonomous region of Madeira, Portugal*, Euro Surveill. 12 (2007), E071115. Available at http://www.eurosurveillance.org/ViewArticle.aspx?ArticleId=3311
6. O. Wichmann, J. Gascon, M. Schunk, S. Puente, H. Siikamaki, I. Gjørup, R. Lopez-Velez, J. Clerinx, G. Peyerl-Hoffmann, A. Sundøy, B. Genton, P. Kern, G. Calleri, M. de Górgolas, N. Mühlberger, and T. Jelinek, *Severe dengue virus infection in travelers: Risk factors and laboratory indicators*, J. Infect. Dis. 195 (2007), pp. 1089–1096.
7. C.A. Devaux, *Emerging and re-emerging viruses: A global challenge illustrated by chikungunya virus outbreaks*, World J. Virol. 1 (2012), pp. 11–22.
8. T. Seyler, F. Grandesso, Y. Le Strat, A. Tarantola, and E. Depoortere, *Assessing the risk of importing dengue and chikungunya viruses to the European Union*, Epidemics 1 (2009), pp. 175–184.

9. G. La Ruche, Y. Souarès, A. Armengaud, F. Peloux-Petiot, P. Delaunay, P. Desprès, A. Lenglet, F. Jourdain, I. Leparc-Goffart, F. Charlet, L. Ollier, K. Mantey, T. Mollet, J. P. Fournier, R. Torrents, K. Leitmeyer, P. Hilairet, H. Zeller, W. Van Bortel, D. Dejour-Salamanca, M. Grandadam, and M. Gastellu-Etchegorry, *First two autochthonous dengue virus infections in metropolitan France, September 2010.* Euro Surveill. 15 (2010): pii = 19676. Available at http://www.eurosurveillance.org/ViewArticle.aspx?ArticleId = 19676
10. E.A. Gould, P. Gallian, X. de Lamballerie, and R.N. Charre, *First cases of autochthonous dengue fever and chikungunya fever in France: From bad dream to reality!*, Clin. Microbiol. Infect. 16 (2010), pp. 1702–1704.
11. Anonymous, *Directive 98/8/EC of the European Parliament and of the Council of 16 February 1998 concerning the placing of biocidal products on the market*, Official Journal of the European Communities, April 24 (1998), L123/1.
12. J. Hemingway and H. Ranson, *Insecticide resistance in insect vectors of human disease*, Ann. Rev. Entomol. 45 (2000), pp. 371–391.
13. G. Munhenga, H.T. Masendu, B.D. Brooke, R.H. Hunt, and L.K. Koekemoer, *Pyrethroid resistance in the major malaria vector* Anopheles arabiensis *from Gwave, a malaria-endemic area in Zimbabwe*, Malar. J. 28 (2008), p. 247.
14. S. Marcombe, R. Poupardin, F. Darriet, S. Reynaud, J. Bonnet, C. Strode, C. Brengues, A. Yébakima, H. Ranson, V. Corbel, and J.P. David, *Exploring the molecular basis of insecticide resistance in the dengue vector* Aedes aegypti: *A case study in Martinique Island (French West Indies)*, BMC Genom. 10 (2009), p. 494.
15. M. Paris, G. Tetreau, F. Laurent, M. Lelu, L. Despres, and J.P. David, *Persistence of* Bacillus thuringiensis israelensis (Bti) *in the environment induces resistance to multiple* Bti *toxins in mosquitoes*, Pest. Manag. Sci. 67 (2011), pp. 122–128.
16. M. Paris, J.P. David, and L. Despres, *Fitness costs of resistance to* Bti *toxins in the dengue vector* Aedes aegypti, Ecotoxicology 20 (2011), pp. 1184–1194.
17. W. Karcher and J. Devillers, *Practical Applications of Quantitative Structure-Activity Relationships (QSAR) in Environmental Chemistry and Toxicology*, Kluwer Academic Publishers, Dordrecht, the Netherlands, 1990, p. 475.
18. N.A. Begum, N. Roy, R.A. Laskar, and K. Roy, *Mosquito larvicidal studies of some chalcone analogues and their derived products: Structure–activity relationship analysis*, Med. Chem. Res. 20 (2010), pp. 184–191.
19. C. Hansch and R.P. Verma, *Larvicidal activities of some organotin compounds on mosquito larvae: A QSAR study*, Eur. J. Med. Chem. 44 (2009), pp. 260–273.
20. S.C. Basak, R. Natarajan, D. Mills, D.M. Hawkins, and J. Kraker, *Quantitative structure-activity relationship modeling of juvenile hormone mimetic compounds for* Culex pipiens *larvae, with a discussion of descriptor-thinning methods*, J. Chem. Inf. Model. 46 (2006), pp. 65–77.
21. F. Darriet, S. Marcombe, and V. Corbel, *Insecticides Larvicides et Adulticides Disponibles pour les Opérations de Lutte contre les Moustiques—Synthèse bibliographique*, IRD, Montpellier, France 2007.
22. Afsset, *La lutte Antivectorielle dans le cadre de L'épidémie de Chikungunya sur l'Ile de la Réunion—Evaluation des risques et de L'efficacité des Produits Adulticides*, Maisons-Alfort, France, 2007.
23. Afsset, *La lutte Antivectorielle dans le cadre de L'épidémie de Chikungunya sur l'Ile de la Réunion—Evaluation des risques et de L'efficacité des Produits Larvicides*, Maisons-Alfort, France, 2007.
24. C. Cox, *Naled (Dibrom)*, J. Pest. Ref. 22 (2002), pp. 16–21.
25. WHOPES, *World health organization pesticide evaluation scheme*. Available at http://www.who.int/whopes/en

26. GCDPP, *Global Collaboration for Development of Pesticides for Public Health*. Available at http://www.who.int/whopes/gcdpp/en (accessed on January 10, 2013)
27. IPHPW, *International public health pesticides workshop*. Available at http://www.iphpw.org (accessed on January 10, 2013)
28. IVCC, *Innovative vector control consortium*. Available at http://www.ivcc.com (accessed on January 10, 2013)
29. V. Belton and T.J. Stewart, *Multiple Criteria Decision Analysis: An Integrated Approach*, Kluwer Academic Publishers, Dordrecht, the Netherlands, 2002. (accessed on January 10, 2013)
30. J. Figueira, S. Greco, and M. Ehrgott, *Multiple Criteria Decision Analysis: State of the Art Surveys*, Springer-Verlag, New York, 2004.
31. M. Ehrgott, J.R. Figueira, and S. Greco, *Trends in Multiple Criteria Decision Analysis*, Springer-Verlag, New York, 2010.
32. R. Baltussen, A.H.A. ten Asbroek, X. Koolman, N. Shrestha, P. Bhattarai, and L.W. Niessen, *Priority setting using multiple criteria: Should a lung health programme be implemented in Nepal*? Health Policy Plan. 22 (2007), pp. 178–185.
33. F. Rakotomanana, R.V. Randremanana, L.P. Rabarijaona, J.B. Duchemin, J. Ratovonjato, F. Ariey, J.P. Rudant, and I. Jeanne, *Determining areas that require indoor insecticide spraying using multi criteria evaluation, a decision-support tool for malaria vector control programmes in the central highlands of Madagascar*, Int. J. Health Geogr. 6 (2007), pp. 1–11.
34. M. Vaillant, J.M. Jouany, and J. Devillers, *A multicriteria estimation of the environmental risk of chemicals with the SIRIS method*, Toxicol. Model. 1 (1995), pp. 57–72.
35. M. Guerbet and J.M. Jouany, *Value of the SIRIS method for the classification of a series of 90 chemicals according to risk for the aquatic environment*, Environ. Impact. Assess. Rev. 22 (2002), pp. 377–391.
36. R. Vincent, F. Bonthoux, and C. Lamoise, *Evaluation du risque chimique. Hiérarchisation des risques potentiels*, Cahiers Notes Doc. Hyg. Sec. Travail 178 (2000), pp. 29–34.
37. AFPMB, Armed forces pest management board. Available at http://www.afpmb.org (accessed on January 10, 2013)
38. ORP, Observatoire des résidus de pesticides. Available at http://www.observatoire-pesticides.gouv.fr (accessed on January 10, 2013)
39. WHO, Collaborating Centre for Drug Statistics Methodology. Available at http://www.whocc.no/atcddd (accessed on January 10, 2013)
40. EudraPharm. Available at http://eudrapharm.eu/eudrapharm/welcome.do (accessed on January 10, 2013)
41. VIDAL. Available at http://www.vidal.fr/fiches-medicaments (accessed on January 10, 2013)
42. ANMV, *Agence nationale du médicament vétérinaire (ANMV/Anses)*. Available at http://www.anmv.anses.fr (accessed on January 10, 2013)
43. CIRCA, Communication & information resource administrator. Available at http://circa.europa.eu/Public/irc/env/bio_reports/library?l=/&vm=detailed&sb=Title (accessed on January 10, 2013)
44. e-phy. Available at http://e-phy.agriculture.gouv.fr (accessed on January 10, 2013)
45. ACTA, *Index Phytosanitaire*, 47th ed., Association de Coordination Technique Agricole, Paris, France, 2011. (accessed on January 10, 2013)
46. C. Tomlin, *The Pesticide Manual*, 11th ed., The Royal Society of Chemistry & The British Crop Protection Council, Farnham, U.K., 2009.
47. W.B. Neely, *Introduction to Chemical Exposure and Risk Assessment*, Lewis Publishers, Boca Raton, FL, 1994.

48. Regulation (EC) No 1272/2008 of the European parliament and of the Council of 16 December 2008 on classification, labelling and packaging of substances and mixtures, amending and repealing Directives 67/548/EEC and 1999/45/EC, and amending Regulation (EC) No. 1907/2006.
49. FOOTPRINT-PPDB (Pesticide properties data base). Available at http://sitem.herts.ac.uk/aeru/footprint/fr/index.htm (accessed on January 10, 2013)
50. CPDB, The carcinogenic potency database. Available at http://potency.berkeley.edu (accessed on January 10, 2013)
51. J. Devillers, E. Mombelli, and R. Samserà, *Structural alerts for estimating the carcinogenicity of pesticides and biocides*, SAR QSAR Environ. Res. 22 (2011), pp. 89–106. (accessed on January 10, 2013)
52. Anonymous, *European workshop on the impact of endocrine disruptors on human health and wildlife,* Report of Proceedings, December 2–4, Weybridge, U.K., 1996.
53. J. Lintelmann, A. Katayama, N. Kurihara, L. Shore, and A. Wenzel, *Endocrine disruptors in the environment. IUPAC technical report*, Pure Appl. Chem. 75 (2003), pp. 631–683.
54. J.M. Porcher, J. Devillers, and N. Marchand-Geneste, *Mechanisms of endocrine disruptions. A tentative overview*, in *Endocrine Disruption Modeling*, J. Devillers, ed., CRC Press, Boca Raton, FL, 2009, pp. 11–46.
55. J. Devillers, N. Marchand-Geneste, A. Carpy, and J.M. Porcher, *SAR and QSAR modeling of endocrine disruptors*, SAR QSAR Environ. Res. 17 (2006), pp. 393–412.
56. H. Kojima, E. Katsura, S. Takeuchi, K. Niiyama, and K. Kobayashi, *Screening for estrogen and androgen receptor activities in 200 pesticides by* in vitro *reporter gene assays using Chinese hamster ovary cells*, Environ. Health Perspect. 112 (2004), pp. 524–531.
57. S. Takeuchi, T. Matsuda, S. Kobayashi, T. Takahashi, and H. Kojima, In vitro *screening of 200 pesticides for agonistic activity via mouse peroxisome proliferator-activated receptor (PPAR)* α *and PPAR* γ *and quantitative analysis of* in vivo *induction pathway*, Toxicol. Appl. Pharmacol. 217 (2006), pp. 235–244.
58. G. Lemaire, W. Mnif, P. Mauvais, P. Balaguer, and R. Rahmani, *Activation of* α- *and* β-estrogen receptors by persistent pesticides in reporter cell lines, Life Sci. 79 (2006), pp. 1160–1169.
59. S. Takeuchi, M. Iida, H. Yabushita, T. Matsuda, and H. Kojima, In vitro *screening for aryl hydrocarbon receptor agonistic activity in 200 pesticides using a highly sensitive reporter cell line, DR-EcoScreen cells, and* in vivo *mouse liver cytochrome P450-1A induction by propanil, diuron and linuron*, Chemosphere 74 (2008), pp. 155–165.
60. H. Kojima, F. Sata, S. Takeuchi, T. Sueyoshi, and T. Nagai, *Comparative study of human and mouse pregnane X receptor agonistic activity in 200 pesticides using* in vitro *reporter gene assays*, Toxicology 280 (2011), pp. 77–87.
61. Endocrine Disruptors Website. Available at http://ec.europa.eu/environment/endocrine/index_en.htm (accessed on January 10, 2013)
62. EPA Endocrine Disruptor Priority Setting Database (EDPSD). Available at http://www.epa.gov/endo/pubs/prioritysetting/finalarch.htm (accessed on 10 January 2013)
63. ANPP, *Méthode CEB no. 95. Méthode de Laboratoire d'Evaluation de la Toxicité Aiguë Orale et de Contact des Produits Phytopharmaceutiques chez l'Abeille Domestique* Apis mellifera *L*, 1st ed., April 1982, revised version November 1996, p. 8. (accessed on January 10, 2013)
64. J. Devillers and M.H. Pham-Delègue, *Honey Bees: Estimating the Environmental Impact of Chemicals*, Taylor & Francis Group, London, U.K., 2002.
65. M.H. Pham-Delègue, A. Decourtye, L. Kaiser, and J. Devillers, *Behavioural methods to assess the effects of pesticides on honey bees*, Apidologie 33 (2002), pp. 425–432.

66. A. Decourtye, C. Armengaud, M. Renou, J. Devillers, S. Cluzeau, M. Gauthier, and M.H. Pham-Delègue, *Imidacloprid impairs memory and brain metabolism in the honeybee* (Apis mellifera *L.*), Pestic. Biochem. Physiol. 78 (2004), pp. 83–92.
67. H. Sasagawa, M. Sasaki, and I. Okada, *Hormonal control of the division of labor in adult honeybees* (Apis mellifera *L.) I. Effect of methoprene on corpora allata and hypopharyngeal gland, and its a-glucosidase activity*, Appl. Entomol. Zool. 24 (1989) 66–77.
68. M.M.G. Bitondi, I.M. Mora, Z.L.P. Simoes, and V.L.C. Figueiredo, *The* Apis mellifera *pupal melanization program is affected by treatment with a juvenile hormone analogue*, J. Insect Physiol. 44 (1998), pp. 499–507.
69. P. Aupinel, D. Fortini, B. Michaud, F. Marolleau, J.N. Tasei, and J.F. Odoux, *Toxicity of dimethoate and fenoxycarb to honey bee brood* (Apis mellifera), *using a new* in vitro *standardized feeding method*, Pest Manag. Sci. 63 (2007), pp. 1090–1094.
70. E. McCauley, W.W. Murdoch, R.M. Nisbet, and W.S.C. Gurney, *The physiological ecology of* Daphnia—*Development of a model of growth and reproduction*, Ecology 71 (1990), pp. 703–715.
71. T.G. Preuss, M. Hammers-Wirtz, U. Hommen, M.N. Rubachd, and H.T. Ratte, *Development and validation of an individual based* Daphnia magna *population model: The influence of crowding on population dynamics*, Ecol. Model. 220 (2009), pp. 310–329.
72. NF EN ISO 6341–1996, *Water quality—Determination of the inhibition of the mobility of* Daphnia magna *Straus (Cladocera, Crustacea)—Acute toxicity test*, ISO, 1996, p. 9.
73. OCDE 202 R 2004, Daphnia *sp. Acute Immobilisation Test, OECD Guidelines for the Testing of Chemicals*, Section 2, OECD Publishing, Paris, France, 2004, p. 12.
74. NF EN ISO 7346–1–1998, *EN-Water quality. Determination of the acute lethal toxicity of substances to a freshwater fish* (Brachydanio rerio *Hamilton-Buchanan (Teleostei Cyprinidae)). Part 1: static method*, ISO, 1998, p. 17.
75. NF EN ISO 7346–2–1998, *EN-Water quality. Determination of the acute lethal toxicity of substances to a freshwater fish* (Brachydanio rerio *Hamilton-Buchanan (Teleostei Cyprinidae)). Part 2: semi-static method*, ISO, 1998, p. 16.
76. NF EN ISO 7346–3–1998, *EN-Water quality. Determination of the acute lethal toxicity of substances to a freshwater fish* (Brachydanio rerio *Hamilton-Buchanan (Teleostei Cyprinidae)). Part 3: Flow-through method*, ISO, 1998, p. 18.
77. NF T 90–305, *Determination of the acute toxicity of a substance to* Salmo gairdneri—*Static and Flow-through methods*, Paris, France, 1985, p. 10.
78. OCDE 203, *Fish, Acute Toxicity Test, OECD Guidelines for the Testing of Chemicals, Section 2,* OECD Publishing, Paris, France, 1992, p. 8.
79. J. Sangster, *LOGKOW Databank. A Databank of Evaluated Octanol-Water Partition Coefficients (log P) on Microcomputer Diskette*, Sangster Research Laboratory, Montreal, Quebec, Canada, 1994.
80. C. Hansch, A. Leo, and D. Hoekman, *Exploring QSAR. Hydrophobic, Electronic, and Steric Constants*, ACS Professional Reference Book, American Chemical Society, Washington, DC, 1995.
81. W.M. Meylan and P.H. Howard, *Atom/fragment contribution method for estimating octanol-water partition coefficients*, J. Pharm. Sci. 84 (1995), pp. 83–92.
82. Anonymous, QSARs in the Assessment of the Environmental Fate and Effects of Chemicals, Tech. Rep. no. 74, ECETOC, Brussels, Belgium, 1998.
83. M. Reinhard and A. Drefahl, *Handbook for Estimating Physicochemical Properties of Organic Compounds*, John Wiley & Sons, New York, 1999.
84. J.B. Unsworth, R.D. Wauchope, A.W. Klein, E. Dorn, B. Zeeh, S.M. Yeh, M. Akerblim, K.D. Racke, and B. Rubin, *Significance of long range transport of pesticides in the atmosphere*, Pure Appl. Chem. 71 (1999), pp. 1359–1383.

85. P.H. Howard, *Handbook of Environmental Fate and Exposure Data for Organic Chemicals. Volume III, Pesticides*, Lewis Publishers, Chelsea, Michigan, 1991
86. K.C. Ma, D. Mackay, S.C. Lee, and W.Y. Shiu, *Handbook of Physical-Chemical Properties and Environmental Fate for Organic Chemicals*, 2nd ed., CRC Press, Boca Raton, FL, 2006.
87. F.C. Grain, *Vapour pressure*, in *Handbook of Chemical Property Estimation Methods*, W.J. Lyman, W.F. Reehl, and D.H. Rosenblatt, eds., American Chemical Society, Washington, DC, 1990.
88. Exposure Assessment Tools and Models. Available at http://www.epa.gov/oppt/exposure/pubs/episuitedl.htm (accessed on January 10, 2013)
89. R. Posthumus, T.P. Traas, W.J.G.M. Peijnenburg, and E.M. Hulzebos, *External validation of EPIWIN biodegradation models*, SAR QSAR Environ. Res.16 (2005), pp. 135–148.
90. R.S. Boethling and J. Costanza, *Domain of EPI suite biotransformation models*, SAR QSAR Environ. Res. 21 (2010), pp. 415–443.
91. R.S. Boethling, P.H. Howard, W. Meylan, W. Stiteler, J. Beaumann, and N.Tirado, *Group contribution method for predicting probability and rate of aerobic biodegradation*, Environ. Sci. Technol. 28 (1994), pp. 459–65.
92. J. Devillers, P. Pandard, and A.M. Charissou, *Sélection multicritère de bioindicateurs de la qualité des sols*, Etude Gestion Sols 16 (2009), pp. 233–242.
93. J. Devillers, P. Pandard, A.M. Charissou, and A. Bispo, *Use of multicriteria analysis for selecting ecotoxicity tests*, in *Ecotoxicology Modeling*, J. Devillers, ed., Springer, New York, 2009, pp. 117–143.

Index

N

O

P

Q

R

S

T

U

V

W

Z

T - #0362 - 101024 - C12 - 234/156/22 - PB - 9781138382206 - Gloss Lamination